T0137025

Advances in Intelligent Systems and Computing

Volume 516

Series editor

Janusz Kacprzyk, Polish Academy of Sciences, Warsaw, Poland
e-mail: kacprzyk@ibspan.waw.pl

About this Series

The series "Advances in Intelligent Systems and Computing" contains publications on theory, applications, and design methods of Intelligent Systems and Intelligent Computing. Virtually all disciplines such as engineering, natural sciences, computer and information science, ICT, economics, business, e-commerce, environment, healthcare, life science are covered. The list of topics spans all the areas of modern intelligent systems and computing.

The publications within "Advances in Intelligent Systems and Computing" are primarily textbooks and proceedings of important conferences, symposia and congresses. They cover significant recent developments in the field, both of a foundational and applicable character. An important characteristic feature of the series is the short publication time and world-wide distribution. This permits a rapid and broad dissemination of research results.

More information about this series at http://www.springer.com/series/11156

Suresh Chandra Satapathy
Vikrant Bhateja · Siba K. Udgata
Prasant Kumar Pattnaik
Editors

Proceedings of the 5th International Conference on Frontiers in Intelligent Computing: Theory and Applications

FICTA 2016, Volume 2

 Springer

Editors
Suresh Chandra Satapathy
Department of Computer Science
  and Engineering
Anil Neerukonda Institute of Technology
  and Sciences
Visakhapatnam, Andhra Pradesh
India

Vikrant Bhateja
Shri Ramswaroop Memorial Group
  of Professional Colleges (SRMGPC)
Lucknow, Uttar Pradesh
India

Siba K. Udgata
SCIS
University of Hyderabad
Hyderabad
India

Prasant Kumar Pattnaik
School of Computer Engineering
KIIT University
Bhubaneswar, Odisha
India

ISSN 2194-5357   ISSN 2194-5365   (electronic)
Advances in Intelligent Systems and Computing
ISBN 978-981-10-3155-7   ISBN 978-981-10-3156-4   (eBook)
DOI 10.1007/978-981-10-3156-4

Library of Congress Control Number: 2016957505

Printed on acid-free paper

This Springer imprint is published by Springer Nature
The registered company is Springer Nature Singapore Pte Ltd.
The registered company address is: 152 Beach Road, #22-06/08 Gateway East, Singapore 189721, Singapore

Preface

The volume is a collection of high-quality peer-reviewed research papers presented at the 5th International Conference on Frontiers in Intelligent Computing: Theory and Applications (FICTA 2016) held at School of Computer Engineering, KIIT University, Bhubaneswar, Odisha, India, during 16–17 September 2016.

The idea of this conference series was conceived by few eminent professors and researchers from premier institutions of India. The first three editions of this conference: FICTA 2012, 2013 and 2014 were organized by Bhubaneswar Engineering College (BEC), Bhubaneswar, Odisha, India. Owing to its popularity and wider visibility in the entire country as well as abroad; the fourth edition of FICTA 2015 has been organized by the prestigious NIT, Durgapur, West Bengal, India. All papers of past FICTA editions are published by Springer AISC Series. Presently, FICTA 2016 is the fifth edition of this conference series which brings researchers, scientists, engineers, and practitioners on to a single platform to exchange and share their theories, methodologies, new ideas, experiences and applications in all areas of intelligent computing theories and applications to various engineering disciplines such as computer science, electronics, electrical, mechanical, and biomedical engineering, etc.

FICTA 2016 had received a good number of submissions from the different areas relating to intelligent computing and its applications, and after a rigorous peer-review process with the help of our program committee members and external reviewers (from the country as well as abroad), good quality papers were selected for publications. The review process has been very crucial with minimum 2 reviews each and in many cases 3–5 reviews along with due checks on similarity and overlaps as well. The number of papers received under FICTA 2016 touched 500 mark including the main track and special sessions. The conference featured eight special sessions in various cutting edge technologies of specialized focus which were organized and chaired by eminent professors. The total number of papers received includes submissions from eight overseas countries. Out of this pool of received papers, only 150 papers were given acceptance yielding an acceptance ratio of 0.4. These papers have been arranged in two separate volumes: Volume 1: 79 chapters and Volume 2: 71 chapters.

The conference featured many distinguished keynote addresses by eminent speakers like Dr. João Manuel and R.S. Tavares, Faculdade de Engenharia da Universidade do Porto, Porto, Portugal on Computational Image Analysis: Methods and Applications; Dr. Swagatam Das, ISI Kolkata on Swarm Intelligence; Dr. Rajib Mall, Information Technology IIT, Kharagpur; Dr. V. Ravi, Centre of Excellence in Analytics Institute for Development and Research in Banking Technology (IDRBT), (Established by RBI, India) Hyderabad; Dr. Gautam Sanyal, Professor, Department of CSE, NIT Durgapur on Image Processing and Intelligent Computing; and a plenary session on 'How to write for and get published in scientific journals?' by Mr. Aninda Bose, Senior Publishing Editor, Springer India.

We thank the General Chairs: Prof. Samaresh Mishra and Prof. Jnyana Ranjan Mohanty at KIIT University Bhubaneswar, India, for providing valuable guidance and inspiration to overcome various difficulties in the process of organizing this conference. We extend our heartfelt thanks to the honorary chairs of this conference: Dr. B.K. Panigrahi, IIT Delhi, and Dr. Swagatam Das, ISI Kolkota, for being with us from the very beginning to the end of this conference, and without their support, this conference could never have been successful. Our heartfelt thanks are due to Prof. P.N. Suganthan, NTU Singapore, for his valuable suggestions on enhancing editorial review process.

We would also like to thank School of Computer Engineering, KIIT University, Bhubaneswar, having coming forward to support us to organize the fifth edition of this conference series. We are amazed to note the enthusiasm of all faculty, staff and students of KIIT to organize the conference in a professional manner. Involvements of faculty coordinators and student volunteers are praiseworthy in every respect. We are confident that in future too we would like to organize many more international level conferences in this beautiful campus. We would also like to thank our sponsors for providing all the support and financial assistance.

We take this opportunity to thank authors of all submitted papers for their hard work, adherence to the deadlines and patience with the review process. The quality of a refereed volume depends mainly on the expertise and dedication of the reviewers. We are indebted to the program committee members and external reviewers who not only produced excellent reviews but also did these in short time frames.

We would also like to thank the participants of this conference, who have considered the conference above all hardships. Finally, we would like to thank all the volunteers who spent tireless efforts in meeting the deadlines and arranging every detail to make sure that the conference can run smoothly. Additionally, CSI Students Branch of ANITS and its team members have contributed a lot to this conference and deserve an appreciation for their contribution.

All efforts are worth and would please us all, if the readers of this proceedings and participants of this conference found the papers and conference inspiring and enjoyable. Our sincere thanks goes to all in press, print and electronic media for their excellent coverage of this conference.

We take this opportunity to thank all keynote speakers, track and special session chairs for their excellent support to make FICTA 2016 a grand success.

Visakhapatnam, India	Suresh Chandra Satapathy
Lucknow, India	Vikrant Bhateja
Hyderabad, India	Siba K. Udgata
Bhubaneswar, India	Prasant Kumar Pattnaik

Organization

Chief Patron

Achyuta Samanta, KISS and KIIT, Bhubaneswar, India

Patron

P.P. Mathur, KIIT University, Bhubaneswar, India

Advisory Committee

Sasmita Samanta, KIIT University, Bhubaneswar, India
Ganga Bishnu Mund, KIIT University, Bhubaneswar, India
Madhabananda Das, KIIT University, Bhubaneswar, India
Prasant Kumar Pattnaik, KIIT University, Bhubaneswar, India
Sudhanshu Sekhar Singh, KIIT University, Bhubaneswar, India

General Chair

Samaresh Mishra, KIIT University Bhubaneswar, India

General Co-chair

Jnyana Ranjan Mohanty, KIIT University, Bhubaneswar, India

Honorary Chairs

Dr. Swagatam Das, ISI Kolkota, India
Dr. B.K. Panigrahi, IIT Delhi, India

Convener

Chittaranjan Pradhan, KIIT University, Bhubaneswar, India

Organizing Chairs

Sachi Nandan Mohanty, KIIT University, Bhubaneswar, India
Sidharth Swarup Routaray, KIIT University, Bhubaneswar, India

Publication Chairs

Suresh Chandra Satapathy, ANITS, Visakhapatnam, India
Vikrant Bhateja, SRMGPC, Lucknow (U.P.), India

Steering Committee

Suresh Chandra Satapathy, ANITS, Visakhapatnam, India
Siba K. Udgata, UoH, Hyderabad, India
Manas Sanyal, University of Kalayani, India
Nilanjan Dey, TICT, Kolkota, India
B.N. Biswal, BEC, Bhubaneswar, India
Vikrant Bhateja, SRMGPC, Lucknow (U.P.), India

Editorial Board

Suresh Chandra Satapathy, ANITS, Visakhapatnam, India
Vikrant Bhateja, SRMGPC, Lucknow (U.P.), India
Siba K. Udgata, UoH, Hyderabad, India
Prasant Kumar Pattnaik, KIIT University, Bhubaneswar, India

Transport and Hospitality Chairs

Harish K. Pattnaik, KIIT University, Bhubaneswar, India
Manas K. Lenka, KIIT University, Bhubaneswar, India
Ramakant Parida, KIIT University, Bhubaneswar, India

Session Management Chairs

Manoj Kumar Mishra, KIIT University, Bhubaneswar, India
Suresh Ch. Moharana, KIIT University, Bhubaneswar, India
Sujoya Datta, KIIT University, Bhubaneswar, India

Registration Chairs

Himansu Das, KIIT University, Bhubaneswar, India
Manjusha Pandey, KIIT University, Bhubaneswar, India
Sharmistha Roy, KIIT University, Bhubaneswar, India
Sital Dash, KIIT University, Bhubaneswar, India

Publicity Chairs

J.K. Mandal, University of Kalyani, Kolkota, India
K. Srujan Raju, CMR Technical Campus, Hyderabad, India

Track Chairs

Machine Learning Applications: Steven L. Fernandez, SCEM, Mangalore, India
Image Processing and Pattern Recognition: V.N. Manjunath Aradhya, SJCE, Mysore, India
Big Data, Web Mining and IoT: Sireesha Rodda, Gitam University, Visakhapatnam, India
Signals, Communication and Microelectronics: A.K. Pandey, MIET, Meerut (U.P.), India
MANETs and Wireless Sensor Networks: Pritee Parwekar, ANITS, Visakhapatnam, India
Network Security and Cryptography: R.T. Goswami, BITS Mesra, Kolkota Campus, India
Data Engineering: M. Ramakrishna Murty, ANITS, Visakhapatnam, India
Data Mining: B. Janakiramaiah, DVR and Dr. HS MICCT, Kanchikacherla, India

Special Session Chairs

SS01: Wireless Sensor Networks: Architecture, Applications, Data Management and Security: Pritee Parwekar, ANITS, Visakhapatnam, India and Sireesha Rodda, Gitam University, Visakhapatnam, India
SS02: Advances in Document Image Analysis: V.N. Manjunath Aradhya, SJCE, Mysore, India
SS03: Computational Swarm Intelligence: Sujata Dash, North Orissa University, Baripada, India and Atta Ur Rehman, (BIIT) PMAS Arid Agricultural University, Rawalpindi, Punjab, Pakistan
SS04: Wide-Area-Wireless Communication Systems-Microwave, 3G, 4G and WiMAX: K. Meena (Jeyanthi), PSNACET, Dindigul, Tamil Nadu, India

SS05: Innovation in Engineering and Technology through Computational Intelligence Techniques: Jitendra Agrawal, RGPV, M.P., India and Shikha Agrawal, RGPV, M.P., India

SS06: Optical Character Recognition and Natural Language Processing: Nikisha B. Jariwala, Smt. Tanuben and Dr. Manubhai Trivedi CIS, Surat, Gujarat, India and Hardik A. Vyas, BMaIIT, Uka Tarsadia University, Bardoli, Gujarat, India

SS07: Recent Advancements in Information Security: Musheer Ahmad Jamia Millia Islamia, New Delhi, India

SS08: Software Engineering in Multidisciplinary Domains: Suma V., Dayananda Sagar College of Engineering, Bangalore, India

Technical Program Committee/International Reviewer Board

A. Govardhan, India
Aarti Singh, India
Almoataz Youssef Abdelaziz, Egypt
Amira A. Ashour, Egypt
Amulya Ratna Swain, India
Ankur Singh Bist, India
Athanasios V. Vasilakos, Athens
Banani Saha, India
Bhabani Shankar Prasad Mishra, India
B. Tirumala Rao, India
Carlos A. Coello, Mexico
Charan S.G., India
Chirag Arora, India
Chilukuri K. Mohan, USA
Chung Le, Vietnam
Dac-Nhuong Le, Vietnam
Delin Luo, China
Hai Bin Duan, China
Hai V. Pham, Vietnam
Heitor Silvério Lopes, Brazil
Igor Belykh, Russia
J.V.R. Murthy, India
K. Parsopoulos, Greece
Kamble Vaibhav Venkatrao, India
Kailash C. Patidar, South Africa
Koushik Majumder, India
Lalitha Bhaskari, India
Jeng-Shyang Pan, Taiwan
Juan Luis Fernández Martínez, California
Le Hoang Son, Vietnam
Leandro Dos Santos Coelho, Brazil

L. Perkin, USA
Lingfeng Wang, China
M.A. Abido, Saudi Arabia
Maurice Clerc, France
Meftah Boudjelal, Algeria
Monideepa Roy, India
Mukul Misra, India
Naeem Hanoon, Malysia
Nikhil Bhargava, India
Oscar Castillo, Mexcico
P.S. Avadhani, India
Rafael Stubs Parpinelli, Brazil
Ravi Subban, India
Roderich Gross, England
Saeid Nahavandi, Australia
Sankhadeep Chatterjee, India
Sanjay Sengupta, India
Santosh Kumar Swain, India
Saman Halgamuge, India
Sayan Chakraborty, India
Shabana Urooj, India
S.G. Ponnambalam, Malaysia
Srinivas Kota, Nebraska
Srinivas Sethi, India
Sumanth Yenduri, USA
Suberna Kumar, India
T.R. Dash, Cambodia
Vipin Tyagi, India
Vimal Mishra, India
Walid Barhoumi, Tunisia
X.Z. Gao, Finland
Ying Tan, China
Zong Woo Geem, USA
M. Ramakrishna Murthy, India
Tushar Mishra, India
And many more…

Contents

About the Editors

Dr. Suresh Chandra Satapathy is currently working as a professor and head, Department of Computer Science and Engineering, Anil Neerukonda Institute of Technology and Sciences (ANITS), Visakhapatnam, Andhra Pradesh, India. He obtained his Ph.D. in Computer Science Engineering from JNTUH, Hyderabad, and master's degree in Computer Science and Engineering from National Institute of Technology (NIT), Rourkela, Odisha. He has more than 27 years of teaching and research experience. His research interest includes machine learning, data mining, swarm intelligence studies and their applications to engineering. He has more than 98 publications to his credit in various reputed international journals and conference proceedings. He has edited many volumes from Springer AISC, LNEE, SIST and LNCS in past, and he is also the editorial board member in few international journals. He is a senior member of IEEE and life member of Computer Society of India. Currently, he is the national chairman of Division-V (Education and Research) of Computer Society of India.

Prof. Vikrant Bhateja is an associate professor, Department of Electronics and Communication Engineering, Shri Ramswaroop Memorial Group of Professional Colleges (SRMGPC), Lucknow, and also the head (Academics and Quality Control) in the same college. His areas of research include digital image and video processing, computer vision, medical imaging, machine learning, pattern analysis and recognition. He has published 100 quality publications in various international journals and conference proceedings. Professor Vikrant has been on TPC and chaired various sessions from the above domain in international conferences of IEEE and Springer. He has been the track chair and served in the core-technical/editorial teams for international conferences: FICTA 2014, CSI 2014, INDIA 2015, ICICT 2015 and ICTIS 2015 under Springer-ASIC Series and INDIACom 2015 and ICACCI 2015 under IEEE. He is an associate editor in International Journal of Synthetic Emotions (IJSE) and International Journal of Ambient Computing and Intelligence (IJACI) under IGI Global. He is also serving in the editorial board of International Journal of Image Mining (IJIM) and International Journal of Convergence Computing (IJConvC) under Inderscience

Publishers. He has been the editor of three published volumes with Springer and another two are in press.

Dr. Siba K. Udgata is a professor of School of Computer and Information Sciences, University of Hyderabad, India. He is presently heading Centre for Modelling, Simulation and Design (CMSD), a high-performance computing facility at University of Hyderabad. He has got his Master's followed by Ph.D. in Computer Science (mobile computing and wireless communication). His main research interests are wireless communication, mobile computing, wireless sensor networks and intelligent algorithms. He was a United Nations Fellow and worked in the United Nations University/International Institute for Software Technology (UNU/IIST), Macau, as research fellow in the year 2001. Dr. Udgata is working as a principal investigator in many Government of India funded research projects mainly for the development of wireless sensor network applications and application of swarm intelligence techniques. He has published extensively in refereed international journals and conferences in India as well as abroad. He has edited many volumes in Springer LNCS/LNAI and Springer AISC Proceedings.

Dr. Prasant Kumar Pattnaik Ph.D. (Computer Science), Fellow IETE, Senior Member IEEE is a professor at the School of Computer Engineering, KIIT University, Bhubaneswar. He has more than a decade of teaching and research experience. Dr. Pattnaik has published numbers of research papers in peer-reviewed International Journals and conferences. His areas of specialization are mobile computing, cloud computing, brain computer interface and privacy preservation.

Information Retrieval for Gujarati Language Using Cosine Similarity Based Vector Space Model

Rajnish M. Rakholia and Jatinderkumar R. Saini

Abstract Based on user query, to retrieve most relevant documents from the web for resource poor languages is a crucial task in Information Retrieval (IR) system. This paper presents Cosine Similarity Based Vector Space Document Model (VSDM) for Information Retrieval in Gujarati language. VSDM is widely used in information retrieval and document classification where each document is represented as a vector and each dimension corresponds to a separate term. Influence and relevancy of documents with user query is measured using cosine similarity under vector space where set of documents is considered as a set of vectors. The present work considers user query as a free order text, i.e., the word sequence does not affect results of the IR system. Technically, this is Natural Language Processing (NLP) application wherein stop-words removal, Term Frequency (TF) calculation, Normalized Term Frequency (NF) calculation and Inverse Document Frequency (IDF) calculation was done for 1360 files using Text and PDF formats and precision and recall values of 78 % and 86 % efficiency respectively were recorded. To the best of our knowledge, this is first IR task in Gujarati language using cosine similarity based calculations.

Keywords Cosine similarity · Document classification · Gujarati language · Information retrieval (IR) · Vector space document model (VSDM)

R.M. Rakholia (✉)
School of Computer Science, R K University, Rajkot, Gujarat, India
e-mail: rajnish.rakholia@gmail.com

J.R. Saini
Narmada College of Computer Application, Bharuch, Gujarat, India
e-mail: saini_expert@yahoo.com

© Springer Nature Singapore Pte Ltd. 2017
S.C. Satapathy et al. (eds.), *Proceedings of the 5th International Conference on Frontiers in Intelligent Computing: Theory and Applications*, Advances in Intelligent Systems and Computing 516, DOI 10.1007/978-981-10-3156-4_1

1 Introduction

To retrieve the most relevant documents from the web is a significant task from the perspective of Natural Language Processing (NLP) research as well as from the perspective of satisfaction of the demands of different users too. It is more difficult for the resource poor languages like Gujarati and Punjabi. Main objective of the current research is to enhance the performance of Information Retrieval (IR) for the Gujarati language. The application areas of the current research are library system, mail classification, sentiment analysis and survey classification, to name a few. In the proposed work, the concept of Cosine similarity has been deployed and measured for documents in the knowledge base and the user's query. In order to be more representative and informative of the pertinent terminologies, we now present the succinct deliberation on Gujarati language, Cosine Similarity Measure and Stop-words.

1.1 Gujarati Language

Gujarati is an official and regional language of Gujarat state in India. It is the 23rd most widely spoken language in the world today, which is spoken by more than 46 million people. Approximately 45.5 million people speak Gujarati language in India and half a million speakers are from outside of India, including territories of countries like Tanzania, Uganda, Pakistan, Kenya and Zambia. Gujarati language belongs to the group of Indo-Aryan languages of the Indo-European language family. Gujarati language is also closely related to the Indian Hindi language.

1.2 Cosine Similarity Measure

It is a measure of similarity between two vectors which is considered as a document in the vector space. Distance computing and cosine measure of angle are two ways to find similarity between two vectors. We have used cosine measure because in vector space distance between two vectors may be too long. In vector space two vectors with the opposed orientation have a cosine similarity of -1 whereas in same orientation they have the cosine similarity of 1.

1.3 Stop-Words

Stop word is a word which has less significant meaning than other tokens. Identification of stop words and its removal process is a basic pre-processing phase in many NLP tasks including Information Retrieval and Data Mining applications. For any NLP tool there is no single universal list of stop words used for a specific language. The stop words list is generally domain specific and dependent on the nature of the application to be developed.

2   Literature Review

Sartori [1] has used terms weighting scheme for text classification and sentiment analysis while Li and Han [2] applied distance weighted cosine similarity measure for text classification.

Al-Talib and Hassan [3] have used TF-IDF weighting for SMS Classification. They stored all SMS into text file and predefined categories (Sales, Friendship, Occasions etc.) were used for experiment purpose. They classified SMS by computing weighting for each term in documents based on its influence. Their experimental results showed that the term weighting method improves the accuracy and efficiency of the system.

Kumar and Vig [4] present WordNet based TF-IDF which is mainly focused on crawling to retrieve most relevant documents for e-library. The performance of the proposed approach shows that retrieved documents are highly related to the user query for a single domain.

Zhang et al. [5] used TF-IDF for information retrieval and document categorization for English and Chinese languages. They presented multi-word characteristic which is more semantic then single terms for document indexing and classification.

Ramos [6] has applied Term-Frequency and Inverse Document Frequency to determine which terms of user query are more relevant to the document. He has also calculated inverse proportion to determine influence of user query to the document based on document length.

Lee et al. [7] proposed Vector Space Model (VSM) for document ranking based on text retrieval query. They advocated that the key word based indexing cannot determine the relevance of the document. They further proposed that by using VSM, it is possible to index most relevant document to the user query.

3   Methodology

3.1   *Preprocessing for Cosine Similarity Measure*

We propose the following algorithm for cleaning the noisy text from documents and create Vector Space Model for Indian Gujarati written script:

INPUT:

A raw text document <d1, d2,..., dn> belonging to the document set D.

OUTPUT:

Document-Term Matrix M (Where, rows represent terms and the columns represent documents), Each element $m_{t,d}$ tabulates the number of times term t occurs in document d. Weights are assigned to each term in a document that depends on the number of occurrences of the term in the document.

A pseudocode representation of our proposed algorithm is presented below:

begin
    variable occurrences=1;
    for each document di in D
    do:

        Remove all punctuation marks, special character and digits from document di;
        Replace each occurrence of consecutive multiple white-space characters by a single white-space character;

        for each word wj in di
        do:

            if(wj is single character or single character with multiple diacritics)
                then

                    remove wj from document di;
                    remove all other words, exact match with wj in document di;
                    continue;

            for each word wj+1 in di
            do:

                if(wj is exact match with other words wj+1 in di OR is a sub word of any other words wj+1 OR wj+1 is sub word of wj)
                then

                    occurrences++;
                    Remove wj+1 from document di;
                end if
            end for
            Assign a frequency of word wj to Mji (M is a Document -term Matrix, number of times term wj occurs in document di);
            Mji=occurrences;
            occurrences=1;

        end for
    end for
end

3.2 *Vector Space Model with Cosine Similarity*

The set of documents is considered a set of vectors in vector space. We have created three documents that contained very less amount of text to demonstrate information retrieval for Gujarati language using Vector Space Model by computing cosine similarity of the user query and documents.

Document-1: કેન્સર ગંભીર બીમારી છે.

Document-2: પ્રકાશ ને હદય ની બીમારી છે.

Documnet-3: કેન્સર નું દવાખાનું સુરત માં છે.

The information retrieval query used by us against the above documents was: કેન્સર બીમારી. It is noteworthy to see that the free order query has been used wherein the word-sequence does not affect the results of search.

This was followed by the calculation of Term Frequency (TF) measure. TF measure is the measure of the number of times a term t occurs in documents. We also calculated Normalized Term Frequency for the words in the documents. The Normalized Term Frequency often referred to also as Normalized Frequency (NF) is defined by the equation presented at (1). NF for each document was calculated to find the accurate influence of each term. Because length of each document might be different, long text documents have a higher probability of more occurrence of a particular word than the short documents. Table 1(a), (b) and (c) present the Term Frequency and the Normalized Frequency for each word of document-1, document-2 and document-3 respectively.

$$\text{Normalized frequency} = \frac{\text{number of occurrences of term}}{\text{total number of terms in document}}. \tag{1}$$

We have considered each term as of an equal importance in the user query. Some terms have more occurrence in document but they carry less importance in document and less occurred terms might carry more influence in document. This problem can be solved by using logarithm function for Inverse Document Frequency (IDF). IDF is defined as presented in Eq. (2).

$$\text{IDF(term t)} = 1 + \log_e \left(\frac{total\ number\ of\ documents\ in\ dataset}{number\ of\ documents\ that\ contained\ term\ t} \right). \tag{2}$$

For instance, "બીમારી" term appeared in Document-1 and Document-2 and the total number of documents is three. Hence, IDF calculation for this term will be carried out as follows:

$$\text{IDF(બીમારી)} = 1 + \log_e \left(\frac{3}{2} \right) = 1.40$$

Table 2 presents IDF for all terms which occurred in all the documents of the corpus. Too frequently occurring terms need to be normalized by lowering down

Table 1 (a) TF and NF for Document-1, (b) TF and NF for Document-2, (c) TF and NF for Document-3

Table-1(a). TF and NF for Document-1

Terms	કેન્સર	ગંભીર	બીમારી	છે
Term frequency	1	1	1	1
Normalized frequency	0.25	0.25	0.25	0.25

Table-1(b). TF and NF for Document-2

Terms	પ્રકાશ	ને	હદય	ની	બીમારી	છે
Term frequency	1	1	1	1	1	1
Normalized frequency	0.17	0.17	0.17	0.17	0.17	0.17

Table-1(c). TF and NF for Document-3

Terms	કેન્સર	નું	દવાખાનું	સુરત	માં	છે
Term frequency	1	1	1	1	1	1
Normalized frequency	0.17	0.17	0.17	0.17	0.17	0.17

Table 2 IDF calculation for each term in document corpus

Term	Term frequency in corpus	IDF	Term	Term frequency in corpus	IDF
કેન્સર	2	1.40546510811	ને	1	2.0986122887
ગંભીર	1	2.0986122887	હદય	1	2.0986122887
બીમારી	2	1.40546510811	ની	1	2.0986122887
છે	3	1.0000000000	નું	1	2.0986122887
પ્રકાશ	1	2.0986122887	દવાખાનું	1	2.0986122887
સુરત	1	2.0986122887	માં	1	2.0986122887

their weightage and while the less frequently occurring terms need to be normalized by giving them higher weightage. This is accomplished using TF-IDF. Tables 3 and 4 present TF-IDF for each document in the corpus and the user query respectively.

We now present the cosine similarity calculation for the user query in context of each of the three documents in the corpus one by one.

Cosine similarity for user query and document1

Here d1 = 0.3513662770275, 0.3513662770275 and q = 0.70273255405, 0.70273255405

Table 3 TF-IDF for each document

Term	Document-1	Document-2	Document-3
કેન્સર	0.3513662770275	0	0.2389290683787
બીમારી	0.3513662770275	0.2389290683787	0

Table 4 TF-IDF for user query

Term	TF	IDF	TFxIDF
કેન્સર	0.5	1.40546510811	0.70273255405
બીમારી	0.5	1.40546510811	0.70273255405

$$\text{Cosine Similarity } (d1, \ q) = dot(d1, q) / \|d1\| \ \|q\|$$

$$dot(d1, \ q) = (0.3513662770275) \times (0.70273255405) + (0.3513662770275) \times (0.70273255405) = \mathbf{0.493833042525}$$

$$\|d1\| = sqrt((0.3513662770275)^{\wedge}2 + (0.3513662770275)^{\wedge}2) = \mathbf{0.496906954333}$$

$$\|q\| = sqrt((0.70273255405)^{\wedge}2 + (0.70273255405)^{\wedge}2) = \mathbf{0.993813908659}$$

$$\begin{aligned}\text{Cosine Similarity } (d1, \ q) &= 0.493833042525 / (0.496906954333) \times (0.993813908659) \\ &= 0.493833042525 / 0.493833042525 \\ &= \mathbf{1}\end{aligned}$$

Same way, we got cosine similarity for user query and document2 (d2, q) = **0.707106781187** and for the user query and document3 (d3, q) = **0.707106781187**.

Table 5 contains cosine similarity measure for each document which can be used for document ranking in search results. Obtained cosine similarity score for document-1 is 1 which is greater than cosine similarity score of document-2 and document-3. Thus, most relevant document for user query is document-1.

Table 5 Document ranking

	Document-1	Document-2	Document-3
Cosine similarity	1	**0.707106781187**	**0.707106781187**

Table 6 Performance evaluation

	Total # documents in corpus	Number of relevant documents	Number of retrieved documents	Number of relevant documents retrieved	Precision (P)	Recall (R)
Sports	**1360**	272	310	242	0.78	0.89
Education		189	201	153	0.76	0.81
Politics		213	240	185	0.77	0.87
Health		173	191	155	0.81	0.90
Entertainment		168	182	142	0.78	0.85
Astronomy		189	212	166	0.78	0.88
Science		156	164	127	0.77	0.82
Average	–	–	–	–	0.78	0.86

4 Results

The experiment was conducted for 1360 different documents in which 722 were prepared in Text file format while 638 documents were prepared in Portable Document Format (PDF). Each document contained more than 800 words. The documents belonged to different seven news categories viz. Sports, Education, Politics, Health, Entertainment, Astronomy and Science. The dataset was collected randomly from the web.

Information Retrieval system returns a set of document which is related to the user query. Recall (R) and Precision (P) are calculated using Eqs. (3) and (4) respectively to evaluate the performance of the system. Table 6 describes randomly selected experiment for each individual category and query.

$$P = \frac{Number\ of\ relevant\ documents\ retrieved}{Number\ of\ retrieved\ documents}. \tag{3}$$

$$R = \frac{Number\ of\ relevant\ documents\ retrieved}{Number\ of\ relevant\ documents}. \tag{4}$$

The average precision and recall values obtained by us are 0.78 and 0.86. The values are promising enough to make the system implementable in real world.

5 Conclusion and Future Work

This work was carried out for Gujarati Information Retrieval by computing cosine similarity of user query and the documents in the knowledgebase of the corpus. We computed TF-IDF followed by the calculation of cosine similarity to find the most relevant documents with the user query. To evaluate the performance of the proposed approach Recall and Precision was measured as well. Experimental results prove that the cosine similarity based approach contributes effectively in Gujarati documents retrieval. We were able to find an efficiency of 78 % and 86 % respectively for the precision and recall values for the system. The present approach could not be applied well when users are looking for most relevant document with semantic relationship between two terms in the document. For, future, we are working on ontology based hybrid approach for information retrieval and document classification for Gujarati language documents.

References

1. C. Sartori, A comparison of term weighting schemes for text classification and sentiment analysis with a supervised variant of tf. idf, in *Data Management Technologies and Applications: 4th International Conference, DATA 2015, Colmar, France, July 20–22, 2015, Revised Selected Papers*, vol. 584 (Springer, 2016), p. 39
2. B. Li, L. Han, Distance weighted cosine similarity measure for text classification, in *Intelligent Data Engineering and Automated Learning–IDEAL 2013* (Springer, Berlin, 2013), pp. 611–618
3. G.A. Al-Talib, H.S. Hassan, A study on analysis of SMS classification using TF-IDF Weighting. Int. J. Comput. Netw. Commun. Secur. (IJCNCS) **1**(5), 189–194 (2013)
4. M. Kumar, R. Vig, e-Library content generation using WorldNet Tf-Idf semantics, in *Proceedings of the International Conference on Frontiers of Intelligent Computing: Theory and Applications (FICTA)* (Springer, Berlin, 2013), pp. 221–227
5. W. Zhang, T. Yoshida, X. Tang, A comparative study of TFx IDF, LSI and multi-words for text classification. Expert Syst. Appl. **38**(3), 2758–2765 (2011)
6. J. Ramos, Using tf-idf to determine word relevance in document queries, in *Proceedings of the First Instructional Conference on Machine Learning* (2003)
7. D.L. Lee, H. Chuang, K. Seamons, Document ranking and the vector-space model. IEEE Softw. **14**(2), 67–75 (1997)

BLDC Motor Drive with Power Factor Correction Using PWM Rectifier

P. Sarala, S.F. Kodad and B. Sarvesh

Abstract Major constraints while using motor drive system are efficiency and cost. Commutation in the conventional DC motors is carried out by commutator which is rotating part placed on the rotor and brushes. Due to these mechanical parts, conventional DC motor consist high amount of losses. Brushless DC (BLDC) Motors are very extensively used motors these days because of its advantages over conventional DC motors. Commutation is carried out with the help of solid-state switches in BLDC motor instead of mechanical commutator as in conventional DC motor. This improves the performance of the motor. BLDC motor draws non-linear currents from the source affecting the loads connected at the source point due to harmonic production. This harmonic production reduces the system efficiency and mainly stresses the loads connected at source point. BLDC drive system with power factor (PF) correction was discussed in this paper. BLDC with normal AC-DC diode bridge rectifier and the performance of BLDC drive with PWM rectifier for power factor correction was discussed. BLDC drive system with PWM rectifier for power factor correction was validated by considering different cases. BLDC motor without power factor correction, BLDC drive with PF correction at starting condition, at steady state and with step-change in DC link voltage models was developed. Torque ripple in BLDC motor drive for these cases were compared. Models were developed and results were obtained using Matlab/Simulink software.

Keywords BLDC · Conventional · DC · Commutator · Solid-state switch · Hall sensor · Power factor (PF)

P. Sarala (✉) · B. Sarvesh
JNTUA, Ananthapur, AP, India
e-mail: dilip1.eee@gmail.com

S.F. Kodad
PESITM, Sivmogga, Karnataka, India

© Springer Nature Singapore Pte Ltd. 2017
S.C. Satapathy et al. (eds.), *Proceedings of the 5th International Conference on Frontiers in Intelligent Computing: Theory and Applications*, Advances in Intelligent Systems and Computing 516, DOI 10.1007/978-981-10-3156-4_2

11

1 Introduction

For a DC motor, supply will be DC type but the EMF should be AC type. This operation is done by commutator and brushes in a conventional DC motor. Commutator is mechanical part placed on the rotor segment for the purpose of commutation. This commutator along with brushes produces wear and tear on the commutator surface and hence commutation might not be effective. Also this mechanical commutator produces high amount of losses. Since both brushes and commutator are good conductors, they produce copper losses. The wear and tear of the commutator surface produces sparks due to uneven current distribution. Sparks produces heat which is a major drawback. The above said disadvantages are mainly due to the presence of commutation process by commutator and brushes. Thus the disadvantages in a conventional DC motor can be overcome by eliminating brushes. This led to the realization of motors without brushes called brushless DC (BLDC) motor [1, 2]. Electrical commutation in BLDC motor is carried out by electronic solid-state switches. Due to the usage of electronic switches for commutation, the drawbacks in conventional DC motor are eliminated thus improving the system performance. DC motors have very good speed control and especially BLDC exhibits many advantages [3, 4] over conventional DC motor like high efficiency, reliability, low acoustic noise, good dynamic response, lighter, improved speed-torque characteristics, higher speed range and requires very less maintenance.

BLDC motor shown in Fig. 1 is typically a combination of permanent magnet (PM) AC machine with electronic commutator. Sensor less operation [5–7] of BLDC is also possible with the help of monitoring back EMF signals. Back EMF is proportional to the speed of the rotor. So, at starting condition of the motor or low speeds, sensor less operation needs additional set-up to control the rotor position. Basically BLDC motor has DC input supply. This input of DC supply needs to be inverted to AC type to drive stator windings of BLDC motor [8–12].

Power factor is a major concern in power system. Power factor should be maintained nearer to unity. Generally drive system draws non-linear currents

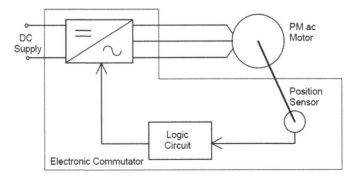

Fig. 1 Typical BLDC motor

affecting the power factor of the system. Many power factor correction techniques [13–16] were discussed. For the operation of BLDC motor, the available AC supply is to be rectified. Elimination of harmonics affecting power factor at the source side due to rectifier is an important task. Effective switching of devices in rectifier can improve power factor. Use of IGBT based pulse width modulation (PWM) rectifier can improve power factor at the source side. The same rectifier rectifies AC-DC and drives supply to converter of BLDC.

In this paper, BLDC motor drive without PWM rectifier was discussed but here in this case a simple diode bridge rectifier was used for AC-DC conversion. Models of BLDC motor drive with PWM rectifier for power factor correction was developed and validated by considering different conditions. BLDC motor with PWM rectifier at starting, at steady-state and with step variation in DC-link voltage was also discussed validating the operation of BLDC motor with PWM rectifier for power factor correction. Simulation work was carried out by Matlab/Simulink and the results were discussed in detail for each case showing the motor characteristics for the respective cases.

2 BLDC Motor Operation

The general available supply is AC. But the supply required for BLDC motor should be DC. So when AC supply is given, this AC supply should be rectified and sent to the BLDC motor. The internal electronic commutation of BLDC motor takes this DC supply through a DC link capacitor. This DC link capacitor maintains constant DC voltage at the input of inverter circuit before actually exciting BLDC motor. By proper switching of inverter switches, the current in the stator windings can be controlled. The rotor position is sensed by using the hall sensors. This is open loop type of BLDC motor without current control (Fig. 2).

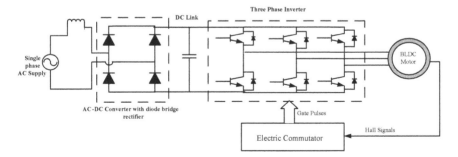

Fig. 2 Diagram of BLDC motor drive with diode bridge rectifier fed from AC supply

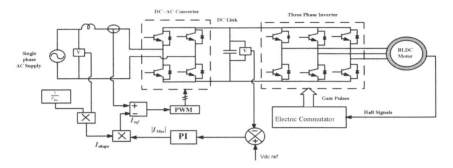

Fig. 3 Diagram of closed-loop speed control of BLDC motor

3 Power Factor Correction for BLDC Motor Drive System

In the operation of BLDC motor, the diode bridge rectifier is replaced with a pulse width modulation (PWM) rectifier which performs a task of rectifier correcting the power factor at source side shown in Fig. 3. This is achieved by switching the IGBT's in rectifier through pulse width modulation.

For power factor correction, the IGBT based rectifier is operated through PWM. The source voltage is measured and is multiplied to inverse of maximum voltage resulting in a current shape. The DC-link voltage is measured and is compared with the reference DC link voltage producing error voltage. This error signal is fed to a PI controller giving out current signal. This current signal is multiplied to current shape obtained earlier producing reference current signal. This reference current signal is measured with the actual current in the line and the error signal obtained is fed to a PWM generator. The PWM generator generates pulses switching ON respective IGBT thus improving power factor at the source side. The IGBT based rectifier performs dual task like rectifying the AC supply to DC feeding converter of BLDC through DC link capacitor and correcting the power factor improving the efficiency of the system when compared to the system employing simple diode bridge rectifier. Diodes are uncontrolled devices and cannot control switching ON or OFF. Replacing diodes with IGBT in bridge rectifier and by switching IGBT's with PWM generator improves system power factor and efficiency as a result.

4 Matlab/Simulink Results and Discussions

Case 1: BLDC motor drive without PFC

Figure 4 shows the model of BLDC motor drive without PFC. Figure 5 shows the source voltage and source current waveforms of BLDC motor drive without

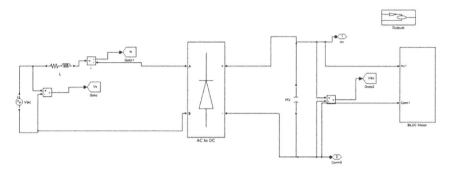

Fig. 4 Simulation model of BLDC motor drive without PFC

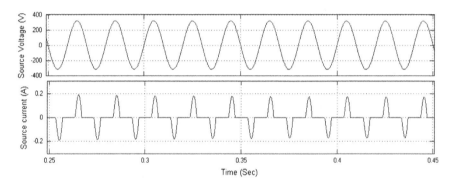

Fig. 5 Source voltage and source current

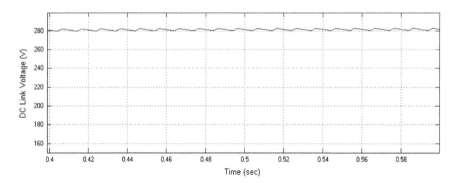

Fig. 6 Simulation result of DC link voltage

PFC. Input source current contains harmonics which can be clearly observed. DC link voltage driving converter of BLDC is shown in Fig. 6.

Figure 7 shows the stator phase currents drawn by BLDC motor. Figure 8 shows the speed of BLDC motor. The torque produced in the BLDC motor was shown in Fig. 9. Source current contains harmonics if BLDC motor drive was operated without PFC. The source current contains 95.89 % of THD as shown in Fig. 10.

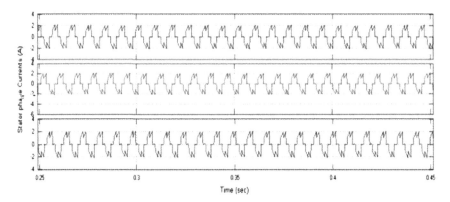

Fig. 7 Simulation result of stator currents

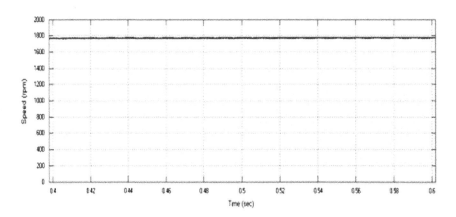

Fig. 8 Simulation result of speed

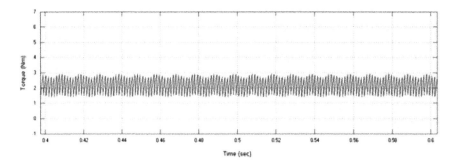

Fig. 9 Torque of BLDC motor without PFC

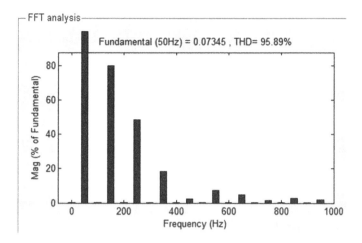

Fig. 10 THD in current of without PFC

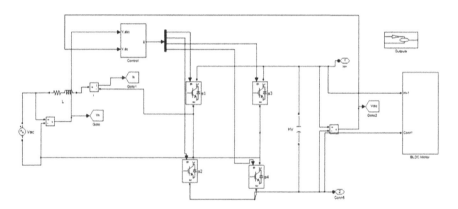

Fig. 11 Simulation model of BLDC motor drive with PFC at starting

Case 2: BLDC motor drive with PFC at starting, $V_{dc} = 50$ V

Figure 11 shows the model of BLDC motor drive with PFC and motor at starting. Figure 12 shows the source voltage and source current waveforms of BLDC motor drive with PFC. Input source current contains very less harmonics which can be clearly observed. DC link voltage driving converter of BLDC is shown in Fig. 13. DC link voltage is maintained constant at 50 V.

Figure 14 shows the stator phase currents drawn by BLDC motor. Figure 15 shows the speed of BLDC motor. The torque produced in the BLDC motor was shown in Fig. 16. Source current contains very less harmonics if BLDC motor drive was operated with PFC.

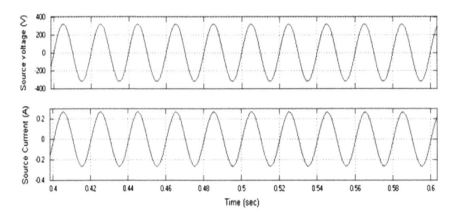

Fig. 12 Source voltage and source current

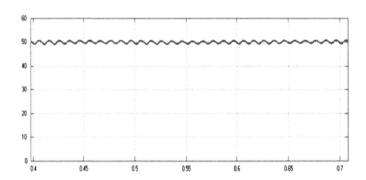

Fig. 13 DC link voltage

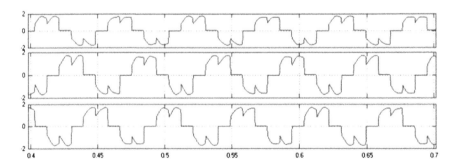

Fig. 14 Simulation result of stator currents

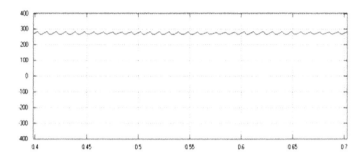

Fig. 15 Simulation result of speed

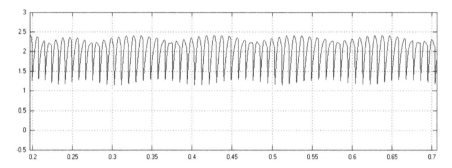

Fig. 16 Simulation result of torque of BLDC motor with PFC at starting

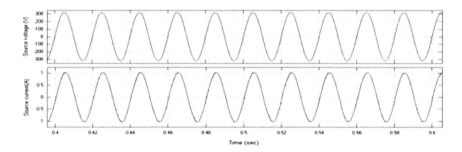

Fig. 17 Source voltage and source current

Case 3: BLDC motor with PFC at steady state condition and $V_{dc} = 200$ V

Figure 17 shows the source voltage and source current waveforms of BLDC motor drive at steady state with PFC. DC link voltage driving converter of BLDC is shown in Fig. 18 maintained constant at 200 V.

Figure 19 shows the stator phase currents drawn by BLDC motor. Figure 20 shows the speed of BLDC motor. The torque produced in the BLDC motor was shown in Fig. 21.

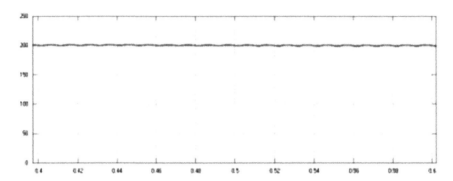

Fig. 18 DC link voltage

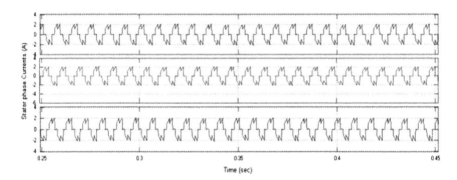

Fig. 19 Simulation result of stator currents

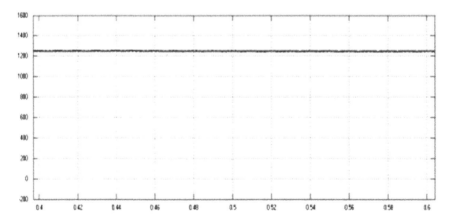

Fig. 20 Simulation result of speed

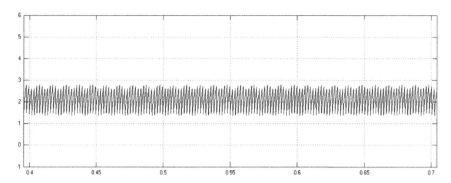

Fig. 21 Simulation result of torque of BLDC motor with PFC at steady-state

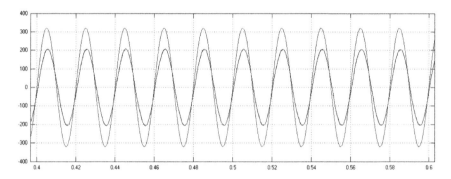

Fig. 22 Simulation result of power factor of system with PFC

Figure 22 shows the power factor of the system with PFC and is maintained nearer to unity. THD in source current is maintained very less at almost 1.5 % as shown in Fig. 23.

Case 4: BLDC motor with PFC with step-change in DC link voltage

Figure 24 shows the source voltage and source current waveforms of BLDC motor drive. The step change is switched at 0.6 s and change in source current can be observed after 0.6 s. Step change in DC link voltage driving converter of BLDC is shown in Fig. 25 maintained constant at 100 V up to 0.6 s and at 0.6 s the DC link voltage is stepped to 150 V.

Figure 26 shows the stator phase currents drawn by BLDC motor. Figure 27 shows the speed of BLDC motor. The torque produced in the BLDC motor was shown in Fig. 28. Step change in DC link voltage is switched at 0.6 s and thus the change in speed and torque characteristics can be observed after 0.6 s when compared to the time before 0.6 s. Table 1 represents the respective THD's in source current with and without PFC.

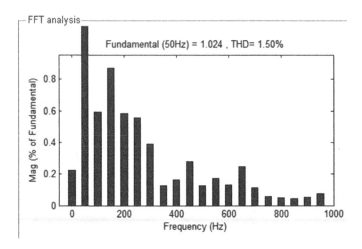

Fig. 23 Simulation result showing THD in source current of system with PFC

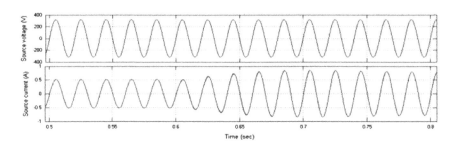

Fig. 24 Source voltage and source current

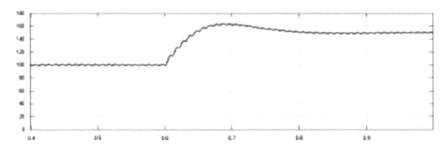

Fig. 25 DC link voltage

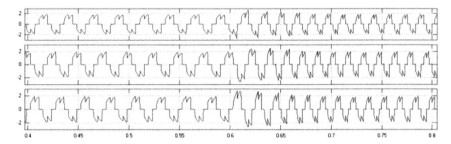

Fig. 26 Simulation result of stator currents

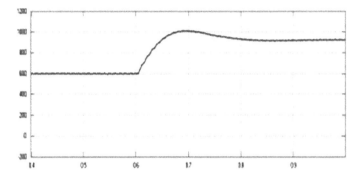

Fig. 27 Simulation result of speed

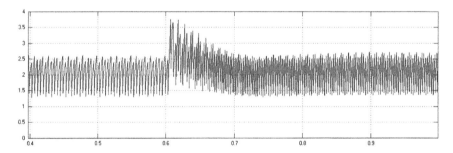

Fig. 28 Simulation result of torque of BLDC motor with PFC with step-change

Table 1 Torque ripple

BLDC motor	THD (%)
System without PFC	95.89
System with PFC	1.5

5 Conclusion

DC motors have very good speed-torque characteristics. DC motors are very much accommodated in many of the industrial drives. Commutation in conventional DC motors was carried out by mechanical parts like brushes and commutator. Presence of brushes for commutation can lead to sparks, losses, reduced efficiency. DC Motors were realized without brushes and mechanical commutator for the commutation purpose called brushless DC (BLDC) motors. BLDC motors eliminate all the disadvantages in conventional DC motors due to the absence of brushes and can give better performance characteristics with smooth speed torque characteristics. BLDC motors use electronic commutator for the purpose of commutation. A converter with solid-state switches was employed to convert DC to AC EMF inside the machine. Models of BLDC motor drive with and without drive system were developed and their respective characteristics were also shown. With power factor correction (PFC) converter, motor was made to run at different operating conditions like motor at starting, motor at steady state and motor with step change in DC link voltage. Characteristics were shown for all cases. Results obtained were simulated using Matlab/Simulink. Results validate the use of BLDC motor drive system with PFC at different motor conditions. THD in source current for the system with and without PFC was compared.

References

1. T.J.E. Miller, Brushless Permanent Magnet and Reluctance Motor Drive (Oxford, 1989)
2. M. Mubeen, Brushless DC Motor Primer (Motion Tech Trends, 2008)
3. P. Yedamale, Hands-on Workshop: Motor Control Part 4 -Brushless DC (BLDC) Motor Fundamentals. Microchip AN885 (2003)
4. D. Liu, Brushless DC Motors Made Easy (Freescale, 2008)
5. D. Arrigo, L6234 Three Phase Motor Driver. ST AN1088 (2001)
6. Sensorless BLDC Motor Control and BEMF Sampling Methods with ST7MC. ST AN1946 (2007)
7. M.R. Feyzi, M. Ebadpour, S.A.K.H. Mozaffari Niapour, A. Feizi, R. Mousavi Aghdam A new single current strategy for high performance brushless dc motor drives, in *International Conference on Electrical and Computer Engineering*, IEEE CCECE 2011, pp. 419–424
8. H.K. Samitha Ransara, U.K. Madawala, A low cost drive for three phase operation of brushless DC motors, in *International Conference on IEEE Industrial Electronics Society*, IECON 2011, pp. 1692–1697
9. R. Krishnan, Permanent Magnet Synchronous and Brushless DC Motor Drives (CRC Press, 2010)
10. B. Akin, M. Bhardwaj, Trapezoidal Control of BLDC Motor Using Hall Sensors (Texas Instruments, 2010)
11. C. Xia, Z. Li, T. Shi, A control strategy for four switch three phase brushless DC motor using single current sensor. IEEE Trans. Ind. Electr. **56**(6), 2058–2066 (2009)
12. P. Kumar, P.R. Sharma, A. Kumar, Simulation and design of power factor correction prototype for BLDC motor control. Eur. Sci. J. **9**(12) (2013)

13. G. Moschopoulos, P. Kain, A novel single-phase soft-switched rectifier with unity power factor and minimal component count. IEEE Trans. Ind. Electron. **51**(3), 566–575 (2004)
14. W. Choi, J. Kwon, E. Kim, J. Lee, B. Kwon, "Bridgeless boost rectifier with low conduction losses and reduced diode reverse-recovery problems. IEEE Trans. Ind. Electron. **54**(2), 769–780 (2007)
15. E.H. Ismail, Bridgeless SEPIC rectifier with unity power factor and reduced conduction losses. IEEE Trans. Ind. Electron. **56**(4), 1147–1157 (2009)
16. B. Lu, R. Brown, M. Soldano, Bridgeless PFC implementation using one cycle control technique, in *Proceedings of IEEE Application Power Electron Conference* (2005), pp. 812–817

Threshold Based Clustering Algorithm Analyzes Diabetic Mellitus

Preeti Mulay, Rahul Raghvendra Joshi, Aditya Kumar Anguria, Alisha Gonsalves, Dakshayaa Deepankar and Dipankar Ghosh

Abstract Diabetes Mellitus is caused due to disorders of metabolism and its one of the most common diseases in the world today, and growing. Threshold Based Clustering Algorithm (TBCA) is applied to medical data received from practitioners and presented in this paper. Medical data consist of various attributes. TBCA is formulated to effectually compute impactful attributes related to Mellitus, for further decisions. TBCAs primary focus is on computation of Threshold values, to enhance accuracy of clustering results.

Keywords Incremental clustering · Knowledge augmentation · Closeness factor based algorithm (CFBA) · Threshold based clustering · Diabetes mellitus · Data mining and TBCA

Please note that the LNCS Editorial assumes that all authors have used the western naming convention, with given names preceding surnames. This determines the structure of the names in the running heads and the author index.

P. Mulay (✉) · R.R. Joshi · A.K. Anguria · A. Gonsalves · D. Deepankar · D. Ghosh
Department of CS and IT, Symbiosis Institute of Technology (SIT),
Symbiosis International University (SIU), Pune, India
e-mail: preeti.mulay@sitpune.edu.in

R.R. Joshi
e-mail: rahulj@sitpune.edu.in

A.K. Anguria
e-mail: aditya.anguria@sitpune.edu.in

A. Gonsalves
e-mail: alisha.gonsalves@sitpune.edu.in

D. Deepankar
e-mail: dakshayaa.deepankar@sitpune.edu.in

D. Ghosh
e-mail: dipankar.ghosh@sitpune.edu.in

© Springer Nature Singapore Pte Ltd. 2017   27
S.C. Satapathy et al. (eds.), *Proceedings of the 5th International Conference on Frontiers in Intelligent Computing: Theory and Applications*, Advances in Intelligent Systems and Computing 516, DOI 10.1007/978-981-10-3156-4_3

1 Introduction

Diabetes is emerged as a major healthcare problem in India and every year it is affecting large number of people. The data science based Knowledge Management System (KMS) in health care industry is getting attention to draw effective recommendations to cure the patient in its early stages [1, 2]. The knowledge augmented through KMS is an asset for society and incremental learning triggers knowledge augmentation [3, 4]. Online interactive data mining tools are available for incremental learning [5]. The threshold acts as a key in incremental learning to investigative formed closeness factors [6]. This approach in a way may change pattern of diabetes diagnosis [6–10]. In this study proposed TBCA is applied on the values of attributes that are collected from patient's medical reports. TBCA implementation unleashes hidden relationships among attributes to extract impactful and non impactful attributes for diabetes mellitus.

In Sect. 2, TBCA is presented. In the following sections i.e., in Sect. 3 the methodology used for its implementation, in Sect. 4 analysis of obtained results, in Sect. 5 concluding remarks and at the last section, references used to carry out this study are listed.

2 TBCA

This section presents a high level pseudo code for TBCA in two parts to show TBCA is an extended version of Closeness Factor Based Algorithm (CFBA).

Input: Data series (DS), Instance (I)
Output: Clusters (K)

CFBA	TBCA

1. Initial cluster count K = 0.
 2. Calculate closeness factor (CF) for series DS(i).
 3. Calculate CF for next series DS(i+1).
 4. Based on CF cluster formation takes place for considered data series (DS).
 5. If not (processed_Flag) then
    CF(newly added cluster) = x_i
    ins_counter(newly added cluster) = 1
    Clusters_CFBA ← Clusters ∪ newly
    added cluster
	6. for all $x_i \in I$
	 7. As processed_Flag = False
	 8. For all clusters ∈ clusters do
	 9. if ‖ x_i - center(cluster)‖ < threshold then
	 10. Update center(cluster)
	 11. ins_counter(cluster)
	 12. As processed_Flag = True
	 13. Exit loop
	 14. end if
	 15. end for

3 Methodology Used to Implement TBCA

TBCA data set considers medical reports of working adult diabetic patients having age group between 35–45 years for the year 2015–2016. TBCA works in three different phases as mentioned below:

(1) In pre-processing input is taken as a CSV file and closeness factor value is calculated by taking into account different possibilities like sum wise, series wise, total weight and error factor for each data series set. The computed values are exported as a CSV file.

(2) In clustering, clusters are formed based on closeness values that are generated through preprocessing for a particular data series and formed clusters are stored in a new CSV file in an incremental fashion.

(3) Post clustering phase is used to extract values of attributes from the formed clusters for further analysis. The attributes related to diabetes mellitus are extracted on the basis of threshold where lower limit is mean of a cluster and upper limit is its higher value. These eight attributes are mentioned in Table 1 where first four are impactful and remaining are non impactful. The following figures represent processing done on 5 K data sets during phases of TBCA in a single and in multiple iterations.

4 TBCA's Analysis

TBCA aims to find out impactful and non impactful attributes and for the same following types of analysis are carried out.

(1) Related attributes analysis: The mean value of each attribute of every cluster is taken into account to analyze related attributes in a single and multiple iterations on data sets as shown in Figs. 1 and 2. The graphs for some of the related attribute analysis are shown below and they depict their behaviour pattern graphically (Fig. 3).

Table 1 Impactful and non impactful attributes for diabetes mellitus

Sr. No.	Name of attribute	Range of attributes in mg/dl
1	Blood glucose fasting	115–210
2	Blood glucose PP	140–250
3	Cholesterol	140–250
4	Triglycerides	140–300
5	HDL cholesterol	40–60
6	VLDL	20–60
7	LDL cholesterol	60–115
8	Non HDL cholesterol	120–170

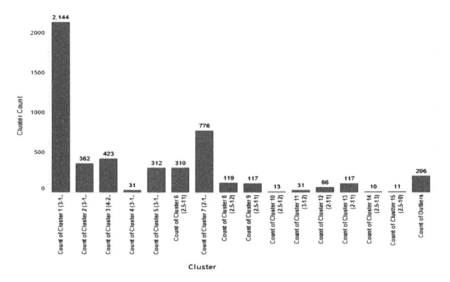

Fig. 1 Processing of 5 K data series in single iteration of TBCA

(2) Outlier analysis to extract impactful attributes: The outlier deviation analysis of datasets with extracted eight attributes is carried out which results in depiction of the deviation of the outlier values from the cluster deviation values. The generated pattern in shown in outlier analysis and it is observed that outlier detection in clustering plays a vital role. The patterns depicted via the statistical graph in Cluster 2 deviation versus outlier deviation for diabetes datasets in Fig. 4. In Fig. 4, after analysis of deviation of each cluster against the outlier deviation, it is observed that attributes BLOOD GLUCOSE FASTING, BLOOD GLUCOSE PP, CHOLESTEROL and TRIGLYCER-IDES are the main factors that are responsible for the generation of the outliers as deviation of the other cluster attributes are overlapping with the outlier deviation. This pattern is cross verified through cluster 2 averages versus outlier average graph shown in another part of Fig. 4.

4.1 Accuracy/Purity of TBCA

The following formula is used for calculation of accuracy or purity of TBCA.

$$= \left(100 - \frac{(\text{Clustering value of multiple iteration} - \text{Clustering value of single iteration})}{\text{Clustering value of multiple iteration}} \times 100 \right)$$

where clustering value = cluster count for cluster that contains maximum clustered data for a particular iteration.

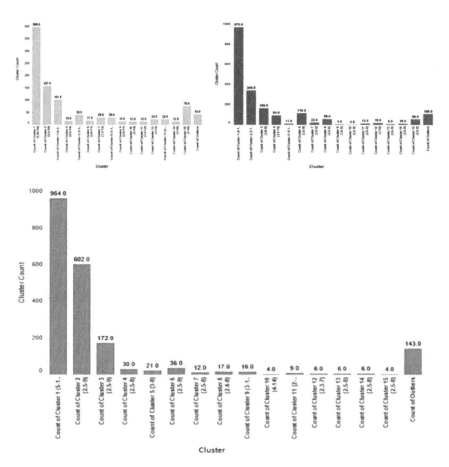

Fig. 2 Processing of 5 K data series in multiple iterations of TBCA

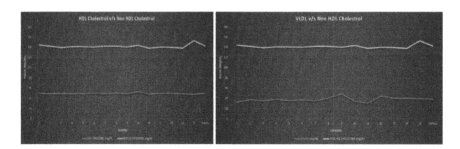

Fig. 3 HDL versus Non HDL Cholesterol, VLDL versus Non HDL Cholesterol analysis

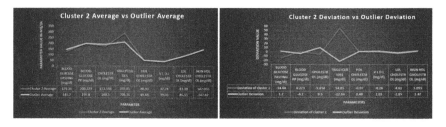

Fig. 4 Clusters, outlier average and clusters deviation, outlier deviation analysis

The accuracy/purity of TBCA is based on clustering value for single iteration and in multiple iterations on same dataset. As shown in Figs. 1 and 2, the first cluster has the maximum weight age (42 and 46 % of the total data resides there) and hence it contains maximum clustered datasets. Therefore, the cluster count or clustering value of this cluster is used calculate the accuracy or purity of TBCA. This accuracy signifies processing of raw datasets and creation precise clusters in single as well as multiple iterations as shown in Figs. 1 and 2 over the same datasets. The multiple iterations on same dataset work in an incremental fashion and confirm cluster members independent of their order, CFBA parameters.

5 Concluding Remarks and Outlook

TBCA proved to be very useful in obtaining inter attribute relationship and outlier value knowledge over various iterations in an accurate manner which eventually triggered towards finding of key attributes related to diabetes mellitus. TBCA has showed 91.9 % of accuracy over single or in several iterations on data set under consideration. It can be effectively used in healthcare domain for prediction of a particular disease like diabetes mellitus. It involves novel mechanism of formation of clusters based on closeness factor and then by using threshold to extract required attributes leading to crisp prediction of impactful set of attributes among them for diabetes mellitus. If a person is suffering from diabetes mellitus properly keeps track of impactful attributes then he/she can manage to cure at early stages. These extracted impactful attributes can act as a catalyst for IT industries for those that are working on medical reports of patients in order to suggest life style management recommendations to cure them from certain diseases. These impactful attributes can also bring revolution in diabetic mellitus patient's treatment in terms of test on a patient for its diagnosis. TBCA algorithm in turn plays a vital role in augmentation of generated knowledge for diabetes mellitus and may also change current way of pathology practices for diagnosis of diabetes mellitus. So, TBCA may prove best in all other disease prediction, being applied across domain, not restricted.

References

1. K.R. Lakshmi, S.P. Kumar, Utilization of data mining techniques for prediction of diabetes disease survivability. Int. J. Sci. Eng. Res. **4**(6), 933–940 (2013)
2. D.S. Vijayarani, M.P. Vijayarani, Detecting outliers in data streams using clustering algorithms. Int. J. Innov. Res. Comput. Commun. Eng. **1**(8), 1749–1759 (2013)
3. P. Mulay, P.A. Kulkarni, Knowledge augmentation via incremental clustering: new technology for effective knowledge management. Int. J. Bus. Inf. Syst. **12**(1), 68–87 (2013)
4. P.A. Kulkarni, P. Mulay, Evolve systems using incremental clustering approach. Evol. Syst. **4**(2), 71–85 (2013)
5. M. Borhade, P. Mulay, Online interactive data mining tool. Proc. Comput. Sci. **50**, 335–340 (2015)
6. P. Mulay, Threshold computation to discover cluster structure: a new approach. Int. J. Electr. Comput. Eng. (IJECE), **6**(1) (2016)
7. R.J. Singh, W. Singh, Data mining in healthcare for diabetes mellitus. Int. J. Sci. Res. (IJSR) **3**(7), 1993–1998 (2014)
8. S.M. Gaikwad, P. Mulay, R.R. Joshi, Attribute visualization and cluster mapping with the help of new proposed algorithm and modified cluster formation algorithm to recommend an ice cream to the diabetic patient based on sugar contain in it. Int. J. Appl. Eng. Res. **10** (2015)
9. M.W. Berry, J.J. Lee, G. Montana, S. Van Aelst, R.H. Zamar, Special issue on advances in data mining and robust statistics. Comput. Stat. Data Anal. **93**(C), 388–389 (2016)
10. M.S. Tejashri, N. Giri, Prof S.R. Todamal, Data mining approach for diagnosing type 2 diabetes. Int. J. Sci. Eng. Technol. **2**(8), 191–194 (2014)

Bio-Inspired Algorithms for Mobile Location Management—A New Paradigm

Swati Swayamsiddha, Smita Parija, Sudhansu Sekhar Singh
and Prasanna Kumar Sahu

Abstract Mobile location management (MLM) has gained a new aspect in today's cellular wireless communication scenario. It has two perspectives: location registration and location search and a trade-off between the two give optimal cost for location management. An outline of the prominent solutions for the cost optimization in location management using various bio-inspired computations is surveyed. For solving complex optimization problems in various engineering applications more and more such bio-inspired algorithms are recently being explored along with incremental improvement in the existing algorithms. This paper surveys and discusses potential approaches for cost optimization using fifteen bio-inspired algorithms such as Artificial Neural Network, Genetic Algorithm to newly developed Flower Pollination Algorithm and Artificial Plant Optimization. Finally, we survey the potential application of these bio-inspired algorithms for cost optimization in mobile location management issue available in the recent literature and point out the motivation for the use of bio-inspired algorithms in cost optimization and design of optimal cellular network.

Keywords Mobile location management · Bio-inspired algorithms · Location update · Paging · Optimization

S. Swayamsiddha (✉) · S.S. Singh
School of Electronics Engineering, KIIT University,
Bhubaneswar, Odisha, India
e-mail: swayamsiddhafet@kiit.ac.in

S.S. Singh
e-mail: ssinghfet@kiit.ac.in

S. Parija · P.K. Sahu
NIT Rourkela, Rourkela, Odisha, India
e-mail: smita.parija@gmail.com

P.K. Sahu
e-mail: pksahu@nitrkl.ac.in

© Springer Nature Singapore Pte Ltd. 2017
S.C. Satapathy et al. (eds.), *Proceedings of the 5th International Conference on Frontiers in Intelligent Computing: Theory and Applications*, Advances in Intelligent Systems and Computing 516, DOI 10.1007/978-981-10-3156-4_4

1 Introduction

In the current times, one of the research focus in the wireless communication domain lies in the design of computationally efficient next generation mobility management system based on bio-inspired algorithms which can provide optimal network configuration by reducing the spectrum utilization as well as the overhead location management cost. Due to rapid progress in technology with multiple applications and exponential rise in mobile users, there is an utmost need to track the mobility pattern of the mobile terminals by the networks. The current location update is required so that an incoming call can be routed to the particular mobile terminal without much delay. Mobile location management (MLM) involves two problems in particular i.e. location update and paging and the ongoing research focuses a tradeoff between the two such that the MLM cost is optimized [1].

Currently, one of the most challenging issues in MLM in mobile communications is selection of suitable strategy for various network environments [2] and choosing appropriate intelligent algorithms from a range of bio-inspired techniques for the optimization of the overhead cost. The developing stages of bio-inspired computing techniques are introduced from its application perspective to MLM by describing the upgrades of its importance in cellular networks design, and the relationship between mobility management technology and bio-inspired optimization.

In course of time the area of bio-inspired computing is getting highlighted. Because of the growing complexity of the optimization problems whose exact solution is not feasible, the bio-inspired techniques prove beneficial. Among the metaheuristics, the bio-inspired algorithms are gaining prominence which can adapt, adjust and accommodate just like the biological entities. Each algorithm is presented for MLM strategies effectively and efficiently.

The remaining paper is structures into the following sections: the second section presents the problem formulation for MLM as an optimization problem. The third section gives an overview of the bio-inspired algorithms. This is followed by the potential scope of application of these algorithms in MLM field. The last section provides useful insights on the exploration of bio-inspired optimization algorithms for location management domain. This paper does not provide implementation details of bio-inspired techniques rather the unexplored areas are identified for future research.

2 Problem Formulation for Mobile Location Management

MLM constitutes two operations: location registration and location inquiry. The mobile terminal updates its current position once it changes the operating cell location which is referred to as location registration whereas, in location search the cellular network traces the cellular position of the residing mobile station so that the arriving call for the mobile user can be directed to the corresponding user.

MLM cost comprises of weighted sum of location registration cost and location search cost given in Eq. (1)

$$Cost = C \times N_{UP} + N_P \tag{1}$$

where, N_{UP} represents the location registration or update cost and N_P represents the paging or search cost over a period of time. The registration cost is generally higher than the search cost by a factor represented as C which is usually taken as a constant set to 10.

Here the fitness function *Cost* is to be optimized (reduced) to obtain best cellular wireless communication network configuration using the heuristic and meta-heuristic optimization techniques which is summarized in the next section.

3 Bio-Inspired Algorithms—An Overview

3.1 Artificial Neural Networks (ANN)

Artificial Neural Networks are a class of machine learning which imitates the functioning of the neurons, the nerve cells of human brain which corresponds to continuous-time nonlinear dynamic system [3]. The strength of ANN lies in the learning and adaptation to the changing environment where numerous processing units are connected to each other in a network in multiple layers. The neural network depicts a layered network structure which works on the inputs to give outputs with learning ability based on inputs and feedbacks which may be positive or negative. It consists of a set of adaptive weights which are tuned by a learning algorithm.

3.2 Genetic Algorithm (GA)

GA was proposed by Holland [4] and was developed by Goldberg and De Jong. GAs is stochastic search mechanisms that utilize a Darwinian criterion of population evolution. The survival of an individual depends on ability to adapt to its environment which reproduces better individuals as well with each growing generation. The population (set of chromosomes) is updated via crossover, mutation and selection which give new set of population called offspring. In the selection stage the individuals having better fitness survive for the future generations following the "survival of the fittest" criterion.

3.3   Differential Evolution (DE)

First introduced by Storn and Price [5] DE is a simple, stochastic and effective tool for global numerical optimization. It is very much similar to GA where the initial population also goes through genetic cycle where differential mutation is performed prior to the crossover operation. The selection of the best individuals for next generation depends on the optimized cost criterion. DE is used for multidimensional real-valued functions optimization which does not require the optimization problem to be differentiable.

3.4   Artificial Immunological Computation (AIC)

AIC came to prominence in mid 1990 where significant work is done using immune networks. It is inspired by the immune system, immune functions, principles and models and is based on human body's immune function of memory and learning [6].

Immunity refers to a condition in which an organism can resist disease which can be innate or acquired. In the immunological computation model there are mechanisms for antigen–antibody recognition. There are various approaches associated with this theory like immune network model, negative selection algorithm, clonal selection algorithm, dendritic cell algorithms, and danger theory which have also been modified for improved solutions.

3.5   Particle Swarm Optimization (PSO)

PSO is swarm-based optimization technique proposed in 1995 by Shi and Eberhart [7]. It is inspired by the food search mechanism of bird flock. The main idea is to simulate the choreography of the flock of birds. The scenario illustrated here suggests the methodology that the PSO follows: A bird flock is searching for food-grains which are in one particular location of that area. None of the birds know the location of the grains but in each iteration they have the knowledge of how far the food is. So, the most effective principle is following the bird which is almost near to the food-grain location. Each bird or particle is considered as a potential solution in the d-dimensional space having some fitness value. The fitness function to be optimized is evaluated for every particle, and velocity and position update is carried out in every cycle where two values called pbest (personal best) and gbest (global best) are also evaluated for every particle. Thus, the optimal solution is achieved by updating the subsequent generations.

3.6 Ant Colony Optimization (ACO)

ACO proposed by an Italian scientist Dorigo is successful in solving many discrete optimization problems [8]. The pheromone, a chemical secreted by the ants influences the travel pattern of its peer ants when they are locating the food. When the ants are confronted to the obstacle on the chosen route, they immediately change their path to the next efficient route to the food source. A higher quantity of pheromone implies a stronger impetus on the path chosen by the ants. It is a very promising research field as it involves derivative free optimization having lot of application areas.

3.7 Artificial Bee Colony (ABC)

ABC algorithm was developed by Karaboga and Basturk [9] is based on the foraging behavior of honey bees swarm. Since then several variants of the basic algorithm have been proposed. The steps of this algorithm are: the employed bees identify the food sources (candidate solutions) with a random stimulus initially. Then the fitness value of the food sources is computed and in the following step better food sources are identified. Meanwhile this knowledge of the food sources is passed on by the employed bees to the onlooker bees who in turn choose the food sources with better fitness values.

3.8 Bacterial Foraging Optimization (BFO)

Developed in 2009, this technique is based on the food gathering behavior of E-Coli bacterium which is replicated as an optimization problem [10]. The principle underlying the foraging theory is that the bacteria find its food in such a mechanism that the energy derived from it per unit time is maximum. The foraging constraints include physiology, predators/prey, environment etc. BFO consists of four prominent stages namely chemotaxis, swarming, reproduction and elimination and dispersion. Chemotaxis involves a tumble followed by a tumble or tumble followed by a run. Swarming means cell to cell signaling via an attractant. In the reproduction stage the population is sorted in ascending order of fitness function. Finally, in the elimination and dispersion cycle the bacterium moves to a new location according to a preset probability.

3.9 Leaping Frog Algorithm (LFA)

LFA technique is inspired from the hunting nature of the frogs [11]. The initial population comprises of virtual frogs generated randomly. First, the whole set of frogs is divided into partitions called memeplexes which in turn is again subdivided into subsets of sub-memeplexes. Here the frog giving worst result will jump towards its prey based on its own memory as well as the experience of the frog giving best results in that memeplex. If the new jump gives better results this process will be repeated otherwise there is generation of a new frog. The sub-memeplexes which gives better results are chosen and again partitioned into new memeplexes and the process continues till the termination criteria is met.

3.10   Cuckoo Search (CS)

Proposed in 2009 Cuckoo search is a bio-inspired technique inspired by the breeding nature of the cuckoo which lay eggs in the host bird's nest [12]. It is a metaheuristic algorithm that can address complex nonlinear optimization problems in practical engineering domain. The cuckoo search is as follows: In the nest the eggs are the solutions and the egg of the cuckoo is a newer solution. At a time, each cuckoo lays one egg in the nest which is randomly selected. The nests of the host birds are fixed in number and with a certain probability the cuckoo's egg is discovered by the host bird. The nests with better solutions are passed on to the future generations.

3.11   Firefly Algorithm (FA)

Yang [13] proposed the firefly algorithm which is based on the flashing nature of the fireflies. The firefly's flash attracts other fireflies by acting as a signal. The intensity of the flashlight corresponds to the fitness value. The bioluminescence signaling indicates other fireflies to deter the predators. This algorithm gives the training for balance between global exploitation and local exploitation. Currently many variants of FA are proposed like adaptive FA, discrete FA, multiobjective FA, chaotic FA, lagrangian FA and modified FA. This algorithm has also been hybridized with bio-inspired algorithms.

3.12   Cat Swarm Optimization (CSO)

CSO, a bio-inspired technique formulated in 2006 by Chu, Tsai and Pan based on the behavior of cats [14]. For most of the time the cats are in their resting period but still active called seeking mode and they take speedy move when awake called tracing mode. In CSO the cats are the candidate solutions having a fitness value. There are two modes: First, the seeking mode which is a resting phase where the cat alertly looks around for its next movement and the second, tracing mode where the cat's velocity and position update takes place. The parameters involved in the seeking mode are seeking memory pool (SMP), counts of dimension to change (CDC) and seeking range of the selected dimension (SRD). Many complex optimisation problems have been solved using CSO which provides satisfactory results.

3.13 Artificial Bat Algorithm (ABA)

ABA is a popular meta-heuristic, which can solve the global optimization tasks. In 2010, Xin-She Yang proposed this algorithm which is based on the echolocation nature of the bats, a flying mammal [15]. It has a very promising application as seen from its implementation and comparison. The control parameters are the pulse frequencies, rate of pulse emission and the loudness which can be fine-tuned to improve the convergence rate of the bat algorithm and control the generation of new solutions close to global optimal solution.

3.14 Flower Pollination Algorithm (FPA)

Yang (2012) proposed the FPA which is based on the process of pollination in the flowering plants. In this algorithm the cross-pollination corresponds to global optimization where a Levy flight is performed by the vectors carrying pollen grains [16]; while the local optimization corresponds to the mechanism of self-pollination. The cross-pollination involves the transfer of the pollens between flowers of two different plants using biotic vectors like insects and birds whereas the self-pollination corresponds to pollination in the flowers of same plant by means of abiotic factors like wind or water. This bio-inspired algorithm can be used for solving real-world applications.

3.15 Artificial Plant Optimization Algorithm (APO)

APO is a bio-inspired algorithm designed to address non-differential, multimodal optimization problems which are based on the growing plant. The branches of the plant represent the potential solution which is initially selected in random. The environment of the growing plant corresponds to the search space. The branches are evaluated for calculating the fitness value and the photosynthesis and phototropism mechanisms determines the new solutions. The photosynthesis operator measures the amount of energy produced whereas, phototropism operator controls the movement of the branches corresponding to the light source [17].

4 Potential Application of Bio-Inspired Algorithms in MLM

In mobile location management, the objective is minimizing the total location registration cost and the search cost which can be potentially performed by the bio-inspired algorithms. The cost reduction in location management has been carried out using various heuristic and meta-heuristic approaches but there is significant scope in applicability of newly developed bio-inspired techniques in this field. The Fig. 1 shows the percentage distribution of publications in Scopus for the location management problem using different bio-inspired algorithms which illustrates that MLM using ANN has 14 % distribution and MLM using GA, PSO, DE, ACO, ABC has less than 4 % distribution in Scopus. So, from the analysis plot it is ascertained that a lot of work can be explored in this area by the scientific community. There are some reviews of bio-inspired algorithms which have focused mainly on the classic algorithms but the newly developed ones have not been captured. Reviews of bio-inspired algorithms for cost minimization in mobile location management are also somewhat sparse. So, this paper significantly overviews the application of bio-inspired optimization tools for reducing the location registration cost and location search cost in mobile location management problem. Such review will definitely add to the knowledge the scope of application in this domain which is still unexplored.

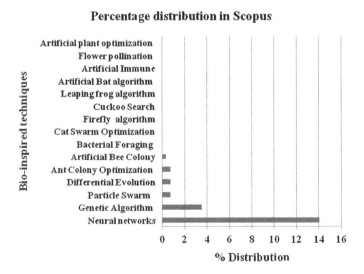

Fig. 1 Percentage distribution of application of bio-inspired algorithms for MLM in Scopus

5 Conclusion and Future Direction

This paper presents an overview of the bio-inspired algorithms which have potential application for cost optimization in location management problem. In recent years, these meta-heuristic optimization techniques have been the focal point of research in this domain except the newly developed ones whose applicability still remains unexplored. While there are different approaches for cost optimization in mobile location management problem, presently the bio-inspired algorithms have captured the attention of the research community for solving the cellular network related issue. The objective of this work is not only to introduce the importance of these algorithms for solving the real time applications but also to provide insights in terms of scope and applicability of these techniques for mobile location management. This survey provides a direction for further research to choose the appropriate bio-inspired algorithms based on their fitment and investigate their performance in comparison to conventional techniques for mobile location management problem.

References

1. R. Subrata, A.Y. Zomaya, A comparison of three artificial life techniques for reporting cell planning in mobile computing. IEEE Trans. Parallel Distrib. Syst. **14**(2), 142–153 (2003)
2. V. Wong, V. Leung, Location management for next generation personal communication networks. IEEE Netw. **14**(5), 18–24 (2009)
3. S. Grossberg, Nonlinear neural networks: principles, mechanisms, and architectures. Neural Netw. **1**(1), 17–61 (1988)
4. J.H. Holland, Adaptation in natural and artificial systems: an introductory analysis with applications to biology, control, and artificial intelligence (University of Michigan Press, 1975)
5. R. Storn, K. Price, Differential evolution—a simple and efficient heuristic for global optimization over continuous spaces. J. Global Optim. **11**, 341–359 (1997)
6. L.N. De Castro, J. Timmis, *Artificial Immune Systems: A New Computational Intelligence Approach* (Springer, London, New York, 2002)
7. Y. Shi, R.C. Eberhart, Parameter selection in particle swarm optimization, *Evolutionary Programming VII* (Springer, Berlin, 1998), pp. 591–600
8. M. Dorigo, C. Blum, Ant colony optimization theory: a survey. Theoret. Comput. Sci. **344**(2), 243–278 (2005)
9. D. Karaboga, B. Basturk, A powerful and efficient algorithm for numerical function optimization: artificial bee colony (ABC) algorithm. J. Global Optim. **39**, 459–471 (2007)
10. S. Das, A. Biswas, S. Dasgupta, A. Abraham, Bacterial foraging optimization algorithm: theoretical foundations, analysis, and applications, in *Foundations of Computational Intelligence vol. 3: 203*, ed. by A. Abraham, et al. (Springer, Berlin, 2009), pp. 23–55
11. J.A. Snyman, The LFOPC leap-frog algorithm for constrained optimization. Comput. Math Appl. **40**(8), 1085–1096 (2000)
12. X.-S. Yang, S. Deb, Cuckoo search via levy flights, in *Proceedings of the World Congress on Nature and Biologically Inspired Computing,* NaBIC 2009 (2009), pp. 210–214)
13. X.-S. Yang, Firefly algorithm, stochastic test functions and design optimisation. Int. J. Bio-Inspired Comput. **2**, 78–84 (2010)

14. S.-C. Chu, P.-W. Tsai, J.-S. Pan, Cat swarm optimization, in *Trends in artificial intelligence*, vol. 4099, ed. by Q. Yang, G. Webb (Springer, Berlin, 2006), pp. 854–858
15. X.-S. Yang, A new metaheuristic bat-inspired algorithm, in *Nature inspired cooperative strategies for optimization (NICSO 2010): 284*, ed. by J. González et al. (Springer, Berlin, 2010), pp. 65–74
16. X.-S. Yang, Flower pollination algorithm for global optimization, in *Unconventional computation and natural computation*, ed. by J. Durand-Lose, N. Jonoska, vol. 7445 (Springer, Berlin, 2012), pp. 240–249
17. Z. Cui, X. Cai, Artificial plant optimization algorithm, in *Swarm Intelligence and Bio-Inspired Computation: Theory and Applications* (2013), (pp. 351–365)

TKAR: Efficient Mining of Top-k Association Rules on Real—Life Datasets

O. Gireesha and O. Obulesu

Abstract Data mining is an important facet for discovering association rules among the biggest scope of itemsets. Association rule mining (ARM) is one of the techniques in data processing with the two sub processes. One is identifying frequent itemsets and the other is association rule mining. Frequent itemset mining has developed as a major issue in data mining and assumes an essential part in various data mining tasks, for example, association analysis, classification, etc. In the structure of frequent itemset mining, the outcomes are itemsets which are frequent in the entire database. Association rule mining is a basic data mining task. Researchers developed many algorithms for finding frequent itemsets and association rules. However, relying upon the choice of the thresholds, present algorithms become very slow and produce a greatly large amount of outcomes or generates few outcomes, omitting usable information. Furthermore, it is well-known that an expansive extent of association rules produced is redundant. This is truly a significant issue because in practice users don't have much asset for analyzing the outcomes and need to find a certain amount of outcomes within a limited time. To address this issue, we propose a one of a kind algorithm called top-k association rules (TKAR) to mine top positioned data from a data set. The proposed algorithm uses a novel technique for generating association rules. This algorithm is unique and best execution and characteristic of scalability, which will be a beneficial alternative to traditional association rule mining algorithms and where k is the number of rules user want to mine.

Keywords Association rules · Data mining (DM) · Frequent itemset mining (FIM) · k · Min_conf · Transactional database

O. Gireesha (✉) · O. Obulesu
Department of Information Technology, Sree Vidyanikethan Engineering College,
Tirupati 517502, Andhra Pradesh, India
e-mail: gireesha93@gmail.com

O. Obulesu
e-mail: oobulesu681@gmail.com

© Springer Nature Singapore Pte Ltd. 2017   45
S.C. Satapathy et al. (eds.), *Proceedings of the 5th International Conference on Frontiers in Intelligent Computing: Theory and Applications*, Advances in Intelligent Systems and Computing 516, DOI 10.1007/978-981-10-3156-4_5

1 Introduction

Data Mining is vital for data analysis in numerous fields as in the references [1–3]. It has pulled in a lot of consideration in the information industry and in the public arena as a whole in recent years because of the wide accessibility of immense measures of data and the impending requirement for transforming such data into valuable information and knowledge. The tasks of data mining are of two types: Descriptive and predictive mining. Descriptive mining characterize the significant features in database. The different techniques of descriptive mining are clustering, association rule mining, summarization and sequential pattern mining. Predictive mining forecast the unknown or hidden values of variables utilizing their existing values. Predictive mining techniques are classification, deviation detection, regression, and prediction (Fig. 1).

Association Rule Mining (ARM) is one of the significant issues in mining prospect which was presented in [4] and extended in [5] to produce association rules. ARM deal with invention of valuable co-occurrence and correlations among itemsets present in vast number of transactions. Rules found fulfill user defined minimum support and confidence thresholds. The generation of association rules can be produced in two parts. Firstly, we can discover frequent itemsets from the datasets, and the secondary deals with the generation of association rules from those frequent patterns [6] (Fig. 2).

Frequent pattern mining is a critical zone of Data mining research. The occurrence of items repeatedly within transactions in a transactional database is called frequent itemsets. Frequent itemset mining will be used to find useful patterns over customer's transaction databases. This kind of finding item set helps organizations to make significant decisions, for example, inventory drawing, cross marketing and customer shopping performance scrutiny. A consumer's transaction database

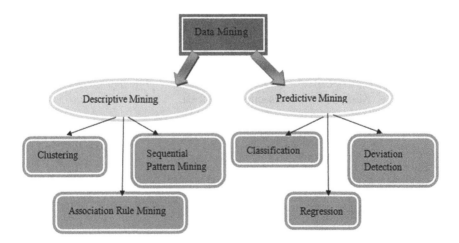

Fig. 1 Categories of data mining tasks [7]

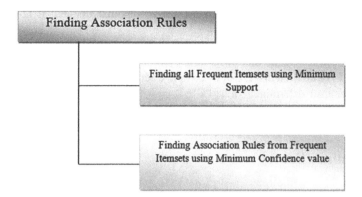

Fig. 2 Generation of Association Rules [8]

comprises of set of transactions, where each and every transaction is an itemset. The purpose of frequent itemset mining is to mine all frequent itemsets in a transactional database. Frequent itemset mining can be ordered into different classifications based upon the type of itemset to be mined. These are discussed in detail in the below [7] (Fig. 3):

(a) **Approximate Frequent Itemset**: When an itemset derive approximate support count from its supersets.
(b) **Closed Frequent Itemset**: An itemset is closed if none of its prompt supersets has the same support as an itemset.
(c) **Constrained Frequent Itemset**: An itemset which satisfy user-specifies set of rules through mining.

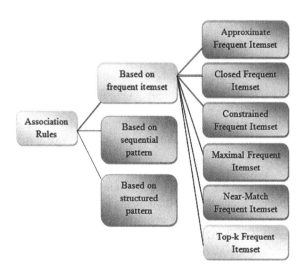

Fig. 3 Kind of pattern to be mined with Association Rules [7]

(d) **Maximal Frequent Itemset**: An itemset is maximal frequent if none of its prompt supersets is frequent.
(e) **Near-Match Frequent Itemset**: An itemset is said to be near-match when its support count is same that of its near-by or similar items.
(f) **Top-k Frequent Itemset**: Out of complete set of frequent itemset, top itemsets that pass user determined value "k".

2 Literature Survey

In the preceding discussion we have studied the fundamental ideas of data mining, Association Rule Mining and frequent itemset mining. This segment will proceed with the literature review of some popular algorithms related to these mining concepts which have been presented from many years in different researches. In this section, we examine closely related past studies and contrast them with our methodology.

The AIS (Agrawal, Imielinski, Swami [4]) algorithm was introduced to mine the association rules [4]. It has some expertise in the standard of databases alongside essential to method call support queries. Through this algorithm one frequent itemsets are produce, which suggests that the following of these guidelines exclusively contain one thing, as a case we tend to exclusively produce rules like A $\cap$ B $->$ C however not those tenets as A $->$ B $\cap$ C. The database is scanned number of times to generate the frequent itemsets in AIS. The bottleneck of this algorithm is several candidate itemsets are produced that requires more space and scans.

The AprioriTID algorithm [5] generates the candidate itemsets. The difference between AIS and AprioriTID is it requires only one database scan and a group of candidate itemsets employed for k > 1. Apriori and AprioriTID use consistent candidate generation methodology and calculation sturdy itemsets. Apriori Hybrid algorithm uses Apriori inside the underlying passes and changes to AprioriTid once it expects that the candidate itemsets at the tip of the pass will be in memory. In the SETM algorithmic guideline, candidate itemsets square measure produced on-the-fly in light because the data is scanned, however counts at the highest point of the pass. The disadvantage of SETM algorithm is similar to the AIS algorithm. Another weakness is that for each candidate itemset, there square measure as a few entries as its support value.

The FP-Growth Algorithm [8], proposed by J. Han requires constructing FP-tree. It is a productive and scalable technique for mining the complete set of frequent patterns through pattern fragment growth, utilizing an amplified prefix-tree structure for storing compacted and pivotal data about frequent patterns (FP) called as Frequent Pattern Tree (FP-tree). This algorithm needs 2 passes irrespective of candidate generation and utilizes divide and conquer approach. The database scanned two times. In the first scan, it maintains a list of frequent items sorted by recurrence

Table 1 Input parameters description

S. no.	Input parameters	Description
1.	TD/Dataset	A transactional database
2.	k	Number of rules user wants to find
3.	Minconf	Minimum confidence threshold value

in dropping order and in the second scan; the FP-tree is generated through the database compression. One issue, which hampers the well known utilization of frequent itemset mining, is that it is troublesome for users to choose a legitimate minimum support threshold. To precisely control the number of the most regular itemsets, is to mine top-k frequent itemsets [9–17] rather than to mine frequent itemsets. The setting of k—value is more instinctive than setting the minimal support threshold because k represents the number of itemsets the user needs to discover while choosing the threshold depends exclusively on database's attributes, which are frequently obscure to users [18]. Despite the fact that these algorithms are different in that they make use of different data structures and pursuit strategies to find top-k frequent itemsets, they follow a same general procedure for finding top-k itemsets (Table 1).

3 Top-k Association Rules Mining (TKAR) Algorithm

The Top-k Association Rules mining algorithm takes as input a transactional database, minimum confidence threshold and k the number of rules user wants to discover. The TKAR algorithm is an advantage to conventional association rule mining algorithms meant for users to control the number of association rules. Table 2 presents an overview of the input parameters for the proposed algorithm.

The proposed TKAR algorithm will work as follows:

Algorithm (dataset, k, minconf)

1. minsup = 1; C = NULL; D = NULL
2. dataset is scanned one time and the sids (sequential ids of each item I is stored in a variable called Tids(i)
3. For every pair of items P & Q Where Tids(P) > = min_sup & Tids (Q) > = min_sup, Set Tids(P => Q) = Null & set Tids(P =>Q) = Null

Table 2 Dataset Characteristics

Datasets	No. of transactions	No. of distinct items	Average transaction size
Chess	3196	75	37
Connect	67557	129	43
Mushrooms	8124	128	23
Pumsb	49046	7116	74

4. For each Tid which belongs to Tids(P) ∩ Tids (Q)
5. If P occurs before Q in the Tid s then Tids(P => Q) = Tids (P => Q) U {s}
    OR If Q occurs before P in Tid s then Tids(Q => P) = Tids (Q => P) U {s}
6. If │ Tids (P => Q) │ / │ Total sequences │ > = minimum support Then
7. Confidence (P => Q) = │ Tids (P =>Q) │ / │ Tids(P) │
8. If confidence (P =>Q) > = minimum confidence then
9. If │ C │ < K then
10. If there exist a rule r2 in C which is similar to the currently generated rule r1 &
    whose support is also similar to the support of r1 then the rule r1 is not added to
    C Otherwise this rule r1 is added to the C. D: D = D U r1.
11. If │ C │ > = K then
12. Remove a rule r from C whose support is equal to the present minimum
    support, Set minimum support = lowest support of rules in C
13. while D is non-empty
14. Do
15. Select a rule r with the highest support in D
16. Expand-Left (rule, N, E, min_sup, min_conf)
17. Expand-Right (rule, N, E, min_sup, min_conf)
18. Remove r from D
19. END

Mining the Top-k association rules has two difficulties:

1. Top-k rules can't depend on minimum support to prune the search space,
    however they may be adjusted to mine frequent itemsets with minimum sup-
    port = 1 to guarantee that all top-k rules can be generated in the generating rules
    step.
2. Top-k association rules mining algorithm can't utilize the two stages procedure
    to mine association rules however they would need to be altered to discover
    top-k rules by generating rules and keeping top-k rules.

4   Experimental Evaluation

For an experimental evaluation of the proposed algorithm (TKAR), we performed
several experiments on real-life datasets. We implemented the proposed algorithm
on a computer equipped with a Core i5 processor running Windows 7 and 4 GB of
main memory running under Java. All datasets (mentioned in the Table 2) were
obtained from the Machine Learning Repository[1] (UCI). The Chess and Connect
datasets are derivative from their respective game steps, the mushroom dataset
contains characteristics of a variety of species of mushrooms, and the Pumsb
consists of the population and housing (census) data. Table 3 shows some

---

[1]http://fimi.ua.ac.be/data/

Table 3 Evaluation for Mushroom Dataset

Min_Conf	Execution time (Milliseconds)				Memory usage (Mega bytes)			
	k = 75	k = 150	k = 600	k = 1500	k = 75	k = 150	k = 600	k = 1500
0.1	6	7	22	35	8.39248	10.28102	12.42887	16.86091
0.2	4	9	22	33	8.61269	9.13397	10.09296	17.32567
0.3	7	10	9	37	9.02098	10.44871	12.13493	17.55789
0.4	7	11	10	38	10.95368	11.74442	13.33350	18.31702
0.5	7	8	6	25	10.10949	8.89583	8.92863	14.42796
0.6	7	10	8	28	11.18904	10.24600	11.28932	24.30187
0.7	8	10	23	36	11.07919	10.73683	13.86193	25.24107
0.8	7	11	14	38	11.14436	11.39779	13.90453	28.53613
0.9	9	6	17	40	11.44678	12.39027	14.01256	34.57773

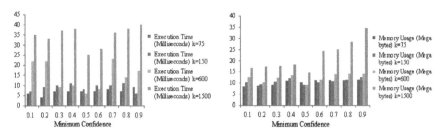

Fig. 4 Detailed results for varying k for the mushroom dataset

characteristics of the real datasets. Typically, these datasets are dense, that is, they generate many frequent itemsets even for very high values of support.

Jeff Schlimmer utilized a database and applied mining techniques with the intention to obtain classifiers which can be used on similar databases. He compiled a mushroom dataset illustrating the records from The Audubon Society Field to North American Mushrooms [6] (1981). The Audubon Society's vision is to preserve and reinstate natural ecosystems. The dataset comprises 8124 instances among them 4208 are classified as edible while 3916 are poisonous and 23 species of gilled mushrooms in the Agaricus and Lepiota families. Each species may recognize as belong to one of the three classes, namely definitely edible, definitely poisonous, or unknown edibility and thus, not recommended. The purpose of choosing the mushroom dataset is to detach edible mushrooms from poisonous ones. It is a classification problem with a target variable of whether mushrooms can be eaten or not. The results generated through TKAR algorithm on mushroom dataset indicates to increase the survival rate when we are in the wild life. If one is capable of distinguishing mushrooms between edible and poisonous by their appearances then we may able to survive in deep mountains by eating mushrooms.

Impact of the k parameter

We have run algorithm by fluctuating estimation of k and considers its values as 75, 150, 600 and 2000 separately. Furthermore we have taken minconf value as 0.1–0.9. As estimation of k increases, both the execution time and memory also increases. This assessment has been appeared underneath in the table and its respective graph. Initially as appeared underneath we consider results of mushroom dataset with various estimation of k. This Table 3 and Fig. 4 indicates memory necessity is directly expanding with estimation of k.

5 Conclusion

The users need to choose the parameter estimations of minconf and minsup in classical algorithm. So amazingly vast number of rules is produced. Due to this the algorithm experience the long execution time and more memory required. We proposed the methodology which produce governs precisely how much user needs.

This proposed algorithm gives preferred execution and performance over others and it is worthwhile to classical algorithms when the user needs to control the number of association rules produced. The top-k association rules mined through the proposed algorithm have the redundancy rules. There is a scope to develop efficient algorithms for mining non-redundant association rules.

References

1. L. Bin, Q. Huayong, S. Yizhen, Realization and application of customer attrition early warning model in security company. TELKOMNIKA Indonesian J. Electr. Eng. **10**(5), 10(5), 1106–1110 (2012)
2. L. Zhou, H. Wang, W. Wang, Parallel implementation of classification algorithms based on cloud computing environment. TELKOMNIKA Indonesian J. Electr. Eng. **10**(5), 10(5), 1087–1092 (2012)
3. A. Ali, M. Elfaki, D. Norhayati, Using Naïve Bayes and bayesian network for prediction of potential problematic cases in tuberculosis. Int. J. Inf. Commun. Technol. (IJ-ICT), **1**(2),1(2), 63–71 (2012)
4. R. Agrawal, T. Imielinski, A. Swami, Mining association rules between sets of items in large databases, in Proceedings of the ACM SIGMOD International Conference on Management of data, vol. 22, no. 2, (ACM, New York, 1993), pp. 207–216
5. R. Agrawal, R. Srikant, Fast algorithms for mining Association Rules in large databases, in Proceedings of the 20th International Conference on Very Large Data Bases (Morgan Kaufmann Publishers Inc., San Francisco, 1994), pp. 487–499
6. D. Jain, S. Gautam, Implementation of apriori algorithm in health care sector: a survey. Int. J. Comput. Sci. Commun. Eng. Nov 2013
7. S. Panwar, Analytical study of frequent Itemset and Utility Mining Approaches. Int. J. Eng. Appl. Technol. 180–187, July 2015. ISSN: 2321-8134
8. Y.L. Cheung, A.W. Fu, Mining frequent itemsets without support threshold: with and without item constraints. IEEE TKDE J. **16**(6), 1052–1069 (2004)
9. T. Le, B. Vo, Bay: An N-list-based algorithm for mining frequent closed patterns. Expert Syst. Appl. **42**(19), 6648–6657 (2015)
10. Q. Huynh-Thi-Le, T. Le, B. Vo, H. Le, H. Bac, An efficient and effective algorithm for mining top-rank-k frequent patterns. Expert Syst. Appl. **42**(1), 156–164 (2015)
11. Z.H. Deng, Fast mining top-rank-k frequent patterns by using node-lists. Expert Syst. Appl. **41**, 1763–1768 (2014)
12. M. Zihayat, A. An, Mining top-k high utility patterns over data streams. Inf. Sci. (2014)
13. K. Chuang, J. Huang, M. Chen, Mining Top-K frequent patterns in the presence of the memory constraint. VLDB J. **17**, 1321–1344 (2008)
14. P. Tzvetkov, X. Yan, J. Han, TSP: Mining top-k closed sequential itemsets. KAIS J. **7**(4), 438–457 (2005)
15. J. Wang, J. Han, Y. Lu, P. Tzvetkov, TFP: an efficient algorithm for mining Top-k frequent closed Itemsets. IEEE TKDE J. **17**(5), 652–664 (2005)
16. J. Han, J. Pei, Mining frequent patterns by pattern-growth: methodology and implications. ACM SIGKDD Explor. Newsl. **2**(2), 14–20 (2000)

17. J. Pei, J. Han, H. Lu, S. Nishio, S. Tang, D. Yang, H-mine: Hyper-Structure mining of frequent Itemsets in large databases, in ICDM, pp. 441–448 (2001)
18. C.W. Wu, B.E. Shie, P. Yu, V.S. Tseng, Mining top-K high utility itemsets, in KDD, pp. 78–86 (2012)

Predicting Defect of Software System

Soumi Ghosh, Ajay Rana and Vineet Kansal

Abstract Any particular study on software quality with all desirable attributes of software products can be treated as complete and perfect provided it is defective. Defects continue to be an emerging problem that leads to failure and unexpected behaviour of the system. Prediction of defect in software system in the initial stage may be favourable to a great extend in the process of finding out defects and making the software system efficient, defect-free and improving its over-all quality. To analyze and compare the work done by the researchers on predicting defects of software system, it is necessary to have a look on their varied work. The most frequently used methodologies for predicting defects in the software system have been highlighted in this paper and it has been observed that use of public datasets were considerably more than use of private datasets. On the basis of over-all findings, the key analysis and challenging issues have been identified which will help and encourage further work in this field with application of newer and more effective methodologies.

Keywords Software defects · Defect prediction · Software datasets · Data mining · Machine learning

S. Ghosh (✉) · A. Rana
Amity University Uttar Pradesh, Sec-125, Noida, India
e-mail: soumighoshphd@gmail.com

A. Rana
e-mail: ajay_rana@amity.edu

V. Kansal
I. T. S. Engineering College, Knowledge Park III, Greater Noida, India
e-mail: dir.engg@its.edu.in

© Springer Nature Singapore Pte Ltd. 2017
S.C. Satapathy et al. (eds.), *Proceedings of the 5th International Conference on Frontiers in Intelligent Computing: Theory and Applications*, Advances in Intelligent Systems and Computing 516, DOI 10.1007/978-981-10-3156-4_6

1 Introduction

In the arena of software engineering, a crucial problem like prediction of defects is often taken into account as a very important step for the purpose of quality improvement obtained in lesser period and by minimum cost. Prediction of defects is highly necessary in order to find out sensitive and defect-prone domains in the stage of software testing, so that it may help in qualitative improvement of the software system with reduced cost. The possibility of detection of potential faults in software system at an early stage may help in effective planning, controlling and execution of software development activities to a considerable extent. In modern days, as the development of software has become very meaningful and keeping pace with the necessity, it may be safely said that hence reviewing and testing of the software system will be very essential and result-oriented in the case of predicting defects. Predicting software defects often involves huge cost and as-such the matter of correction of software defects is altogether a very expensive matter [25]. Those studies which have been carried out in the recent years, reveal the fact that the case of predicting defects assumes more importance compare to testing and reviewing process of software systems [36, 43]. As such, accuracy in predicting software defects is certainly very much helpful in case of improving software testing, minimizing the expenses [10] and improving the software quality [22].

In this paper, although analysis and comparison of various research work (from the year 1992 to 2015) on predicting software defects by using various methodologies have been made but only those unique and most updated methodologies (year 2005–2015) have been highlighted. This paper is having the objective of critically estimate the efficacy of the methods adopted in predicting software defects. Simultaneously, evaluation of the varied systems in prediction of software defects have been made and thus realized the effectiveness and importance of some methodologies like Advance Machine Learning, Neural Network and Support Vector Machine applied most frequently compared to various other techniques for achieving desirable accuracy in predicting defects in the software system. This paper has also highlighted the requirement of further work in this field by applying newer methodologies since the previous ones have not at all been found defect-free or at-least a least defective software system which may finally produce quality software system.

2 Literature Review

In order to perform the analysis, we explored 102 papers (during the period 1992–2015) from various digital library like IEEE Transactions on Software Engineering, ACM, Springer, Elsevier, Science Direct, International Conferences, Reports, Thesis and even technical papers and case studies were also reviewed. After exploring these digital libraries, we found that most of the research work on predicting defects of software system was performed on similar patterns/methodologies/

techniques as well as on nearly same datasets. As such, papers based on similar patterns/methodologies/techniques, datasets were excluded. We included only 49 those papers which are found unique and updated (from the year 2005 to 2015) in this particular field. Since 1992 various methodologies have been applied in predicting defects of software system. But in modern days, various methodologies are basically very favourable in predicting defects in software system. Only those methodologies which were considered unique as well as updated, have been analyzed, compared and the results obtained would help to determine which are the most frequently used and effective methodologies in the field of predicting defects of software system.

2.1 Predicting Defects of Software System Using Data Mining (DM)

Campan et al. [9] experimented with Length Ordinal Association Rule in datasets for searching out any interesting new rules. Song et al. [44] emphasized on Rule Mining methodologies in predicting and correcting software defects. Kamei et al. [26] proposed a methodology combining Logistic Regression analysis with Association Rule Mining for predicting software defects. Chang et al. [12] combined Decision Tree and Classification methodologies-Action Based Defect Prediction (ABDP) along with Association Rule Mining for predicting and discovering software defects pattern with minimum support and confidence. Gray et al. [21] experimented with Support Vector Machine (SVM) classifier based on Static Code Metrics and NASA datasets to maintain defective classes and remove redundant instances. Riquelme et al. [39] applied Genetic Algorithms finding rules featuring subgroups predicting defects and extracted software metrics program dataset from the Promise repository. Gayatri et al. [19] combined Induction methodology with Decision Tree and the new method of feature selection was better as compared to SVM and RELIEF methodologies. Gray et al. [20] analyzed Support Vector Machine (SVM) classifiers based on NASA datasets in such a way that identifies software defects and the basic idea was to classify training data rather than obtaining test datasets. Liu et al. [34] experimented with a new Genetic Programming based search methodology for evaluating the quality of software systems. It found that Validation cum Voting classifier was better than Baseline classifier, Validation classifier. Tao and Wei-Hua [46] found that Multi-Variants GAUSS Naive Bayes methodology was superior as compared to other versions of Naive Bayes methods and J48 algorithm in predicting defects of software system. Catal [11] reviewed different methodologies such as Logistic Regression, Classification Trees, Optimised Set Reduction (OSR), Artificial Neural Networks and discriminate model used during the period 1990 to 2009 on predicting software defects. Kaur and Sandhu [28] found that accuracy level was on higher side in case of software system based on K-Means. Tan et al. [45] attempted prediction of software defects by application of functional cluster of programs vide class or file which significantly improved recall and precision percentage.

Dhiman et al. [15] used a clustered approach in which the software defects will be categorized and measured separately in each cluster. Kaur and Kumar [30] applied clustering methodology for forecasting as well as error forecasting in object-oriented software systems. Najadat and Alsmadi [37] proved Ridor algorithm with other classification approaches on NASA datasets to be an effective methodology for predicting software defects with higher accuracy level. Sehgal et al. [41] focused on application of J48 algorithm of Decision Tree methodology in prediction of defects in software systems. The performance of new methodology was evaluated against the IDE algorithm as well as Natural Growing Gas (NGG) methodology. Banga [7] found that a hybrid architecture methodology called as GP-GMDH or GMDH-GP was more effective as compared to other methodologies on the ISBSG datasets. Chug and Dhall [14] different methodologies were used on different datasets of NASA with both supervised and unsupervised learning methodologies for defect prediction. Okutan and Yildiz [38] for predicting software defects proposed a kernel methodology based on pre-computed kernel metrics. It was observed that the proposed defect prediction methodology was also comparable with other existing methodologies like Linear Regression and IBK. Selvaraj and Thangaraj [42] predicted software defects using SVM and compared its effectiveness with Naive Bayes and Decisions stumps methodologies. Adline and Ramachandran [3] proposed program modules for predicting the fault-proneness when the fault levels of modules are not available. The supervised methodologies like Genetic Algorithm for classification and predicting fault in software were applied. Agarwal and Tomar [4] observed that Linear Twin Support Vector Machine (LTSVM) on the basis of feature selection and F-score methodology was superior to other methodologies. Sankar et al. [40] advocated feature selection methodology using SVM and Naive Baye classifier based on F-mean metrics for predicting and measuring the defects in software system.

2.2   Predicting Defects of Software System Using Machine Learning (MI)

Boetticher [8] analyzed K-Nearest Neighbour (K-NN) algorithm or sampling for predicting software defects and its performance was not effective in case of small datasets. Ardil et al. [5] applied one of the easiest forms of Artificial Neural Network and compared it with other modules of Neural Network. Chen et al. [13] predicted software defects using Bayesian Network and Probabilistic Relational Models (PRM). Jianhong et al. [23] showed that the Resilient Back propagation algorithm based on neural network was superior methodology for predicting software defects. Xu et al. [47] evaluated the effectiveness of software metrics in predicting software defects by applying various Statistical and Machine Learning methodologies. Gao and Khoshgoftaar [17] predicted software defects by use of class-imbalanced and high dimensional database system. In this approach, modelling and feature selection was done on the basis of alternative use of both original and sampled data. Li et al. [32] found that effectiveness of sampled based methodologies like active semi-supervised methodology called as ACoForest was better compared to Random Sampling both with conventional machine learners and semi-supervised learner. Kaur [29] used software metrics along with Neural Network to find out those modules suitable for multiple uses. Abaei and Selamat [1] experimented with the application of various machine learning and artificial intelligent methodologies on different public NASA datasets in connection with predicting software defects. Askari and Bardsiri [6] predicted software defects by using Multilayer Neural Network. Support Vector Machine with the Learning algorithm and Evolutionary methodologies were also used for the purpose of removing the defects. Gayathri and Sudha [18] applied Bell function based Multi-Layer Perceptron Neural Network along with Data Mining for predicting defects in software system and its performance was compared with other Machine Learning methodologies. Jing et al. [24] proposed an efficient model using Advanced Machine Learning methodology-Collaborative representation classification for Software Defect Prediction (CSDP). Kaur and Kaur [27] predicted defects in classes using Machine Learning methodologies with different classifiers. Li and Wang [33] compared various Ensemble Learning methodologies- Ada Boost and Smooth Boost with SVM, KNN, Naive Baye, Logistic and C4.5 for predicting software fault proneness on imbalanced NASA data sets. Malhotra [35] predicted defects and estimated relationship among static code measures, different ML methodologies were applied. Yang et al. [48] used a Learning-to-Rank methodology for predicting defects in software system and also compared its effectiveness with others. Abaei et al. [2] studied the effectiveness of new version of semi-supervised methodology on eight datasets from NASA and Turkish in predicting software defects with high accuracy. Erturk and Sezer [16] proposed a new methodology-Adaptive Neuron Fuzzy Inference System (ANFIS) and compared it with other methodologies (SVM, ANN, ANFIS) using Promise repository for predicting software defects. Laradji et al. [31] Average Probability Ensemble (APE) comprised of seven classifiers was superior to weighted SVM and Random Forest methodologies. Finally, a new version of APE comprised of greedy forward selection was more efficient in removing duplicate and unnecessary features. Zhang et al. [49] predicted software efforts by using methodology based on Bayesian Regression Expectation Maximize (BREM).

3 Methodology

In this paper, a specific methodology was used with the aim of analyzing and comparing only those different, unique and updated methodologies (from the year 2005 to 2015) for predicting defects of software system. Different methodologies were compared on the basis of studies and the results showed that Advance Machine Learning, Neural Network and Support Vector Machine methodologies are the most commonly used techniques for predicting software defects. Summary of major findings are given in Table 1.

Table 1 Summary of major findings of different software defect prediction methodologies

S. no.	Authors	Methodologies	Findings
1.	Boetticher [8]	K-Nearest Neighbour	Its performance was not found satisfactory
2.	Song et al. [44]	Rule Mining, Naive Bayes, PART, C4.5	Accuracy level of proposed method was higher by at least 23 % than other methods
3.	Chang et al. [12]	Association Rule Mining with Action Based Defect Prediction	It was applicable to discover defects as well as to handle continuous attributes of actions
4.	Gray et al. [21]	Support Vector Machine Classifier and Static Code Metrics	It is highly effective and having an accuracy level of about 70 %
5.	Ardil et al. [5]	Artificial Neural Network	Defects were mostly found in the Neural Network modules
6.	Riquelme et al. [39]	Genetic Algorithm (GA) with Promise repository	GA is able to efficiently handle the unbalance datasets that consist of more non-defective than defective sample
7.	Gray et al. [20]	Support Vector Machine	Effectiveness of classifiers in separating training data was experimented
8.	Xu et al. [47]	Neuro-Fuzzy with ISBSG repository	Adapted methodology to improve the accuracy of estimation
9.	Gayatri et al. [19]	Decision Tree Induction method	Its performance is better than all other methodologies
10.	Tao and Wei-Hua [46]	Multi-Variants GAUSS	This method was better than all other Naive Bayes method
11.	Chen et al. [13]	Bayesian Network, Probabilistic Relational Models	Altogether a new model was proposed
12.	Liu et al. [34]	Genetic algorithm-Baseline, Validation, Validation cum Voting classifier	Proposed approach was more effective and accurate when applied with multiple datasets
13.	Jianhong et al. [23]	Resilient Back Propagation Algorithm	Proposed algorithm was applicable to identify the modules having major defects
14.	Catal [11]	Logistic Regression, Classification Trees, Optimised Set Reduction (OSR), Artificial Neural Network and Discriminate model	Performance of OSR methodology was found to be the best among all other methods with an accurate rate of 90 %
15.	Kaur and Sandhu [28]	K-Means clustering approach	It was found effective and having 62.4 % accuracy level in predicting software defects
16.	Gao and Khoshgoftaar [17]	Promise repository based on different criterions	A comparative estimate on effectiveness of all the criterions was made
17.	Li et al. [32]	Semi-supervised model-ACoForest	Defect prediction is not at all effected by size of the datasets

(continued)

Table 1 (continued)

S. no.	Authors	Methodologies	Findings
18.	Tan et al. [45]	Functional cluster approach	Applied for significantly improving recall from 31.6 to 99.2 % and precision from 73.8 to 91.6 %
19.	Najadat and Alsmadi [37]	Ridor algorithm	Better methodology than others and having more accuracy rate
20.	Kaur [29]	Neural Network	Model is effective in improving accuracy level
21.	Sehgal et al. [41]	Decision Tree, J48 algorithm	J48 algorithm is 93.32 % accurate than IDE algorithm as well as Natural Growing Gas (NGG) methodology had an accuracy level of 80 %
22.	Dhiman et al. [15]	Clustered approach	Effective in analyzing defects and integrating software modules
23.	Kaur and Kumar [30]	Clustering methodology	Forecasting error of software system
24.	Chug and Dhall [14]	Classification methods	Different parameters were followed for estimating the effectiveness
25.	Okutan and Yildiz [38]	Linear and RBF kernels	Effective model in case of reducing testing effort and also total project cost
26.	Banga [7]	GP-GMDH	GP-GMDH was superior to all other methodologies
27.	Selvaraj and Thangaraj [42]	SVM	SVM had better performance than Naive Bayes and Decisions stumps methodologies
28.	Adline and Ramachandran [3]	Genetic Algorithm	Software defect prediction under condition of unknown fault level of software
29.	Kaur and Kaur [27]	Machine Learning based on Classifier	J48 and Bagging methodologies were considered most effective than others
30.	Sankar et al. [40]	Support Vector Machine (SVM), Naive Bayes	Naive Bayes was found better than SVM
31.	Abaei and Selamat [1]	Machine Learning, Artificial Immune Systems (AISs)	For, large datasets-Random Forest showed better result. Small datasets-Naive Bayes was more effective. With Feature Selection-Immuons99 outperforms. In absence of Feature Selection-AIRS Parallel was effective
32.	Gayathri and Sudha [18]	Bell function based Multi-Layer Perceptron Neural Network	Proposed approach was effective with an accuracy level of 98.2 %

(continued)

Table 1 (continued)

S. no.	Authors	Methodologies	Findings
33.	Askari and Bardsiri [6]	Multilayer Neural Network	More accurate and précised compared to others
34.	Li and Wang [33]	SVM, KNN, Naive Baye, Logistic, C4.5, Ada Boost and Smooth Boost methods	Smooth Boost method was considered best and most effective as compared to other method
35.	Malhotra [35]	Machine Learning (ML) and Statistical logistic Regression (SLR)	ML methodologies were considered effective and better result oriented than LR methods
36.	Agarwal and Tomar [4]	Linear Twin Support Vector Machine (LTSVM)-F-score	LTSVM performed better than others
37.	Jing et al. [24]	Collaborative Representation Classification for Software Defect Prediction (CSDP)	Experimented with a new methodology for effective defect prediction with low cost
38.	Erturk and Sezer [16]	Adaptive Neuron Fuzzy Inference System (ANFIS), SVM, ANN	Comparative analysis of soft-computing methods found SVM, ANN, ANFIS having 0.7795, 0.8685 and 0.8573 respectively
39.	Abaei et al. [2]	Semi-supervised methodology	Considered as helpful and effective automated methodology for detecting defects
40.	Laradji et al. [31]	Ensemble Learning Method on selected features-Average Probability Ensemble (APE) model with different classifiers	APE model showed better result than SVM, Random Forest method

Figure 1 indicates different methodologies used in software defect prediction from the year 2005 to 2015. This illustrates that these methodologies have been compared on the basis of studies and the results showed that Advance Machine Learning, Neural Network and Support Vector Machine techniques are the most frequently used as compared to other techniques in predicting defects of software system.

The Fig. 2 shows the datasets used in software defect prediction. The research studies using public datasets comprise 64.79 % whereas studies using private datasets cover 35.21 %. In-fact, the public free distributed datasets are mostly connected with PROMISE Repository and NASA Metrics Data Program. Private Datasets are not distributed as public datasets and they basically belong to private companies.

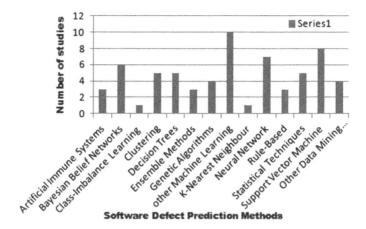

Fig. 1 Methods used in software defect prediction

Fig. 2 Datasets used in
software defect prediction

4 Key Analysis

The analysis of various techniques applied for software defect prediction till date
has brought out the following observations:-

(a) Proper prediction of software defects in the initial phase of design level of
software development lifecycle can improve software quality, provide customer
satisfaction and considerably reduce overall cost, time and initiation of further
work.

(b) In order to minimize efforts in defect prediction with more accuracy and higher
efficiency, it necessitates identifying newer methods and datasets by applying
more sophisticated methodologies which will be appropriate and have adequate
positive and effective impact on prediction of software defects.

(c) Although considerable work has been made so far for prediction of software
defects by applying various parameters, but it may be safely stated that sufficient work had not yet been done in defect prediction of the wave applications
and open source software. As such, there is a need for further research work to
find out more effective methodologies that may produce better result with
higher accuracy in case of predicting software defects.

5 Challenging Issues

After critical analysis, various challenging issues have come to the light that requires immediate attention and timely solution. Owing to various reasons, application of methodologies is not totally problem or defect-free. In-fact, most of the studies implemented open source or public datasets and so, they may not work effectively for private and commercial datasets. Moreover, owing-to privacy issues, the proprietary datasets are not available in public. If availability of proprietary datasets is more, then it may help cross-project defect prediction with higher accuracy. Although various open or public datasets are available for defect prediction but each dataset is not having same number of metrics and similar type of metrics. These metrics are evaluated from different domain and the defect prediction model based object-oriented metrics is not applicable for different metrics or different feature-space. That-is-why, cross-project defect prediction is not very easy and feasibility of cross-project defect prediction model being wide acceptable is very less. It has however been accepted that this model is very useful for the industry. Various defect prediction models that have been proposed so far, could not at all give any guarantee for result of prediction. It is essential to undertake further studies on new metrics, new model or new development process that may be better performance, result-orientated and widely acceptable.

6 Conclusion and Future Work

Defect prediction in software system is truly crucial since, it is considered as an important step for enhancing software quality. Defect prediction in software system with application of proper methodologies is truly significant as it may immensely help in directing test efforts, reducing costs and improving quality and reliability of software. Research work in this field has emerged since 1992 and having huge volume of work done during last 25 years or so, but still it lacks in some areas and needs to solve those issues. However, unique and updated works (from the year 2005 to 2015) have been analyzed separately and the findings reveal that particularly Advance Machine Learning (AML), Neural Network (NN) and Support Vector Machine (SVM) methodologies are the most frequently used techniques as compared to all other techniques for predicting defects of software system. Moreover, it was also an important observation that public datasets used for this purpose comprise 64.79 % where as studies using private datasets cover only 35.21 %. We may conclude by stating that though different methodologies have been applied but no single methodology can be considered as a full proof for predicting software defects. It is highly essential to undertake further work applying newer methodologies in the initial stage for defect prediction with special emphasize on public datasets that are better result-orientated with higher level of accuracy. This work will facilitate further work and make endeavors in designing newer metrics of software that would pave the way and have all the potential to achieve higher prediction accuracy.

References

1. G. Abaei, A. Selamat, A survey on software fault detection based on different prediction approaches. Vietnam J. Comput. Sci. **1**(2), 79–95 (2014)
2. G. Abaeia, A. Selamata, H. Fujita, An empirical study based on semi-supervised hybrid self-organizing map for software fault prediction. Knowl.-Based Syst. **74**, 28–39 (2015)
3. A. Adline, M. Ramachandran, Predicting the software fault using the method of genetic algorithm. Int. J. Adv. Res. Electr. Electron. Instrum. Eng. **3**(2), 390–398 (2014)
4. S. Agarwal, D. Tomar, A feature selection based model for software defect prediction. Int. J. Adv. Sci. Technol. **65**(4), 39–58 (2014)
5. E. Ardil, E. Ucar, P.S. Sandhu, Software maintenance severity prediction with soft computing approach. Int. Sch. Sci. Res. Innov. Proc. World Acad. Sci. Eng. Technol. **3**(2), 253–258 (2009)
6. M.M. Askari, V.K. Bardsiri, Software defect prediction using a high performance neural network. Int. J. Softw. Eng. Appl. **8**(12), 177–188 (2014)
7. M. Banga, Computational hybrids towards software defect predictions. Int. J. Sci. Eng. Technol. **2**(5), 311–316 (2013)
8. G.D. Boetticher, Nearest Neighbour Sampling for Better Defect Prediction. ACM Journal **30**(4), 1–6 (2005)
9. A. Campan, G. Serban, T.M. Truta, A. Marcus, An algorithm for the discovery of arbitrary length ordinal association rules. DMIN **6**, 107–113 (2006)
10. C. Catal, U. Sevim, B. Diri, Practical development of an eclipse-based software fault prediction tool using Naive Bayes algorithm. Expert Syst. Appl. **38**(3), 2347–2353 (2011)
11. C. Catal, Software fault prediction: a literature review and current trends. Expert Syst. Appl. **38**(4), 4626–4636 (2011)
12. C.P. Chang, C.P. Chua, Y.F. Yeh, Integrating in process of software defect prediction with association mining to discover defect pattern. Inf. Softw. Technol. **51**(2), 375–384 (2009)
13. Y. Chen, P. Du, X.H. Shen, P. Du, B. Ge, Research on software defect prediction based on data mining, computer and automation engineering ICCAE, in *The 2nd International Conference*, vol. 1, pp. 563–567 (2010)
14. A. Chug, S. Dhall, Software defect prediction using supervised learning algorithm and unsupervised learning algorithm, confluence 2013, in *The Next Generation Information Technology Summit (4th International Conference)*, pp. 173–179 (2013)
15. P.M. Dhiman, R. Chawla, A clustered approach to analyze the software quality using software defects, advanced computing & communication technologies ACCT, in *2nd International Conference*, pp. 36–40 (2012)
16. E. Erturk, E.A. Sezer, A comparison of some soft computing methods for software fault prediction. Expert Syst. Appl. **42**(4), 1872–1879 (2015)
17. K. Gao, T.M. Khoshgoftaar, Software defect prediction for high-dimensional and class-imbalanced data, in *Proceedings of the 23rd International Conference on Software Engineering & Knowledge Engineering SEKE* (2011)
18. M. Gayathri, A. Sudha, Software defect prediction system using multilayer perceptron neural network with data mining. Int. J. Recent Technol. Eng. **3**(2), 54–59 (2014)
19. N. Gayatri, S. Nickolas, A.V. Reddy, Feature selection using decision tree induction in class level metrics dataset for software defect predictions. Proc. World Congr. Eng. Comput. Sci. **1**, 1–6 (2010)
20. D. Gray, D. Bowes, N. Davey, Y. Sun, B. Christianson, Software defect prediction using static code metrics underestimates defect-proneness, in *International Joint Conference on Neural Network IJCNN*, pp. 1–7 (2010)
21. D. Gray, D. Bowes, N. Davey, Y. Sun, B. Christianson, Using the support vector machine as a classification method for software defect prediction with static code metrics. Eng. Appl. Neural Netw. **43**, 223–234 (2009)

22. T. Hall, S. Beecham, D. Bowes, D. Gray, S. Counsell, A systematic literature review on fault prediction performance in software engineering. IEEE Trans. Softw. Eng. **38**(6), 1276–1304 (2012)
23. Z. Jianhong, P.S. Sandhu, S. Rani, A neural network based approach for modelling of severity of defects in function based software systems. Int. Conf. Electron. Inf. Eng. ICEIE **2**, 568–575 (2010)
24. X.Y. Jing, Z.W. Zhang, S. Ying, Y.P. Zhu, F. Wang, Software defect prediction based on collaborative representation classification, in *Proceedings in ICSE Companion, Proceedings of the 36th International Conference on Software Engineering*, pp. 632–633 (2014)
25. C. Jones, O. Bonsignour, *The Economics of Software Quality* (Pearson Education Inc., 2012)
26. Y. Kamei, A. Monden, S. Morisaki, K.I. Matsumoto, A hybrid faulty module prediction using association rule mining and logistic regression analysis, in *Proceedings of Second ACM-IEEE International Symposium on Empirical Software Engineering and Measurement*, pp. 279–281 (2008)
27. A. Kaur, I. Kaur, Empirical evaluation of machine learning algorithms for fault prediction. Lect. Notes Softw. Eng. **2**(2), 176–180 (2014)
28. J. Kaur, P.S. Sandhu, A K-Means based approach for prediction of level of severity of faults in software systems, in *Proceedings of International Conference on Intelligent Computational Systems* (2011)
29. K. Kaur, Analysis of resilient back-propagation for improving software process control. Int. J. Inf. Technol. Knowl. Manage. **5**(2), 377–379 (2012)
30. S. Kaur, D. Kumar, Software fault prediction in object oriented software systems using density based clustering approach. Int. J. Res. Eng. Technol. IJRET **1**(2), 111–116 (2012)
31. I.H. Laradji, M. Alshayeb, L. Ghouti, Software defect prediction using ensemble learning on selected features. Inf. Softw. Technol. **58**, 388–402 (2015)
32. M. Li, H. Zhang, R. Wu, Z.H. Zhou, *Sample-based Software Defect Prediction with Active and Semi-Supervised Learning, Automated Software Engineering*, vol. 9, no. 2 (Springer Publication, 2011), pp. 201–230
33. R. Li, S. Wang, An empirical study for software fault-proneness prediction with ensemble learning models on imbalanced data sets. J. Softw. **9**(3), 697–704 (2014)
34. Y. Liu, T.M. Khoshgoftaar, N. Seliya, Evolutionary optimization of software quality modelling with multiple repositories. IEEE Trans. Softw. Eng. **36**(6), 852–864 (2010)
35. R. Malhotra, Comparative analysis of statistical and machine learning methods for predicting faulty modules. ELSEVIER J. Appl. Soft Comput. **21**, 286–297 (2014)
36. T. Menzies, Z. Milton, B. Turhan, B. Cukic, Y. Jiang, A. Bener, Defect *Prediction from Static Code Features: Current Results, Limitations, New Approaches, Automated Software Engineering*, vol. 17, no. 4, pp. 375–407 (2010)
37. H. Najadat, I. Alsmadi, Enhance rule based detection for software fault prone modules. Int. J. Softw. Eng. Appl. **6**(1), 75–86 (2012)
38. A. Okutan, O.T. Yildiz, A novel regression method for software defect prediction with kernel methods, in *International Conference on Pattern Recognition Applications and Methods ICPRAM*, pp. 216–222 (2013)
39. J.C. Riquelme, R. Ruiz, D. Rodriguez, J.S. Anguilar-Ruiz, Finding defective software modules by means of data mining methodologies. Latin Am. Trans. IEEE **7**(3), 377–382 (2009)
40. K. Sankar, S. Kannan, P. Jennifer, Prediction of code fault using Naive Bayes and SVM classifiers. Middle-East J. Sci. Res. **20**(1), 108–113 (2014)
41. L. Sehgal, N. Mohan, P.S. Sandhu, Quality prediction of function based software using decision tree approach, in *International Conference on Computer Engineering and Multimedia Technologies*, pp. 43–47 (2012)
42. P.A. Selvaraj, P. Thangaraj, Support vector machine for software defect prediction. Int. J. Eng. Technol. Res. **1**(2), 68–76 (2013)
43. Q. Song, Z. Jia, M. Shepperd, S. Ying, J. Liu, A general software defect-proneness prediction framework. IEEE Trans. Softw. Eng. **37**(3), 356–370 (2011)

44. Q. Song, M. Shepperd, M. Cartwright, C. Mair, Software defect association mining and defect correction effort prediction. IEEE Trans. Softw. Eng. **32**(2), 69–82 (2006)
45. X. Tan, X. Peng, S. Pan, W. Zhao, Assessing software quality by program clustering and defect prediction, in *18th Working Conference on reverse Engineering* (2011)
46. W. Tao, L. Wei-Hua, Naive Bayes software defect prediction model, in *International Conference on Computational Intelligence and Software Engineering*, pp. 1–4 (2010)
47. J. Xu, D. Ho, L.F. Capretz, An empirical study on the procedure to derive software quality estimation models. Int. J. Comput. Sci. Inf. Technol. IJCSIT **2**(4), 1–16 (2010)
48. X. Yang, K. Tang, X. Yao, A learning-to-rank approach to software defect prediction. IEEE Trans. Reliab. **64**(1), 234–246 (2014)
49. W. Zhang, Y. Yang, Q. Wang, Using Bayesian Regression and EM Algorithm with Missing Handling for Software Effort Prediction. Inf. Softw. Technol. **58**, 58–70 (2015)

Fusion of Medical Image Using STSVD

K.N. Narasimha Murthy and J. Kusuma

Abstract The process of uniting medical images which are taken from different types of images to make them as one image is a Medical Image Fusion. This is performed to increase the image information content and also to reduce the randomness and redundancy which is used for clinical applicability. In this paper a new method called Shearlet Transform (ST) is applied on image by using the Singular Value Decomposition (SVD) to improve the information content of the images. Here two different images Positron Emission Tomography (PET) and Magnetic Resonance Imaging (MRI) are taken for fusing. Initially the ST is applied on the two input images, then for low frequency coefficients the SVD method is applied for fusing purpose and for high frequency coefficients different method is applied. Then fuse the low and high frequency coefficients. Then the Inverse Shearlet Transform (IST) is applied to rebuild the fused image. To carry out the experiments three benchmark images are used and are compared with the progressive techniques. The results show that the proposed method exceeds many progressive techniques.

Keywords Magnetic resonance imaging · Positron emission tomography · Medical image fusion · Singular value decomposition · Shearlet transform · Inverse shearlet transform

1 Introduction

Fusing the image is a significant work in many image analysis processes where the images are taken from different generators. The fusion of images is done because to fuse the information into a single image from the information of different types of

K.N.N. Murthy (✉)
Christ University, Bangalore, Karnataka, India
e-mail: murthy_knn@yahoo.co.in

J. Kusuma
Vemana IT, Bangalore, Karnataka, India
e-mail: kusumagowdaj@gmail.com

© Springer Nature Singapore Pte Ltd. 2017   69
S.C. Satapathy et al. (eds.), *Proceedings of the 5th International Conference on Frontiers in Intelligent Computing: Theory and Applications*, Advances in Intelligent Systems and Computing 516, DOI 10.1007/978-981-10-3156-4_7

images [1]. These different types of images which are going to use as the source images are taken from different types of sensors and of varying timings whose physical feature are of different. Because of different features present in the images it is needed to fuse the images to get the exact information which is helpful for both the machine and human sensing. Nowadays there are many techniques has been developed to fuse the medical images of different types [2]. Fusion types are of two: Transform Domain (TD) and Spatial Domain (SD).

The basic method used for fusing the images is averaging the Principal Components Analysis (PCA) which comes under the SD method but it has a limitation that it suffers from reducing contrast and the sharpness [3]. But the pixel based method solves this contrast problem, in this method each pixel value in image is replaced by setting the threshold value. But this method measures inaccurately which reduces the performance of the fusion method. This is solved by the region based method. But in region based method the quality of the image is reduced as because of the artifacts come at edges.

Another method used for fusing is the Wavelet Transform (WT) which comes under TD approach [4, 5]. The WT is developed as a big multi-resolution system; this method permits changes in time extension but in shape. It has many features like sparse representation and multi-scale features of the function [6–8]. But it cannot handle more number of frequency coefficients so to overcome this many methods have attached to the WT such as ridgelet method [9, 10], curvelet method [9] and contourlet method.

Recently ST has been developed by many authors through affine system which is helpful in analysis and synthesis of an image. And another method is the SVD which retrieves the features from the image which is used by many authors for many purposes like in image registration, in face recognition, image fusion and in resolution problems [11, 12]. In this paper a new method is proposed to fuse the images by fusing SVD and ST. The experiments are conducted on three set of images and the proposed method is compared with the existing methods like Discrete Wavelet Transform (DWT), Non-Subsampled Contourlet Transform (NSCT) and Curvelet Transform (CVT). The results show that the proposed method exceeds the existing methods.

The paper is organized as follows: Sect. 2 gives brief description of ST and SVD, Sect. 3 presents the proposed method, the experimental results are illustrated in Sects. 4 and 5 gives the conclusion of the work.

2 Shearlet Transforms

The ST is used to encode the features which are varying in direction in many problems. Its primary steps are localization in directions and decomposition is done in multiple scales [10]. There are two types of ST: continuous and discrete which are defined as follows with the fixed resolution level j.

2.1 Continuous Shearlet Transform

It is defined as the affine systems which are grouped with the complex dilations which are of the form:

$$Sh_{PQ}(\psi) = \left\{ \psi_{a,b,c}(x) = |\det P|^{a/2} \psi(Q^b P^a X - c); a, b \in \mathcal{Z}, c \in \mathcal{Z}^2 \right\} \quad (1)$$

where a represents scale, b is orientation, c is cone for 2 dimensions, P and Q are 2×2 invertible matrices. For the function fn $\in L^2(\mathcal{R}^2)$, is as follows:

$$\sum_{a,b,c} |\langle fn, \psi_{a,b,c} \rangle|^2 = ||fn||^2 \quad (2)$$

For the function fn $\in L^2(\mathcal{R}^2)$, tight frame is formed by the $Sh_{PQ}(\psi)$. The matrix P^a is related with scale transform and the matrix Q^b related with the orientation like shear and rotation.

The continuous ST is of the form:

$$\psi_{a,b,c}(x) = a^{-3/4} \psi\left(B_b^{-1} A_a^{-1}(x-c)\right) \quad (3)$$

where $A_a = \begin{pmatrix} a & 0 \\ 0 & \sqrt{a} \end{pmatrix}$, $B_b = \begin{pmatrix} 1 & b \\ 0 & 1 \end{pmatrix}$, $\psi \in L^2(\mathcal{R}^2)$, which should satisfy the following conditions:

(1) $\hat{\psi}(\xi) = \hat{\psi}(\xi_1, \xi_2) = \hat{\psi}_1(\xi_1)\hat{\psi}_2(\xi_2/\xi_1)$;
(2) $\hat{\psi}_1 \in C^\infty(\mathcal{R})$ supp $\psi_1 \subset [-2, -1/2] \cup [1/2, 2]$;
(3) $\hat{\psi}_2 \in C^\infty(\mathcal{R})$ supp $\psi_2 \subset [-1, -1], \hat{\psi}_2 > 0$, But $||\psi_2|| = 1$

where ψ_1 is a continuous WT, in $\psi_{a,b,c}$ a, b $\in \mathcal{R}$, and c $\in \mathcal{R}^2$, for any function fn $\in L^2(\mathcal{R}^2)$, for different scales of wavelets. Usually, the values of a = 4 and b = 1.

2.2 Discrete Shearlet Transform

There are two steps in discrete shearlet transform: they are directional localization and dividing in different scales. The decomposition of shearlet is shown in Fig. 1.

Let for any scale j, the function fn $\in L(Z_n^2)$. Initially for decomposing an image the Laplacian pyramid is applied. Let consider the image as f_a^j, when Laplacian pyramid is applied on image then it is decomposed into low frequency f_l^j and high frequency f_h^j. If another division is needed then the Laplacian pyramid is applied on

Fig. 1 Frequency tiling by the Shearlet: **a** The frequency plane $\mathcal{R}^2$ is tiled, **b** The frequency size

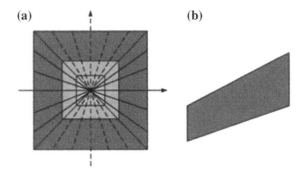

low frequency f_1^j. Then for reconstructing these frequencies into images the inverse fast Fourier Transform is applied.

2.3 Singular Value Decomposition

It is a method which is used to extract the features from the image. When SVD is applied on a matrix it is decomposed into three matrices, where two are singular matrices and one is diagonal matrix [12]. The singular value gives description of the images like scale invariance, feature stability etc. The SVD of the matrix I of size $i \times j$ is as follows:

$$I = U_I \Sigma_I V_I^T \tag{4}$$

where U_I is the matrix with dimension $i \times i$ whose columns are called as left singular vectors, V_I^T is the matrix with dimension $j \times j$ whose rows are called as right singular vectors, and Σ_I is the matrix with dimension $i \times j$ whose diagonal elements are called as singular values. The diagonal elements are arranged in decreasing order; the highest value is at the top left.

3 Proposed Fusion Method

A new fusion method is proposed in this paper named as STSVD to use the advantage of ST that is inter-scale sub-band. The proposed method is divided into two parts which are as low and the high frequency coefficients. The SVD is used by low frequency coefficients; after SVD is applied the max method is used to fuse the low frequencies. The high frequencies are fused by deriving the same and the different levels of high sub-bands. The proposed system structure is shown in Fig. 2.

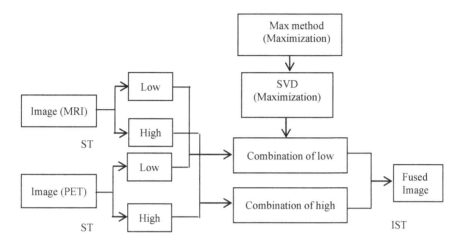

Fig. 2 The structure of the proposed system

3.1 Computing the Low Frequency Sub-Band Coefficients

The ST is applied on the original image which is divided into low and high frequencies. Then the SVD method is applied on the low frequency coefficients. But before the SVD method the covariance matrix is found for the low sub-bands of two input images as follows:

Let L_p (m, n) denote the low sub-band located at (m, n), p = A, B. The covariance matrix C (m, n) for low frequency coefficients is C (m, n) = **Covariance** (L_A (m, n), L_B (m, n)). Then apply SVD on that matrix as SVD (C (m, n)) whose output is [U, D, V], then normalize the U and V matrices as follows:

(1) If V (m, m) $\geq$ V (m + 1, m + 1), then mth column of matrices U, V are normalized.
(2) Otherwise (m + 1)th column of matrices U, V are normalized.

For m = 1,

$$L_z = (U(:, m) + V(:, m)) / (\sum_{m=1}^{x} (U(:, m) + V(:, m))) \qquad (5)$$

$$L_z = (U(:, m+1) + V(:, m+1)) / \sum_{m=1}^{x} (U(:, m+1) + V(:, m+1)) \qquad (6)$$

where x is the last value of the row of the matrix. Then

$$a = [L_z(:,m)/((r *c) \sum L_A)] \tag{7}$$

$$b = [L_z(:,m+1)/((r *c) \sum L_B)] \tag{8}$$

where r is the rows and c is the columns.

$$q = \sqrt{1 + a^2 + b^2} \tag{9}$$

Then the fused low-pass coefficients are obtained as follows:

$$L_F(m,n) = \begin{cases} L_A(m,n) * q \text{ if } L_A(m,n) \geq L_B(m,n) \\ L_B(m,n) * q \text{ if } L_B(m,n) > L_A(m,n) \end{cases} \tag{10}$$

3.2   Computing the High Frequency Sub-Band Coefficients

The edges and corners of the image is obtained from the high frequency coefficients of the ST. The fused coefficients are calculated as follows:

Let $H_p^{k,l}(m,n)$ be the high frequency coefficient at the location (m,n) in the kth sub-band at the lth level, $p = A, B$.

Let $S_{p,h}$ be the summation of the sub-bands $H_p^{k,l}$ and $H_p^{m,l}$ in the same level l call as horizontal sub-bands, which is calculated for each level as follows:

$$S_{p,h} = \sum_{k=1}^{K} \left(H_p^{k,l}, H_p^{k,l} \right) \tag{11}$$

Similarly, $S_{p,v}$ be the sum of $H_p^{k,l}$ and $H_r^{m,n}$ in different level call as vertical sub-band, which is calculated as:

$$S_{p,v} = \sum_{k,m=1}^{L} \sum_{l,n=1}^{K} \left(H_p^{k,l}, H_p^{m,n} \right) \tag{12}$$

Finally, compute the new coefficients $H_{p,new}^{k,l}$ as follows:

$$H_{p,new}^{k,l} = H_p^{k,l} \times \sqrt{1 + S_{p,h}^2 * S_{p,v}^2} \tag{13}$$

Then calculate the fused coefficients $H_F^{k,l}(m,n)$ as follows:

$$H_F(m,n) = \begin{cases} H_{A,new}^{k,l}(m,n), \text{ if } H_{A,new}^{k,l} \geq H_{B,new}^{k,l} \\ H_{B,new}^{k,l}(m,n), \text{ else } H_{B,new}^{k,l} > H_{A,new}^{k,l} \end{cases} \tag{14}$$

Proposed STSVD Algorithm

Input: A and B are the source images which needs to be registered.
Output: Image which is fused.
Step 1: Using ST decompose A and B.
Step 2: Compute low frequency coefficients using the Eqs. (5)–(10).
Step 3: Compute high frequency coefficients using the Eqs. (11)–(14).
Step 4: Then the selected high and low frequency coefficients are fused.
Step 5: Apply IST to reconstruct the image.
Step 6: Fused image is displayed.

4 Experimental Results

The experiments are conducted on three set of images and the proposed method is compared with existing systems i.e. DWT [13–15], NSCT [16, 17] and CVT [18]. The experiment is executed in MATLAB R2013b. The image size selected is 256 × 256 to fusion process.

4.1 Data Set 1

The MRI and PET images are shown in Fig. 3a, b respectively [19]. Figure 3f is the output image of the proposed method. The existing methods DWT, NSCT and CVT outputs are shown in Fig. 3c–e respectively.

4.2 Data Set 2

The MRI and PET images are shown in Fig. 4a, b respectively [19]. Figure 4f is the output image of the proposed method. The existing methods DWT, NSCT and CVT outputs are shown in Fig. 4c–e respectively.

4.3 Data Set 3

The MRI and PET images are shown in Fig. 5a, b respectively [19]. Figure 5f is the output image of the proposed method. The existing methods DWT, NSCT and CVT outputs are shown in Fig. 5c–e respectively.

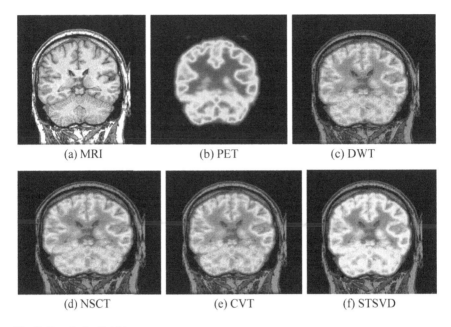

(a) MRI   (b) PET   (c) DWT

(d) NSCT   (e) CVT   (f) STSVD

Fig. 3 Results for Set I image

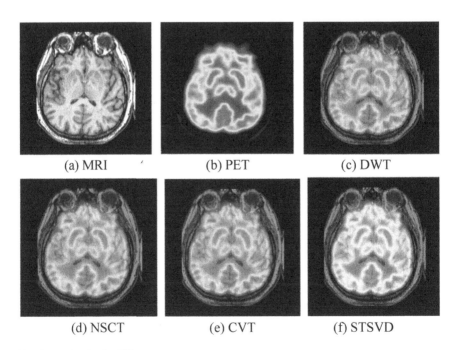

(a) MRI   (b) PET   (c) DWT

(d) NSCT   (e) CVT   (f) STSVD

Fig. 4 Results for Set II image

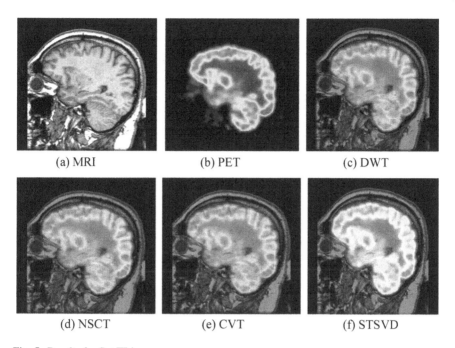

(a) MRI (b) PET (c) DWT

(d) NSCT (e) CVT (f) STSVD

Fig. 5 Results for Set III image

Three popular metrics are used for measuring the performance of proposed method; they are Mutual Information (MI), Cross Correlation (CC) and Edge Indexing (QI). The MI is used to measure the information which is reached to output image from the input image. The CC is used to measure the similarities between the input and the output images. The QI is used to measure the edge information transferred between the inputs to the output. The values of MI, CC and QI are noted in Table 1 for all the three set images and for all the existing and the proposed methods. In each row one value is bolded which is the highest value in that row.

Table 1 Comparative results analysis

Method	Performance Metric	DWT	NSCT	CVT	STSVD
Image-1 (Set I)	CC	0.8787	0.8859	0.8852	**0.9193**
	MI	0.8161	0.8196	0.8321	**0.8576**
	QI	0.8450	0.8512	0.8505	**0.8925**
Image-2 (Set II)	CC	0.8883	0.8935	0.8930	**0.9266**
	MI	0.7432	0.7420	0.7549	**0.8032**
	QI	0.8643	0.8689	0.8688	**0.9084**
Image-3 (Set III)	CC	0.8545	0.8607	0.8600	**0.8955**
	MI	0.6794	0.6765	0.6913	**0.7025**
	QI	0.7877	0.7929	0.7921	**0.8397**

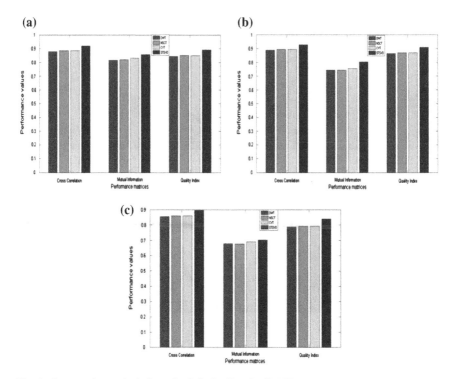

Fig. 6 Comparative analysis for **a** Set-I, **b** Set-II and **c** Set-III

The results of estimated using MI and QI are shown in Table 1. Note that the highest value in each row of Table 1 are showed in bold. The graphs for all the three set images are shown in the Fig. 6 a–c. From the Table 1 and Fig. 6, it is clear that the proposed method exceeds the existing methods for all the performance matrices.

5 Conclusion

In this paper, initially the input image is divided into low and high frequency coefficients by using the ST. Then SVD is applied on low frequency coefficients to retrieve the effective features from the image. The main goal of STSVD is to maintain the significant information from the input images and also to keep the color components of the image as it is even the fusion is performed. This helps in finding the accuracy by bettering the smooth regions of the image. The experiments conducted on the three set of images show that the proposed method exceeds the existing methods.

References

1. A.P. James, B.V. Dasarathy, Medical image fusion: A survey of the state of the art. Inf. Fusion **19**, 4–19 (2014)
2. K.P. Indira, R.R. Hemamalini, Analysis on image fusion techniques for medical applications. IJAREEIE **3**(9), Sept 2014
3. D. Himanshi, V. Bhateja, An improved medical image fusion approach using PCA and complex wavelets, in *IEEE- International Conference on Medical Imaging, m-Health and Emerging Communication Systems*, pp. 442–447 (2014)
4. R.J. Sapkal, S.M. Kulkarni, Image fusion based on wavelet transform for medical application. IJERA **2**(5), 624–627, Sept–Oct 2012. ISSN: 2248-9622
5. P.P. Mirajkar, S.D. Ruikar, Wavelet based image fusion techniques. Int. Conf. Intel. Syst. Sig. Proc. (2013)
6. Y. Liu, J. Yang, J. Sun, *PET/CT Medical Image Fusion Algorithm Based on Multiwavelet Transform* (IEEE, 2010), pp. 264–268
7. L. Tawade, A.B. Aboobacker, F. Ghante, Image fusion based on wavelet transforms. Int. J. Bio-Sci. Bio-Technol. **6**(3), 149–162 (2014)
8. U. Javed, M.M. Riaz, A. Ghafoor, S.S. Ali, T.A. Cheema, MRI and PET image fusion using fuzzy logic and image local features. Sci. World J. Article ID 708075 (2014)
9. F. Ali, I. El-Dokany, A. Saad, F. El-Samie, A curvelet transform approach for the fusion of MR and CT images. J. Mod. Opt. **57**, 273–286 (2010)
10. W.-Q. Lim, The discrete shearlet transform: a new directional transform and compactly supported shearlet frames. IEEE Trans. Image Proces. **19**(5), May 2010
11. D.W. Repperger, A.R. Pinkus, K.A. Farris, R.G. Roberts, R.D. Sorkin, Investigation of image fusion procedures using optimal registration and SVD algorithms, in *Proceedings of the IEEE- National Aerospace and Electronics Conference*, pp. 231 235 (2009)
12. H. Nasir et al. Singular value decomposition based fusion for super resolution image reconstruction. Sig. Process.-Image Commun. (2011)
13. P. Phanindra, J. Chinna Babu, V. Usha Shree, FPGA implementation of medical image fusion based on DWT. Int. J. Adv. Res. Comput. Sci. Softw. Eng. **3**(9), Sept 2013. ISSN: 2277 128X
14. K. Sharmila, S. Rajkumar, V. Vijayarajan, Hybrid method for multimodality medical image fusion using discrete wavelet transform and entropy concepts with quantitative analysis, in *IEEE Advancing Technology for Humanity, International conference on Communication and Signal Processing*, 3–5 Apr 2013
15. N. Nahvi, O.C. Sharma, Implementation of discrete wavelet transform for multimodal medical image fusion. Int. J. Emerg. Technol. Adv. Eng. **4**(7), July 2014, ISSN: 2250–2459
16. G. Bhatnagar, Q.M.J. Wu, Z. Liu, Directive contrast based multimodal medical image fusion in NSCT domain. IEEE Trans. Multimedia **15**(5), Aug 2013
17. V. Savithal, T. Kadhambari, R. Sheeba, Multimodality medical image fusion using NSCT. IJREAT, **1**(6), Dec–Jan 2014. ISSN: 2320–8791
18. N. kaur, M. Bahl, H. Kaur, Review on: image fusion using wavelet and curvelet transform. IJCSIT **5**(2), 2467–2470 (2014)
19. http://www.med.harvard.edu/aanlib/cases/caseNA/pb9.htm

Modified Cuckoo Search Algorithm for Fittest Relay Identification in Microgrid

O.V. Gnana Swathika, Santanab Mukhopadhyay, Yatharth Gupta, Arka Das and S. Hemamalini

Abstract Microgrid is a group of interconnected generating units and loads at the distribution level which operates in two modes—Grid connected mode and Islanded mode. Fault clearance in a microgrid is a key challenge for protection engineers. This paper aims to identify the best fit relay suitable for a microgrid using modified cuckoo search algorithm based on key parameters like current rating, Time Multiplier Setting (TMS), Plug Setting Multiplier (PSM) and time of operation (t_{op}). This algorithm aids in providing suitable relay coordination in microgrid and clears the faulty portion of network effectively from the healthy portion of network.

Keywords Modified cuckoo search algorithm · Microgrid protection · Relay coordination

1 Introduction

A microgrid is a distribution level network with many distributed generator (DG) micro-sources and loads [1]. The microgrid contributes to the main-grid by minimizing congestion, providing additional generation, improving system stability, attending to varying load demands and providing continuity of supply to the

O.V. Gnana Swathika (✉) · S. Mukhopadhyay · Y. Gupta · A. Das · S. Hemamalini
Faculty, School of Electrical Engineering, VIT University, Chennai, India
e-mail: gnanaswathika.ov@vit.ac.in

S. Mukhopadhyay
e-mail: santanab.mukhopadhay2013@vit.ac.in

Y. Gupta
e-mail: yatharth.gupta2013@vit.ac.in

A. Das
e-mail: arka.das2013@vit.ac.in

© Springer Nature Singapore Pte Ltd. 2017
S.C. Satapathy et al. (eds.), *Proceedings of the 5th International Conference on Frontiers in Intelligent Computing: Theory and Applications*, Advances in Intelligent Systems and Computing 516, DOI 10.1007/978-981-10-3156-4_8

consumers [2, 3]. A number of grid synchronization algorithms along with their importance in the control of Distributed Generation on all types of grid conditions are elaborated in [4]. The most frequently occurring faults in a microgrid have their origin in overcurrent issues. Directional overcurrent relays (DOCRs) can be installed at required locations in order to provide protection to the network and this also proves helpful to overcome the aforementioned issues in microgrid. For low voltage applications, the Directional Overcurrent Relay with 2–4 settings is employed [5]. Field-programmable gate array (FPGA) based numerical overcurrent relay [6] protection is provided in islanded microgrid. Reliability and quickness in isolating faults are the key takeaways of this relay. Field Programmable Gate Array (FPGA) based intelligent controller performs the function of overcurrent fault identification and adaptive tripping of suitable breakers during fault clearance [7]. Overcurrent relay coordination maybe optimized using genetic algorithm [8], dual simplex algorithm [9] and particle swarm optimization among other algorithms. In a reconfigurable grid, graph theory algorithms are utilized for radial load flow analysis [10]. Metaheuristic technique based Kruskal-Floyd Warshall algorithm [11] is used in reconfigurable microgrids to identify the shortest path between the location of the fault and the location of the nearest operating source.

Depending on the type of fault in a microgrid, the fault current is calculated and that fault current becomes our actuating quantity to which the relays respond. This is when the identification of the fittest relay becomes important. There maybe a case where the primary relay is not the best fit relay to clear the fault. In such a scenario, using the cuckoo search algorithm, the best fit relay to clear the fault is found out conveniently. This paper aims to identify the fittest relay suitable for a microgrid network using modified cuckoo search algorithm. This ensures that suitable relay coordination maybe achieved in the network and in turn aids in clearing the faulty portion of network conveniently from the healthy portion of network.

2   Modified Cuckoo Search Algorithm

Steps: [12]
M ->MaxLévyStepSize
u ->GoldenRatio
Initialise a population of n nests $x_i(i = 1,2,\ldots,n)$
for all x_i **do**
Calculate fitness $F_i = f(x_i)$
end for
Generation number P ← 1
while NumberObjectiveEvaluations
< MaxNumberEvaluations **do**
P ← P + 1
Sort nests by order of fitness
for all nests to be abandoned do
Current position x_i
Calculate Lévy flight step size α ← M/sqrt P
Perform Lévy flight from x_i to generate new egg x_k
xi ← x_k
Fi ← $f(x_i)$
end for
for all of the top nests **do**
Current position x_i
Pick another nest from the top nests at random x_j
if $x_i = x_j$ **then**
Calculate Lévy flight step size α ← M/P^2
Perform Lévy flight from xi to generate new
egg x_k
$F_k = f(x_k)$
Choose a random nest l from all nests
if $(F_k > F_l)$ **do**
x_l ← x_k
F_l ← F_k
end if
else
dx = $|x_i - x_j|/u$
Move distance dx from the worst nest to the
best nest to find x_k
$F_k = f(x_k)$
Choose a random nest l from all nests
if $(F_k > F_l)$ **then**
x_l ← x_k
F_l ← F_k
end if
end if
end for
end while

3 Application of Modified Cuckoo Search Algorithm for Fittest Relay Identification

The cuckoo search algorithm [12–15] will be used to identify the fittest relay in an electrical network based on certain predefined parameters. If a relay is found to be unsuitable for clearing the fault current in the network, it will be designated as an unfit relay and will be not be considered for future iterations for identification of the best relay to clear the fault current in the network. All remaining relays which are not identified as an unfit relay will be designated as fit relay for the next iteration. Then the next parameter to identify a fit relay is tested. If a relay is found out to be unsuitable for clearing the fault current based on this parameter, it is again added to the list of unfit relays. Relays which are found to be capable of clearing the fault current after this parameter test are added to the list of fit relays. This process continues till all parameters have been tested and a final list of fit and unfit relays are generated. The list of fit relays is now taken for consideration and the relay which performs the best based on the decisive parameter is considered as the fittest relay in the electrical network. The fault current rating and plug setting multiplier are treated as parameters to separate fit and unfit relays. Finally, to identify the fittest relay in the electrical network, the time of operation of the relay acts as the decisive parameter.

3.1 Test Microgrid

A test microgrid system with branches B1.1, B1.2, B1.3 and B1.4 as shown in Fig. 1 is considered for analysis. CPC indicates the central protection center which is responsible for protecting the entire microgrid. Let the positive sequence impedance, negative sequence impedance, zero sequence impedance and fault impedance be Z_1, Z_2, Z_0 and Z_f. The algorithm also considers 3 types of faults in an electric system, namely Single line to ground fault (LG), Line to line fault (LL) and Double line to ground fault (LLG). As the fault current varies based on the type of

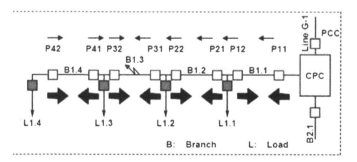

Fig. 1 A sample feeder of microgrid network

fault in the network, test case is analyzed for all the three type of faults so as to arrive at a more comprehensive inference of the fittest relay identification technique.

Let us consider the test system with following specifications:

Power rating: 30 MW, Voltage rating: 11 kV, Base voltage rating: 11 kV, Z_1:0.5Ω, Z_2:0.3 Ω, Z_0:0.6 Ω and Z_f: 0.2 Ω.

3.1.1 Single Line to Ground (SLG) Fault in Test System

Let a single line-to-ground fault occur in the power system network. The secondary windings of current transformer (CT) indicate sequence components of $I_a = 1.5$ A, $I_b = 0$ A and $I_c = 0$ A. Let the fault current measured at secondary of CT be1.5 A. Assume there is availability of four types of relays with specifications of current rating, time multiplier setting (TMS) and plug setting multiplier (PSM) as shown in Table 1. The Cuckoo Search algorithm computes the time of operation as indicated in Table 1.

The relays are first subjected for testing against required current rating parameter. It is evident from Table 1. that Relay 3 and Relay 4 are not capable of clearing the fault because of insufficient fault current rating. At this stage Relay 3 and Relay 4 are categorized as unfit relays in the system. The fit relays are hence Relay 1 and Relay 2. The fit relays are next subjected for testing based on minimum time of operation (t_{op}). Relay 2 with minimum time of operation of 6.28 s passes this test, while Relay 1 with time of operation of 17.03 s joins the unfit list.

3.1.2 Line to Line (LL) Fault in Test System

Let a line-to-line fault occur in the power system network. The secondary windings of current transformer (CT) indicate sequence components of $I_a = 0$ A, $I_b = -0.610500 + j0.456042$ A and $I_c = 0.610500 - j0.456042$ A. Let the fault current measured at secondary of CT be 0.762026 Amps. Assume there is availability of four types of relays with specifications of current rating, time multiplier setting (TMS) and plug setting multiplier (PSM) as shown in Table 1. The Cuckoo Search algorithm computes the time of operation as indicated in Table 1.

The relays are first subjected for testing against required current rating parameter. It is evident from Table 1. that Relay 1 and Relay 4 are not capable of clearing

Table 1 Relay characteristics

Relay number	Current rating (A)	TMS	PSM	t_{op} (s) for SLG fault	t_{op} (s) for LL fault	t_{op} (s) for LLG fault
1	2.5	1	1	17.03	−25.41	68.77
2	1.5	1	0.5	6.28	16.39	8.70
3	1	0.8	0.7	7.25	65.08	12.08
4	0.5	0.5	0.5	3.14	8.19	4.35

the fault because of insufficient fault current rating. At this stage Relay 1 and Relay 4 are categorized as unfit relays in the system. The fit relays are hence Relay 2 and Relay 3. The fit relays are next subjected for testing based on minimum time of operation (t_{op}). Relay 2 with minimum time of operation of 16.39 s passes this test, while Relay 1 with time of operation of 65.08 s joins the unfit list.

3.1.3 Double- Line-to Ground (LLG) Fault in Test System

Let a double-line-to-ground fault occur in the power system network. The secondary windings of current transformer (CT) indicate sequence components of $I_a = 2.702703$ A, $I_b = 1.608973 + j1.848254$ A and $I_c = 1.788973 + j1.713794$ A. Let the fault current measured at secondary of CT be 1.10565 Amps. Assume there is availability of four types of relays with specifications of current rating, time multiplier setting (TMS) and plug setting multiplier (PSM) as shown in Table 1. The Cuckoo Search algorithm computes the time of operation as indicated in Table 1.

The relays are first subjected for testing against required current rating parameter. It is evident from Table 1. that Relay 3 and Relay 4 are not capable of clearing the fault because of insufficient fault current rating. At this stage Relay 3 and Relay 4 are categorized as unfit relays in the system. The fit relays are hence Relay 1 and Relay 2. The fit relays are next subjected for testing based on minimum time of operation (t_{op}). Relay 2 with minimum time of operation of 8.70 s passes this test, while Relay 1 with time of operation of 68.77 s joins the unfit list.

Hence, it is clearly seen from above analysis that only Relay 2 is suitable for clearing the fault current in the given microgrid network irrespective of the type of fault in the system.

A potential advantage of this algorithm is the cost effectiveness, whereby if many types or specifications of relays are available for a microgrid, they can all be replaced by one relay, which is the best fit to clear all types of fault in the network. Another possible application of this algorithm is in the production line of relays. If large number of relays are waiting in the production line, and if the required specifications of the relay for a real-time network is known, then the relay specifications can be designed according to the requirements of the electrical network. As a result of the flexibility and possible cost benefits, this algorithm can have widespread commercial applications for optimizing relay design in electrical networks.

4 Conclusion

The paper proposes the application of modified cuckoo search algorithm in a microgrid network to identify the best fit relays that maybe employed for efficient relay coordination and fault isolation. The proposed technique is validated on a 4-branch microgrid network for all three types of fault namely single- line-ground fault, line- line fault and double line-ground fault. The results are a direct indication of how the modified cuckoo search algorithm is able to identify the best fit relay which is suitable of all fault types. The algorithm generates a list of unfit relays as well. As this algorithm does not set a limit on the predefined parameters, any parameter that an user may consider as a valid parameter to define the fit of a relay is easily incorporated into the system. This reflects the flexibility of the algorithm.

References

1. O.V. G. Swathika, S. Hemamalini, Prims Aided Dijkstra Algorithm for Adaptive Protection in Microgrids. IEEE J. Emerg. Sel. Top. Power Electron. **4**(4), 1279–1286 (2016)
2. O.V. Gnana Swathika, K. Karthikeyan, S. Hemamalini, R. Balakrishnan, Relay coordination in real-time microgrid for varying load demand. ARPN J. Eng. Appl. Sci. **11**, 3222–3227 (2016)
3. O.V. Gnana Swathika, K. Karthikeyan, S. Hemamalini, R. Balakrishnan, PLC based LV-DG synchronization in real-time microgrid network. ARPN J. Eng. Appl. Sci. **11**, 3193–3197 (2016)
4. O.V.G. Swathika, S. Hemamalini, Communication assisted overcurrent protection of microgrid. IEEE International Conference on Computational Intelligence and Communication Networks (CICN), 1472–1475 (2015)
5. O.V.G. Swathika, K. Karthikeyan, S. Hemamalini, Multiple DG synchronization and de-synchronization in a microgrid using PLC. *Advanced Computing and Communication Technologies* (Springer, 2016), pp. 565–572
6. S. Kannuppaiyan, V. Chenniappan, Numerical inverse definite minimum time overcurrent relay for microgrid power system protection. IEEJ Trans. Electr. Electron. Eng. **10**(1), 50–54 (2015)
7. O.V. Gnana Swathika, S. Hemamalini, Adaptive and intelligent controller for protection in radial distribution system, in *Advanced Computer and Communication Engineering Technology* (Springer, 2016), pp. 195–209
8. P. Prashant Bedekar, R. Sudhir Bhide, Optimal coordination of overcurrent relay timing using continuous genetic algorithm, in *Elsevier Science Direct Expert System with Applications*, vol. 38, pp. 11286–11292 (2011)
9. O.V. Gnana Swathika, I. Bose, B. Roy, S. Kodgule, S. Hemamalini, Optimization techniques based adaptive overcurrent protection in microgrids. J Electr. Syst. Spec. (3) (2015)
10. M. Muhammad Aman, B. Ghauth Jasmon, A.H. Abu Bakar et al., Graph theory-based radial load flow analysis to solve the dynamic network reconfiguration problem. Int. Trans. Electr. Energ. Syst. (2015). doi:10.1002/etep. 2108
11. O.V. Gnana Swathika, S. Hemamalini, Kruskal aided floyd warshall algorithm for shortest path identification in microgrids. ARPN J. Eng. Appl. Sci. **10**, 6614–6618 (2015)
12. N.T. Trung, D.N. Vo, Modified cuckoo search algorithm for short-term hydrothermal scheduling. Int. J. Electr. Power Energ Syst. **65**, 271–281 (2015)
13. N.T. Trung, D.N. Vo, Modified cuckoo search algorithm for short-term hydrothermal scheduling. Int. J.Electr. Power Energ Syst. **65**, 271–281 (2015)
14. S.S. Akumalla, S. Peddakotla, S.R.A. Kuppa, A modified cuckoo search algorithm for improving voltage profile and to diminish power loasses by locating multi-type FACTS devices. J. Control Autom. Electr. Syst. **27**, 93–104 (2016)
15. S. Walton, O. Hassan, K. Morgan, M.R. Brown, Modified cuckoo search: a new gradient free optimization algorithm. Chaos, Solitals and Fractals **4**, 710–718 (2011)

Optimization of Overcurrent Relays in Microgrid Using Interior Point Method and Active Set Method

O.V. Gnana Swathika, Arka Das, Yatharth Gupta,
Santanab Mukhopadhyay and S. Hemamalini

Abstract Microgrid is an aggregate of generating units and loads at the distribution level. It operates in two modes: grid connected mode and Islanded mode. Fault clearance in a microgrid is a key challenge for protection engineers. This paper aims to identify the optimized values of time of operation of overcurrent relays in a microgrid network. This paper solves the optimization problem using two methods —Interior Point method and Active Set method. Also three types of relays are used to determine which relay works best in similar constraint environment. These methods aid in providing suitable relay coordination in microgrid and clear the faulty portion of network quickly from the healthy portion of network.

Keywords Microgrid protection · Optimization techniques · Interior point method · Active set method

1 Introduction

A microgrid is a distribution level network with many distributed generator (DG) micro-sources and loads [1]. The microgrid contributes to the main-grid by minimizing congestion, providing additional generation, improving system stability, attending to varying load demands and providing continuity of supply to the consumers [2, 3]. Various grid synchronization algorithms with their role and

O.V. Gnana Swathika (✉) · A. Das · Y. Gupta · S. Mukhopadhyay · S. Hemamalini
School of Electrical Engineering, VIT University, Chennai, India
e-mail: gnanaswathika.ov@vit.ac.in

A. Das
e-mail: arka.das2013@vit.ac.in

Y. Gupta
e-mail: yatharth.gupta2013@vit.ac.in

S. Mukhopadhyay
e-mail: santanab.mukhopadhay2013@vit.ac.in

© Springer Nature Singapore Pte Ltd. 2017   89
S.C. Satapathy et al. (eds.), *Proceedings of the 5th International Conference on Frontiers in Intelligent Computing: Theory and Applications*, Advances in Intelligent Systems and Computing 516, DOI 10.1007/978-981-10-3156-4_9

influence in the control of Distributed Generator penetration on normal and faulty grid condition is elaborated in [4]. Predominant faults in microgrid are due to overcurrent issues. Directional overcurrent relays (DOCRs) maybe employed at required locations to protect the network and prove helpful in overcoming the aforementioned issues in microgrid. In low voltage applications, the DOCR with 2–4 settings are employed [5]. Field-programmable gate array (FPGA) based numerical overcurrent relay [6] protection is provided in islanded microgrid. Reliability and quickness in isolating faults are the key takeaways of this relay. Field Programmable Gate Array (FPGA) based intelligent controller performs the function of overcurrent fault identification and adaptive tripping of suitable breakers during fault clearance [7]. Overcurrent relay coordination maybe optimized using genetic algorithm [8], dual simplex algorithm [9], particle swarm optimization (PSO), and gravitational search algorithm. PSO assures the convergence to the global optimum but possess slow convergence speed. Interior point method is highly capable of obtaining local optima and is widely used in power system applications [10, 11]. In a reconfigurable grid, graph theory algorithms are utilized for radial load flow analysis [12]. From literature, it is evident that graph theory algorithms are not used for microgrid protection. Metaheuristic technique based Kruskal-Floyd Warshall algorithm [13] is used in reconfigurable microgrids to identify the shortest path between the fault and the nearest operating source.

In this paper three types of relays are used to determine which relay works best in similar constraint environment. The TMS of the primary and backup relays of each relay type are determined using two algorithms: the interior point algorithm and the active set algorithm. The result generated by these two algorithms are compared to see which relay is giving better result in terms of minimum operating time. This ensures that suitable relay coordination maybe achieved in the network and in turn aids in clearing the faulty portion of network swiftly from the healthy portion of network.

2 Overcurrent Relay Coordination

A sample three bus portion of a microgrid network is considered in Fig. 1. Fault is taken just beyond relay R2. In this case R2 serves the purpose of primary relay while R1 serves the purpose of backup relay. R1 comes into play when R2 is unable to clear the fault. If the fault had taken place just beyond R1, then R1 would have been the primary relay. In the figure, R2 acts after sometime, 0.2 s after a fault is detected. This time is given so that the relay does not trip the circuit breaker for instantaneous current surges. If the high magnitude current persists beyond 0.2 s in the feeder then it is taken as a fault after which the relay operation takes place. R1 is made to act after a specific amount of time known as Coordination Time Interval (CTI) which is equal to sum of operating time of circuit breaker in Bus 2 and overshoot time of relay R1. CTI is the difference in the operating time of R1 and R2. Following the above stated criteria, the objective function and the related

Fig. 1 A sample feeder of microgrid network

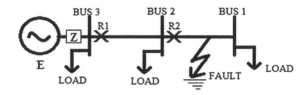

constraint equations are formulated and are solved to obtain the Time Multiplier Setting (TMS) and consequently the Time of Operation (t_{op}) using Interior Point method and Active set method.

3   Problem Formulation

The two main parameters of a directional overcurrent relay are TMS and relay current setting. Relay setting depends on the maximum load on the feeder. The objective is to minimize the total time of operation of the two relays so that they can clear the fault as fast as possible without breaking the constraints applied to the respective relays. The objective function is as follows.

$$Min\, z = \sum_{i=1}^{n} Topi \tag{1}$$

where, T_{opi} is the time of operation of the ith relay when the fault occurs just beyond its point under the following constraints.

- Bounds of operating time

$$topimin \leq topi \leq topimax \tag{2}$$

where,

topimin   minimum operating time of ith relay
topimax   maximum operating time of ith relay

- Coordination Time Interval

$$tbopi - topi \geq CTI \tag{3}$$

where,

tbopi   time of operation of backup relay
topi   time of operation of ith relay

- Relay Characteristics

There are three types of overcurrent relays that have been dealt with in this paper.

1. Inverse Definite Minimum Time (IDMT) Overcurrent Relay whose characteristic equation is

$$topi = TMS * \frac{0.14}{PSM^{0.02} - 1} \tag{4}$$

2. Very Inverse Time (VIT) Overcurrent Relay whose characteristic equation is

$$topi = TMS * \frac{13.5}{PSM - 1} \tag{5}$$

3. Extremely Inverse Time (EIT) Overcurrent Relay whose characteristic equation is

$$topi = TMS * \frac{80}{PSM^2 - 1} \tag{6}$$

topi   time of operation of ith relay
TMS   time multiplier setting
PSM   plug setting multiplier

$$PSM = \frac{If}{CT\ ratio * relay\ setting} \tag{7}$$

If   maximum fault current

4   Interior Point Method

Step 1:   In this method, the problem is transferred into barrier form i.e.,:

$$\begin{array}{ll} min & f(x) \\ x \in R^n & \\ s.t. & c(x)=0 \\ & x \geq 0 \end{array} \qquad \Longrightarrow \qquad \begin{array}{ll} min & f(x)\text{-}\mu \sum_{i=1}^{n} \ln x_i \\ x \in R^n & \\ s.t. & c(x)=0 \end{array}$$

Step 2:   We define $Zi = 1 / Xi$ and solve the above mentioned KKT conditions

$$\Delta f(x) + \Delta c(x)\lambda - Z = 0 \qquad (8)$$

$$c(x) = 0 \qquad (9)$$

$$XZe - \mu e = 0 \qquad (10)$$

We find the solution of the above equations by Newton Rapson Method. i.e.,

$$\begin{bmatrix} W_k & \nabla c(x_k) & -1 \\ \nabla c(x_k) & 0 & 0 \\ Z_k & 0 & X_k \end{bmatrix} \begin{bmatrix} d_k^x \\ d_k^\lambda \\ d_k^z \end{bmatrix} = - \begin{bmatrix} \nabla f(x_k) + \nabla c(x_k) - z_k \\ c(x_k) \\ X_k Z_k e - \mu_j e \end{bmatrix}$$

We solve for d_k^z, after solving for d_k^λ and d_k^x with the explicit solution

$$d_k^z = \mu_k X_k^{-1} e - z_{k-} \sum_k d_k^x \qquad (11)$$

Step 3:   The step size for each iteration is given by

$$x_{k+1} = x_k + \alpha_k d_k^x \qquad (12)$$

$$\lambda_{k+1} = \lambda_k + \alpha_k d_k^\lambda \qquad (13)$$

$$z_{k+1} = z_k + \alpha_k d_k^z \qquad (14)$$

Step 4:   The above steps are repeated until the convergence criteria is met i.e.,

$$\max|\Delta f(x) + \Delta c(x)\lambda - Z| \leq \in_{tol} \qquad (15)$$

$$\max|c(x)| \leq \in_{tol} \qquad (16)$$

$$\max|XZe - \mu e| \leq \in_{tol} \qquad (17)$$

Step 5:   If the above conditions are satisfied then STOP, or repeat steps 1–4, until satisfied.

5   Active Set Method

Step 1:   An initial feasible condition x^1 for Qp is assumed.
If such value does not exist then there exists an empty feasible set of (QP). STOP.
Else assume k = 1.

Step 2: Active Index set $I(x^k)$ and matrix A(k) is assumed. The following set of equations are to be solved.

Step 3: If $a^k = 0$

If $\mu_{active}^k \geq 0$, then

x^k is the optimal solution, RETURN

Else

(a) I(k) corresponding to the most negative Langrage's Multiplier is removed.

(b) $\bar{\alpha}$ is determined such that,

$$\bar{\alpha}_k = \max\{(\alpha|x^k + \alpha d^k \, is \, feasible \, for \, (QP)\} \tag{18}$$

$$= \min\{\frac{A(k,:)x^k - a_0}{(-A(k,:)d^k)}|A(k,:)d^k < 0, k\epsilon N(k)\}. \tag{19}$$

Step 4: Step length α, satisfying $\alpha < \bar{\alpha}$ is calculated,

$$f(x^k + \alpha d^k) < f(x^k)$$

Step 5:

$$\text{Set } x^{k+1} = x^k + \alpha d^k \tag{20}$$

Step 6:

$$\text{Set } k = k + 1. \tag{21}$$

END

6 Application of Active Set and Interior Point Methods to Microgrid Network

A 3 bus feeder of microgrid network is shown in Fig. 2. R2 provides primary protection and R1 provides backup protection when fault occurs just beyond R2. R1 provides primary protection when fault occurs just beyond R1. Assume maximum fault current beyond R1 and R2 are 4000 A and 3000 A. The operating time of each circuit breaker is assumed to be 0.5. The problem can be formulated by taking TMS of R1 as x1 and R2 as x2. Both R1 and R2 act after 0.2 s to prevent network isolation by circuit breaker due to instantaneous current surges. We assume that the overshoot of relay R1 is 10 % of its optimum operating time.

Fig. 2 An example of sample feeder of microgrid with two relays

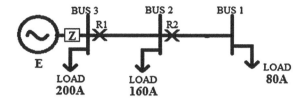

Table 1 Calculation of Coefficients required for formulation of the objective function and constraint equation

Fault position		Relay	
		R1	R2
Just beyond R1	PSM	$\frac{4000}{300*1} = 13.33$	–
	IDMT	$\frac{0.14}{13.33^{0.02} - 1} = 2.6$	
	VIT	$\frac{13.5}{13.33 - 1} = 1.1$	
	EIT	$\frac{80}{13.33^2 - 1} = 0.5$	
Just beyond R2	PSM	$\frac{3000}{300*1} = 10$	$\frac{3000}{100*1} = 30$
	IDMT	$\frac{0.14}{10^{0.02} - 1} = 3$	$\frac{0.14}{30^{0.02} - 1} = 2$
	VIT	$\frac{13.5}{10 - 1} = 1.5$	$\frac{13.5}{30 - 1} = 0.5$
	EIT	$\frac{80}{10^2 - 1} = 0.8$	$\frac{80}{30^2 - 1} = 0.1$

Table 2 The objective function and their respective constraint equations for three different relay types

Relay type	Objective function	Constraint equations
IDMT	$Minz = 2.6 * x1 + 2 * x2$	$3 * x1 - 2 * x2 \geq (0.5 + 0.1 * (0.5 + 2 * x2))$ $2.6 * x1 \geq 0.2$ $2 * x2 \geq 0.2$
VIT	$Minz = 1.1 * x1 + 0.5 * x2$	$1.5 * x1 - 0.5 * x2 \geq (0.5 + 0.1 * (0.5 + 0.5 * x2))$ $1.1 * x1 \geq 0.2$ $0.5 * x2 \geq 0.2$
EIT	$Minz = 0.5 * x1 + 0.1 * x2$	$0.8 * x1 - 0.1 * x2 \geq (0.5 + 0.1 * (0.5 + 0.1 * x2))$ $0.5 * x1 \geq 0.2$ $0.1 * x2 \geq 0.2$

Calculations:

First we choose the CT ratio of the two relays and the plug setting of the relays. At R2, the maximum load current (assuming 25 % overload) = 125 % of 80A = 100 A. Assuming 1 A is the rated current of the relays; CT ratio can be 100:1. Plug setting is assumed to be 100 % i.e., pickup current = 100 % of 1 A = 1 A (Tables 1, 2 and 3).

At R1, the maximum load current (assuming 25 % overload) = 125 % of (160 + 80) A = 300 A. Assuming 1 A is the rated current of the relays, CT ratio

Table 3 The optimized values of x1, x2, objective function and no. of iterations for the two methods

Relay type	Interior point method				Active set method			
	$x1$	$x2$	*Minz*	No. of iterations	$x1$	$x2$	*Minz*	No. of iterations
IDMT	0.257	0.1	0.8673	7	0.257	0.1	0.8673	2
VIT	0.513	0.4	0.7647	8	0.513	0.4	0.7647	2
EIT	0.963	2	0.6813	8	0.962	2	0.6812	2

can be 300:1. Plug setting is assumed to be 100 % i.e., pickup current = 100 % of 1 A = 1 A.

For all types of relay, the following constraint equations on TMS are valid except for $x2$ of EIT:-

$$0.1 \leq x1 \leq 1.2$$

$$0.1 \leq x2 \leq 1.2$$

The constraint on TMS of R2 $(x2)$ in the case of EIT relays is given by:-

$$0.1 \leq x2 \leq 2$$

7 Conclusion

Microgrid has bi-directional power flow and variation in fault current magnitude in both grid connected and islanded modes of operation. Thus it proves challenging to protection engineers. Overcurrent faults are predominant in microgrid and optimization of overcurrent relays in any distribution system is a necessity to clear the fault quickly. From the above result it is clear that using EIT relay is most effective as its total time of operation is minimum. It can also be seen that in between the two optimization algorithms, Active set method is more efficient as the number of iterations it takes to converge to an optimum solution is much less.

References

1. O.V.G. Swathika, S. Hemamalini, Prims aided Dijkstra Algorithm for adaptive protection in microgrids. IEEE J. Emerg. Select. Top. Power Electronics. **4**(4), 1279–1286 (2016)
2. O.V.G. Swathika, K. Karthikeyan, S. Hemamalini, R. Balakrishnan, Relay coordination in real-time microgrid for varying load demand. ARPN J. Eng. Appl. Sci. **11**, 3222–3227 (2016)
3. O.V.G. Swathika, K. Karthikeyan, S. Hemamalini, R. Balakrishnan, PLC based LV-DG synchronization in real-time microgrid network. ARPN J. Eng. Appl. Sci. **11**, 3193–3197 (2016)

4. O.V.G. Swathika, S. Hemamalini, Communication assisted overcurrent protection of microgrid. IEEE Int. Conf. Comput. Intell. Commun. Netw. (CICN). pp.1472–1475 (2015)
5. O.V.G. Swathika, K. Karthikeyan, S. Hemamalini, Multiple DG synchronization and de-synchronization in a microgrid using plc. Springer Adv. Comput. Commun. Technol. 565–572 (2016)
6. S. Kannuppaiyan, V. Chenniappan, Numerical inverse definite minimum time overcurrent relay for microgrid power system protection. IEEE J. Trans. Electr. Electron. Eng. **10**(1), 50–54 (2015)
7. O.V.G. Swathika, S. Hemamalini, Adaptive and intelligent controller for protection in radial distribution system. Adv. Comput. Commun. Eng. Technol. 195–209 (2016)
8. P.P. Bedekar, S.R. Bhide, Optimal coordination of overcurrent relay timing using continuous genetic algorithm. Elsevier Sci. Direct Expert Syst. Appl. **38**, 11286–11292 (2011)
9. O.V.G. Swathika, B. Indranil, R. Bhaskar, K. Suhit, S. Hemamalini, Optimization techniques based adaptive overcurrent protection in microgrids. J. Electr. Syst. (2015) Special Issue 3.
10. J.H. Zhao, F. Wen, Zhao yang dong. Yusheng Xue, Kit Po Wong: Optimal dispatch of electric vehicles and wind power using enhanced particle swarm optimization, IEEE Trans. Ind. Electronics **8**, 889–899 (2012)
11. S. Granville, Optimal reactive dispatch through interior point methods. IEEE Trans. Power System **9**, 136–146 (1994)
12. M.M. Aman, G.B. Jasmon, A.H.A. Bakar, et al., Graph theory-based radial load flow analysis to solve the dynamic network reconfiguration problem, Int. Trans. Electr. Energy Syst. (2015), doi:10.1002/etep.2108
13. O.V.G. Swathika, S. Hemamalini, Kruskal aided floyd warshall algorithm for shortest path identification in microgrids. ARPN J. Eng. Appl. Sci. **10**, 6614–6618 (2015).

Series Smart Wire—Managing Load and Congestion in Transmission Line

Abhishek, Divya Asija, Pallavi Choudekar and Yogasree Manganuri

Abstract Nowadays, congestion management is a major problem in power system deregulation. With the continuous increase in load demand, there is continuous requirement for different new technologies resulting in advanced network operation. This paper presents a solution for congestion management by developing a series smart wire module which operates with the increment in load. The circuitry is bypassed if it does not detects congestion in line else with the detection of congestion series smart wire module is operated. This method improves the reliability of the system by reducing active power losses. The effectiveness of this module is demonstrated in standard IEEE 15 bus system model using MATLAB/Simulink and the results are formulated with graphical representations.

Keywords Series smart wire module · Congestion management · Load flow · Current limiting diode

1 Introduction

With the ever rise in demand and advancement of technologies, the power system network varies from regulated to deregulated power systems. The ever increasing load continuously results in many technical issues. In the modern era, increasing load may cause congestion in transmission line for transmitting more power than its

Abhishek · D. Asija (✉) · P. Choudekar · Y. Manganuri
Amity University, Uttar Pradesh, India
e-mail: divyaasija83@gmail.com

Abhishek
e-mail: abhi.p2947@gmail.com

P. Choudekar
e-mail: pallaveech@gmail.com

Y. Manganuri
e-mail: myogasri@gmail.com

© Springer Nature Singapore Pte Ltd. 2017
S.C. Satapathy et al. (eds.), *Proceedings of the 5th International Conference on Frontiers in Intelligent Computing: Theory and Applications*, Advances in Intelligent Systems and Computing 516, DOI 10.1007/978-981-10-3156-4_10

capacity, as every transmission line has their own capacity for handling power. This extra power flow should be controlled by some technique so that congestion is managed.

There are many problem in modern power system like in deregulated market which includes choosing an appropriate strategy for the electricity, market power issues of the participants, transmission line congestion and related prices, maintaining system reliability, assessing market equilibrium and market efficiency. Out of these issues here congestion management is considered. Congestion is defined as inability of transmission line to transfer power as per market requirements or we can say that the exchanges of power not in control, some lines are overloaded due to rise in demand is also referred as congestion.

The proposed module has been simulated using MATLAB. We have chosen IEEE 15 bus system for testing. The IEEE 15 bus system is basically a substation acting as source of total generation $P = 1.29$ MW and $Q = 1.31$ MVAR, total load of $P = 1.23$ MW and $Q = 1.25$ MVAR, total losses of $P = 0.06$ MW and $Q = 0.06$ MVAR. It has 15 loads connected across 15 buses. The overall system has a base voltage of 11 kV.

For the congestion management, smart wire module has been designed consisting of single turn transformer connection with the conductors of the transmission line [1]. It introduces an additional reactance in the transmission line which limits the over current during congestion. The smart wire module is operated only when there is congestion in the line and it injects inductive impedance in line, which increases the overall reactance and ultimately reduces the congestion of transmission line [2]. Smart wire module is able to raise the value of inductance in the line but it is not able to decrease it. Smart Wire pushes the power to another line, relieves congestion without building another prevailing line.

In the literature it has been found that to control the power flows in transmission line, FACTS devices are used. But, for the large scale adoption, this technology may also suffer through pricing and reliability issues. One of the alternative approaches is to use distributed FACTS devices which are much more cost effective for the control of power in line [3, 4]. In this paper the existing transmission system is converted into smart transmission system which regulates the power flow effectively using Smart Wire module [1]. It is connected directly across the conductor which ultimately pushes the power to alternating low impedance path.

Smart wire is used to reduce congestion which enhances the passive grid asset and reliability with respect to traditional FACTS devices which makes short circuits when it fails and leaving passive assets in operation mode [1]. Minimization of losses through reconfiguration states that system losses have been managed through network reconfiguration so that all loads received power [5].

Congestion management by generic algorithm using FACTS controllers system provides loss minimization, reduction of branch load and voltage stability with optimal location of FACTS devices. For the efficient location with different cases several algorithm is introduced [6].

Optimization of FACTS devices location for congestion management considers that congestion is controlled by the development of simple and effective model for correct location of FACTS devices by its parameters [7]. Low cost function and transmission line monitoring by smart wire studies demonstrate the feasibility of work done and its prototype enhancement along with technology affect [8].

It is also found in literature that for achieving power flow control in line, distributed static series compensator is used. Research shows that DFACTS devices are used for improving the congestion with lower cost. DSSC directly connect across the high voltage line which does not require high voltage insulation [9].

Another study on controlling of power flow in grid through smart wirereveals that line impedance is raised by injecting a pre-defined value of reactance of the single-turn Transformer [10].

2 Reference Model and Parameters

Figure 1. shows the single line diagram of IEEE 15 bus radial distribution having rated line voltage of 11 kV. Table 1 shows the system data which is considered for performing proposed model.

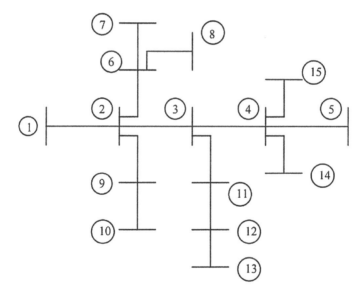

Fig. 1 IEEE 15 bus system [11]

Table 1 IEEE 15 bus system data		P(MW)	Q(Mvar)
	Total generation	1.263815	1.284233
	Total PQ load	1.226329	1.251043
	Total Z shunt	2.71E-13	−4.2E-13
	Total ASM	0	0
	Total losses	0.037485	0.03319

3   Operation of Series Smart Wire Module

From the generation end to load, power flows from generating bus to load bus. The transmitting power between the busses is inversely proportional to Reactance of line and directly proportional to the sine of angle difference of bus voltages.

$$P_{12} = \frac{V_1 V_2 sin\delta}{X_L} \tag{1}$$

V_1 and V_2 are the voltages of bus1 and bus2 and δ is the difference of phases between bus voltages and P_{12} is the power flow between bus1 and bus2. For controlling the power flow in the line, angle between bus voltages have to be changed which is varied by phase shifting transformer that is so expensive. Another method to control power flow is to introduction of additional reactance in the line that indirectly affects the power flow between busses. The reactance increases, power flow reduces due to inverse relation between power and inductive reactance.

Smart wire technology is developed by Georgia tech which introduces an additional reactance to the power line between busses that reduces the power flow and managed congestion problems [8].

Conceptually Smart wire is a system having single turn transformer connected to the current limiting diode and a static switch which is having thyristor connection. The turn ratio of single turn transformer is taken such that it reflects minimum secondary side current even in faulty condition with efficient working of static switch and current limiting diode.

Current limiting diode only conducts for particular limit, if the current exceeds to threshold value than the diode acts as open circuit and the inductance is introduced into the line that increases the overall reactance in line which reduces the power flow. This pushes the extra load requirement to another line. The series connection of this smart wire refers as series smart wire module. Series connection of smart wire building blocks is used here in order to have higher efficiency.

Whenever the load demand increases it require more power flow form the line but line has specific capacity for which it wear power. This overloading of line creates Congestion. Figure 2. shows the series connection of smart wire building block in a transmission line.

Transmission line congestion can be managed through series connected Smart wire building block. If normal flow of power i.e. in no overload condition the Smart wire system is short circuited by current limiting diode but in over loading condition, congestion occur which would be reduced as current raises the value in current limiting diode and it acts as open circuit and overall power flows from single turn transformer that can increases the overall reactance of the line, which in turn reduces the power flow in the line as real power is inversely proportional to reactance.

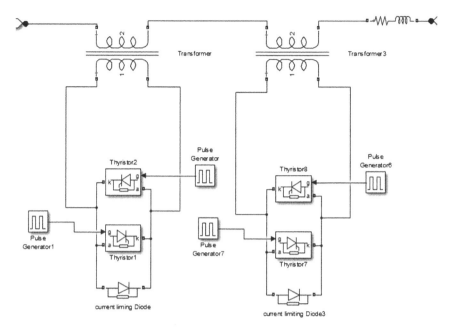

Fig. 2 Series connected smart wire building block

$$P_{12} = \frac{V^2 sin\delta}{X_L + X_{inj}} \tag{2}$$

X_{inj} is the smart wire reactance introduced in the line while congestion. X_{inj} has smaller value which will not affect the transmission line capacity.

4 Results

Implementation of proposed smart wire module is done by connecting line 6–7 in series with the smart wire building block as shown Fig. 3. The system is having 11 kV voltage as base value and 60 Hz frequency. The load 7 connected to transmission line 7 has total requirement of an active power P = 70 KW and reactive power Q = 71.41 KVAR. The change in Load 7 will be analyzed with smart wire module.

Case 1: When Load 7 is unchanged P = 70 KW and Q = 71.41 KVAR

Table 2 shows that with the introduction of series smart wire module reactive power losses have been reduced, while Table 3 shows that voltage profile for line 6–7 has been raised from 0.9780 to 1.094 pu which is required for stability.

Case 2: When Load 7 changes by different percentage level

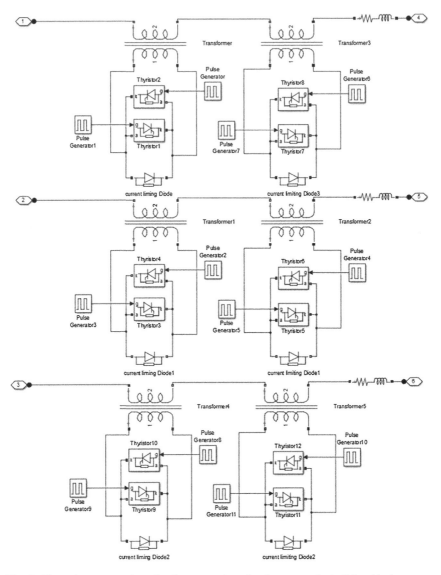

Fig. 3 Three phase connection of series connection of series smart wire building block

Table 2 Variation of current and power losses

Without series smart wire module		With series smart wire module	
Iph (A)	Losses MVAR	Iph (A)	Losses MVAR
5.90	0.06	5.57	−0.07

Table 4 shows the positive impact of smart wire module on reactive power losses with load increment and voltage profile improvement from 0.97 to 1.1 pu which leads to stability enhancement of power system.

Load flow is done for IEEE 15 bus system using Matlab software in which the value of load 7 is raised by 30, 60 and 90 %. Phase current is initially observed as 5.90 A without Smart wire block and along with smart wire block it is 5.57 A. But when the load 7 increases by 30, 60 and 90 % then current suddenly increases such that for 30 % load raised, the phase current is 6.82 A, for 60 % load raised, the phase current observed as 7.83 A and for 90 % load raised phase current is 8.88 A.

Figure 4 shows the graphical representation of current profile with the increment in load. Considering these current referred as beyond the limit value of current limiting diode then it acts as open circuit also having Thyristor not triggered beyond this value. It leads to the introduction of reactance in line 6–7 with series smart wire module and current is drop down for 30 % load increment to 6.14 A, for 60 % load increment phase current observed as 6.33 A and for 90 % load increment the phase current is 6.27 A as shown in Tables 2 and 4. The voltage profile is also improved from the series smart wire module shown in Tables 3 and 4.

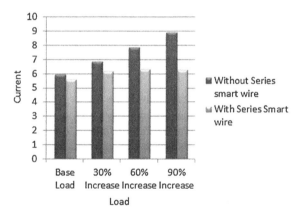

Fig. 4 Variation in current values with incremental load

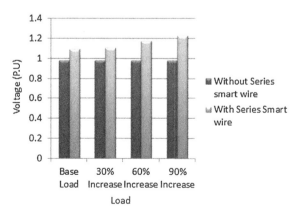

Fig. 5 Improved voltage profile with incremental load

Table 3 Variation of voltages

Without series smart wire module voltage (pu)	With series smart wire module voltage (pu)
0.9780	1.094

Table 4 Variation of current, voltage and power losses with load increment

Load change	Without series smart wire module			With series smart wire module		
Raised by (%)	Iph (A)	Voltage (pu)	Losses (MVAR)	Iph (A)	Voltage (pu)	Losses (MVAR)
30	6.82	0.9773	0.06	6.14	1.1039	−0.11
60	7.83	0.9766	0.06	6.33	1.1746	−0.15
90	8.88	0.9759	0.06	6.27	1.2226	−0.18

Figure 5 shows the graphical representation of improved voltage profile with increment in load. The series smart wire module give high efficiency as voltage profile improves and current is reduced as shown in results. Therefore this technology improves the system reliability and removal of congestion from the transmission line as it forces the power to flow from other low impedance line.

5 Conclusion and Future Scope

From the experimental analysis we are providing a solution having series smart wire module with transmission line for congestion management. The circuitry is bypassed through current limiting diode if it fails to detect congestion in line and if congestion is observed then series smart wire module are operated and current limiting diode acts as open circuit and triggering operation of thyristor are carried out. This improves the reliability and continuous power flow capacity resulting in reduction of transmission line losses by improving voltage and current profile. The effectiveness of this requirement is demonstrated on IEEE 15 bus system using MATLAB/Simulink and the results are formulated with graphical representations which indicate the change in current and voltage values with increase in load. Intelligent techniques along with evolutionary algorithm may be utilized for faster computation by selecting required optimization in various aspects. By considering multi objective differential evolution it will be a good attempt to mitigate congestion.

References

1. S. Pal, S. Sengupta, Congestion management of a multi-bus transmission system using distributed smart wires, in *International Conference on Control, Instrumentation, Energy and Communication (CIEC)*, (2014), pp. 417–420 (January 2014)
2. M. Begovic, in *Electrical Transmission System & Smart Grid, Selected Entries from the Encyclopedia of Sustainability Science and Technology* (Springer, 2013)
3. D. Divan, H. Johal, Distributed FACTS a new concept for realizing grid power flow control, an *International Conference on IEEE Power Electronics Specialist Conference*, vol. 22, pp. 2253–3360 (November 2007)
4. F.A. Rahimi, A. Vojdani, Meet the emerging transmission market segment. IEEE Comput. Appl. Power **12**(1), 26–32 (1999)
5. S. Pal, A. Neogi, S. Biswas, M. Bandyopadhyay, S. Sengupta, Loss minimization and congestion management of a power distribution network through its reconfiguration. Int. J. Electr. **2**, 95–99 (2013)
6. D. Venugopal, A. Jayalaxmi, Congestion management by optimal choice and allocation of facts controllers using genetic algorithm. Int. J. Soft Comput. Eng. **4**, 72–76 (2014)
7. S.N. Singh, A.K. David, Congestion management by optimizing facts device location, in *International Conference on Electric Utility Deregulation and Restructuring and Power Technologies*, 2000. Proceedings. DRPT (2000), pp. 23–28 (April 2000)
8. F. Kreikebaum, D. Das, Y. Yang, F. Lambert, D. Divan, Smart wires a distributed, low-cost solution for controlling power flows and monitoring transmission lines, in *International Conference on Innovative Smart Grid Technologies Conference Europe*, (2010) IEEE PES, pp. 1–8 (October 2010)
9. Deepak M. Divan, William E. Brumsickle, Robert S. Schneider, Bill Kranz, Randal W. Gascoigne, Dale T. Bradshaw, Michael R. Ingram, Ian S. Grant, A distributed static series compensator system for realizing active power flow control on existing power lines. IEEE Trans. Power Deliv. **2**, 654–661 (2004)
10. J. Melcher, Distributed series reactance for grid power flow control IEEE PES chapter meeting, (2012), pp. 3–5 (August 8, 2012)
11. A.K Sharma, T. Manglani, Voltage enhancement in distribution system using voltage improvement factor. IOSR J. Electr. Electron. Eng. **7**(1), 46-48, (2013). e-ISSN: 2278-1676, p-ISSN: 2320-3331

Gene Ontology Based Function Prediction of Human Protein Using Protein Sequence and Neighborhood Property of PPI Network

Sovan Saha, Piyali Chatterjee, Subhadip Basu and Mita Nasipuri

Abstract Predicting functions of protein from its amino acid sequence and inter-acting protein partner is one of the major challenges in post genomic era compared with costly, time consuming biological wet lab techniques. In drug discovery, target protein identification is important step as its inhibition may disturb the activities of pathogen. So, the knowledge of protein function is necessary to inspect the cause of diseases. In this work, we have proposed two function prediction methods FunPred1.1 and FunPred1.2 which use neighbourhood analysis of unknown protein empowered with Amino Acid physico-chemical properties. The basic objective and working of these two methods are almost similar but FunPred1.1 works on the entire neighbourhood graph of unknown protein whereas FunPred1.2 does same with greater efficiency on the densely connected neighbourhood graph considering edge clustering coefficient. In terms of time and performance, FunPred1.2 achieves better than FunPred1.1. All the relevant data, source code and detailed performance on test data are available for download at FunPred-1.

S. Saha (✉)
Department of Computer Science and Engineering, Dr. Sudhir Chandra
Sur Degree Engineering College, Dumdum 700 074, Kolkata, India
e-mail: Sovan.Saha12@gmail.com; sovansaha12@gmail.com

P. Chatterjee
Department of Computer Science and Engineering, Netaji Subhash
Engineering College, Garia 700152, Kolkata, India
e-mail: piyali.gini@gmail.com; chatterjee_piyali@yahoo.com

S. Basu · M. Nasipuri
Department of Computer Science and Engineering, Jadavpur University,
Kolkata 700032, India
e-mail: Subhadip.Basu@cse.jdvu.ac.in; subhadip@cse.jdvu.ac.in

M. Nasipuri
e-mail: Mita.Nasipuri@cse.jdvu.ac.in; mitanasipuri@yahoo.com

© Springer Nature Singapore Pte Ltd. 2017
S.C. Satapathy et al. (eds.), *Proceedings of the 5th International Conference on Frontiers
in Intelligent Computing: Theory and Applications*, Advances in Intelligent Systems
and Computing 516, DOI 10.1007/978-981-10-3156-4_11

Keywords Protein function prediction · Neighbourhood analysis · Protein protein interaction (PPI) network · Functional groups · Relative functional similarity · Edge clustering coefficient · Cellular component · Biological process · Molecular function · Physico-chemical properties · Gene Ontology(GO)

1 Introduction

Proteins executes vital functions in essentially all biological processes. Computational methods like gene neighborhood, sequence and structure, protein-protein interactions (PPI) etc. have naturally created a larger impact in the field of protein function prediction than the biological based experimental methods. Unknown protein function predicted from protein interaction information is an emerging area of research in the field of bioinformatics. In this approach functions of unannotated proteins are determined by utilizing their neighborhood properties in PPI network on the basis of the fact that neighbors of a particular protein have similar function.

In the work of Schwikowski [1] at first most frequent occurrence of k functional labels are identified. Then a simple counting technique is used to assign k functions to the unannotated protein based on the identification. Though the entire methodology is not too much complex in execution but the fact that the entire network has not been considered cannot be denied. Besides confidence score also play an important role in predicting functional annotations which is also missing in this work. This deficiency of assignment of confidence score has been erased in the work of Hishigaki et al. [2]. Here annotations of k functions to the unannotated protein P is dependent on k largest $chi - square$ scores which is defined as $\frac{(n_f - e_f)^2}{e_f}$, where n_f is the count of proteins belonging to the n-neighborhood of the protein P that have the function f and e_f is the expectation of this number based on the number of occurrences of f among all proteins available in the entire network. While on the other hand, the exploitation of the neighborhood property of PPI network up to the higher levels has been executed in the work of Chen et al. [3]. Whereas Vazquez et al. [4] annotate a protein to a function in such a way that the connectivity of the allocated protein to that function is maximum. An identical technique on a collection of PPI data as well as on gene expression data is applied by Karaoz et al. [5]. Nabieva et al. [6] applies a flow based approach considering the local as well as global properties of the graph. This approach predicts protein function based on the amount of flow it receives during simulation. It should be noted here that each annotated protein acts as the source of functional flow. While the theory of Markov random field has been reflected in the work of Deng et al. [7] where the posterior probability of a protein of interest is estimated. Letvsky and Kasif [8] use totally a different approach by the application of binomial model in unknown protein function prediction. Similarly, Wu et al. [9] includes the summation of both protein structure and probabilistic approach in this field of study. In the work of Samanta et al. [10], a network based statistical algorithm is proposed, which assumes that if two proteins share significantly larger number of common inter-acting partners they share a common functionality. Arnau et al. [11] proposed another application named as UVCLUSTER which is based on bi-clustering. This application iteratively explored distance datasets. In the early stage, Molecular Complex Detection (MCODE) is executed by Bader and Hogue [12] where iden-tification of dense regions takes place according to some heuristic parameters. Altaf-ul-Amin et al. [13] also use a clustering approach. It starts from a single node in a graph and clusters are gradually grown until the similarity of every added node within a cluster and density of clusters reaches a certain limit. Graph clustering approach is used by Spirin and Mirny [14] where they detect densely connected modules within themselves as well as sparsely connected with the rest of the network based on super paramagnetic clustering and Monte Carlo algorithm. Theoretical graph based approach is observed in the work of Pruzli et al. [15] where clusters are identified using Leda's routine components and those clusters are analyzed by Highly Connected Sub-graphs (HCS) algorithm. While the application of Restricted Neighborhood Search Clustering algorithm (RNCS) is highlighted in the work of King et al. [16]. The interaction networks are partitioned into clusters by this algorithm using a cost weightage mechanism. Filtering of clusters is then carried out based on their properties like size, density etc.

This survey highlights the fact that there is an opportunity for inclusion of domain as well as some other related specific knowledge like protein sequences to enhance the performance of protein function prediction from protein interaction network. Motivated by this fact, a neighborhood based method has been proposed for predicting function of an unannotated protein by computing the neighborhood scores on the basis of protein functions and physico-chemical properties of amino acid sequences of proteins. The unannotated protein is associated with the function corresponding to highest neighborhood score.

1.1 Dataset

We have used the Gene Ontology (GO) dataset of human obtained from UniProt. The dataset is available at FunPred-1. Three categories: Cellular-component, Molecular-function and Biological-process are involved in the GO system. In this system, each protein may be annotated by several GO terms (like GO: 0000016) in each category. So, here, at first we have ranked every GO terms of 3 categories based on the maximum number of occurrences in each of them. Then 10 % of proteins belonging to the top 15 GO terms in each of three categories are selected as unan-notated while the remaining 90 % proteins are chosen as training samples using random sub-sampling technique. Since we have considered both *Level*-1 and *Level*-2 neighbors, the protein interaction network formed for each protein in any functional group is large and complex. Therefore, in the current experiment we consider only 10 % of available proteins in each functional group as test set. Table 1 show the detailed statistics of the train-test dataset for the three GO categories. While overall

Table 1 Distribution of proteins and protein pairs in 3 functional categories in GO based Human dataset, considered under the current experiment

Organism	Number of proteins	No. of interactions	GO terms	Cellular component	Molecular function	Biological process
Human	2577	3329	3730	522	717	2491
	Cellular component		*Molecular function*		*Biological process*	
	Selected unannotated proteins	Annotated proteins	Selected unannotated proteins	Annotated proteins	Selected unannotated proteins	Annotated proteins
	846	1731	765	1812	1216	1361
	Total selected unannotated proteins in entire GO : 846 + 765 + 1261 = 2872					
	Total annotated proteins in entire GO: 1731 + 1812 + 1361 = 4904					

protein interaction network of the three functional categories along with known (marked blue) and unannotated proteins (marked yellow) with their respective result comparison by FunPred 1 has been highlighted here.

2   Related Terminologies

In both FunPred 1.1 and FunPred 1.2, we have used four scoring techniques: Protein Neighborhood Ratio Score $\left(\text{Pscore}^{l(=1,2)}\right)$[17], Relative functional similarity $\left(W_{u,v}^{l(=1,2)}\right)$ [17, 18], Proteins path-connectivity score $\left(Q_{u,v}^{l(=1,2)}\right)$ [17, 19] and physico-chemical properties score $\left(\text{PCP}_{\text{score}}^{l(=1,2)}\right)$ [20]. $\text{PCP}_{\text{score}}^{l(=1,2)}$ is incorporated since sequences of amino acid of each protein also plays a vital role in unknown protein function prediction. While in FunPred 1.2, we have used one additional feature Edge Clustering Coefficient $\left(\text{ECC}_{u,v}^{l(=1,2)}\right)$ [21] to find densely populated region in the network. All the other relevant graphical terms and properties are described in our previous work [17, 22].

3   Proposed Method

Two methods [17] have been proposed for unannotated protein function prediction. Uniqueness can be defined in the aspect that the selection of the neighborhood of the unannotated proteins in both these two methods differs over the different aspects of neighborhood properties defined in the previous section. The first method FunPred 1 is described below:

3.1　FunPred 1.1

FunPred 1.1 [17] uses the combined score of neighborhood ratio, proteins path connectivity, physico-chemical property score and relative functional similarity. Now, this method always focuses in identifying the maximum of the summation of four scores thus obtained in each level and assign the unannotated protein to the corresponding functional group (GO term) of the protein having the maximum value. Given G_p', a sub graph consisting of any proteins (nodes) of set $FC = \{FC_1, FC_2, FC_3\}$; where, FC_i represents a particular functional category, this method annotates proteins belonging to the set of un-annotated proteins P_{UP} to any GO term of set FC. Steps of FunPred 1.1 are described as Algorithm 1.

Algorithm 1 Basic methodology of FunPred 1.1

```
Input:Unannotated protein set P_UP.
Output:The proteins of the set P_UP gets annotated to any
functional group (GO term) ofset FC.
Step 1: Any protein from set P_UP is selected.
Step 2: Count Level −1 and Level −2 neighbors of that
protein in G'_p associated with set FC.
```
Step 3: Compute $P_{FC_{i(=1,...,3)}}^{l(=1,2)}$ for each GO term in set FC and assign this score to eachprotein
$\left(\text{Pscore}^{l(=1,2)}\right) \in P_A$, belonging to the respective functionalcategory.
Step 4: Compute $Q_{u,v}^{l(=1,2)}$, $W_{u,v}^{l(=1,2)}$ for each edge in Level −1 and Level −2.
Step 5:Obtain neighborhood score i.e.

$$N_{(FC_K)}^l = Max((\max(\text{Pscore}^1 + Q_{u,v}^1 + W_{u,v}^1 + ECC_{u,v}^1 + PCP_{score}^1)),$$

$$\left(\max(\text{Pscore}^2 + Q_{u,v}^2 + W_{u,v}^2 + ECC_{u,v}^2 + PCP_{score}^2)\right)))$$

```
Step 6: The unannotated protein from the set P_UP is
assigned to the GO term belonging to FC_K.
```

3.2　FunPred 1.2

In FunPred 1.1, all *Level*-1 neighbors and *Level*-2 neighbors belonging to any GO term of 3 functional categories are considered for any unannotated protein. Neighborhood property based prediction is then carried out, the computation of which considers all *Level*-1 or *Level*-2 neighbors. But if the computation is confined only on significant neighbors who have maximum neighborhood impact on the target protein then exclusion of non-essential neighbors may substantially reduce the computational time which is the basis of our heuristic adopted in FunPred 1.2 [17]. So this method looks for the promising regions instead of calculating neighborhood ratios for all of them and only then the calculation of $N^l_{(FC_K)}$ is done. Here, at first edge clustering coefficient (ECC) of each edge in $Level-1$ and $Level-2$ (as mentioned in the earlier section) is calculated. Edges having relatively low edge clustering coefficient gets eliminated and thus the original network gets reduced upon which we will apply our previous method. Now the original FunPred-1.1 algorithm is applied on this reduced PPI network (renaming the entire modified method as FunPred 1.2). The computational steps associated with FunPred 1.2 are described as Algorithm 2.

Algorithm 2 Basic methodology of FunPred 1.2

```
Input:Unannotated protein set P_UP.
Output: The proteins of the set P_UP gets annotated to any
functional group (GO term) of set FC.
Step 1:Any protein from set P_UP is selected.
Step 2:Protein interaction network of the selected
protein has been constructed detecting its
Level −1 and Level −2 neighbors.
Step 3: Compute ECC^l(=1,2)_u,v for each edge in Level −1 and
Level −2.
Step 4: Eliminate non-essential annotated proteins
(neighbors) associated with edges having lower
values of ECC^l(=1,2)_u,v both in Level −1 and Level −2 thus
generating a densely connected reduced protein
interaction network.
Step 5:Count Level −1 and Level −2 neighbors of that
protein in G'_p associated with set FC.
Step 6:Compute P^l(=1,2)_FC_i(=1,...,3) for each GO term in set FC and
assign this score to each protein (Pscore^l(=1,2)) ∈ P_A,
belonging to the respective functional group.
Step 7: Compute Q^l(=1,2)_u,v, W^l(=1,2)_u,v for each edge in Level −1
and Level −2.
Step 8: Obtain neighborhood score i.e.
```

$$N^l_{(FC_K)} = \text{Max}((\text{max}(\text{Pscore}^1 + Q^1_{u,v} + W^1_{u,v} + ECC^1_{u,v} + PCP^1_{score})),$$

$$(\text{max}(\text{Pscore}^2 + Q^2_{u,v} + W^2_{u,v} + ECC^2_{u,v} + PCP^2_{score})))$$

```
Step 9: The unannotated protein from the set P_UP is
assigned to the GO term belonging to FC_K.
```

4 Results and Discussion

We have used standard performance measures, such as Precision (P), Recall (R) and F-Score (F) values for evaluating the training results for the ith functional category as described in our previous work [17]. The detailed analysis of FunPred 1.1 and FunPred 1.2 with respect to Precision, Recall and F-score values has been shown in Table 3. Functional category-wise Precision, Recall and F-scores of the two methods are given in Table 2. The average Precision of FunPred 1.2 is estimated as 0.743 (see Table 3). Although we observe relatively low values of Recall for the two methods, high Precision scores indicate that our algorithm has succeeded in generating more significant results. High F-score values have been retrieved in one functional category i.e. Molecular function. Ten percent of proteins from each of the high ranking GO terms in the three functional categories are considered as unannotated proteins using random sub-sampling in both of our methods.

The performance of FunPred 1.1 has been significantly improved in FunPred 1.2 as FunPred 1.2 reduces the neighborhood network. For example, from Table 2, it can be observed that a Precision improvement of 5.2 and 9.8 % occurs in the Cellular component and Molecular function respectively in FunPred 1.2 over FunPred 1.1. In our experiment, Biological process performs worst in comparison to the other functional category. Except this category, in almost all other cases we have either achieved good prediction performance in FunPred 1.1 or obtained significant hike in performance in FunPred 1.2 in comparison to its predecessor.

To compare the performance of the current method with the other existing neighborhood analysis methods, we have identified four relevant methods and

Table 2 Evaluated results of FunPred 1.1 and FunPred 1.2 for three functional categories of GO based human dataset

Functional categories	Methods	Precision	Recall	F-Score
Cellular component	FunPred-1.1	0.662	0.602	0.631
	FunPred-1.2	0.714	0.650	0.680
Molecular function	FunPred-1.1	0.722	0.725	0.724
	FunPred-1.2	0.820	0.823	0.821
Biological process	FunPred-1.1	0.660	0.625	0.642
	FunPred-1.2	0.695	0.657	0.676

Table 3 Recall, Precision, F-Score for FunPred 1.1 and FunPred 1.2 in accordance to Mean and standard deviation

Methods	Mean/SD	Precision	Recall	F-Score
FunPred-1.1	Mean	0.681	0.651	0.665
	Standard deviation	0.035	0.065	0.050
FunPred-1.2	Mean	0.743	0.710	0.726
	Standard deviation	0.067	0.097	0.082

compared the performances of the same on our Human dataset. More specifically we compared our work with the neighborhood counting method [1], Chi-square method [2], a recent work on Neighbor Relativity Coefficient (NRC) [19] and FS-weight based method [23].

The best performance among the four methods is the work of Moosavi et al. [19]. The NRC method generates average Precision, Recall and F-score values of 0.374, 0.434 and 0.368 respectively. The detailed result analysis of our method as highlighted in Table 3 over 15 functional groups clearly reveals the fact that our method is relatively better than the NRC based method in terms of average prediction scores. This betterment is achieved since both Level-1 and Level-2 neighbors have been considered along with the exploration of a variety of scoring techniques in the human PPI network. Not only that we have also included protein sequences, successors as well as the ancestors of a specific unannotated protein while estimating neighborhood score for unannotated protein function prediction.

The result obtained in all Chi-square methods [2] is comparatively lower than the other methods because it only concentrates only on the denser region of the interaction network. The neighborhood counting method though performs well but fails when compared to NRC, FS-weight#1 (only direct neighbors are considered) and FS-weight #1 and #2 (both direct and indirect neighbors are considered) methods since it does not consider any difference between direct and indirect neighbors. Figure 1 shows a comparative detailed analysis of the four methods (taken into consideration in our work) along with our proposed systems.

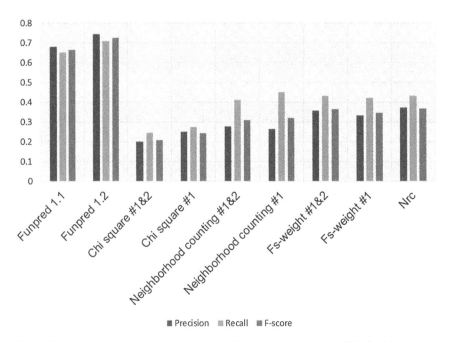

■ Precision ▨ Recall ■ F-score

Fig. 1 Comparative analysis of other methods with our developed method FunPred 1

All these analysis show that our proposed FunPred-1 software, relatively performs much better than the other existing methods in unannotated protein function prediction. But this work is limited to only 15 high ranking GO terms/functional groups in the human PPI network, which we would like to extend for other significant GO terms as well. Simultaneously, the function prediction of our method can be well enhanced in our future work if domain-domain affinity information [24] and structure related information [25] can be incorporated.

Acknowledgments Authors are thankful to the "Center for Microprocessor Application for Training and Research" of the Computer Science Department, Jadavpur University, India, for providing infrastructure facilities during progress of the work.

References

1. B. Schwikowski, P. Uetz, S. Fields, A network of protein-protein interactions in yeast. Nat. Biotechnol. **18**, 1257–1261 (2000)
2. H. Hishigaki, K. Nakai, T. Ono, A. Tanigami, T. Takagi, Assessment of prediction accuracy of protein function from protein-protein interaction data. Yeast (Chichester, England) 18, 523–31 (2001)
3. J. Chen, W. Hsu, M.L. Lee, S.K. Ng, Labeling network motifs in protein interactomes for protein function prediction, in *IEEE 23rd International Conference on Data Engineering* (2007), pp. 546–555
4. A. Vazquez, A. Flammini, A. Maritan, A. Vespignani, Global protein function prediction from protein-protein interaction networks. Nat. Biotechnol. **21**, 697–700 (2003)
5. U. Karaoz, T.M. Murali, S. Letovsky, Y. Zheng, C. Ding, C.R. Cantor, S. Kasif, Whole-genome annotation by using evidence integration in functional-linkage networks. Proc. Natl. Acad. Sci. U. S. A. **101**, 2888–2893 (2004)
6. E. Nabieva, K. Jim, A. Agarwal, B. Chazelle, M. Singh, Whole-proteome prediction of protein function via graph-theoretic analysis of interaction maps. Bioinform. **21**, i302–i310 (2005)
7. M. Deng, S. Mehta, F. Sun, T. Chen, Inferring domain–domain interactions from protein–protein interactions. Genome Res. 1540–1548 (2002)
8. S. Letovsky, S. Kasif, Predicting protein function from protein/protein interaction data: a probabilistic approach. Bioinform. **19**, i197–i204 (2003)
9. D.D. Wu, An efficient approach to detect a protein community from a seed, in *IEEE Symposium on Computational Intelligence in Bioinformatics and Computational Biology* (2005), pp. 1–7
10. M.P. Samanta, S. Liang, Predicting protein functions from redundancies in large-scale protein interaction networks. Proc. Natl. Acad. Sci. U. S. A. **100**, 12579–12583 (2003)
11. V. Arnau, S. Mars, I. Marín, Iterative cluster analysis of protein interaction data. Bioinform. **21**, 364–378 (2005)
12. G.D. Bader, C.W.V. Hogue, An automated method for finding molecular complexes in large protein interaction networks. BMC Bioinform. **27**, 1–27 (2003)
13. M. Altaf-Ul-Amin, Y. Shinbo, K. Mihara, K. Kurokawa, S. Kanaya, Development and implementation of an algorithm for detection of protein complexes in large interaction networks. BMC Bioinform. 7, doi:10.1186/1471-2105-7-207 (2006)
14. V. Spirin, L.A. Mirny, Protein complexes and functional modules in molecular networks. Proc. Natl. Acad. Sci. U. S. A. **100**, 12123–12128 (2003)

15. A.D. King, N. Przulj, I. Jurisica, Protein complex prediction via cost-based clustering. Bioinform. **20**, 3013–3020 (2004)
16. S. Asthana, O.D. King, F.D. Gibbons, F.P. Roth, Predicting protein complex membership using probabilistic network reliability. Genome Res. **14**, 1170–1175 (2004)
17. S. Saha, P. Chatterjee, S. Basu, M. Kundu, M. Nasipuri, Funpred-1: protein function prediction from a protein interaction network using neighborhood analysis cell. Mol. Biol. Lett. (2014). doi:10.2478/s11658-014-0221-5
18. X. Wu, L. Zhu, J. Guo, D.Y. Zhang, K. Lin, Prediction of yeast protein-protein interaction network: insights from the Gene Ontology and annotations. Nucleic Acids Res. **34**, 2137–2150 (2006)
19. S. Moosavi, M. Rahgozar, A. Rahimi, Protein function prediction using neighbor relativity in protein-protein interaction network. Comput. Biol. Chem. 43, doi:10.1016/j.compbiolchem. 2012.12.003 (2013)
20. S. Saha, P. Chatterjee, Protein function prediction from protein interaction network using physico-chemical properties of amino acid. Int. J. Pharm. Bio. Sci. **24**, 55–65 (2014)
21. W. Peng, J. Wang, W. Wang, Q. Liu, F.X. Wu, Y. Pan, Iteration method for predicting essential proteins based on orthology and protein-protein interaction networks. BMC Syst. Biol. 6, doi:10.1186/1752-0509-6-87 (2012)
22. S. Saha, P. Chatterjee, S. Basu, M. Kundu, M. Nasipuri, Improving prediction of protein function from protein interaction network using intelligent neighborhood approach, in *International Conference on Communications, Devices and Intelligent Systems* (IEEE, 2012), pp. 604–607
23. H.N. Chua, W.K. Sung, L. Wong, Exploiting indirect neighbours and topological weight to predict protein function from protein-protein interactions. Bioinform. **22**, 1623–1630 (2006)
24. P. Chatterjee, S. Basu, M. Kundu, M. Nasipuri, D. Plewczynski, PSP_MCSVM: brainstorming consensus prediction of protein secondarystructures using two-stage multiclass support vector machines. J. Mol. Model. **17**, 2191–2201 (2011)
25. P. Chatterjee, S. Basu, J. Zubek, M. Kundu, M. Nasipuri, D. Plewczynski, PDP-CON: prediction of domain/linker residues in protein sequences using a consensus approach. J. Mol. Model. doi:10.1007/s00894-016-2933-0 (2016)

PLoc-Euk: An Ensemble Classifier for Prediction of Eukaryotic Protein Sub-cellular Localization

Rajkamal Mitra, Piyali Chatterjee, Subhadip Basu,
Mahantapas Kundu and Mita Nasipuri

Abstract Protein Sub-Cellular Localization is very important information as they play a crucial role in their functions. Thus, prediction of protein Sub-Cellular Localization has become very promising and challenging problem in the field of Bioinformatics. Recently, a number of computational methods based on amino acid compositions or on the functional domain or sorting signal. But, they lack of contextual information of the protein sequence. In this paper, an ensemble classifier, PLoc-Euk is proposed to predict sub-cellular location for the eukaryotic proteins which uses multiple physico-chemical properties of amino acid along with their composition. PLoC-Euk aims to predict protein Sub-Cellular Localization in eukaryotes across five different locations, namely, *Cell Wall, Cytoplasm, Extracellular, Mitochondrion,* and *Nucleus.* The classifier is applied to the dataset extracted from http://www.bioinfo.tsinghua.edu.cn/~guotao/data/ and achieves 73. 37% overall accuracy.

Keywords Sub-cellular localization · Physico-chemical properties of amino acid · Ensemble classifier

R. Mitra
Rate Integration Software Technologies Pvt. Ltd.,
213 A, A.J.C. Bose Road, Kolkata 20, India
e-mail: rajkamal.mitra@evolving.com

P. Chatterjee (✉)
Department of Computer Science & Engineering,
Netaji Subhash Engineering College, Garia 152, Kolkata, India
e-mail: chatterjee_piyali@yahoo.com

S. Basu (✉) · M. Kundu · M. Nasipuri
Department of Computer Science & Engineering,
Jadavpur University, Kolkata 700032, India
e-mail: subhadip@cse.jdvu.ac.in

M. Kundu
e-mail: mahantapas@gmail.com

M. Nasipuri
e-mail: mitanasipuri@yahoo.com

© Springer Nature Singapore Pte Ltd. 2017
S.C. Satapathy et al. (eds.), *Proceedings of the 5th International Conference on Frontiers in Intelligent Computing: Theory and Applications*, Advances in Intelligent Systems and Computing 516, DOI 10.1007/978-981-10-3156-4_12

1 Introduction

With the deluge of gene products in the post genomic age, the gap between the newly found protein sequences and their cellular location is growing larger. To use these newly found protein sequences for drug discovery it is desired to develop an effective method to bridge such a gap. In real life, it is found that proteins may simultaneously exist at or move between two or more different Sub-Cellular locations. Thus, localization of proteins is very challenging problem in Bioinformatics. The annotations of protein Sub-Cellular localization can be detected by various biochemical experiments such as cell fraction, electron microscopy and fluorescent microscopy. These accurate experimental approaches are time consuming and expensive which necessitates the computational techniques to predict protein Sub-Cellular Localization which will be useful for protein function prediction. A number of in-silico Sub-Cellular Localization methods have been proposed. Most of the prediction methods can be classified into various categories which are based on the recognition of protein N-terminal sorting signals, amino acid composition, functional domain, homology and fusion. Sorting signals are short sequence segments that localize proteins to intra or extra cellular environments. These include signal peptides, membrane-spanning segments, lipid anchors, nuclear import signals and motifs that direct proteins to organelles such as Mitochondria, Lysosomes etc. [1]. Nakai and Kanehisa [2] took pioneering attempt to propose a computational method, named PSORT, based on sequence motifs and amino acid composition by exploiting a comprehensive knowledge of protein sorting. Reinhardt and Hubbard [3] used amino acid composition information to predict protein subcellular location in neural network based system. Chou and Elrod [4, 5] also used amino acid composition in prediction of subcellular location applying covariant discriminant algorithm. They got better prediction accuracies when they used correlations of residue pairs and acid composition. A work based on Signal based information [6] has been proposed by Emanuelsson and co-authors where individual sorting signals e.g. signal peptides, mitochondrial targeting peptides chloroplast transit peptides are identified [14]. Then they proposed an integrated prediction system using neural network based on the prediction of individual sorting signals. The reliability of the method is based on the quality of the genes 5′-region or protein N-terminal sequences assignment. However, the assignment of 5′-regions are usually not reliable using gene identification methods. Inadequate information of signals may give inaccurate results which results in low accuracy. Hua and Sun [7] used a radial Basis kernel SVM based prediction system using Amino Acid composition. Another voting scheme based work using amino acid composition for prediction of 12 Sub-Cellular locations is done by Park and Kanehisa [8] where a set of SVMs was trained based on its amino acid, amino acid pair and gapped amino acid pair compositions. MultiLoc [9] is an SVM based approach which integrates N-terminal targeting sequences, amino acid composition and protein sequence motifs. It predicts eukaryotic proteins very well. Hortron et al. [10] proposes extension to PSORT-II which is a sorting signal composition based method called WOLF PSORT where amino acid content, sequence length, sorting signals are used. The use of feature sets increased the prediction accuracy of PSORT II with the same classifier k-nearest neighbor. In the work of Chou and Shen [11] proposed an ensemble classifier with kNN basic classifier which uses the concept of pseudoAA (pseAA) composition. Mer and his co-author proposed a novel approach [12] exploiting amino acid composition and different levels of amino acid exposure. The concept was based on that differently exposed residues have different evolutionary pressures to mutate towards specific amino acid types whose side chains have physicochemical properties that agree to the Sub-Cellular location where the protein performs its better activity. To predict singleplex or multiplex protein siLoc-Euk [13] uses multi-label classifier over 22 location sites. APSLAP [14] uses adaptive boosting technique empowered with physicochemical descriptor, Amino acid composition and CTD. From the above mentioned methods, it can be observed that some predictors have experimented with different feature sets for a particular classifier [2–7] or some predictors have taken a voting scheme or ensemble classifier from set of classifiers [8, 11]. In this work, these facts motivate us to use multiple physico-chemical properties weighted by AAC and ensemble classifier of different classifier.

2   Materials and Methods

In this work, an attempt has been taken to use combination of amino acid composition and their physicochemical properties for prediction of five different eukaryotic Sub-Cellular locations, i.e. Cell wall, Cytoplasm, Mitochondrion, Extracellular and Nucleus. Here, whole experiment is conducted in two stages. In the first stage, four different types of classifiers, namely, PART, Multi-Layer Perceptron (MLP), Adaboost and RBF neural network are taken and their performance are observed for prediction. In the second stage of experiment, an ensemble classifier is constructed on the basis of two well performed classifier (in this case, PART and Adaboost Classifier) to achieve better prediction accuracy.

2.1   The Feature Set

The Amino Acid Composition (AAC) of a protein specifies the occurrence (sometimes percentage) for each of the 20 amino acids. AAC of a protein for location is based on the hypothesis that differences in AAC associate with different locations [12]. On the other hand, use of appropriate physico-chemical properties of amino acids also determines its location of activity. Relevant physico-chemical properties of amino acids can be mentioned in this respect, namely, hydropathy, charge, solubility, pKa value, LP value, hydrophilicity and Isoelectric point value. According to the theory of Lim (1974), amino acid residue hydrophilic patterns incline to occur in secondary structure of a protein sequence. The hydrophobic value of amino acid residue represents the major driving force behind protein folding and protein has activity only in specific folding pattern. As proteins take different functions in different part of cellular location it can be concluded that the Hydropathy and Hydrophilicity feature of amino acid have a great influence in protein Sub-Cellular localization. Charge is also important in this field, e.g., it has been seen that the most nucleus protein consists of much more amino acid residues which are positively charged [15]. On the other hand, LP [16] values of amino acids are basically used for protein function prediction as the function and location of a protein is highly correlated, LP value can be used as a feature for protein Sub-Cellular Localization. Studies say that the solubility of a protein is highly related with its function [3] and is a major property of proteins that determines their function and location within a cell. Isoelectric points or pKa value of amino acids are changed according with their location environment. So proteins which reside in particular location of a cell may have identical isoelectric point and pKa value.

In this work, every protein sequence is represented by seven elements vector where each element in the vector represents a particular physicochemical property weighted by AAC. It is mathematically represented as $P = [P_1, P_2, P_3, P_4, P_5, P_6, P_7]$ of any protein P refers to occurrence of any residue a_i of 20 amino acids and is calculated using the Eq. 1. Finally it is normalized in the range [0, 1].

$$AACa_i = \frac{Occurrence\ of\ a_i}{length\ of\ protein\ sequence}. \tag{1}$$

The feature indices of Charge, Hydrophilicity, LP value, Hydropathy were taken from AAindex dataset [17]. The physicochemical properties are weighted by AAC using Eqs. 2–8.

$$P_1 = \sum_{n=1}^{20} AAC_i \times hydropathy\,(a_i) \tag{2}$$

$$P_2 = \sum_{n=1}^{20} AAC_i \times charge\,(a_i) \tag{3}$$

$$P_3 = \sum_{n=1}^{20} AAC_i \times solubility\,(a_i) \tag{4}$$

$$P_4 = \sum_{n=1}^{20} AAC_i \times isoelectricpoint\,(a_i) \tag{5}$$

$$P_5 = \sum_{n=1}^{20} AAC_i \times pK\,(a_i) \tag{6}$$

$$P_6 = \sum_{n=1}^{20} AAC_i \times hydrophilicity\,(a_i) \tag{7}$$

$$P_7 = \sum_{n=1}^{20} AAC_i \times LP\,(a_i) \tag{8}$$

2.2 Design of the Classifier

As previously mentioned, four different classifiers, namely, PART, RBF NN, Adaboost and MLP are taken and their individual performance is observed. Prediction decisions of two well performed classifiers are combined to construct an ensemble classifier PLoc-Euk to boost up its prediction accuracy. The basis of ensemble classifier is to accept prediction decision from one of its component classifier which classifies a protein at higher confidence. Ensemble classifier PLoc-Euk is constructed from two component classifiers PART and Adaboost as they are found to have better prediction accuracy compared to MLP and RBFNN.

2.3 Experimentation and Results

Data Set. We have taken 1001 Eukaryotic protein sequences with five Sub-Cellular locations extracted from (http://www.bioinfo.tsinghua.edu.cn/~guotao/data/) where 750 protein sequences serve as training data and remaining 251 sequences act as test data. For training data 150 protein sequences are taken from each Sub-Cellular location and 50-51 protein sequences are taken as test data for every five locations.

Performance Measure. The performance of classifiers is evaluated using two performance measures: Matthews Correlation Coefficient and Accuracy which is described as follows:

Matthews Correlation Coefficient (MCC)

It is used in machine learning as a measure of quality of binary (two class) classifications. It takes into account true and false positives and negatives and is generally regarded as a balanced measure which can be used even if the classes are of different sizes. The MCC is a correlation coefficient between the observed and predicted binary classifications. It returns a value between -1 and $+1$. A coefficient of $+1$ represents a perfect prediction, 0 an average random prediction and -1 an inverse prediction. Here, when considering a particular class as positive (here location i.e., cell wall) then all other locations are considered to be negative class. Thus, TP, FP, TN, FN for every class or location are calculated and used in computation of MCC.

$$MCC = \frac{(TP \times TN - FP \times FN)}{\sqrt{((TP+FN)(TP+FP)(TN+FP)(TN+FN))}}. \tag{9}$$

Accuracy

It is calculated to measure the performance of a predictor system and defined by

$$Accuracy = \frac{(TP+TN)}{(TP+TN+FP+FN)}. \tag{10}$$

where, TP, TN, FP, FN have their usual meanings.

Performance Evaluation. The whole experiment is conducted in two stages. Initially, four classifiers are applied for the prediction of Sub-Cellular location of test proteins. In the second stage, two best classifiers are taken as component classifier for constructing an ensemble classifier PLoc-Euk. As two classifiers are taken as component classifier, so PLoc-Euk takes prediction decisions from them which classify the test sample at higher confidence. In this section, performance of four classifiers, namely, PART, RBFNN, MLP and Adaboost classifier are observed in prediction of subcellular location i.e., cell wall, extracellular, mitochondrion, nucleus and cytoplasm. Tables 1, 2, 3 and 4 show MCC scores and Accuracy measures of our classifiers. From this table, it is evident, the average accuracies of PART classifier and Adaboost classifier are comparatively better than MLP and RBFNN. Finally, Table 5 shows performance of PLoc-Euk where in most of the cases it performs well compared to component classifiers. The comparison of the performances of PLoc-Euk and its component classifiers are graphically presented in Fig. 1.

Comparison of PLoc-Euk with existing Predictors. We have taken Cello v 2.5 [18] and WOLF-PSORT [10] as existing methods for comparison because they are freely available though they are not too recent but they are based on machine learning method. To compare the performance of the present work PLoc-Euk, 251 test proteins are tested with Cello v2.5 and WOLF -PSORT. From Fig. 2. It can be explained that for cytoplasmic protein, mitochondrion and Nucleus proteins PLoc-Euk performs better than WOLF-PSORT. In case of mitochondrion protein it performs better than two predictors. But, for extracellular proteins, PLoc-Euk does not achieve well.

Table 1 Performance measures of PART classifier

Location	MCC	Accuracy (%)
Cell wall	0.6536	86
Cytoplasm	0.5846	58
Extra cellular	0.5312	50
Mitochondrion	0.7252	78
Nucleus	0.794	90
Average	0.66	72.51

Table 2 Performance measures of MLP classifier

Location	MCC	Accuracy (%)
Cell wall	0.393	66
Cytoplasm	0.5629	70
Extra cellular	0.3824	44
Mitochondrion	0.6276	62
Nucleus	0.74	68.6
Average	0.5412	62.154

Table 3 Performance measures of RBFNN classifier

Location	MCC	Accuracy (%)
Cell wall	0.47	64
Cytoplasm	0.47	66
Extra cellular	0.46	48
Mitochondrion	0.52	52
Nucleus	0.72	78
Average	0.53	61.753

Table 4 Performance measures of Adaboost classifier

Location	MCC	Accuracy (%)
Cell wall	0.6326	78
Cytoplasm	0.5233	66
Extra cellular	0.49	54
Mitochondrion	0.7510	72
Nucleus	0.704	78
Average	0.62	69.32

Table 5 Performance measures of PLoc-Euk classifier

Location	MCC	Accuracy (%)
Cell wall	0.65	84
Cytoplasm	0.58	60
Extra cellular	0.61	64
Mitochondrion	0.72	74
Nucleus	0.78	86
Average	0.67	73.37

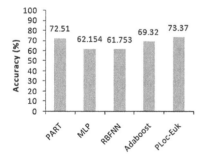 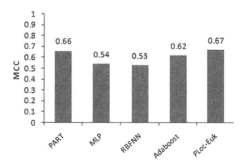

Fig. 1 Performance comparison of ensemble classifier PLoc-Euk and other classifiers

Conclusion. Sub-Cellular localization information of any protein gives proper insight of its function. Thus it has become very challenging task in Bioinformatics. Previously signal based, amino acid composition based, structural based approaches were taken for computation prediction approach. In this work we have combined

Fig. 2 Comparison of
PLoc-Euk, CelloV-2.5 and
Wolf-PSORT on test proteins
of five locations

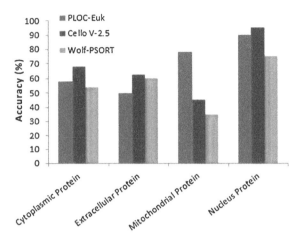

weighted physicochemical based properties of amino acids and their composition as input vector. We have taken 7 relevant physicochemical properties and represented them according to their amino acid composition. Thus weighted properties indicate their intensity and dominance over the protein thereby making the predictor to predict their Sub-Cellular location properly. In addition to these physicochemical properties the performance of the different classifiers has been observed and it is found that we get good performance in PART and Adaboost classifier and also from PLOC-Euk classifier which was designed upon PART and Adaboost classifier. We also compare our work with some existing prediction system. Signal based information can be added with the physicochemical properties to strengthen the prediction power of this classifier. Individual physicochemical properties also have its own influence on a protein to be in a particular location within the cell. So, a number of physicochemical properties can be taken and any feature optimization technique can be employed to reduce the dimension of the input feature vector physicochemical properties, more cellular location also can be included to increase the number of classes and it will also make our system reliable. From further analysis of our work, we can also create a relationship between the Sub-Cellular location and Protein-Protein Interaction [19, 20] and domain information [21] of protein which may be a further research of Bioinformatics.

References

1. R. Mott, J. Schultz, P. Bork, C.P. Ponting, Predicting protein cellular localization using a domain projection method. Genome Res. **12**, 1168–1174 (2002)
2. K. Nakai, M. Kanehisa, A knowledge Base for predicting protein localization sites in Eucaryotic cells. Genomics **14**, 897–911 (1992)
3. A. Reinhardt, T. Hubbard, Using Neural networks for prediction of the subcellular location of proteins. Nucleic Acids Res. **26**, 2230–2236 (1998)

4. K.C. Chou, D.W. Elrod, Protein subcellular location prediction. Protein Eng. **12**, 107–118 (1999)
5. K.C. Chou, D.W. Elrod, Prediction of membrane types and subcellular locations. Proteins **34**, 137–153 (1999)
6. O. Emanuelsson, H. Nielson, S. Brunak, G. von Heijne, Predicting subcellular localization of proteins based on their N-terminal amino acid sequence. J. Mol. Biol. **300**, 1005–1016 (2000)
7. S. Hua, Z. Sun, Support vector machine approach for protein subcellular localization prediction. Bioinformatics **17**, 721–728 (2001)
8. K. Park, M. Kanehisa, Prediction of protein subcellular locations by support vector machines using compositions of amino acids and amino acid pairs. Bioinformatics **19**, 1656–1663 (2003)
9. A. Hoglund, P. Donnes, T. Blum, H.W. Adolph, O. Kohlbacher, MultiLoc: prediction of protein subcellular localization using N-terminal targeting sequences, sequence motifs and amino acid composition. Bioinformatics **22**, 1158–1165 (2006)
10. P. Horton et al., WoLF PSORT: protein localization predictor. Nucleic Acids Res. **35**, W585–W587 (2007)
11. K.C. Chou, B. Shen, Cell-PLoc: a package of Web servers for predicting subcellular localization of proteins in various organisms. Nat. Protoc. **3**, 153–162 (2008)
12. A.S. Mer, M.A. Andrade-Navarro, A novel approach for protein subcellular location prediction using amino acid exposure. BMC Bioinformatics **14** (2013), doi:10.1186/1471-2105-14-342
13. K.C. Chou, Z.C. Wu, X. Xiao, iLoc-Euk: a multi-label classifier for predicting the subcellular localization of singleplex and multiplex eukaryotic proteins. PLOS One **6**, e18258 (2011)
14. V. Saravanan, P.T. Lakshmi, APSLAP: an adaptive boosting technique for predicting subcellular localization of apoptosis protein. ActaBiotheoretica **61**, 481–497 (2013)
15. H. Nakashima, K. Nishikawa, Discrimination of intracellular and extracellular proteins using amino acid composition and residue-pair frequencies. J. Mol. Biol. **238**, 54–61 (1994)
16. M.A. Andrade, S.I. O'Dnoghue, B. Rost, Adoption of protein surfaces to sub-cellular locations. J. Mol. Biol. **276**, 517–525 (1998)
17. S. Kawashima, H. Ogata, M. Kanehisa, AAindex: amino acid index database. Nucleic Acids Res. **27**, 368–369 (1999)
18. C.S. Yu, Y.C. Chen, C.H. Lu, J.K. Hwang, Prediction of protein subcellular localization. Proteins **64**, 643–651 (2006)
19. P. Chatterjee, S. Basu, M. Kundu, M. Nasipuri, D. Plewczynski, PPI_SVM: prediction of protein-protein interactions using machine learning, domain-domain affinities and frequency tables. Cell Mol. Biol Lett. **16**, 264–278 (2011). doi:10.2478/s11658-011-0008-x
20. S. Saha, P. Chatterjee, S. Basu, M. Kundu, M. Nasipuri, Funpred-1: protein function prediction from a protein interaction network using neighborhood analysis cell. Mol. Biol. Lett. (2014). doi:10.2478/s11658-014-0221-5
21. P. Chatterjee, S. Basu, J. Zubek, M. Kundu, M. Nasipuri, D. Plewczynski, PDP-CON: prediction of domain/linker residues in protein sequences using a consensus approach. J. Mol. Model. (2016). doi:10.1007/s00894-016-2933-0

Drive-by-Download Malware Detection in Hosts by Analyzing System Resource Utilization Using One Class Support Vector Machines

Prabaharan Poornachandran, S. Praveen, Aravind Ashok, Manu R. Krishnan and K.P. Soman

Abstract Drive-by-Download is an unintentional download of a malware on to a user system. Detection of drive-by-download based malware infection in a host is a challenging task, due to the stealthy nature of this attack. The user of the system is not aware of the malware infection occurred as it happens in the background. The signature based antivirus systems are not able to detect zero-day malware. Most of the detection has been performed either from the signature matching or by reverse engineering the binaries or by running the binaries in a sandbox environment. In this paper, we propose One Class SVM based supervised learning method to detect the drive-by-download infection. The features comprises of system RAM and CPU utilization details. The experimental setup to collect data contains machine specification matching 4 user profiles namely Designer, Gamer, Normal User and Student. The experimental system proposed in this paper was evaluated using precision, recall and F-measure.

Keywords Malware detection · Drive-by-Download · Anomaly detection · Hardware performance metrics · System resources · One class SVM

P. Poornachandran (✉) · S. Praveen · A. Ashok · M.R. Krishnan
Center for Cyber Security Systems and Networks, Amrita Vishwa
Vidyapeetham, Amrita University, Amritapuri, India
e-mail: praba@am.amrita.edu

S. Praveen
e-mail: praveen_sbsh@yahoo.co.in

A. Ashok
e-mail: aravindashok@am.amrita.edu

M.R. Krishnan
e-mail: manurk@am.amrita.edu

K.P. Soman
Center for Computational Engineering and Networking, Amrita Vishwa
Vidyapeetham, Amrita University, Coimbatore, India
e-mail: kp_soman@amrita.edu

© Springer Nature Singapore Pte Ltd. 2017
S.C. Satapathy et al. (eds.), *Proceedings of the 5th International Conference on Frontiers in Intelligent Computing: Theory and Applications*, Advances in Intelligent Systems and Computing 516, DOI 10.1007/978-981-10-3156-4_13

1 Introduction

In the present information age, computers play an essential role in almost all the possible fields one could think of. At the same time, there has been a substantial growth in the number of cyber-attacks involving data theft from these computers and using them for personal gain. Criminals use multiple stealthy approaches to hide the presence of malwares in the end-users system. These stealthy approaches are so advanced that they are not only invisible to the end-user but also bypass antivirus and other signature based software. One such stealthy attack approach is drive-by-download. Drive-by Download attacks is one of the most common and effective methods for spreading malware nowadays [1]. In this attack, the user unknowingly downloads malware by visiting a malicious website. The download and execution of the malware happens in the background and hence goes unnoticed. Presence of such malwares also affects the performance of the end-user machine and makes it comparatively slower. This research is based on the assumption that any unwanted activity happening in this system, will reflect on the system resources, like increased clock speed, power consumption, CPU usage etc. By comparing the system resource data of an infected machine with that of a clean system and by applying machine learning techniques, presence of any malware in the system could be detected. In this paper, we propose to use One Class SVM technique to detect any anomalous malicious activities happening on the system based on the hardware metrics (RAM, CPU usage) collected from the system.

2 Literature Survey

There have been a lot of articles regarding drive-by-downloads in the past few years. Provos et al. in [2] and Seifert et al. in [3] provided the much needed awareness regarding the threat posed by drive-by downloads. Several articles discusses the problems caused by drive-by-downloads, its detection and mitigation techniques [4–7]. But these work focus on detecting drive-by-downloads from the javascript code by performing static and dynamic analysis. In [8], Moshchuk et al. used client-side techniques to study nature of spyware threat on the web. They also performed a study over 5 months to examine the drive-by-download attacks and came with a conclusion that in their crawled dataset, 1 in every 62 domains contained at least one scripted drive-by-download attack. Tools as in [9] and [10] classify a web-page as malicious, if it is having bad reputation in any of the third party applications, anti-virus, malware analysis tools, or domain blacklists.

The idea of detecting malwares on Smartphones based on its power consumption was proposed in [11]. They used signatures and these signatures are based on power consumptions of a program. A tool PowerScope was built by Flinn et al. a first of its kind to profile energy usage of mobile applications [12]. In [13], a tool called Virus Meter was proposed which monitors and audits power consumption on mobile devices with a behavior-power model. In [14], an on-line power estimation and model generation framework is described which informs the smartphone developers and users of the power consumption implications of decisions about application design and use. In [15], a full system was simulated by characterizing and predicting the performance behavior of Operating System services. During the learning period, samples are captured and characterized from all important behavior points of operating system services. These characteristics are then clustered and given a signature. The future occurrence of Operating System services are then profiled to get its signature. This signature is then compared with the signature of the clusters collected during the learning period and its corresponding performance characteristics are predicted from it. In [16], a research was done to find the relationship between usage of computer resources and performance of computer system. They chose CPU used rate, physical memory, used swap space and process blocked number as the key resources. From the study conducted by them, it was found out that the performance degradation in computer happens with the usage of system resources. For predicting the value of the resource during software degradation, state-space model is used. It's a flexible model and can be used for the modeling of many scenarios and requires only two parameters.

In the system proposed in [17] UML2 and SystemC was combined for Platform modeling for performance evaluation of real time embedded systems using a Y-chart model.

3 Proposed System

Figure 1 describes the proposed system architecture. An agent (or System Resource Data Collector) is installed in the computer and is used to collect all available system resource data. The data collected are then saved on to a flat database. Relationship between these metrics can be unveiled using machine learning techniques. For Classification we use One Class Support Vector Machine.

3.1 One Class Support Vector Machine

One class SVM is little different from the conventional SVM algorithm [18]. Conventional SVM algorithm is devised to solve a two-class or multi-class problem. In two-class SVM, there are two classes, a positive class and a negative class. However, in one class SVM, training is done by considering only positive class information (or one class classification problem). This means that, in training data, information about the target class is available and no details regarding outliers are present. In one class SVM, the task is to estimate a boundary which accepts as much data as possible as well as try not to include the outliers [19]. One class SVM requires large amount of data to fix an accurate nonlinear boundary as it is

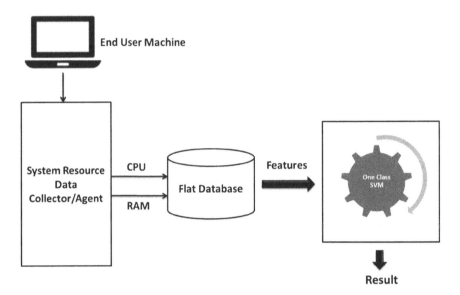

Fig. 1 Proposed system architecture

determined only from the positive class data. One class SVM uses a kernel function $\phi(x)$ to map the data into a higher dimensional space. This algorithm learns the boundary by considering most of the data into the positive class and hence making the number of outliers very less, which is illustrated in (Fig. 2).

The terminologies such as outlier detection, novelty detection or concept learning are widely used for one class SVM. These terms originates w.r.t different applications of one class SVM. The data points are mapped to a hyper sphere using a kernel function with a small error is taken into account. Consider the training data $x_1, x_2, \ldots, x_m \in R^m$ where m is the no of training samples and let $\phi(.)$ be a function which maps to a feature space $R^m \rightarrow F$. In this formulation [18], data points are mapped to a higher dimensional space and using a kernel function, a hyper sphere is

Fig. 2 One class SVM boundary

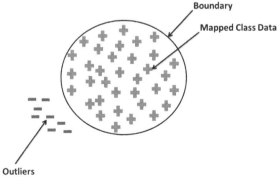

found which contains most of the mapped data points. The objective function can be written in the form

$$\min_{R\in\mathfrak{R},\zeta\in\mathfrak{R}^m,c\in F} R^2 + \frac{1}{\nu m}\sum_i \zeta_i \tag{1}$$

Subject to,

$$\|\phi(x_i) - c\|^2 \le R^2 + \zeta_i,\ \zeta_i \ge 0,\ for\ i \in [m] \tag{2}$$

where R is the radius of hyper sphere, c is the center of the sphere and $\nu \in [0, 1]$ is a control parameter which determines the size of the ball which contains the positive class. Slack variables ζ_i are introduced to create a soft margin. Since nonzero slack variables ξ_i are penalized in the objective function, we can expect that if w and ρ solve this problem, and then the decision function will be,

$$f(x) = \text{sgn}\left(\left(w^T \phi(x)\right) - \rho\right) \tag{3}$$

The negative sign represents the outlier data.

4 Results and Discussion

The usage of system hardware resources varies depending upon the type of user. Based on user activity, 4 profiles were created to collect data namely—Gamer, Student, Designer and Normal user. The specifications of the system used for this work is specified in Table 1.

The RAM and CPU usage (hardware resource usage) details were collected from all the computers under these profiles. An agent collect the system resource details, is run on different systems under these profiles to collect hardware data over a period of time. The data collected is stored on to a flat database and is then used to train the model. We ran the agent, both before and after executing the malware as an attempt to collect the normal hardware resource usage and the hardware resource usage after infection. The data collected is then used to train the model. From the experiments performed, it was observed that hardware data (CPU and RAM usage) of the computers after malware infection is more than that of the hardware resource

Table 1 System resource specifications

Resources	Specifications
Operating System	Windows 7 Professional 64 bit
Processor	Intel Core i7-3770 CPU @ 3.40 GHz
RAM	4 GB DDR3
Hard disk	120 GB
Browser	Mozilla Firefox, Google Chrome

usage before executing the malware. The CPU and RAM usage of the test profiles before and after executing the malware is shown in Figs. 3, 4, 5 and 6. From the aforementioned figures it can be observed that the CPU and RAM usage details capture the change occurring due to the execution of the downloaded malwares. With this reason, the CPU and RAM usage details are chosen as features to detect the presence of drive-by-download malwares using One Class SVM classifier.

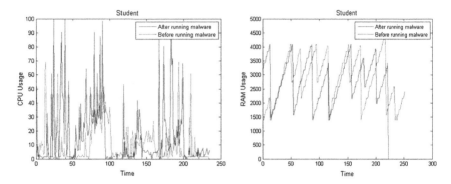

Fig. 3 CPU and RAM usage for a student user profile

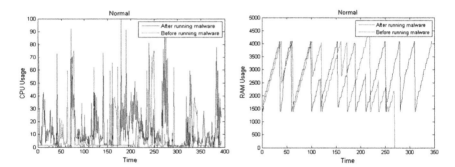

Fig. 4 CPU and RAM usage for a normal user profile

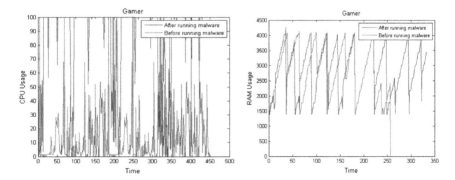

Fig. 5 CPU and RAM usage for a gamer user profile

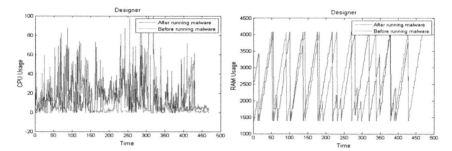

Fig. 6 CPU and RAM usage for a designer user profile

The classification is evaluated using precision, recall and F-measure [20].

$$Precision = \frac{True\_Positives}{True\_Positives + False\_Positives} \tag{4}$$

$$Recall = \frac{True\_Positives}{True\_Positives + False\_Negatives} \tag{5}$$

$$F - Measure = \left(1 + \beta^2\right) \frac{precision \times recall}{\left(\beta^2 \times precision\right) + recall} \tag{6}$$

The value of β is kept as 1 in this research. True Positive is the number of data items which is correctly identified as belonging to the positive class. False Positive is the number of data items that are incorrectly labelled as belonging to the positive class. False Negatives are data items which belong to positive class but not labelled as positive class.

From Table 2 it can be observed that designer profile has its measures around 69 % detection, Gamer profile with nearly 84 % detection, Normal User profile with nearly 76 % detection and Student profile with a nearly detection of 62 %.

4.1 Challenges Faced

The idea of predicting the presence of malware from the RAM and CPU usage details has been new and gaining vast interests among researchers. However, this

Table 2 Performance measures

Profile/measures	Precision	Recall	F-measure
Designer	0.695	0.693	0.69
Gamer	0.845	0.844	0.844
Normal user	0.763	0.764	0.762
Student	0.629	0.622	0.622

task is very daunting as the hardware performance may vary according to the programs run by the user and on how different operating systems handle the programs running on it. Listed below are some of the challenges faced during the experiment.

- Opening too many browser tabs or running a high-end game and execution of a malware may have the same kind of memory usage
- A malware running in a high-end machine may not produce the same spike it generated in a normal day-to-day user system. This motivates to use good machine learning models which are intelligent enough to predict malwares for different processors and systems with different configuration
- A program running on Windows may not use the same amount of CPU/RAM while running in Unix/Linux platform
- Any interrupt caused by the user may create spikes in hardware usage which will increase the number of false positives

5 Conclusion

This paper proposes a method for detecting malwares via drive-by-download in systems with the help of supervised learning technique known as One Class SVM method. For the experiments, four different user profiles were created and collected system hardware details such as CPU and RAM usage details as features. The drive-by-download malicious malware detection task is casted as an outlier detection problem and a supervised model is learned using One Class SVM technique. The experimental results show that the One Class SVM technique can capture the malicious malware information via drive-by-download scenario. Since very limited work has been done on this area, this motivates to extend this work to improve the system in terms of features and machine learning methods to obtain improved detection accuracy.

References

1. M. Egele et al., Defending browsers against drive-by downloads: mitigating heap-spraying code injection attacks, in *Detection of Intrusions and Malware, and Vulnerability Assessment.* (Springer, Berlin, Heidelberg, 2009), pp. 88–106
2. N. Provos et al., The ghost in the browser: analysis of web-based malware. HotBots, **7**, 4–4 (2007)
3. C. Seifert et al., Know your enemy. Malicious web servers. The Honeynet Project (2007)
4. M. Cova, K. Christopher, V. Giovanni, Detection and analysis of drive-by-download attacks and malicious JavaScript code, in *Proceedings of the 19th International Conference on World Wide Web* (ACM, 2010)

5. K. Rieck, K.T. Krueger, A. Dewald. Cujo: efficient detection and prevention of drive-by-download attacks in *Proceedings of the 26th Annual Computer Security Applications Conference* (ACM, 2010)
6. L. Lu et al., Blade: an attack-agnostic approach for preventing drive-by malware infections, in *Proceedings of the 17th ACM Conference on Computer and Communications Security*, (ACM, 2010)
7. N.P.P. Mavrommatis, M.A.R.F. Monrose, All your iframes point to us in *USENIX Security Symposium* (2008)
8. A. Moshchuk et al., A Crawler-based Study of Spyware in the Web, in *NDSS* vol. 1 (2006)
9. A. Ikinci, T. Holz, F.C. Freiling, Monkey-Spider: detecting Malicious Websites with Low-Interaction Honeyclients, in *Sicherheit* vol. 8 (2008)
10. N. Provos, SpyBye—Finding Malware (2016), http://www.monkey.org/~provos/spybye Accessed 15 June 2016
11. H. Kim, J. Smith, K.G. Shin, Detecting energy-greedy anomalies and mobile malware variants, in *Proceedings of the 6th International Conference on Mobile Systems, Applications, and Services* (ACM, 2008)
12. J. Flinn, M. Satyanarayanan, Powerscope: A tool for profiling the energy usage of mobile applications. Mobile computing systems and applications, in *Proceedings Second IEEE Workshop on WMCSA'99* (IEEE, 1999)
13. L. Lei et al., Virusmeter: preventing your cellphone from spies in *Recent Advances in Intrusion Detection* (Springer, Berlin, Heidelberg, 2009)
14. L. Zhang et al., Accurate online power estimation and automatic battery behavior based power model generation for smartphones. in *Proceedings of the Eighth IEEE/ACM/IFIP International Conference on Hardware/Software Codesign and System Synthesi* (ACM, 2010)
15. S. Kim et al., Accelerating full-system simulation through characterizing and predicting operating system performance, in *2007 IEEE International Symposium on Performance Analysis of Systems & Software, ISPASS* (IEEE, 2007)
16. S. Rui et al., The relationship research between usage of resource and performance of computer system, in *WRI World Congress on Software Engineering, 2009 WCSE'09*, vol. 3 (IEEE, 2009)
17. J. Kreku et al., Combining UML2 application and SystemC platform modelling for performance evaluation of real-time embedded systems. EURASIP J. Embed. Syst. **1**, 1–18 (2008)
18. K.P. Soman, R. Loganathan, V. Ajay, *Machine Learning with Svm and Other Kernel Methods* (PHI Learning Pvt. Ltd, 2009)
19. R. Perdisci, G. Gu, W. Lee, Using an ensemble of one-class SVM classifiers to harden payload-based anomaly detection systems, in *Sixth International Conference on Data Mining ICDM'06* (IEEE, 2006)
20. DM. Powers, Evaluation: from precision, recall and F-measure to ROC, informedness, markedness and correlation (2011)

Privacy Preserving Data Mining: A Parametric Analysis

Darshana Patel and Radhika Kotecha

Abstract With technological revolution, a huge amount of data is being collected and as a consequence the need of mining knowledge from this data is triggered. But, data in its raw form comprises of sensitive information and advances in data mining techniques have increased the privacy breach. However, due to socio-technical transformations, most countries have levied the guidelines and policies for publishing certain data. As a result, a new area known as Privacy Preserving Data Mining (PPDM) has emerged. The goal of PPDM is to extract valuable information from data while retaining privacy of this data. The paper focuses on exploring PPDM in different aspects, such as types of privacy, PPDM scenarios and applications, methods of evaluating PPDM algorithms etc. Also, the paper shows parametric analysis and comparison of different PPDM techniques. The goal of this study is to facilitate better understanding of these PPDM techniques and boost fruitful research in this direction.

Keywords Data mining · Privacy · Privacy preserving data mining techniques

1 Introduction

Internet has brought a drastic change in today's world. The Internet today is a widespread information infrastructure presently allowing many people all over the globe to communicate and better understand each other. Such information is stored in huge databases. Knowledge discovery from such databases is the aim of data mining and is attracting researchers vastly. This knowledge discovery has resulted

D. Patel (✉) · R. Kotecha
V.V.P. Engineering College, Rajkot, Gujarat, India
e-mail: darshana.h.patel@gmail.com

R. Kotecha
e-mail: kotecha.radhika7@gmail.com

© Springer Nature Singapore Pte Ltd. 2017
S.C. Satapathy et al. (eds.), *Proceedings of the 5th International Conference on Frontiers in Intelligent Computing: Theory and Applications*, Advances in Intelligent Systems and Computing 516, DOI 10.1007/978-981-10-3156-4_14

in a remarkable increase in the disclosure of private information about individuals. As a result, preserving privacy has become an active research area. Privacy preservation implies that an access to the published data should not enable the adversary to learn anything extra about any target victim as compared to having no access to the database, even with the presence of any adversary's background knowledge obtained from other sources [1]. Thus, Privacy preserving data mining (PPDM) deals with hiding sensitive information of individuals like names, age, zip code etc. without compromising the usability of data [2, 3]. An efficient privacy preserving data mining technique must ensure that any information disclosed should not: (1) be traced to a specific individual, and, (2) form an intrusion.

1.1 Types of Privacy

Privacy is a valuable characteristic of a person. It is the fundamental right of every human being and needs to be preserved. Following are the types of privacy [4]:

(a) Information Privacy: Entails the establishment of rules governing the gathering and managing of private data such as credit information and medical records.
(b) Bodily privacy: Concerns the fortification of people's physical identity in opposition to invasive procedures such as drug testing and cavity searches.
(c) Privacy of communications: Protections of messages, telephones, email and other forms of communication.
(d) Territorial privacy: Involves the setting of confinement on infringing of the domestic and other environments such as the place of work or civic space.

1.2 Different Scenarios in PPDM

With the advance of the information age, data gathering and data investigation have exploded out both in magnitude and complications. Consequently, there arises a need of data sharing amongst stakeholders. For sharing data in privacy preserving, considering in broader aspect, following two different scenarios exist specifically in PPDM [2]:

(a) Central Server Scenario: Firstly, data owner conceals the micro-data by applying various PPDM techniques before publishing it to the data miner which then performs different data mining tasks on such concealed data. In this scenario, data owners/data miners are independent of managing privacy issues. It is also referred as Data Publishing scenario.

(b) Distributed Scenario: The data owners can also be the miners and get collective outcomes on the amalgamation of their records. This is a situation where the privacy is ensured on results of data mining. Distributed scenario can further be classified into three following different models [3, 5]:

  (1) Trust Third Party Model: In such kind of model, each and every party gives the data to a trusted third party keeping blind faith on it. On the other hand, this trusted third party carries out the computation and conveys only the outcomes. However, any such reliable third party does not exist, so this is a superlative model.

  (2) Semi-honest Model: In the semi-honest model, every party pursue the policy of the etiquette using its truthful put in, but may try to interpret facts from the data interchange process.

  (3) Malicious Model: In this sort of model, no restrictions are placed on any of the contributors. As a consequence, any party is completely free to treat in whatever way it wants. Thus, it becomes quite complicated to conduct data mining under the malicious model.

2 Applications of Privacy Preserving Data Mining

Privacy-preserving data mining studies techniques for assembling the potentially contradictory objectives with regard to human privileges and allowing applicable organization to bring together and excavate it for massive data sets. This technique has been consumed in plentiful application areas, some are listed below [6, 7]:

(a) Clinical Records: Registration of patients in clinics leads to collection of enormous amount of clinical records. Such kinds of data are shared with the scientific and government organizations for research and development. These clinical records can be anonymized before publishing the records with other firms so that privacy of an individual is preserved.

(b) Cyber Terrorism/Social Networking: Internet is very famous at the present time for pleasing people with a variety of services related to different fields. But, people are oblivious of the privacy issues. The published data in social network can be anonymized such that it does not infer to any one individual.

(c) Homeland security applications: Some examples of such applications are as follows: identity theft, credential validation problem, web camera surveillance, video-surveillance, and watch list problem.

(d) Government and Financial companies: The government and financial companies are required to publish the data to civilians. If such data is sanitized then no one could reveal any identity and lead to self humiliation.

(e) Business: Enormous data is generated by piling up data from daily shopping of customers. Also, the retailers may share data with other retailers for mutual benefit. In such cases, various anonymization methods can be applied on the data to ensure the privacy of each individual.

(f) Surveillance: Raw samples gathered from surveillance may end up in the hands of adversary. These adversaries could forecast the description of biometric qualities of an individual stored in the database and takes benefit of the existing feature. Privacy preserving algorithms can be adapted and deployed on the databases storing raw data of each individual.

3 Evaluation of Privacy Preserving Data Mining Algorithms

Privacy preserving data mining, the recognition of appropriate evaluation principles and the development of related standards is an important aspect in the expansion and evaluation of algorithms. It is thus vital to provide abusers with a set of metrics which will facilitate them to select the most suitable privacy preserving technique for the data at furnish; with respect to some specific consideration they are interested in optimizing. The chief list of evaluation parameters that are used for evaluation of privacy preserving data mining algorithms are given below [3, 6]:

(a) Performance: Performance of PPDM algorithms is evaluated in terms of time taken to preserve the private attributes.

(b) Data Utility: Data utility is basically a measure of information loss or loss in the functionality of data in supplying the results, which could be created without PPDM algorithms.

(c) Uncertainty level: It measures uncertainty with which the sensitive information that has been concealed can still be forecasted.

(d) Resistance: It depicts the measure of acceptance shown by PPDM algorithm against various data mining algorithms.

4 Related Work

4.1 Different Privacy Preserving Techniques

Varied forethoughts from an attacker could lead to an information disclosure of a particular user. Therefore, one needs to limit this disclosure risks to an acceptable level while maximizing data efficiency before releasing original data for research and analysis purposes. To limit such risks, modification of data is done by applying variety of operations to the original stuff [8, 9]. They can broadly be classified into three categories mainly: Anonymization, Randomization and Cryptography.

(a) Anonymization: Anonymization aims to make individual record impossible to differentiate among a group of records by using following techniques:

  (1) Generalization and Suppression: Replaces definite values from the original data [10, 11].
  (2) Anatomization and Permutation: De-associates the correlation between attributes [1].
  (3) Perturbation: Distorts the data by adding noise [12].
  (4) Bucketization: Separates the SAs (Sensitive attributes) from the QIs (Quasi-Identifiers) by arbitrarily permuting the SA values in each bucket [13].
  (5) Microaggregation: Unified approach consisting of partition and aggregation [14, 15].

(b) Randomization: Applied generally to provide estimates for data distributions rather than single point estimates. It distorts values of each attribute in a sample independently and doesn't use information about other samples [16].

(c) Cryptography: Cryptographic techniques are ideally meant for multiparty scenarios where most frequently used protocol is SMC (Secure Multiparty Computation) which mainly constitutes secure sum, secure union, secure intersection, etc. operations [17, 18].

4.2 Types of Attack Models

Privacy preserving data mining is a significant asset that any mining system must satisfy. User's data is considered to be effectively protected when an opponent could not fruitfully recognize a particular user's data through linkages between an record owner to sensitive attribute in the published data. Thus, these linkage attacks can be classified broadly into three types of attack models namely Record Linkage, Attribute Linkage and Table Linkage [1, 19]. In all these three types of attacks or linkage, it is assumed that opponent knows the QIs (Quasi-identifiers) of the victim.

(a) Record Linkage: If an opponent is capable to link a record holder to a record in a published data table then such kind of privacy risk is known as record linkage. It is assumed that the adversary can make out the victim's record is in the released table and tries to recognize the victim's record from the table. To prevent record linkage privacy models such as k-Anonymity, MultiR k-Anonymity, l-Diversity, (α, k)-Anonymity, (X, Y)-privacy, (c, t)-Isolation, etc. can be used [1].

(b) Attribute Linkage: If an opponent is capable to link a record holder to a sensitive attribute in a published data table then such kind of privacy risk is known as attribute linkage. It is assumed that the adversary can make out the victim's record is in the released table and tries to recognize the victim's sensitive information from the table. To prevent attribute linkage privacy models such as l-Diversity, Confidence Bounding, (α, k)-Anonymity, (X, Y)-privacy, (k, e)-Anonymity, (ε, m)-Anonymity, Personalized Privacy, t-closeness, etc. can be used [1].

(c) Table Linkage: If an opponent is capable to link a record holder to the published data table then such kind of privacy risk is known as table linkage. In this scenario, the attacker seeks to determine the occurrence or nonappearance of the victim's record in the released table. To prevent table linkage privacy models such as δ-presence, ε-Differential privacy, (d, γ)-privacy and distributional privacy can be used.

4.3 Data Mining Tasks

Privacy preserving [20] has been extensively studied by the Data Mining community in recent years. Currently, the PPDM algorithms are mainly used on the functionality of data mining such as Classification, Association and Clustering [21–23].

(a) Classification: Classification is the process of finding a set of model that differentiates data classes such that the model can be used to predict the class of objects whose class label is unknown. The common algorithms used for Classification are Decision Tree, Bayesian classification, etc.

(b) Association Rule: It is a technique in data mining that recognizes the regularities found in bulky amount of data. Such a technique may identify and reveal hidden information that is private for an individual or organization. The common approaches used for Association are a priori, FP-Growth etc.

(c) Clustering: Preserving the privacy of individuals when data are shared for clustering is a complex problem. An important task is to protect the primary data values without interpretation of the similarity between objects under analysis in clustering functionality. The common algorithms used for Clustering are Partitioning methods, Hierarchical methods, Grid methods etc.

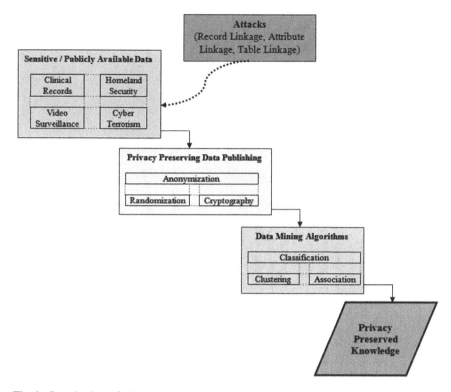

Fig. 1 Generic view of privacy preserving data mining

As a summary of the above details, Fig. 1 shows the generic view of privacy preserving data mining.

Table 1 shows the comparative and parametric analysis of various privacy preserving data mining techniques considering advantages and disadvantage of each privacy model, possible attacks on the data and different data mining tasks.

In Table 1, the correct mark in attack models represent that the technique prevents data from the particular attack and the correct mark in data mining task represents that the functionality can be applied to perform the said data mining method.

Table 1 Parametric analysis of privacy preserving data mining techniques

Privacy Models		Advantages	Disadvantages	Attack models[a]			Data mining tasks[a]		
				RL	AL	TL	1	2	3
Anonymization	Generalization and suppression	• Simple Technique • Protects Identity Disclosure • More flexible	• Suffers from homogeneity attack and background knowledge attack • Significant loss of granularity • Not applicable for continuous data. • Suppression complicates analysis	✓			✓	✓	✓
	Anatomization and Permutation	• More Accurate than Generalization • Certain aggregate computations can exactly be performed without violating the privacy of the data	• Linking Attack • Cannot be applied to High dimensional data without complete loss of utility • Lacks a formal framework for providing how much privacy is guaranteed	✓			✓	✓	✓
	Perturbation	• Attributes are preserved independently • Direct protection for privacy of data is possible due to statistical nature of data mining	• Does not reconstruct the original data values but only data distributions. • Loss of information • Need to develop distribution based algorithm every time	✓			✓	✓	✓
	Bucketization	• Used for high dimensional data • Better data utility than Generalization	• Does not prevent membership disclosure • Requires a clear separation between Quasi-identifiers and sensitive attributes	✓	✓		✓	✓	✓
	Micro-aggregation	• Unified approach unlike Suppression and Generalization • Reduces the impact on outliers • Reduces data distortion	• Finding an optimal partition in multidimensional micro-aggregation is NP-Hard problem	✓			✓	✓	✓

(continued)

Table 1 (continued)

Privacy Models	Advantages	Disadvantages	Attack models[a]			Data mining tasks[a]		
			RL	AL	TL	1	2	3
Randomization	• Simple and easily implemented at data collection phase • Efficient as compared to Cryptography • Doesn't require knowledge of distributions of other records of data • Doesn't require trusted server	• High Information Loss • Cannot be used for multiple attribute databases • Treats all the record equally and reduces the utility of the data	√			√		
Cryptography	• Offers a well-defined model for privacy • Better privacy as compared to Randomization • Vast toolset of cryptographic algorithms for implementing PPDM	• Complexity increases when more parties are involved • Does not address the question of whether the disclosure of the final data mining result may break the privacy of distinct records • Long Process	√	√	√	√	√	√

[a] RL Record Linkage, AL Attribute Linkage, TL Table Linkage
[b] 1 Classification, 2 Clustering, 3 Association

5 Conclusion and Future Work

With proliferation in data mining techniques, the privacy of the individuals or organizations is being disclosed which has fuelled the field of privacy preserving data mining. In this paper, a broad study on privacy preserving data mining has been conducted with respect to various parameters. The field of PPDM mainly considers information privacy and the paper highlights various scenarios and applications that emphasize the importance of PPDM. Further, a tabular analysis depicting the relation between privacy models, attack models and data mining tasks is presented. Also, the merits and demerits of some of the popular PPDM techniques have been described that would help beginners to carry out research considering different dimensions of PPDM. From the study, it has been concluded that a single algorithm, in its naive form, is not efficient enough for effectively protecting data privacy since the process of preserving privacy leads to decrease in accuracy of the final data mining result. Thus, the future work can be to attempt applying variants of privacy models and their enhancements to improve the accuracy of data mining tasks while preventing the data from various attack models.

References

1. P. Tan, M. Steinbach, V. Kumar, *Introduction to Data Mining*, 1st edn. (Addison Wesley Longman Publishing, Co. Inc., 2005)
2. J. Panackal, A. Pillai, Privacy preserving data mining: an extensive survey, in *ACEEE, Proceedings of International Conference on Multimedia Processing, Communication and Information Technology* (2013)
3. M. Dhanalakshmi, S. Sankari, Privacy preserving data mining techniques-survey, in *Proceedings of information communication & embedded systems* (2014)
4. K. Babu, Utility-Based Privacy Preserving Data Publishing. Ph.D. thesis, National Institute of Technology Rourkela (2013)
5. X. Ge, J. Zhu, *New Fundamental Technologies in Data Mining*, ch. Privacy preserving data mining (In Tech Publishing, 2011)
6. S. Gokila, P. Venkateswari, A survey on privacy preserving data publishing. Int. J. Cybern. Inf. **3**(1) (2014)
7. C. Aggarwal, P. Yu, *Advances in Database Systems*, ch. A general survey of privacy-preserving data mining models and algorithms (Springer, 2008)
8. G. Nayak, S. Devi, A survey on privacy preserving data mining: approaches and techniques. Int. J. Eng. Sci. Technol. (2011)
9. V. A-Rivera, P. McDonagh, T. Cerqueus, L. Murphy, A systematic comparison and evaluation of k-anonymization algorithms for practitioners. Trans. Data Privacy, 337–370 (2014)
10. P. Samarati, Protecting Respondents' identities in Microdata Release. IEEE Trans. Knowl. Data Eng. (2001)
11. L. Sweeney, Achieving K-anonymity privacy protection using generalization and suppression. Int. J. Uncertain. Fuzziness Knowl.-Based Syst. **10**(5), 571–588 (2002)
12. R. Brand, Microdata protection through noise addition, in *Inference Control in Statistical Databases*. Lecture Notes in Computer Science, vol. 2316, pp. 97–116 (2002)

13. X. Xiao, Y. Tao Anatomy: simple and effective privacy preservation, in *Proceedings of the 32nd International Conference on Very Large Data Bases*, pp. 139–150 (2006)
14. J. D-Ferrer, V. Torra Ordinal, continuous and heterogeneous k-anonymity through Microaggregation, in *Data Mining Knowledge Discovery*, vol. 11, no. 2 (2005)
15. P. Samarati, L. Sweeney, Generalizing data to provide anonymity when disclosing information, in *Proceedings of ACM Symposium on Principles of Database Systems* (1998)
16. C. Aggarwal, P. Yu, *Advances in Database Systems*, ch. A survey of randomization methods for privacy-preserving data mining (Springer, 2008)
17. Y. Lindell, B. Pinkas, Privacy preserving data mining. J. Cryptol. **15**(3), 177–206 (2002)
18. L. Vasudevan, D. Sukanya, N. Aarthi, Privacy preserving data mining using cryptographic role based access control approach, in *Proceedings of the International Multi-Conference of Engineers and Computer Scientists*, Hong Kong, pp. 19–21 (2008)
19. B. Fung, K. Wang, R. Chen, P. Yu, Privacy-preserving data publishing: a survey of recent developments. ACM Comput. Surv. **42**(4) (2010)
20. R. Kotecha, S. Garg, Data streams and privacy: two emerging issues in data classification, in *Nirma University International Conference on Engineering* (IEEE, 2015)
21. K. Saranya, K. Premalatha, S. Rajasekar, A survey on privacy preserving data mining, in *International Conference on Electronics & Communication System* (IEEE, 2015)
22. J. Han, M. Kamber, *Data Mining: Concepts and Techniques* (Morgan Kaufmann Publishers Inc., 2005)
23. R. Kotecha, V. Ukani, S. Garg, An empirical analysis of multiclass classification techniques in data mining, in *Nirma University International Conference on Engineering* (IEEE, 2011)

Low Power 14T Hybrid Full Adder Cell

Chauhan Sugandha and Sharma Tripti

Abstract The performance of the adder entirely influenced by the performance of its basic modules. In this paper, a new hybrid 1-bit 14 transistor full adder design is proposed. The proposed circuit has been implemented using pass gate as well as CMOS logic hence named hybrid. The main design objective for this circuit is low power consumption and full voltage swing at a low supply voltage. As a result the proposed adder cell remarkably improves the power consumption, power-delay product and has less parasitic capacitance when compared to the 16T design. It also improves layout area by 7–8 % than its peer design. All simulations are performed at 90 & 45 nm process technology on Synopsys tool.

Keywords Power consumption · Delay · Parasitic capacitance · Area · Power-delay product

1 Introduction

Presently in electronics industry low power has come up as a prime theme. The requirement for low power has become a major factor as essential as area and performance [1]. In all digital circuits full adder is used as a vital building block so, definitely, improving the performance will improving the overall design performance [2]. In arithmetic and logical functions, the vast use of this operation has attracted many researchers to present various types of unique logic styles for designing 1-bit Full Adder cell. Power consumption, speed, and area are the three most important parameters of VLSI system. Designing low power VLSI systems have become an essential goal because of the rapid increasing technology in the

C. Sugandha · S. Tripti (✉)
Department of Electronics & Communication Engineering,
Chandigarh University Gharuan, Mohali, Punjab, India
e-mail: tripsha@gmail.com

C. Sugandha
e-mail: Sugandharajput27@gmail.com

© Springer Nature Singapore Pte Ltd. 2017
S.C. Satapathy et al. (eds.), *Proceedings of the 5th International Conference on Frontiers in Intelligent Computing: Theory and Applications*, Advances in Intelligent Systems and Computing 516, DOI 10.1007/978-981-10-3156-4_15

151

field of communication and computing during the recent years. By using less number of transistors in order to implement any logic function has become profitable in reduction of parasitic capacitance and chip area, resulting into low power consumption or high speed [3] by means of exponential relationship between number of transistors and area. Hence, low-power circuits have become a top priority in modern VLSI design.

To accommodate the growing demands, we propose a new power efficient full adder using 14 transistors that produces very promising results, in terms of power, area coverage and threshold loss in contrasting with a range of existing full adders having varied transistor count of 28, 32, 16, 10 and so on [4–16]. The rest paper has been formulated as follows: Sect. 2 illustrates the design and functioning of existing 16T hybrid full adder [16]. Section 3 presents the detail explanation of the proposed 14T full adder. Pre and Post layout simulation and analysis are included in Sect. 4 and in the closing stage Sect. 5 concluded the paper.

2 Previous Work

After reviewing, various adders design approaches like CMOS [4], CPL [5], TFA [6], 10T [7], HPSC [8], TGA [9], 8T [10], GDI Based [11] and other hybrid designs [12–16]. All full adders have various parameter values, no single adder have less delay, power and Power-Delay product. According to this, there is a tradeoff between these all parameters. Out of these all adders, we choose 16T Hybrid CMOS 1-bit full adder, which is recent published in IEEE transaction 2015 [16]. The reason behind selecting 16T Hybrid full adder is that it offered improved PDP compared with the above mentioned full adders.

The schematic of 16T full adder [16] is shown in Fig. 1 given below. It includes three modules where SUM signal is implemented by Module 1 and 2, representing XNOR logic function. Module 3 is used to implement COUT signal which includes transmission gate.

Module 1 and 2: Both modules are liable for most of the power dissipation in the entire circuit. Hence this module is designed in such a way that degradation of power is avoid which is removed purposefully by inverter comprised of transistor P1 and N1, output of this is $\overline{B}$. Transistors P2 and N2 (controlled inverter) are used to generates XNOR output with low logic swing. Further full swing of output levels is fixed with transistor P3 and N3. Similarly Module 2 includes pMOS (P4, P5 and P6) and nMOS (N4, N5 and N6).

Module 3: Output carry signal is implemented using (P6, P7, N7 and N8). Aspect ratio of transistor made large for deficiency in propagation delay of COUT signal.

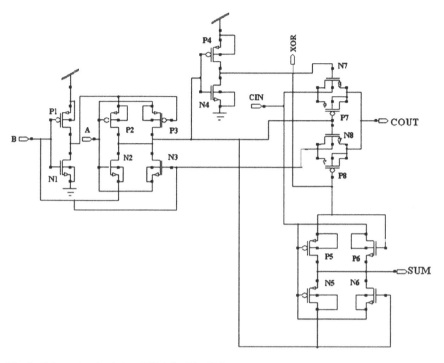

Fig. 1 Schematic of existing 16T full adder [16]

After analyzing the truth table of design, the C_{OUT} generation has been alleviated as following condition below:

$$\text{If, } A = B. \text{ then } C_{OUT} = B; \text{ else, } C_{OUT} = C_{IN}$$

Further the proposed 14T Full Adder included in the Sect. 3 has been compared with 16T Full Adder [16] as shown in Fig. 1 as this adder is already proved its superiority among the others [4–15].

3 Proposed 14T Hybrid Full Adder

In this paper, the new design of 14 transistor full adder has been proposed to fulfil the requirements of low power electronic devices. The schematic of 14T full adder is shown in Fig. 2. In this circuit pMOS transistors (P1, P2 and P3) and nMOS transistor (N1, N2 and N3) are used to form XNOR gate. The aspect ratio (W/L) of respective transistors is given in Fig. 2. Transistor P6 and N3 are used to obtain output SUM signal where P7 and N4 are used to obtain output carry signal. Analyzing the truth table of a basic full adder the following conditions have been founded as below (Tables 1 and 2).

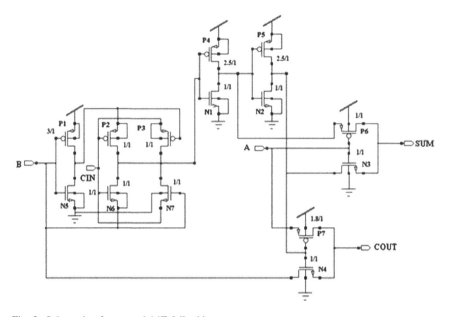

Fig. 2 Schematic of proposed 14T full adder

Table 1 Conditions for SUM (Sum Output)

B ⊕ C	A	SUM
1	0	B ⊕ C
1	1	0

Table 2 Conditions for COUT (Carry Output)

B⊙C	COUT
0	A
1	CIN

The new design has several modifications as one 2T multiplexer is used to implement SUM and other for COUT signal. The selector terminal of a multiplexer used to produce COUT is XNOR signal and input A for the generation of SUM output.

As noted in Fig. 2 the aspect ratio (W/L) of pMOS P2 and P3 have been reduced to 1/1 from 4/1. When P2 is in "active" regime then P3 is in "cut-off" regime or vice versa as both transistors work opposite to each other. Hence at any time either will suffer from sub threshold leakage problem which will provoke the increment in power dissipation. Hence to discard the leakage up to some limit the aspect ratio of both the pMOS (P2 and P3) have been reduced to one-fourth than prior. According to the Eqs. (1) and (2) sub threshold current will decrease as aspect ratio (W/L) is directly proportional to sub threshold current (I_{sub}).

$$I_{sub} = I_x \left[1 - exp\left(\frac{V_{DS}}{V_{thermal}}\right) \right] . exp\left(\frac{V_{GS} - V_T}{\eta V_{thermal}}\right) \tag{1}$$

$$I_x = \frac{W}{L} \mu_n C_{ox} V_T^2 \tag{2}$$

Further, the substrate terminal of nMOS N3 has been connected to GND as compared to A in existing design. Because A is variable and when A = 1 the bulk of N3 is forward bias mode will lead to increase in threshold voltage (V_T) as well as power consumption will increase. The modification is done in such a way alternatively connecting the bulk to A, it must be for all time connected to GND to pause this issue, so that all the time of operation the bulk of N3 will be reversed bias preventing rise of threshold voltage as well as power consumption.

Authors have stated [16] that strong transmission gates have been used to implement the carry generation module so as to assured reduced propagation delay but it will increase the power consumption as well as on-chip area due to the switching of extra added transistors while compared to 14T full adder cell and 2T multiplexer has been used to generate COUT output signal by immolating the delay by some extent the power consumption of the circuit is enormously improved which result into much reduced power-delay product.

By using two cascaded CMOS inverters the 26–30 % voltage swing has been improved, for logic combinations "100" and "111" of the SUM signal, the threshold level of proposed design is 0.95 V and full swing for rest combinations has been achieved at the output.

The input-output waveform of 16T and 14T hybrid adder is shown in Fig. 3 that declares certain changes in output signal (SUM and COUT).

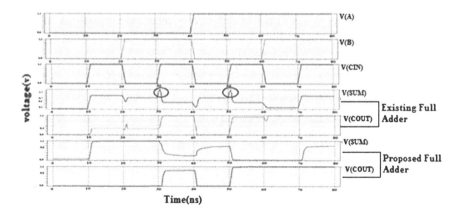

Fig. 3 Input-Output waveform of existing and proposed full adders

(1) The kink shown in SUM signal of 16T full adder (highlighted by circle) has been removed in 14T design as a result of changes in the design as discussed above

(2) By reducing aspect ratio of transistors, sub threshold leakage has been overcome and now in proposed circuit the level of SUM output signal is 1.2 while compare to existing is 0.8.

4 Simulation Results and Analysis

4.1 Pre Layout Simulations

Pre layout simulation of existing and proposed full adder for extracting the power consumption by the design have been performed using 90 nm (45 nm) at 1.2 V (0.8 V) on Synopsys tool and results are shown in Table 3 and Table 4. In order to prove that proposed design is consuming low power and have better performance at various temperature and frequency of operation, simulation are carried out on varying temperature and frequency indicated in Table 5 and Table 6.

Table 5 depicts that the power consumption increases with escalating frequency, as frequency and power consumption has linear relationship with each other [2]. The data shown in Table 5 reveals that the average power consumption of proposed design surmount its peer design over the range of high frequencies. The simulation has been performed using 45 nm technology.

Table 6 shows the effect of varying temperature on the power consumption of both the circuits. The transistor characteristics influenced with temperature variation; also Fermi-potential (φ_F) which is inversely proportional to temperature will decrease by rising temperature and consequently threshold voltage will cause to go down [17], hence resulting into augmented drain current as well as power consumption of the device. Due to the less device count, active transistors in the proposed design at any instant will be less than the 16T adder and therefore ensures low power consumption when compared.

Table 3 Simulation results for 90 nm @ 1.2 V and 25 °C

Full adder design	Average power consumption (W)	Power-delay product (W s)
16T (Existing)	1.18×10^{-6} [16]	0.12×10^{-15} [16]
14T (Proposed)	0.46×10^{-6}	0.005×10^{-15}

Table 4 Simulation results at 45 nm @ 25 °C

Full adder design	Average power consumption (W)	
	@ 0.8 V	@ 1 V
16T (Existing)	3.41×10^{-6}	1.512×10^{-5}
14T (Proposed)	7.68×10^{-8}	1.94×10^{-7}

Table 5 Simulation results for Frequency versus Average power consumption @ 0.8 V

Operating frequency (MHz)	Average power consumption (W)	
	16T hybrid full adder	14T proposed full adder
100	8.61×10^{-8}	5.85×10^{-8}
300	4.67×10^{-7}	2.30×10^{-7}
500	3.86×10^{-7}	2.02×10^{-7}

Table 6 Average power consumption with varying temperature @ 0.8 V

Temperature (°C)	Average power consumption (W)	
	16T hybrid full adder	14T proposed full adder
−10	1.41×10^{-6}	7.32×10^{-8}
30	3.72×10^{-6}	7.74×10^{-8}
60	5.59×10^{-6}	8.01×10^{-8}

4.2 Post- Layout Simulations

The post layout simulations of both designs have been performed on 90 nm process technology at 1 and 1.2 V input as well as supply voltage using EDA Tool. The designed layout of 16T 1-bit full adder and proposed design 14T 1-bit full adder are shown in Fig. 4a, b respectively. It can be seen that the 14T full adder required less area which can be fruitful in increasing the chip density. In both the layouts multiple active contacts are used for reduction of diffusion resistance; also the Source and Drain diffusion regions are filled with maximum number of contacts possible to reduce the resistance of the connection from metal to diffusion, and to maximize the amount of current flow through the contacts. Multiple contacts are not implemented for the MOSFETs having small aspect ratio as it violates the design rule for optimum distance between the contacts. The two layouts make it obvious that the less number of vias are used to implement 14T rather than 16T which may be a source for increase in power results. Moreover, the parasitic capacitance of the proposed design is much less than the 16T hybrid adder, reason behind, 75 % of the 16T design is implemented using the parallel combination of transistor which lead to increase in capacitance which are eliminated in proposed one. It reduces the capacitance by 70–82 % which become a prime cause of extraordinary reduction in the power consumption. Table 7 illustrates the extracted results after post-layout simulation. It confirms the supremacy of the proposed design over its existing peer design.

(a)

(b)

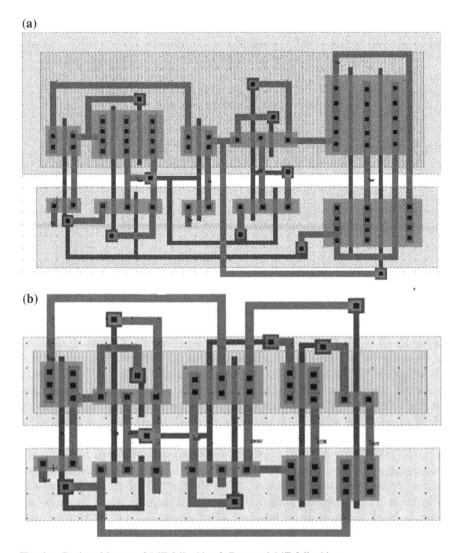

Fig. 4 **a** Designed layout of 16T full adder. **b** Proposed 14T full adder

Table 7 Extracted Parasitic capacitance and Average power consumption at 1 V and 1.2 V

Full adder design	Layout area (μm^2)	Output capacitance (Farad)	Average power consumption (W)	
	90 nm	90 nm	@ 1 V	@ 1.2 V
16T (Existing)	25.84	1.792×10^{-15}	1.84×10^{-5}	4.09×10^{-5}
14T (Proposed)	23.91	0.371×10^{-15}	1.44×10^{-7}	1.91×10^{-7}

5 Conclusion

In this paper a low-power 14T hybrid 1-bit full adder has been proposed. The simulations were carried out using Synopsys (Custom-Designer) EDA Tool at 90 nm and 45 nm technology and compared with existing 16T design. The simulation results proved that designed adder is found to give superior performance than most adder mentioned in literature in Sect. 1. The proposed design has remarkably low power consumption at various voltages, temperature and frequency of operation than its peer design. Moreover the post layout performance adds an extra weather to the winning cap of the proposed full adder by justifying the results of pre-layout simulation. The proposed circuit has been examined to have extreme less parasitic capacitance and approximately 8 % reduction in layout area. Hence the proposed design can be a viable option for low power high frequency applications.

References

1. N. Weste, K. Eshraghian, *Principles of CMOS Digital Design. A System Perspective* (Addison Wesley, Massachusetts, MA, USA, 1993)
2. S. Kang, Y. Leblebici, *CMOS Digital Integrated Circuit Analysis and Design*, 3rd edn. (McGraw-Hill, 2005)
3. K. Roy, S.C. Prasad, *Low Power CMOS VLSI Circuit Design*. ISBN 0471-11488 (2000)
4. R. Zimmermann, W. Fichtner, Low-power logic styles: CMOS versus pass-transistor logic. IEEE J. Solid-State Circuits **32**(7), 1079–1090 (1997)
5. D. Radhakrishnan, Low-voltage low-power CMOS full adder. IEEE Proc. Circuits Devices Syst. **148**(1), 19–24 (2001)
6. A.M. Shams, T.K. Darwish, M. Bayoumi, Performance analysis of low-power1-bit CMOS full adder cells. IEEE Trans. Very Large Scale Integr. (VLSI) Syst. **10**(1), 20–29 (2002)
7. H.T. Bui, Y. Wang, Y. Jiang, Design and analysis of low power 10-transistor full adders using novel XOR-XNOR gates. IEEE Trans. Circuits Syst. II, Analog Digit. Signal Process. **49**(1), 25–30 (2002)
8. M. Zhang, J. Gu, C.-H. Chang, A novel hybrid pass logic with static CMOS output drive full-adder cell, in *Proceedings of the International Symposium on Circuits and Systems*, pp. 317–320 (2003)
9. C.H. Chang, J.M. Gu, M. Zhang, A review of 0.18-μm full adder performances for tree structured arithmetic circuits. IEEE Trans. Very Large Scale Integr. (VLSI) Syst. **13**(6), 686–695 (2005)
10. S.R. Chowdhury, A. Banerjee, A. Roy, H. Saha, A high speed 8 transistor full adder design using novel 3 transistor XOR gates. Int. J. Electron. Circuits Syst. **2**, 217–223 (2008)
11. A. Bazzazi, B. Eskafi, Design and Implementation of Full Adder Cell with the GDI Technique Based on 0.18 μm CMOS Technology, in P*roceedings of the International Multiconference of Engineers and Computer Scientist*, vol. II, IMECS Hong Kong (2010)
12. S. Goel, A. Kumar, M. Bayoumi, Design of robust, energy-efficient full adders for deep-sub micrometer design using hybrid-CMOS logic style, in IEEE Trans. Very Large Scale Integr. (VLSI) Syst. **14**(12), 130–1320 (2006)
13. M. Aguirre-Hernandez, M. Linares-Aranda, *CMOS full-adders for energy-efficient arithmetic applications*, in *Proceedings of the 4th IEEE VLSI Systems*, vol. 19, pp. 718–721(2011)

14. B. Sathiyabama, Dr. S. Malarkkan, Novel low power hybrid adders using 90 nm technology for DSP applications. **1**(2) (2012). ISSN:2278–067X
15. C.-K. Tung, S.-H. Shieh, C.-H. Cheng, Low-power high-speed full adder for portable electronic applications. Electron. Lett. **49** (2013)
16. P. Bhattacharyya, B. Kundu, S. Ghosh, V. Kumar, A. Dandapat, Performance analysis of a low-power high-speed hybrid 1-bit full adder circuit. 10th IEEE Trans. Very Large Scale Integr. Syst. **23** (2015)
17. I.M. Filanovsky, A. Allam, Mutual compensation of mobility and threshold voltage temperature effects with applications in CMOS circuits. IEEE Trans Circuits Syst. I: Fund. Theory Appl., 876–884 (2001)

Improved Dynamic Time Warping Based Approach for Activity Recognition

Vikas Tripathi, Sharat Agarwal, Ankush Mittal
and Durgaprasad Gangodkar

Abstract Dynamic Time Warping (DTW) has been a very efficient tool in matching two time series and in past much work has already done in modifying DTW so as to enhance its efficiency and further broadening its application areas. In this paper we are proposing an enhanced version of DTW by calculating mean and standard deviation of the minimum warping path because of which the efficiency of DTW increased in detecting different human activities. We also introduce a new fusion of DTW with Histogram of Gradients (HOG) as it helped in extracting both temporal and spatio information of the activity and this fusion has worked very effectively to depict human activities. We used Random Forest as a classification tool giving highest accuracy of 88 % in weizMan dataset.

Keywords DTW · HOG · Random forest · Activity recognition

1 Introduction

During the past decade there has been a lot of research and development in the field of Computer Vision. Many technologies are existing and many more are yet to come in this field. The main aim of Human Computer Interaction is to provide an interface where the user can interact with the system and in parallel do other work. Human Computer Interaction can be in two ways, either the system responding to

V. Tripathi (✉)
Uttarakhand Technical University, Dehradun, India
e-mail: vikastripathi.be@gmail.com

S. Agarwal · A. Mittal · D. Gangodkar
Graphic Era University, Dehradun, India
e-mail: sharat29ag@gmail.com

A. Mittal
e-mail: dr.ankush.mittal@gmail.com

D. Gangodkar
e-mail: dgangodkar@yahoo.com

© Springer Nature Singapore Pte Ltd. 2017
S.C. Satapathy et al. (eds.), *Proceedings of the 5th International Conference on Frontiers in Intelligent Computing: Theory and Applications*, Advances in Intelligent Systems and Computing 516, DOI 10.1007/978-981-10-3156-4_16

our voice commands or to our actions [1, 2]. Our main focus in on recognizing different actions and performing particular function in respond to that action, like while doing work you can command your system with some action [1] and your system respond to it with a particular task. We all want to make our life easier and simpler but at the same time we want to be multitasking, and for that smart systems have to be created. Much research is going on in this field and in future we would see that this whole working experience will be different, our computer will just do all the tasks by our voice commands or by recognizing our gestures. An improved algorithm with better accuracy is the main focus nowadays. Human Activity recognition does not focus only on creating a HCI system, it is also most useful in video surveillance. Automated systems are being developed where computer can detect some abnormal activities of human. This type of technique is very useful and helpful in many fields. For example if the cameras in ATM can detect some abnormal human activity [3], then in response to that some immediate actions can be taken. With this kind of technology anyone can protect his private property with automated system, without much human efforts. Many more applications can be there, like cameras in chemical plants can keep an eye to any kind of leakage or gas emissions [4] because it's always been very risky to work in chemical plants. Likewise, in medical field also it has played a major role in performing operations or operating from distant doctor without being physically present with the patient.

In this paper we are mainly focusing on improved DTW and enhancing Human Activity Recognition combining it with HOG. We have used WEIZMAN, ATM existing dataset [3] and also we have created our own dataset with 5 classes of different activity with 10 video in each class, and further vindicating our result by using Random Forest filter. This paper is organized in 5 sections Sect. 2 contains literature related work and background behind this approach, our proposed methodology is presented in Sect. 3, further depicting our result in Sect. 4 result and conclusion and finally Sect. 5 shows final conclusion and future work.

2 Literature Review

DTW is a popular algorithm with its extreme feature of detecting patterns in time series. It is highly used in detecting different pattern and shapes. DTW aims at comparing and aligning two time series. DTW efficiently work in O(mn) as it uses dynamic programming as an efficient tool, but further more experiments have been done to reduce its time complexity by using threading and multi scaling techniques [5]. Originally DTW was used in speech recognition but by time it was further used in bioinformatics, econometric and handwriting detection, fingerprint verification [5]. Many variations were already brought up to original DTW, legendary work was done by Muller and Meinard [6] in their Book Information Retrieval for Music and Motion in that book they brought three different variations of introducing step size condition, local weights and global constraints their work was appreciated a lot and that new variation definitely broaden the application area of DTW, thus giving better results in Motion Comparison and Retrieval. Some or the other experiments are done on DTW. Earlier landmark work done by G.A. ten Holt of introducing Multi Dimensional DTW in Gesture Recognition [7, 8] which substantially enhanced their results even in noisy conditions. Also Eamonn J. Keogh scaled up the DTW for massive dataset [9] and experimented their approach on medical, astronomical and sign language data. Thus finally in 2011 Samsu Sempena proposed Human action Recognition using Dynamic Time Warping [10] and used body part detecting using depth camera to recover human joints in 3D world coordinate system. We have also used HOG as a descriptor to fuse with HOG getting inspiration from Dalal and Trigs [11] proposed new descriptor HOG in their pedestrian detection [12].

By rapid growing field of computer vision in human activity recognition motivates us to experiment with new techniques and bring up a new system. Vikas Tripathi [3] also proposed a framework for ATM booth detecting abnormal activities. There are many existing feature to detect human activity but the need of the hour is for an efficient way which is useful in every background and situation with best accuracy.

3 Methodology

The proposed methodology employs DTW and its improved versions for extracting different descriptors for activity recognition. Following phases are created for feature extraction. Firstly the frames are extracted applying DTW and its different versions give 5 attributes further combining these with 72 values of HOG creates a testing and training dataset of 77 attributes, which are classified using random forest (Fig. 1).

DTW has earned its popularity by being very accurate and efficient in measuring similarity between time series and allowing elastic transformation in time series to detect different shapes and patterns. DTW uses dynamic programming as a tool and evaluates the warping path in O(mn), in given two time series X = (x1, x2, x3, ..., xn) and Y = (y1, y2, y3, ..., ym). First of all a local distance matrix is calculated using these two given time series using the Eq. (1) local distance matrix signifies

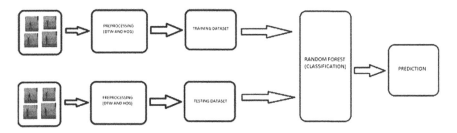

Fig. 1 Proposed framework for action recognition

the difference in distance between two sequences. Further creating a local cost matrix including using algorithm [13]. After the local cost is constructed the main part of constructing a minimum cost warping path is done, which is to found by following three criteria of monotonicity, continuity and boundary. Calculation of warping path is done according to the algorithm proposed by Pavel Senin [13].

$$\text{Dist\_matrix } (i, j) = |X_i - Y_j|; \tag{1}$$

Algorithm 2 returns the min cost path and also its minimum cost. So changing to this original algorithm we calculated mean of that minimum path which is min_path vector in Algorithm 1 which stores all the values of that warping path. Also the second change was to find the standard deviation of the warping path found. And after testing these changes to DTW we found significantly rise up in the accuracy for the weizman dataset as shown in Table 1. As shown in Table 1 DTW mean has shown top most accuracy by using DTW only. While applying DTW on video we used the concept of sliding window, and patch in frames. We extracted frames of the video and divided into equal patches of 20×20 sizes and calculated the mean of that patch. In weizman dataset frame were 144×180 sizes thus creating total 24 patches in the frame, forming a vector of 24 mean values which represents the whole frame. Starting from frame number 15 ignoring first 15 frames, a mean matrix is created and DTW is calculated for every nth frame with respect to $(n - 3)$, $(n - 6)$, $(n - 9)$, $(n - 12)$, $(n - 14)$ as depicted in Algorithm 3. Thus we were getting 5 descriptor values from DTW for every frame.

Algorithm 1 For Calculating DTW of Frame

```
%% mean_mat is the mean matrix for every video each row representing one frame
for i = 15:NumberOfFrames
vect1 = mean_mat(i,:);
vect2 = mean_mat(i-3,:);
vect3 = mean_mat(i-6,:);
vect4 = mean_mat(i-9,:);
vect5 = mean_mat(i-2,:);
vect6 = mean_mat(i-14,:);
DTW_train(count,1) = DTW_major1_mean(vect1,vect2);
DTW_train(count,2) = DTW_major1_mean(vect1,vect3);
DTW_train(count,3) = DTW_major1_mean(vect1,vect4);
DTW_train(count,4) = DTW_major1_mean(vect1,vect5);
DTW_train(count,5) = DTW_major1_mean(vect1,vect6);
count = count + 1;
end
```

In Algorithm 1 DTW_train matrix creates the training dataset for all the training videos and similarly DTW_test will create testing dataset. Originally DTW used to return total minimum cost of the warping path but we calculated mean and standard deviation because in calculating the total cost by adding then the range is not fixed it varies, and also when there is set of known numbers then the arithmetic mean is best in the sense of minimizing the squared deviation from the typical value. And for standard deviation the motive was that it helps to identify the variation of set of values from the average. And as Table 1 depicts results were better in both the variations.

As DTW gives us only temporal information about the set of values so it was combined with HOG as it gives spatio information, and to detect any activity both temporal and spatio information is required and thus this fusion as in Table 3 shows worked efficiently and accuracy went up. We used Random Forest as a classification technique because it is a stable algorithm and also it works best in noisy environment. Andy Liaw [14] proposed many applications of Random Forest one was that it is best for classification problems where the class frequencies are unbalanced.

4 Result and Conclusion

The algorithm has been trained and tested using MATLAB 2015(b) on a computer having xenon processor with 16 GB RAM and 2.8 Ghz, video resolution varied from 144 × 180 in weizman to 240 × 320 in ATM dataset. We used classes bend, jack, jump, pjump, wave1, wave2 of weizman with 8 videos in each class.ATM dataset was classified into two categories of normal and abnormal, for multiple and single attendees 10 videos in each class with total of 40 videos. We used 7 videos for training and 3 for testing.

Table 1 shows the accuracy with using random Forest as the classifier with 100 trees. Table 2 shows the accuracy of Activity recognition when only HOG was used. We used inbuilt HOG function MATLAB 2015(b) returning 72 values for each frame.

Table 3 depicts very clearly that combining DTW mean and standard deviation with HOG gives a very good result, and also all the combination have performed better with an average accuracy of 85.88. Algorithm was further tested on ATM dataset and Table 4 shows the comparison of different techniques performed and their corresponding accuracy.

Table 1 Comparative analysis between normal DTW and its variations

Technique	Dataset	Accuracy
DTW NORMAL	WEIZMAN	41.958
DTW MEAN	WEIZMAN	48.8112
DTW STANDARD DEVIATION	WEIZMAN	43.2168

Table 2 Analysis for only HOG

Technique	Dataset	Accuracy
HOG	WEIZMAN	78.8811

Table 3 Comparative analysis with different combination, of HOG and DTW

Technique	Dataset	Accuracy
DTW NORMAL COMBINED HOG	WEIZMAN	81.9891
DTW MEAN COMBINED HOG	WEIZMAN	87.4126
DTW STANDARD DEVIATION COMBINED HOG	WEIZMAN	85.8741
DTW MEAN COMBINED WITH STANDARD DEVIATION AND HOG	WEIZMAN	88.2517

Table 4 Accuracy analysis in ATM

Technique	Dataset	Accuracy
DTW NORMAL	ATM	45.0766
DTW MEAN	ATM	46.3213
DTW STANDARD DEVIATION	ATM	47.0766
HOG COMBINED DTW STANDARD DEVIATION	ATM	66.2523
HOG COMBINED WITH DTW MEAN AND STANDARD DEVIATION	ATM	66.9007
HOG COMBINED DTW MEAN NORMAL AND STANDARD DEVIATION	ATM	67.3449

5 Conclusion

In this paper we have represented a new and much efficient technique for human activity recognition, and also the new fusion of DTW with HOG have showed better results than being testing them individually. We tested on 2 dataset of different features and video qualities with lot more dynamic background in ATM dataset. Technique is very simple to implement not much costly in terms of space and time without much drastic changes in the technique it has showed very efficient results. Further in future work many more changes can be brought to this technique. While this approach is open to fuse with histogram of optical flow, oriented gradients and also other kind of changes can be brought in extracting values from frames and applying to DTW for further increase in accuracy.

References

1. V.I. Pavlovic, R. Sharma, T.S. Huang, Visual interpretation of hand gestures for human-computer interaction: a review. IEEE Trans. Pattern Anal. Mach. Intell. **19**(7), 677–695 (1997)
2. S.S. Rautaray, A. Agrawal, Vision based hand gesture recognition for human computer interaction: a survey. Artif. Intell. Rev. **43**(1), 1–54 (2015)
3. V. Tripathi, D. Gangodkar, V. Latta, A. Mittal, Robust abnormal event recognition via motion and shape analysis at ATM installations. J. Electr. Comput. Eng. (2015)
4. A. Safitri, X. Gao, M.S. Mannan, Dispersion modelling approach for quantification of methane emission rates from natural gas fugitive leaks detected by infrared imaging technique. J. Loss Prev. Process Ind. **24**(2), 138–145 (2011)

5. Z.M. Kovács-Vajna, A fingerprint verification system based on triangular matching and dynamic time warping. IEEE Trans. Pattern Anal. Mach. Intell. **22**(11), 1266–1276 (2000)
6. M. Müller, *Information Retrieval for Music and Motion* (Springer, Berlin, 2007)
7. M. Wöllmer, M. Al-Hames, F. Eyben, B. Schuller, G. Rigoll, A multidimensional dynamic time warping algorithm for efficient multimodal fusion of asynchronous data streams. Neurocomputing **73**(1), 366–380 (2009)
8. G.A. ten Holt, M.J. Reinders, E.A. Hendriks, Multi-dimensional dynamic time warping for gesture recognition, in *Thirteenth Annual Conference of the Advanced School for Computing and Imaging*, vol. 300 (2007)
9. E.J. Keogh, M.J. Pazzani, Scaling up dynamic time warping to massive datasets, in *Principles of Data Mining and Knowledge Discovery* (Springer, Berlin, 1999), pp. 1–11
10. S. Sempena, N.U. Maulidevi, P.R. Aryan, Human action recognition using dynamic time warping, in *Electrical Engineering and Informatics (ICEEI)* (IEEE Press, 2011), pp. 1–5
11. N. Dalal, B. Triggs, Histograms of oriented gradients for human detection. Comput. Vision Pattern Recognit. **1**, 886–893 (IEEE Press, 2005)
12. N. Dalal, B. Triggs, C. Schmid, Human detection using oriented histograms of flow and appearance, in *Computer Vision–ECCV* (Springer, Berlin, Heidelberg, 2006), pp. 428–441
13. P. Senin, Dynamic time warping algorithm review. Information and Computer Science Department University of Hawaii at Manoa Honolulu, pp 1–28, USA (2008)
14. A. Liaw, M. Wiener, Classification and regression by random forest. R News **2**(3), 18–22 (2002)

STRIDE Based Analysis of the Chrome Browser Extensions API

P.K. Akshay Dev and K.P. Jevitha

Abstract Chrome browser extensions have become very popular among the users of Google Chrome and hence they are used by attackers to perform malicious activities which lead to loss of user's sensitive data or damage to the user's system. In this study, we have done an analysis on the security of the Chrome extension development APIs. We have used the STRIDE approach to identify the possible threats of the Chrome specific APIs which are used for extension development. The analysis results show that 23 out of the 63 Chrome specific APIs are having various threats as per the STRIDE approach. Information disclosure is the threat faced by many APIs followed by tampering. This threat analysis result can be used as reference for a tool which can detect whether the extension is malicious or not by deeply analysing the ways in which the APIs having threats are used in the extension code.

Keywords Chrome extensions · STRIDE analysis · Chrome specific APIs · Malicious extensions

1 Introduction

Google Chrome is the most popular web browser available today and many extensions are being developed to enhance the features and functionality of the browser. Extensions are basically small programs developed by browser vendors or third party developers to improve the functionality of browsers. Due to the wide

P.K. Akshay Dev (✉)
TIFAC CORE in Cyber Security, Amrita School of Engineering,
Amrita Vishwa Vidyapeetham, Amrita University, Coimbatore, India
e-mail: akshaydev313@gmail.com

K.P. Jevitha
Department of Computer Science and Engineering,
Amrita School of Engineering, Coimbatore, Amrita Vishwa Vidyapeetham,
Amrita University, Coimbatore, India
e-mail: kp_jevitha@cb.amrita.edu

© Springer Nature Singapore Pte Ltd. 2017   169
S.C. Satapathy et al. (eds.), *Proceedings of the 5th International Conference on Frontiers in Intelligent Computing: Theory and Applications*, Advances in Intelligent Systems and Computing 516, DOI 10.1007/978-981-10-3156-4_17

acceptance of the browser extensions by the users, it has also become the major target of the attackers [1]. Extensions created by third-party developers may not have enough security considerations and become vulnerable to attacks. Extensions once compromised can result in different malicious activities which can affect the user's sensitive data and system. Also, malicious extensions are being developed to attack the user's system and data. Chrome extensions can have the privilege to perform many sensitive actions in the browser using the different permissions and chrome specific APIs available to them. This makes them highly dangerous when used for malicious purposes.

STRIDE is a threat analysis approach developed by Microsoft for analysing the possible threats of web applications or components [2]. STRIDE approach can be used to map the possible threats into six categories namely Spoofing, Tampering, Repudiation, Information disclosure, Denial of service, Elevation of privilege. Spoofing is impersonating something or someone else. Tampering is malicious modification of code, data or settings. Repudiation is denial of truth or validity of something. Information disclosure is exposing information to someone not authorized to see it. Denial of service is denying or degrading service to users. Elevation of privilege is gaining capabilities without proper authorization [3].

The rest of the paper is organized as follows. Section 2 briefs the related work done in this area of research. Section 3 describes the security model of the chrome extensions and Sect. 4 describes the security analysis of the Chrome specific APIs based on STRIDE. Section 5 explains the statistical results of the analysis and Sect. 6 concludes the work.

2 Related Work

Different research work has been done in the area of browser extension security over the past few years. Due to the wide acceptance of extensions by the users of popular web browsers like Google Chrome and Mozilla Firefox, the security of those extensions has become a major concern and also gained attention of the security researchers. An in-depth analysis of the Chrome's extension security model was done [1] and they conclude that its vulnerabilities are rooted from the violation of the principles of least privilege and privilege separation. Using a series of bot-based attacks, the authors demonstrate that malicious chrome extensions pose serious threat to Chrome browser. A thorough evaluation of how well the Chrome extension's security mechanism defends against the vulnerabilities has been done in [4] and they found out that developers are not always following the security mechanisms for developing the extensions which leads to vulnerabilities. A framework to analyse the security of Chrome extensions has been proposed in [5]. The framework proposed uses the permission feature of Chrome extensions and flow of data through the extension's JavaScript code to detect vulnerabilities and malicious behaviour of extensions. A systematic review of the different research work done in the area of browser extension security has been done and the different security measures that can be taken for browser extension vulnerabilities have been tabulated in [6].

Various solutions have been proposed to protect the users from different vulnerabilities in extensions and from attacks of malicious extensions. A framework called LvDetector was proposed, which combines static and dynamic program analysis techniques for automatic detection of information leakage vulnerabilities in legitimate browser extensions [7]. A dynamic analysis system called Hulk detects malicious extensions by monitoring the actions and by creating a dynamic environment that adapts to the extension's needs to trigger the intended malicious behaviour [8]. Researchers from Google analysed the use of Chrome web store by criminals to spread malicious extensions and came up with a system called WebEval that broadly identifies malicious extensions [9]. Apolicy enforcer for Firefox browser called Sentinel, gives the control to the users over the actions of existing JavaScript extensions and prevents common attacks on the extensions [10]. A system called Expector automatically inspects and identifies extensions that inject ads and classifies those ads as malicious or benign based on their landing pages [11].

3 Security Model of Chrome Extensions

The security model of Chrome extensions is based mainly on three principles. They are least privilege, privilege separation and strong isolation [1]. Extensions will get only those privileges requested in their manifest. Least privilege does not protect users from malicious extensions. Instead, least privileges helps if a benign extension has a vulnerability. If somehow an extension is compromised, the attacker's privileges will be limited. For privilege separation, instead of dumping all the extension code into a single file, developers divide it into two parts, a background page and a content script [12]. The background page is given the major share of the extension's privileges, but it is isolated from direct contact with webpages. On the other hand, content scripts has direct contact with webpages, but has fewer privileges. For strong isolation, extension's content script is run in a separate environment from that of the websites so that websites can't access content script's variables or functions [4].

Google Chrome is concerned with the security of users of Chrome extensions. As part of Chrome's continuing security efforts, from 2013, it is no more possible to silently install extensions into Chrome on windows using the windows registry mechanism for extension deployment [13]. Also, from 2014, Chrome supports only installation of extensions that are hosted in the Chrome web store, in the windows stable and beta channels [13]. But they continue to support local extension installs in developer mode as well as installs via enterprise policy. In Chrome web store, all extensions go through an automated review process, but in most cases, will be published without any further manual review. Chrome web store will block those extensions that violate the Chrome's malware policy. Despite all these security checks, malicious extensions are being hosted in the Chrome web store.

4 Security Analysis of Chrome Specific APIs

This section explains the security analysis of the Chrome specific APIs based on STRIDE approach. Google Chrome provides specific APIs for the extensions to use, in addition to giving access to all the APIs that web pages and apps can use. They are often called chrome.* APIs [12]. There are a total of 63 chrome.* APIs. These APIs allow tight integration with the browser. These chrome.* APIs can be used to perform specific functions with extensions. These APIs have different method, events and properties which can be used to do specific tasks. For example, bookmarks API can be used to create, organize and otherwise manipulate bookmarks [14]. These chrome specific APIs are so powerful in terms of the ways they can be used to access or modify user data or perform different actions in the browser. There are different ways in which these APIs can be used in extensions for performing malicious actions. So, we have used the STRIDE threat analysis approach to analyse the possible threats of using chrome specific APIs.

4.1 STRIDE Threat Analysis

STRIDE approach, developed by Microsoft is used to identify the possible threats of each Chrome specific API. In the case of extensions, spoofing can occur in the form of phishing, clickjacking etc. For example, update method of bookmarks API can be used to update an existing bookmark with a new URL which can be used for phishing, a form of spoofing. Tampering can occur in the form of XSS, clickjacking etc. For example, executeScript method of tabs API can be used to perform cross site scripting attack which is a form of tampering. Repudiation can occur when the extension hides some activity it is doing. For example, erase method of downloads API can be used to erase the download information from the browser which can hide the fact that the extension has downloaded something. Any activity of extensions that accesses user's private data can come under information disclosure. For example, getVisits method of history API can be used to retrieve the history information from the browser. Any activity of extensions that deny users normal browsing experience comes under Denial of service. For example, images property of contentSettings API can be used to block images in webpages. Elevation of privilege involves doing things without the proper consent of users or doing things the user thinks the extension can never do. For example, create method of bookmarks API can be used to create new bookmarks and remove method of bookmarks API can be used to remove existing bookmarks from the web browser.

4.2   Threat Analysis of Chrome Specific APIs

We have used the STRIDE approach to analyse the possible threats of using chrome specific APIs. Chrome specific APIs have different methods, events and properties which can be used for different purposes. We have analysed the specific use of different methods, events and properties of each API and identified the possible threats based on the STRIDE approach. We have analysed the functionality of particular methods, events and properties of each API to identify the possible threats as per STRIDE approach and did not consider the combination of methods and events or two or more methods to identify the threats. Table 1 lists the STRIDE analysis results for the Chrome specific APIs that were identified as having threats.

Out of the 40 APIs that are not in the table, 27 were found to have no threats as per STRIDE approach, 11 APIs works only on the Chrome OS which is not so popular yet. We have concentrated on the users of Chrome browser. The APIs that work only on Chrome OS include accessibilityFeatures, documentScan etc. [14]. There are 3 APIs which are called the devTools APIs. They are devtools.inspectedWindow, devtools.network and devtools.panels. They work only in the devTools page which is called only when the devTools window is opened. The rest

Table 1   STRIDE analysis of APIs

Chrome.* APIs	Methods/events/properties	S	T	R	I	D	E
bookmarks	methods—get, getChildren, getRecent, getTree, getSubTree & search	✗	✗	✗	✓	✗	✗
	methods—move, create, remove & removeTree	✗	✓	✗	✗	✗	✓
	method—update	✓	✓	✗	✗	✗	✓
	events—onCreated, onRemoved, onChanged, onMoved, onChildrenReordered, onImportBegan & onImportEnded	✗	✗	✗	✗	✗	✗
browsingData	method—settings	✗	✗	✗	✓	✗	✗
	methods—remove, removeAppcache, removeCache, removeCookies, removeDownloads, removeFileSystems, removeFormData, removeHistory, removeIndexedDB, removeLocalStorage, removePluginData, removePasswords & removeWebSQL	✗	✓	✗	✗	✗	✓
contentSettings	properties—cookies, images, javascript, location, plugins, popups, notifications, fullscreen, mouselock, unsandboxedPlugins&automaticDownloads	✗	✓	✗	✗	✓	✓
cookies	methods—get, getAll & getAllCookieStores	✗	✗	✗	✓	✗	✗
	methods—set, remove	✗	✓	✗	✗	✗	✓
	event—onChanged	✗	✗	✗	✗	✗	✗

(continued)

Table 1 (continued)

Chrome.* APIs	Methods/events/properties	S	T	R	I	D	E
debugger	method—attach	✗	✓	✗	✗	✓	✓
	method—sendCommand	✗	✓	✗	✗	✗	✓
	methods—detach & getTargets	✗	✗	✗	✗	✗	✗
	events—onEvent & onDetach	✗	✗	✗	✗	✗	✗
desktopCapture	method—chooseDesktopMedia	✗	✗	✗	✓	✗	✗
	method—cancelChooseDesktopMedia	✗	✗	✓	✗	✗	✗
downloads	methods—download, pause, resume, cancel, open & removeFile	✗	✓	✗	✗	✗	✓
	method—search	✗	✗	✗	✓	✗	✗
	methods—getFileIcon, show & showDefaultFolder	✗	✗	✗	✗	✗	✗
	method—erase	✗	✗	✓	✗	✗	✓
	methods—acceptDanger, drag & setShelfEnabled	✗	✗	✗	✗	✗	✗
	events—onCreated, onErased, onChanged & onDeterminingFilename	✗	✗	✗	✗	✗	✗
gcm	methods—register & unregister	✗	✗	✗	✗	✗	✗
	method—send	✗	✗	✗	✓	✗	✗
	events—onMessage, onMessagesDeleted & onSendError	✗	✗	✗	✗	✗	✗
history	methods—search & getVisits	✗	✗	✗	✓	✗	✓
	methods—addUrl, deleteUrl, deleteRange & deleteAll	✗	✓	✗	✗	✗	✓
	events—onVisited & onVisitRemoved	✗	✗	✗	✗	✗	✗
management	methods—getAll, get, getPermissionWarningsById & getPermissionWarningsByManifest	✗	✗	✗	✓	✗	✗
	method—setEnabled	✗	✓	✗	✗	✓	✓
	methods—uninstall & launchApp	✗	✓	✗	✗	✗	✓
	methods—getSelf, uninstallSelf, createAppShortcut, setLaunchType & generateAppForLink	✗	✗	✗	✗	✗	✗
	events—onInstalled, onUninstalled, onEnabled & onDisabled	✗	✗	✗	✗	✗	✗
power	method—requestKeepAwake	✗	✓	✗	✗	✗	✗
	method—releaseKeepAwake	✗	✗	✗	✗	✗	✗
privacy	property—network	✗	✗	✗	✗	✗	✗
	properties—services & websites	✗	✓	✗	✗	✗	✓
proxy	property—settings	✗	✓	✗	✗	✓	✓
	event—onProxyError	✗	✗	✗	✗	✗	✗
sessions	methods—getRecentlyClosed & getDevices	✗	✗	✗	✓	✗	✗
	method—restore	✗	✗	✗	✗	✗	✗
	event—onChanged	✗	✗	✗	✗	✗	✗

(continued)

Table 1 (continued)

Chrome.* APIs	Methods/events/properties	S	T	R	I	D	E
storage	property—sync	✗	✗	✗	✓	✗	✗
	properties—local & managed	✗	✗	✗	✗	✗	✗
	event—onChanged	✗	✗	✗	✗	✗	✗
system.cpu	method—getInfo	✗	✗	✗	✓	✗	✗
system.memory	method—getInfo	✗	✗	✗	✓	✗	✗
system.storage	method—getInfo	✗	✗	✗	✓	✗	✗
	methods—ejectDevice & getAvailableCapacity	✗	✗	✗	✗	✗	✗
	events—onAttached & onDetached	✗	✗	✗	✗	✗	✗
tabCapture	method—capture	✗	✗	✗	✓	✗	✗
	method—getCapturedTabs	✗	✗	✗	✗	✗	✗
	event—onStatusChanged	✗	✗	✗	✗	✗	✗
tabs	methods—connect, sendRequest, sendMessage, getSelected, getAllInWindow, create, duplicate, query, highlight, move, reload & detectLanguage	✗	✗	✗	✗	✗	✗
	methods—update & remove	✗	✗	✗	✗	✓	✓
	methods—get, getCurrent & captureVisibleTab	✗	✗	✗	✓	✗	✗
	method—executeScript	✗	✓	✗	✗	✗	✗
	method—insertCSS	✓	✓	✗	✗	✗	✗
	methods—setZoom, getZoom, setZoomSettings & getZoomSettings	✗	✗	✗	✗	✗	✗
	events—onCreated & onActivated	✗	✗	✗	✗	✓	✗
	events—onUpdated, onMoved, onSelectionChanged, onActiveChanged, onHighlightChanged, onHighlighted, onDetached, onAttached, onRemoved, onReplaced & onZoomChange	✗	✗	✗	✗	✗	✗
topSites	method—get	✗	✗	✗	✓	✗	✗
webRequest	method—handlerBehaviorChanged	✗	✗	✗	✗	✗	✗
	event—onBeforeRequest	✓	✓	✗	✗	✓	✓
	events—onBeforeSendHeaders & onHeadersReceived	✗	✓	✗	✗	✓	✓
	events—onSendHeaders, onAuthRequired, onResponseStarted, onBeforeRedirect & onCompleted	✗	✓	✗	✗	✗	✓
	event—onErrorOccurred	✗	✗	✗	✗	✗	✗
windows	method—get, getLastFocused & getAll	✗	✗	✗	✓	✗	✗
	method—getCurrent	✗	✗	✗	✓	✓	✗
	method—create	✗	✗	✗	✗	✗	✗
	methods—update & remove	✗	✓	✗	✗	✗	✓
	event—onCreated	✗	✗	✗	✗	✓	✗
	events—onRemoved & onFocusChanged	✗	✗	✗	✗	✗	✗

3 APIs are events, extensionTypes and types. events API is basically a namespace containing common types used by APIs dispatching events to notify something. extensionTypes API contain type declarations for Chrome extensions and types API contains type declarations for Chrome.

5   Results

This section describes the statistical results obtained from our threat analysis of the Chrome specific APIs based on STRIDE. Chrome provides JavaScript APIs specifically for developing extensions. These APIs are very powerful in terms of the activities they can perform. There are a total of 63 Chrome specific APIs that we have analysed using the STRIDE approach. We have identified that 23 APIs are having various threats based on the STRIDE approach. Table 1 lists the STRIDE analysis results for the Chrome specific APIs that were identified as having threats. In summary, APIs that have threats as per STRIDE approach are: bookmarks, browsingData, contentSettings, cookies, debugger, desktopCapture, downloads, gcm, history, management, power, privacy, proxy, sessions, storage, system.cpu, system.memory, system.storage, tabCapture, tabs, topSites, webRequest and windows. Table 2 lists the count of the APIs, methods, events and properties having various threats as per STRIDE approach. It is the total count of APIs, methods, events and properties affected by each threat as per STRIDE approach.

Table 2 Summary of API threat analysis	Threats (STRIDE)	APIs	Methods	Events	Properties
	Spoofing	3	2	1	0
	Tampering	14	40	8	14
	Repudiation	2	2	0	0
	Information disclosure	17	33	0	1
	Denial of service	7	5	6	12
	Elevation of privilege	13	42	8	14

6 Conclusion

Chrome extensions are getting wide acceptance among the users of Google Chrome and hence are being widely used for malicious purposes by attackers. The research work has revealed the potential dangers the users of Chrome extensions are facing when they use the extensions which can use all the Chrome specific APIs for its own purposes. Table 1 shows the various threats different Chrome APIs are having and the attackers can use any of these APIs in the extensions to perform malicious activities. The APIs having threats are so powerful that when used in a malicious way can lead to loss of user's sensitive data or damage to user's system. This threat analysis results can help in the identification of the malicious behaviour of chrome extensions. This threat analysis has only considered the functionality of each Chrome specific API for identifying the threats. This work can be extended by analysing the possible threats considering the different combinations of Chrome specific APIs, as it can serve different purposes and can be used for more malicious activities in the extension code.

References

1. L. Liu, X. Zhang, G. Yan, S. Chen, Chrome extensions: threat analysis and countermeasures, in *NDSS* (2012)
2. Microsoft STRIDE threat model, https://msdn.microsoft.com/en-us/library/ee823878%28v= cs.20%29.aspx
3. S.F. Burns, Threat modeling: a process to ensure application security, in *GIAC Security Essentials Certification (GSEC) Practical Assignment* (2005)
4. N. Carlini, A. Porter Felt, D. Wagner, An evaluation of the google chrome extension security architecture, in *Presented as Part of the 21st USENIX Security Symposium (USENIX Security 12)*, pp. 97–111 (2012)
5. V. Aravind, M. Sethumadhavan, A framework for analysing the security of chrome extensions. Adv. Comput. Netw. Inf. **2**, 267–272 (2014)
6. J. Arunagiri, S. Rakhi, K.P. Jevitha, A systematic review of security measures for web browser extension vulnerabilities, in *Proceedings of the International Conference on Soft Computing Systems* (Springer India, 2016)
7. R. Zhao, C. Yue, Q. Yi, Automatic detection of information leakage vulnerabilities in browser extensions, in *Proceedings of the 24th International Conference on World Wide Web* (International World Wide Web Conferences Steering Committee, 2015)
8. A. Kapravelos, et al., Hulk: eliciting malicious behavior in browser extensions, in *23rd USENIX Security Symposium (USENIX Security 14)* (2014)
9. N. Jagpal, et al., Trends and lessons from three years fighting malicious extensions, in *24th USENIX Security Symposium (USENIX Security 15)* (2015)
10. K. Onarlioglu, et al., Sentinel: securing legacy firefox extensions. Comput. Secur. **49**, 147–161 (2015)

11. X. Xing, et al., Understanding malvertising through ad-injecting browser extensions, in *Proceedings of the 24th International Conference on World Wide Web* (International World Wide Web Conferences Steering Committee, 2015)
12. Chrome extension developer guide, https://developer.chrome.com/extensions/overview
13. Chromium blog, http://blog.chromium.org/
14. Chrome extension specific API index, https://developer.chrome.com/extensions/api_index

Some Properties of Rough Sets on Fuzzy Approximation Spaces and Applications

B.K. Tripathy and Suvendu Kumar Parida

Abstract The notion of Rough sets introduced by Pawlak has been extended in many directions to enhance its modelling power. One such approach is to reduce the restriction of the base relation being an equivalence relation. Adding the flavour of fuzzy sets to it a fuzzy proximity relation was used to generate a fuzzy approximation space by De et al. in 1999 and hence the rough sets on fuzzy approximation spaces could be generated. These are much more general than the basic rough sets and also the rough sets defined on proximity relations. However, some of the results established in this direction by De et al. have been found to be faulty. In this paper we show through examples that the results are actually faulty and provide their correct versions. Also, we establish some more properties of these rough sets. A real life application is provided to show the application of the results.

Keywords Rough sets · Fuzzy sets · Fuzzy relations · Fuzzy proximity relations · Fuzzy approximation spaces

1 Introduction

In order to handle datasets with uncertainty the rough set model was proposed by Pawlak in 1982 [1]. In an attempt to extend the basic rough set model, which depends upon equivalence relations, several attempts have been made and using fuzzy proximity relations, which are fuzzy reflexive and fuzzy symmetric only. The effort to define rough sets on fuzzy approximation spaces, which are defined by taking fuzzy approximation relations instead of equivalence relation is due to De

B.K. Tripathy (✉)
School of Computer Science and Engineering, VIT University, Vellore,
Tamil Nadu 632014, India
e-mail: tripathybk@vit.ac.in

S.K. Parida
SCS Autonomous College, Puri 752001, Odisha, India
e-mail: paridasuvendukumar@gmail.com

© Springer Nature Singapore Pte Ltd. 2017
S.C. Satapathy et al. (eds.), *Proceedings of the 5th International Conference on Frontiers in Intelligent Computing: Theory and Applications*, Advances in Intelligent Systems and Computing 516, DOI 10.1007/978-981-10-3156-4_18

et al. in [2] and [3]. Attempts have been made to define generalized approximation spaces instead of knowledge bases in [4] and [5]. After the introduction of this new notion, many applications have been studied. Notable among them are these studies are in [6–12].

The structure of the paper henceforth is as follows. We present the definitions to be used and the notations to be followed in the paper in Sect. 2. The main results for rough sets on fuzzy approximation spaces will be discussed in Sect. 3. An application of the results is to be provided in Sect. 4. In Sect. 5 we provide concluding remarks and finally end up with a list of works consulted during the preparation of the paper in the reference section.

2 Definitions and Notions

The most successful uncertainty based model is perhaps fuzzy sets introduced by Zadeh in 1965 [11]. The formal definition is as follows.

Definition 2.1 A fuzzy set A defined over U uniquely defined by its membership function μ_A, given by

$$\mu_A : U \to [0, 1], \tag{2.1}$$

that is, for each $x \in U$, $\mu_A(x)$, is called the grade of membership of x in A is any real number in [0, 1].

The non-membership function ν_A of A is defined as

$$\nu_A(x) = 1 - \mu_A(x), \text{ for all } x \in U \tag{2.2}$$

The notion of rough sets was introduced by Pawlak in 1982 [1].

Let U be a universe of discourse and R be an equivalence relation over U. Then R induces a classification over U comprising of equivalence classes of elements x, denoted by $[x]_R$. For any subset X of U we associate two crisp sets with respect to R called the lower and upper approximations of X with respect to R, denoted by $\underline{R}X$ and $\overline{R}X$ respectively. These are defined as $\underline{R}X = \{x \in U \,|\, [x]_R \subseteq X\}$ and $\overline{R}X = \{x \in U \,|\, [x]_R \cap X \neq \phi\}$. Also, we define $BN_R(X) = \overline{R}X \backslash \underline{R}X$ called the boundary of X with respect to R. We call X to be rough with respect to R iff $\underline{R}X \neq \overline{R}X$ or equivalently $BN_R(X) \neq \phi$. Otherwise, X is said to be R-definable.

Definition 2.3 A fuzzy relation over U is said to be a fuzzy proximity relation over U if and only if

$$R \text{ is fuzzy reflexive that is } \mu_R(x, x) = 1, \forall x \in U. \tag{2.3}$$

$$R \text{ is fuzzy symmetric if and only if } \mu_R(x, y) = \mu_R(x, y), \forall x, y \in U. \tag{2.4}$$

Definition 2.4 For any $\alpha \in [0, 1]$, the α –*cut* R_α of R is a subset of $U \times U$ is given by

$$R_\alpha = \{(x, y) | \mu_R(x, y) \geq \alpha\}. \tag{2.5}$$

For any fuzzy proximity relation R on U and $\alpha \in [0, 1]$, if $(x, y) \in R_\alpha$ then we say that x and y are α – *similar* and we denote it by $xR_\alpha y$.

Definition 2.5 Two elements x and y are said to be α *identical* denoted by $xR(\alpha)y$ if either $xR_\alpha y$ or there exists a sequence of elements $u_1, u_2, \ldots u_n$ in U such that $xR_\alpha u_1 R_\alpha u_2 \ldots u_n R_\alpha y$

It may be noted that the relation $R(\alpha)$ is an equivalence relation for each $\alpha \in [0, 1]$. Here (U, R) is called a fuzzy approximation space. Also, it may be noted that for any $\alpha \in [0, 1]$, $(U, R(\alpha))$ is an approximation space in the same sense as that used by Pawlak.

For any x in U we denote the equivalence class of x with respect to $R(\alpha)$ by $[x]_{R(\alpha)}$.

Definition 2.6 Let U be a universal set and R be a fuzzy proximity relation on U. Then for any $\alpha \in [0, 1]$, we define the lower and upper approximations [13, 14] of a subset X in U as

$$\underline{R(\alpha)}X = \left\{x \in U | [x]_{R(\alpha)} \subseteq X\right\} and \tag{2.6}$$

$$\overline{R(\alpha)}X = \left\{x \in U | [x]_{R(\alpha)} \cap X \neq \phi\right\}. \tag{2.7}$$

We say that X is $R(\alpha)$ discernible if and only if $\underline{R(\alpha)}X = \overline{R(\alpha)}X$. Else, X is said to be $R(\alpha)$-rough.

3 Results on Rough Sets on Fuzzy Approximation Spaces

Out of the several properties for rough sets on fuzzy approximation spaces established in [13, 14], we show that some are technically incorrect. We show this through counter examples, state the correct forms and prove them. It is easy to see that

Lemma 3.1 If R and S are fuzzy proximity relations over U then $R \cup S$ and $R \cap S$ are fuzzy proximity relations over U.

The first result incorrectly stated in [13, 14] is that

Proposition 3.1 Let R and S be two fuzzy proximity relations on U. Then

$$(R \cup S)(\alpha) \subseteq R(\alpha) \cup S(\alpha) \tag{3.1}$$

$$(R \cap S)(\alpha) \supseteq R(\alpha) \cap S(\alpha) \tag{3.2}$$

Counter Example 3.1:

The following example shows the incorrectness of property (3.1). Suppose $U = \{x_1, x_2, x_3, x_4, x_5, x_6\}$ is a universe. Two fuzzy proximity relations R and S are defined over U are given in their matrix form in Table 1 and Table 2 respectively.

From the above two tables we obtain the union and intersection and is represented in the following tables. Table 3 represents $R \cup S$ and Table 4 represents $R \cap S$.

Let us consider $\alpha = 0.7$.

Therefore from Tables 1 and 2 we obtain the following:

$$R(\alpha) = \{(2,3),(2,4),(3,4),(5,6),(3,2),(4,2),(4,3),(6,5),(1,1),(2,2),(3,3),(4,4),(5,5),(6,6)\} \tag{3.3}$$

$$S(\alpha) = \{(3,5),(5,3),(1,1),(2,2),(3,3),(4,4),(5,5),(6,6)\} \tag{3.4}$$

And from Tables 3 and 4 we obtain the following:

$$(R \cup S)(\alpha) = \{(2,3),(2,4),(3,4),(3,5),(5,6),(2,5),(2,6),(3,2),(4,2),(4,3),(5,3),$$
$$(6,5),(1,1),(2,2),(3,3),(4,4),(5,5),(6,6)\} \tag{3.5}$$

$$(R \cap S)(\alpha) = \{(1,1),(2,2),(3,3),(4,4),(5,5),(6,6)\} \tag{3.6}$$

From (3.3) and (3.4) we obtain

Table 1 Fuzzy proximity relation R

R	x_1	x_2	x_3	x_4	x_5	x_6
x_1	1	0.3	0.3	0.3	0.3	0.3
x_2	0.3	1	0.7	0.7	0.3	0.3
x_3	0.3	0.7	1	0.7	0.3	0.3
x_4	0.3	0.7	0.7	1	0.3	0.3
x_5	0.3	0.3	0.3	0.3	1	0.7
x_6	0.3	0.3	0.3	0.3	0.7	1

Table 2 Fuzzy proximity relation S

S	x_1	x_2	x_3	x_4	x_5	x_6
x_1	1	0.1	0.1	0.1	0.1	0.1
x_2	0.1	1	0.1	0.1	0.1	0.1
x_3	0.1	0.1	1	0.1	0.8	0.1
x_4	0.1	0.1	0.1	1	0.1	0.1
x_5	0.1	0.1	0.8	0.1	1	0.1
x_6	0.1	0.1	0.1	0.1	0.1	1

Table 3 Fuzzy proximity relation $R \cup S$

$R \cup S$	x_1	x_2	x_3	x_4	x_5	x_6
x_1	1	0.3	0.3	0.3	0.3	0.3
x_2	0.3	1	0.7	0.7	0.3	0.3
x_3	0.3	0.7	1	0.7	0.8	0.3
x_4	0.3	0.7	0.7	1	0.3	0.3
x_5	0.3	0.3	0.8	0.3	1	0.7
x_6	0.3	0.3	0.3	0.3	0.7	1

Table 4 Fuzzy proximity relation $R \cap S$

$R \cap S$	x_1	x_2	x_3	x_4	x_5	x_6
x_1	1	0.1	0.1	0.1	0.1	0.1
x_2	0.1	1	0.1	0.1	0.7	0.1
x_3	0.1	0.1	1	0.1	0.3	0.1
x_4	0.1	0.1	0.1	1	0.1	0.1
x_5	0.1	0.7	0.3	0.1	1	0.1
x_6	0.1	0.1	0.1	0.1	0.1	1

$$R(\alpha) \cup S(\alpha) = \{(2,3), (2,4), (3,4), (5,6), (3,5), (5,3), (3,2), (4,2), (4,3), (6,5),$$
$$(1,1), (2,2), (3,3), (4,4), (5,5), (6,6)\}$$

$$(3.7)$$

$$R(\alpha) \cap S(\alpha) = \{(1,1), (2,2), (3,3), (4,4), (5,5), (6,6)\} \qquad (3.8)$$

It follows from (3.5) and (3.7) that $(R \cup S)(\alpha) \subseteq R(\alpha) \cup S(\alpha)$ is not true.
A similar example can be constructed to show that (3.2) is erroneous.

Theorem 3.1 Let R and S be two fuzzy proximity relations on U. Then for any $\alpha \in [0, 1]$,

$$(R \cup S)(\alpha) \supseteq R(\alpha) \cup S(\alpha) \qquad (3.9)$$

$$(R \cap S)(\alpha) \subseteq R(\alpha) \cap S(\alpha) \qquad (3.10)$$

Proof of (3.9):

$$(x, y) \in R(\alpha) \cup S(\alpha) \Rightarrow (x, y) \in R(\alpha) \text{ or } (x, y) \in S(\alpha)$$

Now 4 cases can occur.
Case (i): $xR_\alpha y$ or $xS_\alpha y$
Then $\mu_{R(x,y)} \geq \alpha$ or $\mu_{S(x,y)} \geq \alpha$. So, $\max\{\mu_{R(x,y)}, \mu_{S(x,y)}\} \geq \alpha$

$$\Rightarrow \mu_{(R \cup S)(x,y)} \geq \alpha \Rightarrow (x, y) \in (R \cup S)(\alpha)$$

Case (ii): $xR_\alpha y$ or $\exists$ a sequence u_1, u_2, ... u_n such that $xS_\alpha u_1$, $u_1 S_\alpha u_2$, ... $u_n S_\alpha y$
If $xR_\alpha y$ then $\mu_{R(x,y)} \geq \alpha$. So, $\mu_{(R \cup S)(x,y)} = \max\{\mu_{R(x,y)}, \mu_{S(x,y)}\} \geq \alpha$.

Again, suppose $\exists$ a sequence u_1, u_2, ... u_n such that $xS_\alpha u_1$, $u_1 \ S_\alpha u_2$, ... $u_n S_\alpha y$. Then $\mu_{S(x,u_1)} \geq \alpha, \mu_{S(u_1,u_2)} \geq \alpha, ..., \mu_{S(u_n,y)} \geq \alpha$. So, as above $\mu_{(R \cup S)(x,u_1)} \geq \alpha$, $\mu_{(R \cup S)(u_1,u_2)} \geq \alpha, ..., \mu_{(R \cup S)(u_n,y)} \geq \alpha$. This implies that $x(R \cup S)_\alpha u_1, u_1$ $(R \cup S)_\alpha u_2, ..., u_n(R \cup S)_\alpha y$. Thus, $x(R \cup S)(\alpha)y$.

Case (iii): $\exists$ a sequence u_1, u_2, ... u_n such that $xR_\alpha u_1$, $u_1 R_\alpha u_2$, ... $u_n R_\alpha y$ or $xS_\alpha y$. The proof is similar to case (ii) above.

Case (iv): $\exists$ a sequence u_1, u_2, ... u_n such that $xR_\alpha u_1$, $u_1 R_\alpha u_2$, ... $u_n R_\alpha y$ or

$$\exists \text{ a sequence } v_1, v_2, ... v_m \text{ such that } xS_\alpha v_1, v_1 S_\alpha v_2, ... v_m S_\alpha y$$

In any one of the cases we apply the same argument as in case (ii) 2nd part and conclude that $x(R \cup S)(\alpha)y$.

Hence in all the cases the proof follows. Proof of (3.10) is similar.

We denote the equivalence classes generated by a fuzzy proximity relation R over U with respect to a grade $\alpha \in [0, 1]$ by R_α^*. We have the following result from [13, 14].

Proposition 3.2 Let R and S be two fuzzy proximity relations on U. Then

$$(R \cup S)_\alpha^*(\alpha) \subseteq R_\alpha^* \cup S_\alpha^* \tag{3.11}$$

$$(R \cap S)_\alpha^*(\alpha) \supseteq R_\alpha^* \cap S_\alpha^* \tag{3.12}$$

We show below that the result is incorrect. For this we provide two counter examples where the results fail to be true.

Counter Example 3.3:

We continue with the above example where the fuzzy proximity relations R and S and their union and intersection are given in Tables 1, 2, 3 and 4. We have,

$$R_\alpha^* = \{\{1\}, \{2,3,4\}, \{5,6\}\} \tag{3.13}$$

$$S_\alpha^* = \{\{1\}, \{2\}, \{4\}, \{6\}, \{3,5\}\} \tag{3.14}$$

$$(R \cup S)_\alpha^* = \{\{1\}, \{2,3,4,5,6\}\} \tag{3.15}$$

$$(R \cap S)_\alpha^* = \{\{1\}, \{2\}, \{3\}, \{4\}, \{5\}, \{6\}\} \tag{3.16}$$

$$R_\alpha^* \cup S_\alpha^* = \{\{1\}, \{2\}, \{4\}, \{6\}, \{2,3,4\}, \{3,5\}, \{5,6\}\} \tag{3.17}$$

$$R_\alpha^* \cap S_\alpha^* = \{\{1\}\} \tag{3.18}$$

It is clear from (3.17) and (3.18) that (3.11) holds true.

Similarly, a counter example can be constructed to show that (3.12) is incorrect.

Theorem 3.2 Let R and S be two fuzzy proximity relations on U. Then

$$(R \cup S)^*_\alpha(\alpha) \subseteq R^*_\alpha \cup S^*_\alpha \tag{3.19}$$

$$(R \cap S)^*_\alpha(\alpha) \supseteq R^*_\alpha \cap S^*_\alpha \tag{3.20}$$

Proof of (3.19):

Let $[x] \in R^*_\alpha \cup S^*_\alpha$. Then $[x] \in R^*_\alpha$ or $[x] \in S^*_\alpha$.

So, for any $y \in [x]$, $(x, y) \in R(\alpha)$ or $(x, y) \in S(\alpha) \Leftrightarrow (x, y) \in (R \cup S)(\alpha)$ (by (3.9))

Hence $[x] \in (R \cup S)^*_\alpha$. This proves (3.19). Proof of (3.20) is similar.

4   An Application of Rough Sets on Fuzzy Approximation Spaces

Let us consider a situation where the computer needs to match two images based on the different matching properties of images.

Suppose $V = \{i_1, i_2, i_3, i_4, i_5\}$ be a set of images that needs to be matched based on the following criteria

- Based on edge matching
- Based on color matching
- Based on shape and surface area matching
- Based on texture matching

After the feature extraction process of the images we obtain the matching percentages of properties of two images. R, S, T, and W are the relations based on the above properties on the set V which can be used to match images based on the required criteria. To find matching images based on two or more properties, we find the union or intersection of the relations based on the requirements. Following tables represent the relations R, S, T, and W (Tables 5, 6, 7 and 8).

If we fix $\alpha = 0.85$ as the percentage of matching property, we obtain the following:

R $(\alpha) = \{(2, 4)\}$, S $(\alpha) = \{(1, 3), (4, 5)\}$, T $(\alpha) = \{(4, 5)\}$ and W $(\alpha) = \{(1, 3)\}$

$R^*_\alpha = \{\{1\}, \{3\}, \{5\}, \{2, 4\}\}$, $S^*_\alpha = \{\{2\}, \{4, 5\}, \{1, 3\}\}$, $T^*_\alpha = \{\{1\}, \{2\}, \{3\}, \{4, 5\}\}$ and $W^*_\alpha = \{\{2\}, \{4\}, \{5\}, \{1, 3\}\}$

Table 5 Matching edges

R	i_1	i_2	i_3	i_4	i_5
i_1	1	0.30	0.15	0.60	0.50
i_2	0.30	1	0.40	0.87	0.20
i_3	0.15	0.40	1	0.52	0.05
i_4	0.60	0.87	0.52	1	0.66
i_5	0.50	0.20	0.05	0.66	1

Table 6 Matching colours

S	i_1	i_2	i_3	i_4	i_5
i_1	1	0.30	0.90	0.60	0.75
i_2	0.30	1	0.23	0.41	0.50
i_3	0.90	0.23	1	0.60	0.41
i_4	0.60	0.41	0.60	1	0.96
i_5	0.75	0.50	0.41	0.96	1

Table 7 Matching shape and surface area

T	i_1	i_2	i_3	i_4	i_5
i_1	1	0.46	0.30	0.70	0.25
i_2	0.46	1	0.50	0.35	0.80
i_3	0.30	0.50	1	0.10	0.03
i_4	0.70	0.35	0.10	1	0.95
i_5	0.25	0.80	0.03	0.95	1

Table 8 Matching texture

W	i_1	i_2	i_3	i_4	i_5
i_1	1	0.09	0.87	0.43	0.67
i_2	0.09	1	0.15	0.33	0.54
i_3	0.87	0.15	1	0.63	0.70
i_4	0.43	0.33	0.63	1	0.30
i_5	0.67	0.54	0.70	0.30	1

Table 9 $R \cup T$

$R \cup T$	i_1	i_2	i_3	i_4	i_5
i_1	1	0.46	0.30	0.70	0.50
i_2	0.46	1	0.50	0.87	0.80
i_3	0.30	0.50	1	0.52	0.05
i_4	0.70	0.87	0.52	1	0.95
i_5	0.50	0.80	0.05	0.95	1

Inference: From the above results we can conclude that based on the edge matching property images i_2 and i_4 are matched maximum. Similarly, based on colour matching images i_1 and i_3, and images i_4 and i_5 are matching. Images i_4 and i_5 are matched based on shape and surface area, and images i_1 and i_3 on texture matching.

Now if we want to find the matching images based on edge matching or shape and surface area matching, we find the union of relations R and T. The union of the two relations is as shown in Table 9.

From Table 9 we find

$$(R \cup T)(\alpha) = \{(2,4),(4,5),(2,5)\} \text{ and } (R \cup T)^*_\alpha = \{\{1\},\{3\},\{2,4,5\}\}$$

Table 10 $S \cap W$

$S \cap W$	i_1	i_2	i_3	i_4	i_5
i_1	1	0.09	0.87	0.43	0.67
i_2	0.09	1	0.15	0.33	0.50
i_3	0.87	0.15	1	0.60	0.41
i_4	0.43	0.33	0.60	1	0.30
i_5	0.67	0.50	0.41	0.30	1

Inference: We can say that images i_2, i_4 and i_5 form a group of matching images based on edge matching and shape and surface area matching

Next, if we want to find the matching images based on colour matching as well as texture matching, we find the intersection of relations A_2 and A_4. The intersection of the two relations is as shown in Table 10.

From Table 10 we find

$$(S \cap W)(\alpha) = \{(1,3)\} \text{ and } (S \cap W)_\alpha^* = \{\{2\}, \{4\}, \{5\}, \{1,3\}\}$$

Inference: We can say that images i_1 and i_3 is a set of images that are matched both in terms of colour and texture similarity.

5 Conclusions

Rough sets on fuzzy approximation spaces are generalizations of basic rough sets introduced in 1999 [13, 14]. However, some of the properties established in the first paper were erroneous. In this chapter, we first established that the results are incorrect through counter examples and then established the correct versions of the properties. Finally, we illustrated through an application in computer vision as how the results are applicable in real life situations.

References

1. Z. Pawlak, Rough sets. Int. J. Inf. Comp. Sci. **11**, 341–356 (1982)
2. A. Skowron, J. Stepaniuk, Tolerance approximation spaces. Fund. Inf. **27**(2–3), 245–253 (1996)
3. R. Slowinski, D. Vanderpooten, Similarity Relation as a Basis for Rough Approximations, in *Advances in Machine Intelligence & Soft-Computing*, vol. IV, ed. by P.P. Wang (Duke University Press, Durham, 1997), pp. 17–33
4. A. Skowron, J. Stepaniuk, Generalized approximation spaces, in *Soft Computing: Rough Sets, Fuzzy Logic, Neural Networks, Uncertainty Management*, ed. by T.Y. Lin, A.M. Wild Berger (San Diego Simulation Councils Inc., 1995), pp. 18–21
5. R. Slowinski, D. Vanderpooten, A generalized definition of rough approximations based on similarity. IEEE Trans. Data Knowl. Eng. **12**, 331–336 (2000)

6. B.K. Tripathy, Rough sets on fuzzy approximation spaces and application to distributed knowledge systems. National Conference on Mathematics and its applications, Burdwan (2004)

7. B.K. Tripathy, Rough sets on fuzzy approximation spaces and intuitionistic fuzzy approximation spaces, in *Rough Set Theory: A True Landmark in Data Analysis* (Springer, Berlin, Heidelberg (2009), pp. 3–44

8. B.K. Tripathy, S.S. Gantayat, Rough sets on fuzzy similarity relations and applications to information retrieval. Int. J. Fuzzy Syst. Rough Syst. (IJFSRS) **3**(1),1–13 (2010)

9. B.K. Tripathy, S.S. Gantayat, Rough sets on generalized fuzzy approximation spaces, in *International Conference on Multidisciplinary Information Technology (INSCIT - 2006)*, Spain (2006)

10. B.K. Tripathy, S.S. Gantayat, D. Mohanty, Properties of rough sets on fuzzy approximation spaces and knowledge representation, in *Proceedings of the National Conference on Recent Trends in Intelligent Computing*, 17–19 Nov 2006, Kalyani Govt. Engineering College (W. B), pp. 3–8 (2006)

11. L.A. Zadeh, Fuzzy sets. Inf. Control **8**(3), 338–353 (1965)

12. D.P. Acharjya, B.K. Tripathy, Rough sets on fuzzy approximation spaces and applications to distributed knowledge systems. Int. J. Artif. Intell. Soft Comput. **1**(1), 1–14 (2008)

13. S.K. De, Some aspects of fuzzy sets, rough sets and intuitionistic fuzzy sets. Ph.D. Thesis, I.I.T. Kharagpur (1999)

14. S.K. De, R. Biswas, A.R. Roy, Rough sets on fuzzy approximation space. J. Fuzzy Math. **11**, 1–15 (2003)

Design and Optimization of Fractal Antenna for UWB Application

Arati Behera, Sarmistha Satrusallya and Mihir Narayan Mohanty

Abstract UWB application increases day-by-day along with various applications. As the communication is mostly dependent on wireless based, so a suitable antenna design is a major challenge for the researchers. In this paper we've taken an attempt to partially meet the challenge. The antenna is microstrip type based on Fractal Geometry. Initially the fractal antenna design has been made so its performance is in terms of bandwidth. Further the parameter of the antenna have been optimized to have better performance as compared to Un-optimized antenna. It shows output as shown in result section. Particle Swarm Optimization (PSO) which is a viable developmental improvement strategy is used for optimizing the proposed antenna. This method provides better results in the design of the antenna and it also analyzed the effect of the various design parameters like ground plane, feed line width, middle triangle radius. The improvement in the results has been included in the design. It is found to be suitable for UWB communication application.

Keywords Microstrip antennas · Particle swarm optimization (PSO) · Return loss

1 Introduction

The geometries of fractal antenna methodology approach become put sent by Nathen Cohen in 1995 [1]. The fractal antenna performs a most essential function in area of wireless communications due to its wideband and multi-band traits. Because

A. Behera · S. Satrusallya · M.N. Mohanty (✉)
ITER, S 'O' A University, Bhubaneswar, Odisha, India
e-mail: mihir.n.mohanty@gmail.com

A. Behera
e-mail: arati.kisd@gmail.com

S. Satrusallya
e-mail: sarmisthasatrusalya@soauniversity.ac.in

© Springer Nature Singapore Pte Ltd. 2017
S.C. Satapathy et al. (eds.), *Proceedings of the 5th International Conference on Frontiers in Intelligent Computing: Theory and Applications*, Advances in Intelligent Systems and Computing 516, DOI 10.1007/978-981-10-3156-4_19

of this, the usage of microstrip fractal antenna geometry has been turnout to be latest topic for the researchers inside the global [2]. In remote correspondence wideband and multiband attributes is attractive to outline greater successful antennas. Fractal implies the uneven fragments, which might be the association of complex geometries which have the self-likeness systems [3]. Fractal geometries likewise diminish the overall length of antenna and it create different resonant bands [4]. The principle points of interest of fractal antennas are that they have low profile, linear and circular polarization, less fabrication cost, light weight [5]. Fractal antennas are composed by way of applying the boundless number of times an iterative calculation, for example, MRCM (Multiple Reduction Copy Machine) [2]. In the patch of fractal antenna geometry discontinuities presented to expand the radiation proficiency of antennas [6].

2 Proposed Design of Antenna

The different stages to develop the antenna is shown in Fig. 1. At 1st stage the triangular patch is considered as shown in Fig. 1a. Next to it an equilateral triangle is integrated in the middle of the patch as shown in Fig. 1b. The middle triangle (equilateral) can be integrated into patch in 2nd stage. The center triangle was expelled from the equilateral radio wire by leaving all three sides just as measured triangles. By then Koch-fractal is delivered from the design by substituting the portion of the Fig. 1a–c [7].

The main phase of Sierpinski fractal (S1) was accomplished as shown in Fig. 1b. The 3rd stage was the presentation of two Koch, such as bends at both sides of the initially repeated Sierpinski fractal patch Fig. 1c [8]. The 4th stage was the addition of Koch curves whose iteration factor is $\frac{1}{3}$ and is consider Fig. 1d. This structure is simulated by utilizing High Frequency Structure Simulated (HFSS v13.0) software. Proposed antenna's geometry is formed on a minimal effort FR4_epoxy substrate having thickness 1.6 mm with relative permittivity 4.4 and loss tangent 0.02 (0.019), at resonant frequency of 6 GHz. The equilateral triangular patch monopole radio wire's side length is computed by utilizing the mathematical statement and is found to be 15.89 mm.

$$W_p = \frac{2 \times c}{3 \times f_o \times \sqrt{\varepsilon_r}} \, in \, mm$$

where,

$f_0 =$ Operating frequency (GHz)
$c =$ Speed of light in free space (m/s)
$\varepsilon_r =$ relative permittivity

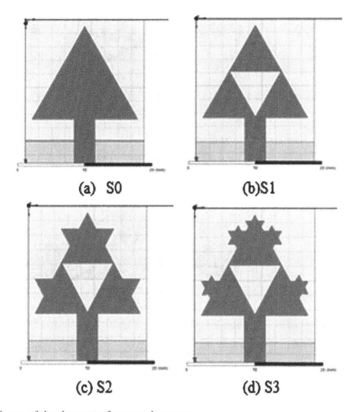

Fig. 1 Stages of development of proposed antenna

A substrate of low dielectric constant is selected to make the proposed antenna meet the bandwidth requirements. Measurement of microstrip feed dimension is done by using transmission line mathematical statement [9, 10] is found for fulfilment of characteristics impedance of 50 Ω (Fig. 2; Table 1).

3 Optimization Using Particle Swarm Optimization Method

Particle-swarm optimization (PSO) become offered by James Kennedy and Russell Eberhart in 1995. It is based on a random probability distribution algorithm inspired by social behavior of bird flocking [11]. When a swarm of bees fly in a garden, they generally attempt to discover a place having large number of flowers [12]. This techniques utilizes various particles constituting a group transferring around within the search space looking for an optimized answer. In this method, each particle is considered as a point in an N-dimensional space. The particles adjust their position

Fig. 2 Geometry of the
proposed antenna

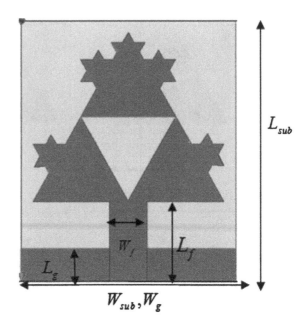

Table 1 Un-optimized
values of the antenna

Length of substrate (L_{sub})	21.45 mm
Width of substrate (W_{sub})	17.89 mm
Height of substrate (H_{sub})	1.6 mm
Feed line width (L_f)	6.9 mm
Feed width (W_f)	3.2 mm
Ground plane length (L_g)	19.7 mm
Ground plane width (W_g)	21.45 mm
Patch length	15.89 mm
Thickness of patch	0.1 mm

according to the present positions and velocities, the distance between the present position and the personal best position (*pbest*), the distance between the present position and the global best position (*gbest*). The velocity expression of the particle is represented as [12]:

$$V(t) = W * V(t-1) + A1 * x1(pbest - S(t-1)) + A2 * x2(gbest - S(t-1))$$

The position of the particle is given by

$$S(t) = S(t-1) + V(t)$$

where V(t) refers to the particle velocity along the 't' dimension and S(t) is the particle co-ordinate in that dimension. W is the internal weight which is taken

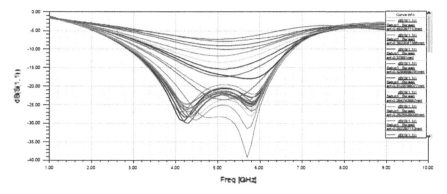

Fig. 3 Impact of ground length variation on return loss characteristics

between 0 and 1. It decides that extent the particle remains in its original course unaffected by the pull of *pbest* and *gbest*. A1 & A2 are the scaling factors that determine the relative 'pull' of *pbest* and *gbest*. A1 decides how lots the particle is prompted by means of ibest and A2 decides how the particle is inspired by way of the relaxation of the particle swarm.

Therefore, it is valuable to optimize using different parameters such as ground plane, feedline width (W_f), middle triangle radius. Here the optimization sweeps were carried out by utilizing PSO and the steps are:

(1) Enter the operating frequency, thickness and dielectric constant of the substrate in patch calculator which is programmed by using MATLAB tool.
(2) Utilizing the outputs length (L) and width (W) of the patch respectively to design the antenna using HFSS.
(3) Inspect the execution of the proposed patch antenna plan as far as return loss for bandwidth calculation.
(4) Check if the return loss is below than −20 dB, then the prospective patch antenna is optimized, else go to Step-3 and repeat these steps.

4 Result and Discussion

The results of proposed antenna has shown from Fig. 4 through Fig. 7 and the comparison for antenna parameter as shown in Table 2 (Figs. 3, 5, 6).

Return Loss: It is an essential parameter of antenna, that's the evaluation inside the middle of ahead and reflected power in dB. The satisfactory estimation of return loss is not exactly −10 dB for the antenna to work proficiently. Below figures demonstrates the simulated dB plot versus frequency plot of Optimized and Un-optimized value of proposed antenna. It shows that the resonant frequency is not exactly at 6 GHz.

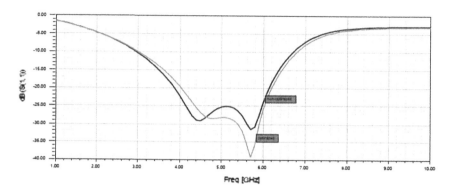

Fig. 4 Impact of optimized and un-optimized ground length variation on return loss characteristics

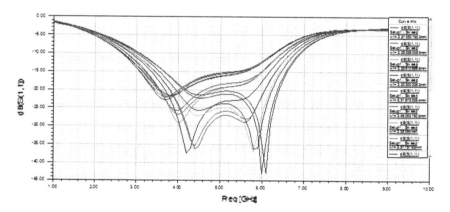

Fig. 5 Impact of feed line width variation on return loss characteristics

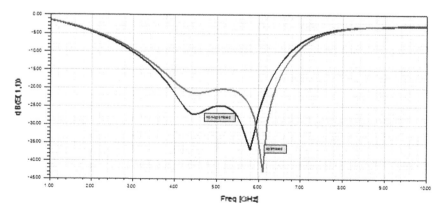

Fig. 6 Impact of optimized and un-optimized feed line width variation on return loss characteristics

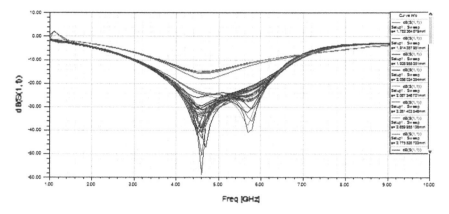

Fig. 7 Impact of middle triangle radius variation on return loss characteristics

Table 2 Comparison of parameters among un-optimized and optimized antenna

Antenna	Ground plane length (mm)	Bandwidth (GHz)	Return loss (dB)
Un-optimized antenna	2.95	3.7028	−31.28
Optimized antenna	2.7	3.7638	−39.2661

4.1   Ground Plane Optimization

The height gap (h_{gap}) between the ground plane and the patch was varied from 0 to 4 mm. measurement of return loss is completed and the impedance bandwidth alteration with feed gap was shown in Fig. 3. It is found that the accomplished bandwidth is maximum when the gap is between 2–3 mm.

4.2   Feed Line Width Optimization

The feed line width of microstrip was varied from 1 to 4 mm. This represents varying the feed line's characteristic impedance. It can be seen that the bandwidth is moderately steady between 3–4 mm, inferring that the bandwidth is not exceptionally delicate to feed line width (Table 3).

4.3   Middle Triangle Radius Optimization

See Fig. 8 and Table 4.

Table 3 Comparison of parameters among un-optimized and optimized antenna

Antenna	Feed width length (mm)	Bandwidth (GHz)	Return loss (dB)
Un-optimized antenna	3.1	3.7725	−36.9495
Optimized antenna	3.4	3.8335	−42.9358

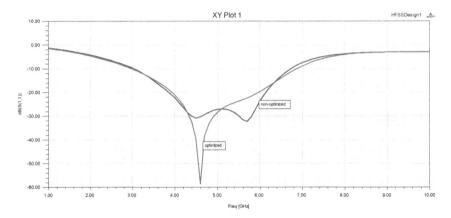

Fig. 8 Impact of optimized and un-optimized middle triangle radius variation on return loss characteristics

Table 4 Comparison of parameters among un-optimized and optimized antenna

Antenna	Middle triangle radius (mm)	Bandwidth (GHz)	Return loss (dB)
Un-optimized antenna	4.557	3.6652	−31.9037
Optimized antenna	2.2814	3.8726	−58.7156

5 Conclusion

While optimizing the antenna parameter, the overlapping problems most often encountered using HFSS. The best possible optimization is done with the calibration and redesigning of the antenna in HFSS simulation tool. Now the results of both the un-optimized antenna as well as the optimized antenna are compared. These results are presented in the tables above. From the table it is observed that the achievement of the optimized antenna using PSO is superior to the Un-optimized in terms of the return loss.

References

1. B.K. Jeemon, K. Shambhavi, Z.C. Alex, Design and analysis of multi-fractal antenna for UWB applications, in *IEEE International Conference on Emerging Trends in Computing*
2. V.P. Meshram, P. Sangare, P. Wanjari, I. Chintawar, Design and fabrication of wide band fractal antenna for commercial applications, in *IEEE International Conference on Machine Intelligence and Research Advancement*, 978-0-7695-5013-8/13 (2013)
3. Z.H. Feng, M.C. Liang, The study of Koch fractal applied to isosceles triangle patch antenna, in *Proceedings of ISAP,* Kaohsiung, Taiwan (2014)
4. S. Dhar, K. Patra, R. Ghatak, B. Gupta, D.R. Poddar, A dielectric resonator loaded minkowski Fractal shaped slot loop hepta band antenna. IEEE Trans. Antenna Propag. (2015)
5. J.S. Sivia, S.S. Bhatia, Design of fractal based microstrip rectangular patch antenna for multiband applications, in *IEEE International Advanced Computing Conference (IACC)* (2015)
6. S.R. Best, A comparison of the resonant properties of small space-filling fractal antennas. IEEE Antennas Wirel. Propag. Lett. **2** (2003)
7. S.R. Best, On the resonant properties of the koch fractal and other wire monopole antennas. IEEE Antennas Wirel. Propag. Lett. **1** (2002)
8. Z.W. Yu, G.M. Wang, X.J. Gao, K. Lu, A novel small size single patch antenna based on Koch and Sierpinski fractal shapes. Progr. Electromagn. Res. Lett. **17**, 95–103 (2010)
9. C.A. Balanis, *Antenna Theory Analysis and Design*, 2nd ed. (Wiley India (P.) Ltd., 2007)
10. R. Garg, P. Bartia, I. Bhal, A microstrip antenna design hand book. Artech House-2001
11. I. Kennedy, R.C. Eberhart, Particle swarm optimization, in *Proceedings of IEEE International Conference on Neural Networks*, vol. IV (IEEE service center, Piscataway, NJ, 1995), pp. 1942–1948
12. Y. Shi, R.C. Eberhart, A modified particle swarm optimizer, in *Proceedings of the IEEE International Conference on Evolutionary Computation* (IEEE Press, Piscataway, NJ, 1998), pp. 69–73

Multi-objective IT Professionals' Utilization Problems Using Fuzzy Goal Programming

R.K. Jana, Manas Kumar Sanyal and Saikat Chakrabarti

Abstract This paper presents a fuzzy goal programming approach to solve IT professionals' utilization problems for software firms. These problems involve multiple objectives and binary decision variables. The fuzzy goal programming approach helps to quantify uncertainness of the objectives of the problem. With the help of membership functions, the problem is converted to its equivalent deterministic form. A case study demonstrates the effectiveness of the approach.

Keywords Multi-objective programming · Fuzzy goal programming · Human resource planning · Information technology

1 Introduction

Human resources (HR) are the most precious resources for any organization. Software firms are no exception to this. The success and failure of software firms greatly depend on proper utilization of information technology (IT) professionals who are highly paid. It is a challenge for the firms to ensure the best utilization of their human resources. Two major factors of utilization of HR in software projects are cost involved in employing professionals and timely completion of the projects.

R.K. Jana (✉)
Indian Institute of Social Welfare & Business Management,
Kolkata 700073, WB, India
e-mail: rkjana1@gmail.com

M.K. Sanyal
University of Kalyani, Kalyani 741235, WB, India
e-mail: manas_sanyal@klyuniv.ac.in

S. Chakrabarti
Institute of Engineering & Management, Ashram Campus,
Kolkata, India
e-mail: studymat.saikat@gmail.com

© Springer Nature Singapore Pte Ltd. 2017   199
S.C. Satapathy et al. (eds.), *Proceedings of the 5th International Conference on Frontiers in Intelligent Computing: Theory and Applications*, Advances in Intelligent Systems and Computing 516, DOI 10.1007/978-981-10-3156-4_20

A technique is presented in [1] for selecting professionals for software projects by utilizing the concept of Taguchi's parameter design. The processes and principles of managing projects related to IT domain are discussed in [2]. The literature reports techniques [3, 4] for optimal utilization of IT professionals' in software projects.

IT professionals' utilization problems consist of multiple and conflicting objectives. There exist various techniques for solving multiple objective problems (MOP) [5]. Goal programming (GP) [6] is one of the most prominent techniques. Over the last few decades, GP has been used for solving MOP problems [7–9]. Surprisingly, GP has not been used much in the field of IT professionals' utilization problems. In [10], a mathematical model is proposed for utilizing IT professionals for software projects. In spite of the simplicity and popularity of GP techniques, they suffer from a major drawback. The goals need to be defined precisely. In a decision making situation, this assumption might not always hold. It may not be easy for the decision maker to come up with a precise target value of the goals. In the year 1980, fuzzy logic [11] was combined with GP and fuzzy goal programming (FGP) [12] was proposed to overcome this drawback. Since then FGP has been applied to solve MOP problems from various domains [13–15]. On the contrary, FGP has been applied for solving multi-objective IT professionals' utilization problems only in few studies [16, 17].

This paper focuses on first proposing a mathematical model of the IT professionals' utilization problems for software firms. To accommodate uncertainty of the decision making process, the objectives are converted to fuzzy goals. Membership functions corresponding to the fuzzy goals are defined to derive the deterministic equivalent model. The Branch-and-Bound solver of LINGO 10 solves the model to obtain the most effective allocation of IT professionals so that the cost is minimized and the effort required is maximized simultaneously.

The paper is organized as follows: FGP model formulation is presented in Sect. 2. The multi-objective IT professionals' utilization model is presented in Sect. 3. The case study is presented in Sect. 4. Results and discussions are presented in Sect. 5. Concluding remarks are presented in Sect. 6.

2 FGP Model Formulation

In this section, the FGP model is presented. The general form of a binary multi-objective programming problem [5] can be stated as follows:

$$
\begin{aligned}
&\text{min: } z_1(x) \\
&\text{min: } z_2(x) \\
&\quad \vdots \\
&\text{min: } z_M(x) \\
&\text{subject to} \quad h_r(x) \le 0, r = 1, 2, \ldots, R \\
&\qquad\qquad\ x = 0 \text{ or } 1.
\end{aligned} \tag{1}
$$

where x is the vector of $n$binary decision variables $(x_1, x_2, \ldots, x_n)$; $z_i(x), (m = 1, 2, \ldots, M)$ are individual objectives, $h_r(x)(r = 1, 2, \ldots, R)$ are constraints.

If the information related to the objectives is not precise in a decision making situation, then FGP comes as a handy tool to model the problem. In the considered problem, there are different sources of uncertainties. As a result, the objectives are defined as fuzzy goals. If b_m be the target value of the m-th objective $z_m(x)$, then the FGP formulation of (1) can be represented as follows:

Find x such that the following goals and constraints are satisfied:

$$
\begin{aligned}
& z_m(x) \gtrsim b_m, m = 1, 2, \ldots, m_1 \\
& z_m(x) \lesssim b_m, m = m_1 + 1, m_1 + 2, \ldots, M \\
\text{subject to} \quad & h_r(x) \leq 0, r = 1, 2, \ldots, R \\
& x = 0 \text{ or } 1.
\end{aligned}
\tag{2}
$$

Membership function [11]: Let $\tilde{A}$ be a fuzzy set, U is the universal set, and $u \in U$ is an element of U. A fuzzy set $\tilde{A}$ in U is defined as:

$$
\tilde{A} = \left\{ \left(u, \mu_{\tilde{A}}(u) \right) \right\} | u \in U
\tag{3}
$$

where $\mu_{\tilde{A}}(u) \rightarrow [0, 1]$ is called the membership function of $\tilde{A}$ and $\mu_{\tilde{A}}(u)$ is the degree of membership in which $u \in U$. Therefore, the fuzzy set $\tilde{A}$ associates each point in U with a nonnegative real number in $[0, 1]$. The value of $\mu_{\tilde{A}}(u)$ at u represents the membership grade of u in U. As the value of $\mu_{\tilde{A}}(u)$ increases, the grade of 'belongingness' of u in $\tilde{A}$ also increases.

The fuzzy goals in (2) are characterized by their membership functions. Depending up on the fuzzy restrictions given to a fuzzy goal, the lower or upper tolerance limits are defined. The membership function corresponding to the fuzzy goal $z_m(x) \gtrsim b_m, (m = 1, 2, \ldots, m_1)$ can be constructed based on the lower tolerance limit t_m^l for the achievement of the target value b_m as follows:

$$
\mu_{z_m}^l(x) = \begin{cases}
1 & \text{if } z_m(x) \geq b_m \\
\frac{z_m(x) - (b_m - t_m^l)}{t_m^l} & \text{if } b_m - t_m^l \leq z_m(x) < b_m \\
0 & \text{if } z_m(x) < b_m - t_m^l
\end{cases}
\tag{4}
$$

Proceeding in a similar manner, the membership function corresponding to the fuzzy goal $z_m(x) \lesssim b_m (m = m_1 + 1, m_1 + 2, \ldots, M)$ can be constructed based on the upper tolerance limit t_m^u for the achievement of the aspiration level b_m as follows:

$$
\mu_{z_m}^u(x) = \begin{cases}
1 & \text{if } z_m(x) \leq b_m \\
\frac{(b_m + t_m^u) - z_m(x)}{t_m^l} & \text{if } b_m < z_m(x) \leq b_m + t_m^u \\
0 & \text{if } z_m(x) > b_m + t_m^u
\end{cases}
\tag{5}
$$

FGP model: The deterministic equivalent form of the FGP problem (2) can be obtained using membership functions corresponding to both types of fuzzy goals given in (4) and (5). Following [18], it can be written as follows:

$$\max: \sum_{m=1}^{m_1} \mu^l_{z_m}(x) + \sum_{m=m_1+1}^{M} \mu^u_{z_m}(x)$$

$$\text{subject to} \quad \mu^l_{z_m}(x) \le \frac{z_m(x) - (b_m - t^l_m)}{t^l_i}, \quad m = 1, 2, \cdots, m_1$$

$$\mu^u_{z_m}(x) \le \frac{(b_m + t^u_m) - z_m(x)}{t^u_m}, m = m_1 + 1, m_1 + 2, \ldots, M \qquad (6)$$

$$h_r(x) \le 0, r = 1, 2, \ldots, R$$

$$x = 0 \text{ or } 1$$

$$0 \le \mu^l_{z_m}(x) \le 1, \quad m = 1, 2, \ldots, m_1$$

$$0 \le \mu^u_{z_m}(x) \le 1, \quad m = m_1 + 1, m_1 + 2, \ldots, M.$$

3 Multi-objective Model for IT Professionals' Utilization

The multi-objective mathematical model for utilization of IT professionals' for the software firms is presented here. The model assumptions are mentioned first. Notations, objectives and constraints of the model are presented next.

3.1 Assumptions

To model is formulated based on the following assumptions

(i) There are three categories of IT professionals—Architect, Developer and Quality Engineer.
(ii) There are five phases of each project—Study, design; development; quality check and maintenance.
(iii) The company successfully bids for all the phases of the projects.
(iv) The projects are fixed time projects, i.e., the time of delivery is fixed beforehand.

3.2 Notations

Index

i Index for professionals
j Index for professionals' category $(j = 1, 2, 3)$
I_j Number of professionals in category j
I Total number of professionals $(I = I_1 + I_2 + I_3)$
k Index for phase in a project $(k = 1, 2, \ldots, 5)$
p Index for projects $(p = 1, 2, \ldots, P)$

Coefficient

c_{ijkp} Cost of employing the professional i of category j in phase k of project p
e_{ijkp} Effort of the professional i of category j in phase k of project p
M_{kp} Minimum number of professionals of specific category required to complete the task in time $(k = 1, 2, \ldots, 5; p = 1, 2, \ldots, P)$
A_{kp} Available number of professionals of appropriate category for completing the task k of project p

Decision Variable

x_{ijkp} Professional i of category j involved in phase k of project p, a binary decision variable

3.3 Objectives of the Model

The objectives of the model are presented next:

Cost Minimization: The objective of the firm is to complete the projects in time by appointing the right professionals in right phases of the right projects so that the total cost is minimized. Therefore, the cost minimization objective can be written as:

$$Min: z_1(x) = \sum_{p=1}^{P} \sum_{k=1}^{5} \sum_{j=1}^{3} \left\{ \sum_{i=1}^{I_j} c_{ijkp} x_{ijkp} \right\}. \tag{7}$$

Effort Maximization: The efforts of the employed professionals must be maximized in order to complete all the projects in time. Therefore, the effort maximization objective can be written as:

$$Max: z_2(x) = \sum_{p=1}^{P} \sum_{k=1}^{5} \sum_{j=1}^{3} \left\{ \sum_{i=1}^{I_j} e_{ijkp} x_{ijkp} \right\}. \tag{8}$$

3.4 Constraints of the Model

Requirement of minimum number of IT professionals: To successfully complete each task of the projects, a minimum number of professionals must be employed. So, the corresponding constraint can be written as:

$$\sum_{i=1}^{I_j} x_{ijkp} \geq M_{kp}, \ k = 1, 2, \ldots, 5; \ p = 1, 2, \ldots, P. \tag{9}$$

Requirement of maximum number of IT professionals: Each task of the projects needs involvement of a maximum number of professionals. So, the corresponding constraint can be written as:

$$\sum_{i=1}^{I_j} x_{ijkp} \leq A_{kp}, \ k = 1, 2, \ldots, 5; \ p = 1, 2, \ldots, P. \tag{10}$$

IT Professionals' involvement: If professional i of category j is not employed in phase k of project p, then the value of the corresponding decision variable is zero. So,

$$x_{ijkp} = 0, \text{ for all } i, j, k, p. \tag{11}$$

If professional i of category j is involved in phase k of project p, then the value of the corresponding decision variable is one. So,

$$x_{ijkp} = 1, \text{ for all } i, j, k, p. \tag{12}$$

4 A Case Study

This case study is taken from [17]. It is developed based on data collected from a software firm located in the Electronic Complex of Salt Lake, Sector—V, Kolkata, India. Apart from the consulting works, the firm develops a variety of in-house software. IT professionals of the firm can be classified into three primary categories. They are architect (C_1), developer (C_2) and quality engineer (C_3). The phases of the projects can also be classified into five primary categories. They are study (P_1), design (P_2); development (P_3); quality check (P_4), and maintenance (P_5). The involvement matrix of three categories of professionals in five phases of projects is shown in the following matrix:

Table 1 Requirement of IT professionals

Project	P_1	P_2	P_3	P_4	P_5
1	1	1	2	1	1
2	1	1	6	2	1
3	1	1	3	2	1
4	1	1	4	2	1

Table 2 Target and tolerance for cost objective

Objective	Target (Rs '00000)	Tolerance (Rs '00000)
Cost	60	5

Table 3 Target and tolerance for effort objective

Objective	Target	Tolerance
Effort	290	10

$$\begin{array}{c} \\ C_1 \\ C_2 \\ C_3 \end{array} \begin{array}{ccccc} P_1 & P_2 & P_3 & P_4 & P_5 \\ \left(\begin{array}{ccccc} 1 & 1 & 0 & 0 & 0 \\ 0 & 0 & 1 & 0 & 1 \\ 0 & 0 & 0 & 1 & 1 \end{array} \right) \end{array}$$

In the involvement matrix, a value '1' indicates that a particular category of IT professional is involved in a specific phase of a project, and a value '0' indicates that a particular category of IT professional is not involved in a specific phase of a project.

There are nearly 50 employees in the software firm. Out of the 50 employees, 28 are IT professionals. The remaining are non-technical staffs. In the technical category, there are 4 are architects, 16 developers, and 08 quality engineers. The firm plans to execute 4 upcoming projects. The requirement of professionals for the upcoming projects is shown in Table 1.

The target and tolerance values of the cost minimization and effort maximization objectives are shown in Tables 2 and 3, respectively.

Due to space and certain other limitations, the remaining data are not shown here.

5 Results and Discussion

Based on the data described in the previous section, the FGP model given in (7) is first formulated. The model becomes a binary integer linear programming problem involving 160 binary decision variables. The problem is solved using LINGO 10 in a Desktop (Processor: Intel® Core™ i3-2120 CPU @ 3.30 GHz; RAM: 4 GB). Due to the presence of large number of binary variables, Branch-and-Bound solver in LINGO is used to find the solution. The obtained solution is global optimal. LINGO finds the solution only at the 15th iteration. The time taken by the solver is less than one second to find the solution. The non-zero variables of the solution are presented in Table 4.

The corresponding assignment of IT professionals is shown in Table 5.

The above assignment of professionals is achieved by spending Rs. 59.8808 lac. The corresponding value of effort is 285.5.

For the purpose of comparing the results, the solution obtained in [17] is presented in Table 6.

Table 7 presents the summary of the present technique and the technique used in [17].

The present technique takes less number of iteration to obtain the results. The computational time is less and uses less memory compared to [17]. Also, the

Table 4 Non-zero decision variables obtained from FGP model

Non-zero decision variables							
X1112	X2114	X2351	X4113	X4353	X7344	X10231	X16232
X1122	X2124	X3111	X4123	X5232	X7354	X11233	
X1232	X2234	X3121	X4232	X6234	X8232	X12232	
X1342	X2341	X3233	X4342	X6343	X8342	X14233	
X1352	X2342	X3344	X4343	X7231	X9234	X15234	

Table 5 Proposed solution

Project	P_1	P_2	P_3	P_4	P_5
1	C_{1-3}	C_{1-3}	C_2-7,10	C_3-2	C_3-2
2	C_{1-1}	C_{1-1}	C_2-1, 4, 5, 8, 12, 16	C_3-1, 2, 4,8	C_3-1
3	C_{1-4}	C_{1-4}	C_2-3, 11, 14	C_3-4, 6	C_3-4
4	C_{1-2}	C_{1-2}	C_2-2, 6, 9, 15	C_3-3, 7	C_3-7

C_j: Employee category j; P_k: Phase of project k

Table 6 Solution obtained in [17]

Project	P_1	P_2	P_3	P_4	P_5
1	C_1-3	C_1-3	C_2-6, 8, 10	C_3-4, 6	C_3-5, 6
2	C_1-1	C_1-1	C_2-4, 12, 13, 14, 15, 16	C_3-7, 8	C_3-3
3	C_1-4	C_1-4	C_2-1, 2, 3	C_3-1, 2	C_3-4
4	C_1-2	C_1-2	C_2-5, 7, 9, 11	C_3-3, 5	C_3-5

C_j: Employee category j; P_k: Phase of project k

Table 7 Comparison of the techniques

Method	Iteration	Run time (s)	Memory used (K)	Cost (lac)
Proposed solution	15	$\ll 1$	84	59.88
Jana et al. (2016b)	2591	1	104	60

solution suggests that the cost of executing all the projects can be reduced using the present technique. As a result, the present technique is more suitable than technique presented in [17].

6 Conclusions

This paper develops a multi-objective model of IT professionals' utilization for software firms. The problem is very challenging as proper utilization of HR is the key to success for any firms. The problem involves multiple objectives. Also, there are many binary decision variables in the problem. The technique used provides global optimal solution to the problem. The technique is compared with a binary FGP technique. It is observed that the technique used in this paper is better than that used in [17]. This technique may be used to solve HR management problems of similar types from the other industries and domains.

References

1. H.-T. Tsai, H. Moskowitz, L.-H. Lee, Human resource selection for software development projects using taguchi's parameter design. Eur. J. Oper. Res. **151**, 167–180 (2003)
2. M. Cotterall, B. Hughes, *Software Project Management*, 4th edn. (McGraw Hill Education, Berkshire, UK, 2006)
3. J. Xiao, Q. Wang, M. Li, Y. Yang, F. Zhang, L. Xie, Constraint-driven human resource scheduling method in software development and maintenance process, in *Proceedings of IEEE International Conference on Software Maintenance*, (IEEE, 2008), pp. 17–26. doi:10. 1109/ICSM.2008.4658050
4. L. Zhou, A Project Human Resource Allocation Method Based on Software Architecture and Social Network, in *Proceedings of 4th International Conference on Wireless Communications, Networking and Mobile Computing*, (IEEE, 2008), pp. 1–6. doi:10.1109/WiCom.2008. 174910.1109/ WiCom.2008.1749
5. Y.-J. Lai, C.-L. Hwang, *Fuzzy Multiple Objective Decision Making: Methods and Applications* (Springer, Berlin, 1994)
6. A. Charnes, W.W. Cooper, *Management Models and Industrial Applications of Linear Programming* (Wiley, New York, 1961)
7. D. Jones, M. Tamiz, *Practical Goal Programming* (Springer, New York, 2010)
8. S.M. Lee, *Goal Programming for Decision Analysis* (Auerbach Publishers, Philadelphia, 1972)
9. W.T. Lin, A survey of goal programming applications. Omega **8**, 115–117 (1980)
10. R.K. Jana, S. Chakrabarti, A goal programming approach to human resource planning for software projects, Emerging Challenges for HR: VUCA Perspective, ed. by U.K. Bamel, A. Sengupta, P. Singh (Emerald, 2016), pp. 151–158 (ISBN: 978-09-9268-008-4)
11. L.A. Zadeh, Fuzzy sets. Inf. Control **8**, 338–353 (1965)
12. R. Narasimhan, Goal programming in a fuzzy environment. Decis. Sci. **11**, 325–336 (1980)
13. R.K. Jana, Dinesh K. Sharma, Daniel I. Okunbor, Nutrient management system for sustainable sugarcane production—An application of fuzzy goal programming and decision theoretic approach. Int. J. Appl. Manag. Sci. **8**(3), 230–245 (2016)

14. H.P. Sharma, D.K. Sharma, R.K. Jana, A varying domain fuzzy goal programming approach to mutual fund portfolio selection. Int. J. Finance **22**(2), 6394–6412 (2010)
15. D.K. Sharma, R.K. Jana, A. Gaur, Fuzzy goal programming for agricultural land allocation problems. Yugosl. J. Oper. Res. **17**(1), 31–42 (2007)
16. I. Giannikos, P. Polychroniou, A fuzzy goal programming model for task allocation in teamwork, Int. J. H.R. Dev. Manag. **9**(1), 97–115 (2009)
17. R.K. Jana, M.K. Sanyal, Saikat Chakrabarti, Binary fuzzy goal programming for effective utilization of IT professionals. Adv. Intell. Syst. Comput. **458**, 469–480 (2016)
18. R.N. Tiwari, S. Dharmar, J.R. Rao, Fuzzy goal programming—an additive model. Fuzzy Sets Syst. **24**, 27–34 (1987)

Convex Hyperspectral Unmixing Algorithm Using Parameterized Non-convex Penalty Function

K. HariKumar and K.P. Soman

Abstract Unmixing of hyperspectral data is an area of major research because the information it provides is utilized in plethora of fields. The year of 2006 witnessed the emergence of Compressed Sensing algorithm which was later used to spearhead research in umixing problems. Later, the notion of ℓ_p norms $0 < p < 1$ and other non-smooth and non-convex penalty function were used in place of the traditional convex ℓ_1 penalty. Dealing with optimization problems with non-convex objective function is rather difficult as most methodologies often get stuck at local optima. In this paper, a parameterised non-convex penalty function is used to induce sparsity in the unknown.The parameters of penalty function can be adjusted so as to make the objective function convex, thus resulting in the possibility of finding a global optimal solution. Here ADMM algorithm is utilized to arrive at the final iterative algorithm for the unmixing problem. The algorithm is tested on synthetic data set, generated from the spectral library provided by US geological survey. Different parametric penalty functions like log and arctan are used in the algorithm and is compared with the traditional ℓ_1 penalties, in terms of the performance measures RSNR and PoS. It was observed that the non-convex penalty functions out-performs the ℓ_1 penalty in terms of the aforementioned measures.

Keywords Unmixing · Linear model · Parameterised non-convex penalties · ADMM

K. HariKumar (✉)
Amrita School of Engineering, Centre for Computational Engineering
and Networking (CEN), Coimbatore, India
e-mail: hari.kumar748@gmail.com

K.P. Soman
Amrita Vishwa Vidyapeetham, Amrita University, Coimbatore, India
e-mail: kp_soman@amrita.edu

© Springer Nature Singapore Pte Ltd. 2017
S.C. Satapathy et al. (eds.), *Proceedings of the 5th International Conference on Frontiers in Intelligent Computing: Theory and Applications*, Advances in Intelligent Systems and Computing 516, DOI 10.1007/978-981-10-3156-4_21

209

1 Introduction

Most materials on Earth can be uniquely characterised by their spectral reflectance curves and serves as their identity [1], but in many cases the spectral measurements may contain information pertaining to more than one material. This may be due to the low spatial resolution of the measuring device, or the substance inherently being a mixture of two or more materials. The notion of hyperspectral unmixing addresses this issue and aims at identifying the materials contributing to the mixed pixel [1] and serves a myriad of applications like high precision farming [2], mapping of soil burn severity after a fire [3] and in military for target detection [4]. Till date, there are many algorithms designed to carry out the unmixing process and the crux of all is presented in the survey paper [5]. This paper discuss linear mixing model where the resultant mixed pixel is a linear combination of all the spectral signature with appropriate weights called the abundance vector. For all practical scenarios, the abundance vector should be sparse, as too many materials are unlikely to contribute the mixed pixel [5]. This closely resembles the famous compressed sensing problem [6] and many algorithms takes this approach to arrive at sparse solution to the set of linear equations. Other constrains imposed on the abundance vector is that its elements should be nonnegative and should sum up to 1. Compressed sensing regime asks for the solution that minimizes ℓ_0 for recovery and papers [7, 8] discusses its approximation via ℓ_1 minimization. The approximation of ℓ_0 norm using the concept of ℓ_p norm with $0 < p \leq 1$, inspired form the works in [9], is used in [10] to perform hyperspectral unmixing. In [11], an arctan penalty function is used to drive the abundance vector towards a more sparser solution compared to the traditional ℓ_1 penalties. The arctan function being non-convex, there is a high chance that the algorithm gets stuck in a local optima. This paper introduces a new parameter into the algorithm which changes its dynamics in such a way that, with slow variation of this supplementary parameter, the solution is taken through a set of local optima to global optima.

In this paper, parameterised non-convex penalty functions introduced in [12] are used to achieve sparsity in the solution. The ADMM framework is utilized to arrive at an algorithm for the formulated optimization problem, inclusive of the ANS and ANC conditions [11]. The objective function in the ADMM update step, featuring the non-convex penalty function, is made convex by properly adjusting the parameters of penalty function. The algorithm is tested on simulated data from a standard dataset, in the presence of AWGN noise. Reconstruction signal to noise ratio (RSNR) and probability of success (POS) [11] are used to measure the algorithm performance.

2 Problem Definition

Let $y \in \mathbb{R}^M$ be an arbitrary pixel vector from a hyperspectral image of a particular locality. Assuming a linear model, the vector y can be expressed as

$$y = Lx + y_n \tag{1}$$

where the matrix $L = [L_1 L_2 \ldots L_N]$ known as the spectral library contains spectral signatures of N materials, pertaining to the region, along its columns as $L_i \in \mathbb{R}^M$ where $i \in Z_N = \{1, 2, \ldots, N\}$ and $y_n \in \mathbb{R}^M$ models noise from the sensors. $x \in \mathbb{R}^N$ is the unknown abundance vector which has to be determined from (1) such that $x = [x_1 x_2 \ldots x_N]^T$ is sparse. Additionally there are two more constraints [11] that are imposed on the abundance vector x, the first being the non-negativity constrain $x(i) > 0$ for all $i \in Z_N$ and the second is a constraint on the sum of the abundance values $e^T x = 1$ where $e = [11 \ldots 1]_{1 \times N}^T$. The following requirement can be translated to an optimization framework and is formulated as

$$\begin{aligned} \min_{x \in \mathbb{R}^N} \quad & \|y - Lx\|_2^2 + \sum_{i=1}^N \phi(x(i)) \\ s.t \quad & e^T x = 1 \\ & x \geq 0 \end{aligned} \tag{2}$$

where $\phi : \mathbb{R} \to \mathbb{R}$ is a penalty function which induces sparsity to a vector the x. The nature of ϕ discussed in this article follows from papers [12, 13], where all its properties are clearly stated. They are basically parameterised non-convex functions which are cleverly designed in such a way that it combines with an appropriate convex quadratic term to become convex as a whole [12].

3   ADMM Algorithm

Alternate Direction Method of Multipliers, presents a generic iterative procedure to solve a large class of optimization problems [14]. Initially, it introduces a surrogate variable $z \in \mathbb{R}^N$, apart from x, and doubles the number of unknown. It then splits the main objective function into two $f_1, f_2 : \mathbb{R}^N \to \mathbb{R}$, which are preferably convex, and optimizes those separately in an iterative fashion, connected together by a linear constraint on x and z namely $A_{xz} [x^T z^T]^T = C_{xz}$ where $A_{xz} \in \mathbb{R}^{P \times 2N}$ and $C_{xz} \in \mathbb{R}^P$ and $P \in \mathbb{N}^+$ [11, 14]. To solve for the formulation (2), domain constraint functions $f_1(x) = \{\|y - Lx\|_2^2 : x \in \mathbb{R}, e^T x = 0\}$ and $f_2(x) = \left\{ \sum_{i=1}^N \phi(x(i)) : x \in \mathbb{R}, x \geq 0 \right\}$ are introduced. The formulation conditioned for applying ADMM algorithm and its corresponding lagrangian along with the augmented lagrangian term are

$$\begin{aligned} \min_{x, z \in \mathbb{R}^N} \quad & f_1(x) + f_2(z) \\ s.t \quad & x - z = 0, \end{aligned} \tag{3}$$

$$l(x, z, u) = f_1(x) + f_2(z) + \frac{\mu}{2} \|x - z - u\|_2^2. \tag{4}$$

where $\mu \in \mathbb{R}$ is a constant and $u \in \mathbb{R}^N$ is the lagrangian multiplier. The solution is obtained in an iterative procedure [14] as shown in Algorithm 1.

3.1 Solving for x^{j+1} and z^{j+1}

The expansion of formulation in step 3, of the Algorithm 1 is given by

$$
\begin{array}{c}
\min_{x \in \mathbb{R}^N} \ \|Lx - y\|_2^2 + \frac{\mu}{2}\|x - z^j - u^j\|_2^2 \\
s.t \quad e^T x = 1
\end{array}
\tag{5}
$$

Algorithm 1 Generic ADMM Solution

1: Initialize u^1, z^1 and j = 1
2: **repeat**
3:     $x^{j+1} \leftarrow \arg\min_{x \in \mathbb{R}^N} l(x, z^j, u^j)$
4:     $z^{j+1} \leftarrow \arg\min_{z \in \mathbb{R}_+^N} l(x^{j+1}, z, u^j)$
5:     $u^{j+1} \leftarrow u^j - (x^{j+1} - z^{j+1})$
6:     $j \leftarrow j + 1$
7: **until** Convergence

The solution for formulation (5) is discussed in [11] and is expressed as

$$
x^{j+1} = A^{-1}B - A^{-1}e\left(e^T A^{-1}e\right)^{-1}\left(e^T A^{-1}B - 1\right)
\tag{6}
$$

where $A = (L^T L + \mu I)$ and $B = (L^T y + \mu(z^j + u^j))$.

The objective function for z update in ADMM algorithm step 4 consists of a non-convex function ϕ. As per papers [12, 13], the objective function can be made convex by setting the parameter of the penalty function to be in range $[0, \frac{1}{\lambda}]$. Majorizaton-minimisation (MM) method [15] is used to determine that z_s^{j+1} which solves formulation in step 4, of Algorithm 1, whose ϕ is replaced with majorizer $g(z, c) = \frac{2\phi'(c)}{c}z^2 + b(c)$ [12]. The vector z_s^{j+1} comes out as

$$
z_s^{j+1} = \frac{x^{j+1} - u^j}{1 + \frac{\lambda}{\mu}\frac{\phi'(c)}{c}}.
\tag{7}
$$

For simplicity the parameter c can be replaced with the previous value of z, that is z^j as mentioned in [11, 13]. The constraint mentioned in the definition of function f_2, is applied by projecting the z_s^{j+1} onto the set $\{z : z \in \mathbb{R}^N, z > 0\}$. The final expression for the z update is of the form $\left(z_s^{j+1}\right)+$ where the $(.)+$ represents a function that will put 0 in place of any negative element in the vector [11]. The whole ADMM algorithm pans out in the form shown in Algorithm 2.

Algorithm 2 ADMM Algorithm for Unmixing

1: $u^1, z^1 \leftarrow e/N$
2: **repeat**
3:      $x^{j+1} \leftarrow A^{-1}B - A^{-1}e \left(e^T A^{-1} e\right)^{-1} \left(e^T A^{-1} B - 1\right)$
4:      $z^{j+1} \leftarrow \left(\dfrac{x^{j+1} - u^j}{1 + \frac{\lambda}{\mu} \frac{\phi'(z^j)}{z^j}}\right) +$
5:      $u^{j+1} \leftarrow u^j - (x^{j+1} - z^{j+1})$
6:      $j \leftarrow j + 1$
7: **until** Convergence

---

4 Experiment and Observation

Experiments with Algorithm 2, conducted in this paper, follows from the ones conducted in [11]. The data for this purpose is created from a standard spectral library provided by the United States Geological Survey (USGS) [16]. The library contains spectral signatures corresponding to 1365 different materials in $M = 224$ bands which ranges from 0.4 to 2.5 μm. As per [11], a set of $N = 240$ spectral signatures are chosen from the original spectral library to form the library matrix L in such a way that

$$\arccos_{\forall i,j \in Z_N, i \neq j} \left(\frac{L_i L_j}{\|L_j\|_2 \|L_j\|_2}\right) \geq 4.4 \frac{pi}{180}. \tag{8}$$

The pixel vector y is generated by first choosing a $K \in Z_{10} = \{1, 2, \ldots, 10\}$ sparse abundance vector x. Let the positions of non-zero entries of x be in the set $Z_P \subset Z_N$ such $|Z_P| = K$, where $|.|$ represents the cardinality of a set. The abundance vector x, satisfying the Dirichlets distribution [11], is defined as

$$x(k) = \frac{\log \left(c_k\right)}{\sum\limits_{j=1}^{N} \log \left(c_j\right)}, \tag{9}$$

where c_k follows a uniform distribution in the range $[0, 1]$ when $k \in Z_P$ and 0 when $k \in Z_N \setminus Z_P$. The pixel vector y is generated using Eq. (1) where $y_n \in \mathbb{R}^M$ is the AWGN noise from the sensors. $T \in N^+$ fractional abundance vectors, all having same sparsity K, along with their corresponding pixel vectors are generated and

Table 1 Table showing different penalty functions $\phi(d)$ and their corresponding derivatives $\phi'(d)$ used the algorithm, where $d \in \mathbb{R}$

Name	$\phi(d)$	$\phi'(d)$
ϕ_{ℓ_1}	$\lvert d \rvert$	$\frac{\lvert d \rvert}{d}$
ϕ_{log}	$\frac{1}{a}\log(1 + a\lvert d\rvert)$	$\frac{1}{1+a\lvert d\rvert}$
$\phi_{\arctan}$	$\frac{2}{a\sqrt{3}}\tan^{-1}\left(\frac{1+2a\lvert d\rvert}{\sqrt{3}} - \frac{\pi}{6}\right)$	$\frac{1}{1+a\lvert d\rvert+a^2\lvert d\rvert^2}$

Algorithm 2 is used to find the reconstructed abundance fraction $\hat{x}$. The performance metrics used to evaluate the algorithm are

$$RSNR = \frac{E(\|x\|_2^2)}{E(\|x - \hat{x}\|_2^2)}, \tag{10}$$

$$PoS = Pr\left(\frac{\|x - \hat{x}\|_2}{\|x\|_2} \leq \zeta\right) \tag{11}$$

where $Pr(.)$ represents the probability of the statement [11]. The experiment is repeated for all $K \in Z_{10}$ and for different penalty functions ϕ shown in Table 1. The algorithm is also tested under noise levels of SNR 15 and 50 dB as show in Fig. 1, were SNR is defined as

$$SNR = 10\log_{10}\left(\frac{x^T L^T L x}{y_n^T y_n}\right) dB. \tag{12}$$

Experiments are conducted over data set consisting of $T = 2500$ synthetically generated pixel vectors. The ADMM algorithm is run for 100 iterations with the following initialization: $x^1 \leftarrow \frac{e}{N}, u^1 \leftarrow \frac{e}{N}, z^1 \leftarrow \frac{e}{N}, \mu \leftarrow 10, a \leftarrow 0.9\frac{1}{\lambda}$. The parameter λ is given according to the experiment conducted.

Figure 1 shows that the newly introduced penalty function outperforms the classical ℓ_1 penalty in some cases. A study of how the RSNR and PoS varies with respect to λ, for arctan penalty function, on a data set with fixed SNR, is shown in Fig. 2. It can be seen that as λ increases beyond 3 the performance drops considerably and $\lambda < 3$ gives better results for sparsity values beyond 3 compared to $\lambda = 3$. Figure 3 shows how performance varies when SNR of the data set is varied for arctan penalty function with constant $\lambda = 3$.

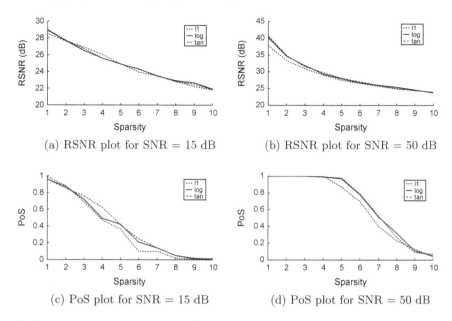

Fig. 1 Graphs representing of RSNR and Probability of Success (PoS) against sparsity for different values of SNRs at $\lambda = 3$ and $\mu = 10$

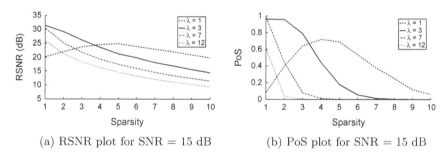

Fig. 2 Graphs representing RSNR and Probability of Success (PoS) against sparsity considering a constant $SNR = 15$ dB and $\mu = 10$ for different values of λ with 'tan' penalty function

5 Conclusion

This paper presents a convex way of formulating hyperspectral unmixing problem using parameterised non-convex penalty functions. The convexity is brought about by adding a convex quadric term with the parametric non-convex function and setting an apt value for the parameter. The algorithm for unmixing is then derived with the help of ADMM framework. Experimental analysis of the algorithm with different penalty functions on synthetically generated data sets suggests heightened performance compared with the traditional ℓ_1 penalty with respect to the performance

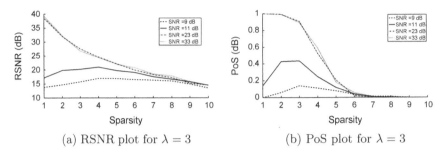

(a) RSNR plot for $\lambda = 3$ (b) PoS plot for $\lambda = 3$

Fig. 3 Graphs representing RSNR and Probability of Success (PoS) against sparsity for constant $\lambda = 3$ and $\mu = 10$ for different SNRs with 'tan' penalty function

measures RSNR and PoS. It was observed that the sparsity of the abundance vector and parameter λ are related, and choosing the apt λ in case of unknown sparsity is yet to be explored.

Acknowledgements The authors would like to thank Rohith Mohan and Jocelyn Babu M., MTech students of CEN department, Amrita University, for their assistance with coding in Matlab and also for improving and editing the work and manuscript. The authors would like to thank all the students and staff of CEN department who have contributed to build up this ideology.

References

1. J.B. Campbell, R.H. Wynne, *Introduction to Remote Sensing* (Guilford Press, 2011)
2. H. McNairn, J. Deguise, J. Secker, J. Shang, Development of remote sensing image products for use in precision farming, in *Submitted to Third European Conference on Precision Farming, Montpellier, France* (2001), pp. 18–20
3. P.R. Robichaud, S.A. Lewis, D.Y. Laes, A.T. Hudak, R.F. Kokaly, J.A. Zamudio, Postfire soil burn severity mapping with hyperspectral image unmixing. Remote Sens. Environ. **108**(4), 467–480 (2007)
4. D. Manolakis, C. Siracusa, G. Shaw, Hyperspectral subpixel target detection using the linear mixing model. IEEE Trans. Geosci. Remote Sens. **39**(7), 1392–1409 (2001)
5. N. Keshava, A survey of spectral unmixing algorithms. Lincoln Lab. J. **14**(1), 55–78 (2003)
6. D.L. Donoho, Compressed sensing. IEEE Trans. Inf. Theory **52**(4), 1289–1306 (2006)
7. M.-D. Iordache, J.M. Bioucas-Dias, A. Plaza, Sparse unmixing of hyperspectral data. IEEE Trans. Geosci. Remote Sens. **49**(6), 2014–2039 (2011)
8. J.M. Bioucas-Dias, M.A. Figueiredo, Alternating direction algorithms for constrained sparse regression: application to hyperspectral unmixing, in *2010 2nd Workshop on Hyperspectral Image and Signal Processing: evolution in Remote Sensing (WHISPERS)*. IEEE (2010), pp. 1–4
9. R. Chartrand, Exact reconstruction of sparse signals via nonconvex minimization. IEEE Signal Proces. Lett. **14**(10), 707–710 (2007)
10. F. Chen, Y. Zhang, Sparse hyperspectral unmixing based on constrained lp-l 2 optimization. IEEE Geosci. Remote Sens. Lett. **10**(5), 1142–1146 (2013)
11. Y. Esmaeili Salehani, S. Gazor, I.-M. Kim, S. Yousefi, 0-norm sparse hyperspectral unmixing using arctan smoothing. Remote Sens. **8**(3), 187 (2016)

12. P.-Y. Chen, I.W. Selesnick, Group-sparse signal denoising: non-convex regularization, convex optimization. IEEE Trans. Signal Process. **62**(13), 3464–3478 (2014)
13. W. He, Y. Ding, Y. Zi, I.W. Selesnick, Sparsity-based algorithm for detecting faults in rotating machines. Mech. Syst. Signal Proces. **72**, 46–64 (2016)
14. S. Boyd, N. Parikh, E. Chu, B. Peleato, J. Eckstein, Distributed optimization and statistical learning via the alternating direction method of multipliers. Found. Trends Mach Learn **3**(1), 1–122 (2011)
15. K. Soman, R. Ramanathan, Digital signal and image processing-the sparse way (Isa Publication, 2012)
16. R.N. Clark, G.A. Swayze, R. Wise, K.E. Livo, T.M. Hoefen, R.F. Kokaly, S.J. Sutley, USGS digital spectral library splib06a (2007)

Novel Techniques for Detection of Anomalies in Brain MR Images

K. Bhima and A. Jagan

Abstract With the significant growth in the field of medical imaging, the analysis of brain MR images is constantly evolving and challenging filed. MR Images are widely used for medical diagnosis and in numerous clinical applications. In brain MR Image study, image segmentation is mostly used for determining and visualizing the brain's anatomical structures. The parallel research results articulated the enhancement in brain MR image segmentation by combining varied methods and techniques. Yet the precise results are not been proposed and established in the comparable researches. Thus, this work presents an analysis of accuracy for brain disorder detection using most accepted Watershed and Expectation Maximization-Gaussian Mixture Method. The bilateral filter is employed to the Watershed and Expectation Maximization-Gaussian Mixture Method to improve the image edges for better segmentation and detection of brain anomalies in MR images. The comparative performance of the Watershed and EM-GM method is also been demonstrated with the help of multiple MR image datasets.

Keywords Brain MR images · T1 images · Watershed method · EM-GM method · Bilateral filter

1 Introduction

With the enormous growths in the brain imaging technique have explored the extended possibilities of brain anatomy analysis based on MR images. The MR image presents the noninvasive extensive visualization of inner anatomical arrangement of the brain. In order to improve the accuracy of detection for brain anatomy, the imaging techniques have extended the quality of the images. Thus the

K. Bhima (✉) · A. Jagan
B.V. Raju Institute of Technology, Narsapur, Telangana, India
e-mail: bhima.mnnit@gmail.com

A. Jagan
e-mail: jagan.amgoth@bvrit.ac.in

© Springer Nature Singapore Pte Ltd. 2017
S.C. Satapathy et al. (eds.), *Proceedings of the 5th International Conference on Frontiers in Intelligent Computing: Theory and Applications*, Advances in Intelligent Systems and Computing 516, DOI 10.1007/978-981-10-3156-4_22

analysis of this complex and high quality images became the most tedious task for the technicians [1]. Moreover, due to the human intervention the investigations are bound to be erroneous. Also these manual analyses are often time-consuming and limited in finding difficulties in brain data analysis compared to the computerized methods for anomalies detection [1]. The broadly used method for investigation of medical images is segmentation based imaging in clinical analysis. Image segmentation is generally used for measuring and visualizing the brain's anatomical structures, for analyzing brain changes, for delineating pathological regions, and for surgical planning and image-guided interventions. The limitations identified from the study demonstrates various segmentation techniques [2] are restricted in generating high accuracy and mostly focused in brain tumor detection. The recent researches also fail to achieve the unsurpassed accuracy [3]. MR Image is mainly used for analysis and detection of brain disorder. In MR images have diverse series such as T1 image and T2 image Fig. 1. This work presents T1 Image [4], In T1 image the time those protons within a tissue necessitate to the appearance to the original magnetization state, which is particular by the static magnetic field. The common T1 image provides superior anatomical details than T2 images. In this work we analysis the accuracy of major accepted methods such as watershed and expectation maximization of Gaussian mixture are normally used for brain MR image segmentation and detection of brain disorder.

Henceforth the rest of the work is organized with the focus to demonstrate the enhancement in accuracy of disorder detection for T1 Images and this work presents the comparative performance of the watershed and expectation maximization of Gaussian mixture segmentation techniques with a Bilateral filtering to improve the accuracy of the brain disorder detection.

The rest of the paper is organized as follows, in Sect. 2 this paper discuss about the Presented Techniques for Brain Anomalies Detection using Watershed and EM-GM method, in Sect. 3 Presents the comparative performance of the Watershed and EM-GM method tested on multiple MRI datasets and in Sect. 4 presents the conclusions.

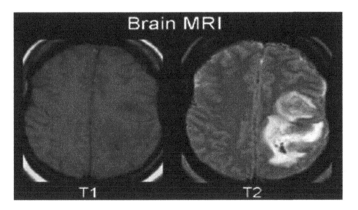

Fig. 1 T1 and T2 types of brain MR image

2 Presented Techniques for Brain Anomalies Detection

The main purpose behind this work is to the find the relative performance of watershed and expectation maximization of Gaussian mixture segmentation for the detection of brain anomalies in MR Images. The magnetic resonance techniques for producing the visual depiction of brain images result in two different types of images such as T1 and T2 image. The studies exhibits the accuracy of T1 images is elevated for detecting the anomalies [4]. Therefore in this work we focus on T1 images to carry out the presented method. The presented methods of Watershed and Expectation Maximization-Gaussian Mixture are equipped with the bilateral filter for the better segmentation of MR images and for detection of brain anomalies in brain. Bilateral filtering is most popular techniques used to smooth MR images though preserving edges. The applications of bilateral filter has grown-up drastically and it's now used for MR image denoising, MR image enhancement etc. The bilateral filter is frequently used for improving the input image variance and standard deviation [5]. The core framework of presented approach has been demonstrated in Fig. 1.

Proposed Method for Brain Anomaly Detection:

- Bilateral Filter is applied to MR Image to smooth regions.
- Watershed method is applied on MR Image dataset for detection of the anomalous regions.
- EM-GM Method is applied on Brain MR Image dataset for detection of the anomalous regions.
- Compute the accuracy for the detection of the anomalous region using Watershed and EM-GM Method.
- Evaluate the performance of the Watershed and EM-GM method on MR Image dataset.

2.1 Watershed Algorithm

The watershed [4, 6–9] method is described as morphological gradient based segmentation for this work and the minimal watershed method is illustrated in Fig. 2. The objective of the watershed algorithm [1, 3, 5, 8] is to improve the accuracy of the image segmentation. A markers [3, 5, 6] are connected component belong to an image, Marker overcome the over-segmentation difficulty in the watershed method and markers consist of the internal markers and external markers, the internal marker related among objects of interest and the external markers related among the background.

The marker usually consists of two steps that is preprocessing and find the condition those markers essential to persuade and detection of anomalies in Brain MR Image using Marker based watershed method.

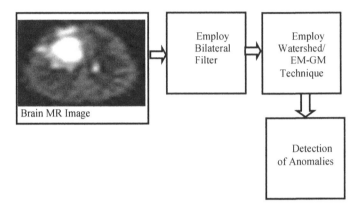

Fig. 2 Presented Framework for brain anomaly detection

- Employ the internal markers to acquire watershed lines of the gradient of the Brain MR image to be segmented.
- Employ the obtained watershed lines as external markers.
- Each region defined by the external markers contains a single internal marker and part of the background
- The problem is reduced to partitioning each region into two parts i.e. object such as containing internal markers and a single background such as containing external markers.

The brief discuss about watershed technique basic notations and algorithms are employed and the various universal parameters of watershed algorithm are defined below.

- Let RM_1, RM_2, RM_3…, RM_a are represents the set of coordinates in the regional minima of image f(i, j).
- The Coordinate catchment basin CB (RM_x) is associated with regional minimum RM_x.
- The set of coordinates (S, Y) for which f(S, Y) < N is defined as

$$Y[N] = \{(S, Y)|f(S, Y) < N\} \tag{1}$$

Hence the procedure of watershed method is illustrated below.

(a) Determine the Min and Max pixel value of f(i, j) and Min coordinate is assigned to RM_x. The topography will be flooded in integer flood increments to N = Min + 1. The coordinate CB(RM_x) in the catchment basin associated with Min RM_x that are flooded at point N.

(b) Calculate

$$CB_N(RM_x) = CB(RM_X) \bigcap Y[N] \qquad (2)$$

(c) If $(i, j) \in CB(RM_x)$ and $(i, j) \in Y[N]$, $CB_n(RM_x) = 1$
at location (i, j); otherwise $CB_N(RM_i) = 0$.

(d) Let CB[N] represents the union of the flooded catchment basin at point N.

$$CB[N] = \bigcup_{x=1}^{a} CB_N(RM_x) \qquad (3)$$

Put N = N + 1.

(e) Obtain the set of connected components in Y[N] defined as CM. Here three
situations for every connected component $cm \in CM[N]$.

   i. If $cm \bigcap CB[N-1]$ is empty and cm is connected component added to
CB[N − 1] to form CB[N] since it define new minimum is come across.

   ii. If $cm \bigcap CB[N-1]$ holds one connected component of CB[N − 1] to
form CB[N], cm connected component is added into CB[N − 1] to form
CB[N] since it means cm lies within the catchment basin of some
regional minimum.

   iii. If $cm \bigcap CB[N-1]$ holds more than one connected component of CB
[N − 1], it represents all or part of a ridge separating two or catchment
Basin is encountered to find the points of ridges and presented as dam.

(f) Create CB [N] as per to Eqs. (2) and (3). Set N = N+1.

(g) Repeat step 3 and 4 until N reaches Max + 1.

Consequently with the Min and Max pixel value and edges presently the
watershed method can recursively recognize the segments.

2.2 Expectation Maximization and Gaussian Mixture (EM-GM) Method

The Expectation maximization (EM) method [10, 11] is an iterative two phase
method mostly used to locate maximum likelihood factor of a given statistical
model and the name of the method is specified by the names of the two phases such
as Expectation and Maximization. The EM method is representation of distribution
of intensities by mixture of weighted Gaussians. The EM method is an iterative
process used to estimate the maximum-likelihood when the observations are
incomplete and EM method has two phases such as Expectation (E) and Maximization (M), these phases are carries out iteratively till the end results converge.
The as Expectation (E) phase estimates an expectation of the likelihood by
including the latent variables as if they were observed and a Maximization
(M) phase, which calculate the maximum likelihood estimates of the parameters by

maximizing the expected likelihood found on the previous Expectation (E) phase. The parameters found on the Maximization (M) phase are subsequently used to initiate a further Expectation (E) phase and the procedure is repeated until convergence.

The Gaussian mixture method [10, 11] for each pixel value for input image is described as:

$$GMM(p) = GND(p|\mu, \Sigma) \tag{4}$$

where GMM denotes the Gaussian mixture method, GND denotes the Gaussian normal distribution, μ denotes the mean and Σ denotes the variance. The inclusion of the multiple neighbouring pixels will result in

$$GMM(p) = GND(p|\mu_1, \Sigma_1) + GND(p|\mu_2, \Sigma_2) \tag{5}$$

Henceforth, the applicability of the mixing coefficient will result in

$$GMM(p) = \pi_1 GND(p|\mu_1, \Sigma_1) + \pi_2 GND(p|\mu_2, \Sigma_2) \tag{6}$$

Therefore the Gaussian mixture for the entire image will result in

$$GMM(p) = \sum_{x=1}^{N} \pi_x GND(p|\mu_x, \Sigma_x) \tag{7}$$

where N indicates the total number of pixels in the input image, subsequent to the application of Gaussian mixture method, the expectation maximization needs to be applied Expectation step as score for each pixel:

$$p: \gamma_x \tag{8}$$

Then, the Gaussian parameters to be mapped into the score point:

$$\pi_x, \mu_x, \Sigma_x \rightarrow \gamma_x \tag{9}$$

Lastly the likelihood to be calculated to converge.

3 Results and Discussions

In order to demonstrate the results and theoretical construction presented in this work provides the MATLAB implementation of the framework to test the visual advantages of Watershed and EM-GM Method for detection of brain disorder in MR Images. The presented application has been tested for ten dissimilar dataset's of brain MR image and furnishes better results with proposed approach.

Table 1 The comparative performance of the watershed and EM-GM method

MR image dataset	Input file name	Truth file name	Accuracy (%)	
			EM-GM method	Watershed method
Dataset 1	MRI01_T1	MRI01_truth	96.87	98.13
Dataset 2	MRI02_T1	MRI02_truth	98.12	98.02
Dataset 3	MRI03_T1	MRI03_truth	92.53	98.00
Dataset 4	MRI04_T1	MRI04_truth	91.67	98.07
Dataset 5	MRI05_T1	MRI05_truth	95.63	95.36
Dataset 6	MRI06_T1	MRI06_truth	93.04	92.69
Dataset 7	MRI07_T1	MRI07_truth	94.25	97.98
Dataset 8	MRI08_T1	MRI08_truth	92.03	97.98
Dataset 9	MRI09_T1	MRI09_truth	86.43	85.43
Dataset 10	MRI10_T1	MRI10_truth	97.02	97.54

In our presented approach, we have analyzed the efficiency of Watershed and EM-GM methods for detection of anomalies in brain MR image and for this analysis; we have used the images from the most popular brain MR image datasets that contained the brain scan MR images along with their ground truth image. The accuracy of Watershed and EM-GM methods is measured by finding the comparison between the anomalies extracted from input Brain MR Image and the ground truth image of the parallel input image that is presented in the dataset. The relative performance of the Watershed and EM-GM method is presented in Table 1 and Fig. 3.

Thus this work exhibits the relative performance of the Watershed and EM-GM method with the help of ten MR Image datasets for detection of brain anomalies. The testing results clearly demonstrate that the Watershed method presents better accuracy for the majority of the tested MR image datasets.

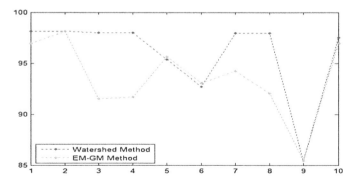

Fig. 3 Comparison of watershed method and EM-GM method

4 Conclusion

The significant amount of analysis has been done to compare the accuracy in results of the brain MR Images for watershed and Expectation Maximization–Gaussian Mixture method. The bilateral filter applied on MR Image dataset to improve the image edges for better segmentation and detection of brain. This work also concludes the comparative analysis for medical image segmentation and detection of brain anomalies using Watershed and EM-GM techniques. Evaluated with the existing research outcomes, this work presented the improvement in detection of brain anomalous regions. In this work, Watershed method presents an elevated value of average comparison thus Watershed method is accurate in extracting the anomalies in brain MR images. With the final outcome of accuracy improvement, this work certainly and satisfyingly extends the possibilities of better segmentation of MR images.

References

1. N. Van Porz, Multi-modalodal glioblastoma segmentation: man versus machine. PLoS ONE **9**, e96873 (2014)
2. J. Liu, M. Li, J. Wang, F. Wu, T. Liu, Y. Pan, A survey of MRI-Based brain tumor segmentation methods, vol. 19, no. 6 (Tsinghua Science and Technology, 2014)
3. S. Bauer, R. Wiest, L.-P. Nolte, M. Reyes, A survey of MRI-based medical image analysis for brain tumor studies. Phys. Med. Biol. **58**(13), R97–R129 (2013)
4. E. Ilunga-Mbuyamba, J.G. Avina-Cervantes, D. Lindner, J. Guerrero-Turrubiates, C. Chalopin, Utomatic brain tumor tissue detection based on hierarchical centroid shape descriptor in Tl-weighted MR images, in *IEEE International Conference on Electronics, Communications and Computers (CONIELECOMP)*, pp. 62–67, 24–26 Feb 2016
5. M. Stille, M. Kleine; J. Hagele; J. Barkhausen; T. M. Buzug, Augmented likelihood image reconstruction, IEEE Trans. Med. Imag. **35**(1)
6. C.C Benson, V.L Lajish, R. Kumar, Brain tumor extraction from MRI brain images using marker based watershed algorithm, in *IEEE international Conference on Advances in Computing, Communications and Informatics* (ICACCI), pp. 318–323, 10–13 Aug 2015
7. D.W. Shattuck, G. Prasad, M. Mirza, K.L. Narr, A.W. Toga, Online resource for validation of brain segmentation methods. Neuroimage **45**(2), 431–439 (2009)
8. J.B.T.M. Roerdink, A. Meijster, The watershed transform: definitions, algorithms and parallelization strategies. Fundam. Inform. **41**, 187–228 (2000)
9. G. Li, Improved watershed segmentation with optimal scale based on ordered dither halftone and mutual information, in 3rd *IEEE international conference, computer science and information technology* (ICCSIT 2010), pp. 296–300, 9–11 July 2011
10. G.Biros Gooya, C. Davatzikos, Deformable registration of glioma images using EM algorithm and diffusion reaction modeling. IEEE Trans. Med. Imag. **30**(2), 375–390 (2011)
11. L. Weizman, Automatic segmentation, internal classification, and follow-up of optic pathway gliomas in MRI. Med. Image Anal. **16**(1), 177–188 (2012)

Cryptanalysis of Secure Routing Among Authenticated Nodes in MANETs

Rajeev Ranjan

Abstract Secure routing (SR) is one of the most important issues in Mobile Ad hoc Networks (MANETs). Recently, in 2013, Zhao et al. proposed an efficient routing integrated framework for MANETs. They claimed that their proposed scheme distributes the system parameter only to the authenticate nodes before network set up phase. However, based on cryptanalysis, we have found that an unauthenticated nodes are also be able to get that original system parameter and behave like a malicious node in the network. Thus, their scheme fails to provide an authenticate distribution mechanism in real life application. As a counter measurement, this paper aims to present an efficient authenticated distribution mechanism that can be incorporated very efficiently in their scheme. Our proposed technique is found to be secure under the hardness of Computational Diffie-Hellman (CDH) assumption.

Keywords Secure routing · MANETs · Security attacks · Authentication technique · Routing protocols

1 Introduction

In the current decade, mobile ad hoc networks (MANETs) have received more and more attention because of their capabilities of self-maintenance and configuration. MANETs is a system of wireless mobile nodes which dynamically self-organize in arbitrary and temporary network topologies. In MANETs nodes may be mobile phones, computer, laptop, personal digital assistants (PDA) and handheld digital devices etc. MANETs doesn't have any fixed infrastructure, i.e. there is no base station. Mobile ad hoc network structure shown in Fig. 1. Nodes arbitrarily change their own position resulting in a highly dynamic topology causing wireless links to be broken and re-established on the fly. The deployment of such networks faces

R. Ranjan (✉)
Department of Computer Science and Engineering,
Indian Institute of Technology (ISM), Dhanbad 826004, India
e-mail: rajeev.macet@gmail.com

© Springer Nature Singapore Pte Ltd. 2017 227
S.C. Satapathy et al. (eds.), *Proceedings of the 5th International Conference on Frontiers in Intelligent Computing: Theory and Applications*, Advances in Intelligent Systems and Computing 516, DOI 10.1007/978-981-10-3156-4_23

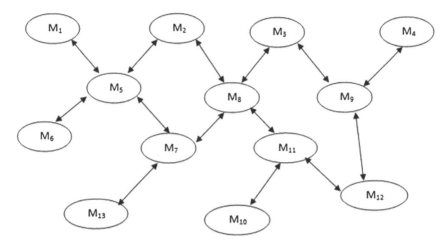

Fig. 1 Mobile ad hoc network (MANETs)

many challenges such as limited physical security, node mobility, low processing power, low bandwidth and limited memory storage capacity. MANETs is used in such field like Emergency search operation, rescue operations, disaster relief effort, mine site operations, military operations in battle fields, electronic classrooms, conferences, convention centers etc. [1]. In MANETs, nodes can directly communicate with other neighbor nodes within radio-ranges; whereas nodes that are not in the direct communication range use the intermediate node to communicate with each other. In both situations, all the participated nodes in the communication automatically form a wireless network, therefore this kind of wireless mobile network is called as a mobile ad hoc network. Compared to the wired networks and MANETs, MANETs are much more vulnerable for security attacks.

This is mainly due to its features of not closed medium, active topology, cooperative algorithms, lack of centralized supervise and management point. Secure routing is a progressive research area in ad hoc network and recent years, several routing protocol has been proposed for MANETS [2–5]. Current research is going on to secure routing and communication in MANETs. For security reasons MANETS should be expected to meet the following different security requirements [6].

Confidentiality: Only the knowing receivers should be able to execute the transmitted data.

Integrity: Before and after the transmission process data should be modified. It is ensured that data integrity must be maintained. i.e. during the transmission process the data should be entact.

Availability: The term availability refers that information should be available for legitimate nodes or authorized parties when needed.

Authentication: Authentication is a method to check the message coming from authentic node or authorized party or not. Nodes are able to authenticate the data has been sent by the authorized node.

Non-repudiation: Non repudiation is a process by which sender of a message cannot deny the sending the message and also receiver cannot deny the receipt after receiving the message. In recent years, many routing protocols have been proposed by researchers for MANETs. These routing protocols are proactive (Table driven) [7, 8], reactive (on demand) [9, 10] and Hybrid routing protocols [11]. The lists of well-known routing protocols are categories in Fig. 2.

Reactive routing protocol: It is a on demand routing protocol, when route is required node flooded the route request message and find the route. After that source route sends the message to the destination node.

Proactive routing protocol: In proactive or table driven routing protocols, every node maintains the network topology information in the form of routing tables by sending HELLO messages periodically. If the link is not broken, then both the nodes exchange their routing information.

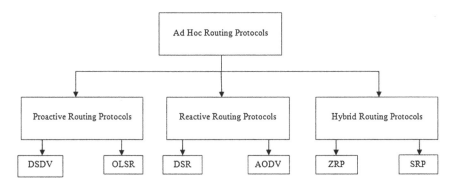

Fig. 2 Classification of routing protocols

Table 1 List of notations used

Symbol	Description
Z	Finite integer set
Z_n	Set of integer after modulo n
F_q	q elements with finite fields
Z_q^*	The multiplicative group of integers modulo prime number q
E/F_p	Elliptic curve over finite field (F_p)
d_{id}	Private key corresponding to node identity
Q_{id}	Public key corresponding to node identity
S	Secret key (Master)
P_{pub}	Public key (System)
H_i	An i, integer usage as subscript, in the system multiple hash function are used
G_1	The additive group, sub group of points on E/F_p
G_2	The multiplicative group, subgroup of the finite field
P, Q	Two points on elliptic curve

Hybrid routing protocol: Hybrid routing protocols are protocols that combine the best features of both reactive and proactive routing protocols.

In this paper we used IBE technique and same system parameter used in Boneh-Franklin's IBC scheme [12, 13]. The notations used in this scheme are summarized in Table 1. For the setup of system parameter a PKG is required for IBC cryptosystems. The system public key (P_{pub}) setup by PKG. $P_{pub} = sP$, where s is any number belongs to Z_q^* and P is a point on elliptic curve (E/F_p).

2 Review of Authentication Phase in Zhao et al. Routing Protocol

The authentication technique used in Zhao et al. protocol [14] is as follows:

- The PKG selects a random $s_0 \in Z_q^*$ and $P_0 \in E/F_p$ of order q, and distributes pseudo system parameters $params = < p, q, P_0, P_{pub0}, H, H_0, H_1 >, (P_{pub0} = s_0 P_0)$ through public channels as traditional IBC schemes do.
- Nodes demands to private keys and real system parameters. Node A selects a random x_A in Z_q^*, encrypts its identity ID (with proof) with $K_A = e(x_A.P_{pub0}, P_0)$ and sends the encrypted identity C_{ID} with $x_A P_0$ to the PKG: $< C_{ID}, x_A P_0 >$.
- The PKG decrypts ID with $k_0 = e(s_0.(x_A P_0), P_0) = K_A$, authenticate identity of node A. The parameter P, P_{pub} and As private key with k_0 encrypts updated system if that gets through it and through public channels, it is sent back to the respective node. Then updated system parameters and its private key is decrypted by node A. Other node which is not aware of x_A cannot learn that information.

3 Attack on Authentication Phase in Zhao et al. Routing Protocol

It is found that Zhao et al. scheme is insecure under Key compromisation attack. That is, a node is able to generate the key and read and/or modify the message during the communication. Other node which is not aware of x_A cannot learn that information. The process is as follows:

- Sender node(A) is sending $< C_{ID}, x_A P_0 >$
- Attacker interrupt the message and collect $V = x_A P_0$ and can able to compute the symmetric secret key $k_A = e(V, P_{pub0}) = e(P_0, P_0)^{x_A s_0}$

Thus, not only authenticated nodes, but also attacker are able to compute the secret key, and thus the whole process is under attack.

4 Proposed Authentication Technique

The process is similar to the previous system. Additionally, we choose another point Q in E/F_P and make $params = <p, q, P_0, Q, P_{pub0}, H, H_0, H_1>$, $(P_{pub0} = s_0 P_0)$ through public channels as did in previous system. Nodes apply to real system parameters and private keys. A node A selects a random r in Z_q^*, encrypts its identity ID (with proof) with $k = e(P_{pub0}, rQ)$ and sends the encrypted identity C_{ID} with rP_0 to the PKG: $<C_{ID}, rP_0>$

$$K = e(P_{pub0}, rQ) = e(s_0 P_0, rQ) = e(P_0, Q)^{rs_0} \tag{1}$$

$$C_{ID} = AES(K, ID_A) \tag{2}$$

The PKG decrypts ID with $K_0 = e(s_0.Q, (rP_0)) = K$, authenticate identity of node A. The parameter P, P_{pub} and private key of node A with k_0 encrypts updated system if that gets through it and through public channels, it is sent back to the respective node. Then updated system parameters and its private key are decrypted by node A.

$$K_0 = e(rP_0, s_0 Q) = e(P_0, Q)^{rs_0} \tag{3}$$

If $K = K_0$, verifies identity of node A after that PKG updated system parameter P, P_{pub} in place of P_0 and P_{pub0} it is sent back to the respective nodes through public channels. The updated new system parameter is $<p, q, P, Q, P_{pub}, H, H_0, H_1>$.

Now nodes are authenticated for secure communication. We choose proactive routing protocol OLSR because it has high efficacy, flexibility and extensibility for secure communication in MANETs.

5 Security Analysis

Definition (*Computational Diffie-Hellman Assumption*) For an algorithm A, the computation of $Z = b.Q$ in polynomial time t from the given tuple $<P, Q, bP>$ is very hard, where b is chosen at random.

Theorem *The proposed authentication scheme is secure under the hardness of CDH assumption, i.e., if the scheme is breakable then the underline hard problem is solvable in polynomial time.*

Proof Here we show that due to the unavailability of r other nodes cannot learn the corresponding information due to the hardness of CDH assumption which was leaked in Zhao et al. authentication technique. Unlike their scheme, we have introduced another point Q in E/F_p and resist the previous attack, i.e., a malicious node can also behave as original node and read as well as modified the message due to the valid secret key. Here, in attacker point of view, the hardness of finding a solution is to compute $Z = s_0 Q$ from the given tuple $< Q, P_0, s_0 P_0 >$ for an unknown $s_0 \in Z_q^*$. If it is solvable then an attacker is also able to generate the symmetric key $K = e(P_0, Q)^{s_0}$ and decrypt any message by acting as an authentication node. Since, there is no solution exists in polynomial time for the CDH problem, so, our authentication technique is secure as long as CDH is hard.

6 Conclusion

Recently, Zhao et al. proposed an efficient routing technique. For this, they have given a proposal of an authentication technique. However, we show that their proposed authentication technique is not secure in real time attack. So, as an enhancement, we have proposed an efficient authentication technique which is suitable for Zhao et al. routing scheme. The proposed technique is found to be secured under the hardness of CDH assumption.

References

1. P. Michiardi, R. Molva, Core: a collaborative reputation mechanism to enforce node cooperation in mobile ad hoc networks, in *Advanced communications and multimedia security* (Springer, 2002), pp. 107–121
2. S. Buchegger, J.-Y. Le Boudec, Performance analysis of the confidant protocol, in *Proceedings of the 3rd ACM international symposium on Mobile ad hoc networking & computing* (ACM, 2002), pp. 226–236
3. D. Coppersmith, M. Jakobsson, Almost optimal hash sequence traversal, in *Financial Cryptography* (Springer, 2002), pp 102–119
4. Ralf Hauser, Tony Przygienda, Gene Tsudik, Lowering security overhead in link state routing. Comput. Netw. **31**(8), 885–894 (1999)
5. D.B. Johnson, Routing in ad hoc networks of mobile hosts, in *First Workshop on Mobile Computing Systems and Applications, WMCSA 1994.* (IEEE, 1994), pp. 158–163
6. W. Su, M. Gerla, Ipv6 flow handoff in ad hoc wireless networks using mobility prediction, in*Global Telecommunications Conference, 1999. GLOBECOM'99*, vol. 1 (IEEE, 1999), pp 271–275
7. T. Clausen, P. Jacquet, C. Adjih, A. Laouiti, P. Minet, P. Muhlethaler, A. Qayyum, L. Viennot, Optimized link state routing protocol (olsr) (2003)
8. C.E. Perkins, P. Bhagwat, Highly dynamic destination-sequenced distance-vector routing (dsdv) for mobile computers, in textitACM SIGCOMM Computer Communication Review, vol. 24 (ACM, 1994), pp. 234–244
9. D.B Johnson, The dynamic source routing protocol for mobile ad hoc networks. *draft-ietf-manet-dsr-09.txt* (2003)
10. E.M. Royer, C.E. Perkins, Multicast operation of the ad-hoc on-demand distance vector routing protocol, in *Proceedings of the 5th annual ACM/IEEE International Conference on Mobile Computing and Networking* (ACM, 1999), pp. 207–218
11. Z.J. Haas, M.R. Pearlman, P. Samar, The zone routing protocol (zrp) for ad hoc networks (2002)
12. D. Boneh, M. Franklin, Identity-based encryption from the weil pairing, in *Advances in Cryptology CRYPTO 2001* (Springer, 2001), pp. 213–229

13. Muhammad Bohio, Ali Miri, Efficient identity-based security schemes for ad hoc network routing protocols. Ad hoc Netw. **2**(3), 309–317 (2004)
14. S. Zhao, R. Kent, A. Aggarwal, A key management and secure routing integrated framework for mobile ad-hoc networks. Ad Hoc Netw. **11**(3), 1046–1061 (2013)

EEG Based Oscitancy Classification System for Accidental Prevention

Jay Sarraf, Satarupa Chakrabarty and Prasant Kumar Pattnaik

Abstract Drowsiness and alcohol consumption has always been the root cause of the road mishaps that takes place. Excessive consumption of alcohol gives rises to many complications such as it prevents healthy thinking and slows down reflex actions. So in order to determine a person's capability to do a job, his oscitancy tracking is very much important. In this paper, we classify the EEG signal taken from 50 drunk and 50 non drunk people. Various band decomposition of the data was done using the DWT (Discrete Wavelet transformation) and further trained by ANN (Artificial Neural Network) approach. Further we suggest an intelligence system which monitors and decide whether the driver should be allowed to drive the vehicle or not based on his drowsiness classification which can prevent accidents with drunken drivers.

Keywords EEG · ANN · Drunk · Alcohol · DWT

1 Introduction

In our daily lives, we can see a lot of traffic accidents that takes place and takes the lives of many. In most of the cases or rather in 95 % cases the reason for such kind of accidents is drunken driving and drowsiness and sleepiness while driving. A person can sustain his daily life using a convinced volume of alcohol, but when the same is taken in an excess quantity, the catastrophe happens. According to the

J. Sarraf · S. Chakrabarty · P.K. Pattnaik (✉)
School of Computer Engineering, KIIT University, Bhubaneswar, Odisha, India
e-mail: patnaikprasantfcs@kiit.ac.in

J. Sarraf
e-mail: Jaysarraf596@gmail.com

S. Chakrabarty
e-mail: rimpa06@gmail.com

© Springer Nature Singapore Pte Ltd. 2017
S.C. Satapathy et al. (eds.), *Proceedings of the 5th International Conference on Frontiers in Intelligent Computing: Theory and Applications*, Advances in Intelligent Systems and Computing 516, DOI 10.1007/978-981-10-3156-4_24

amount of alcohol consumed, it reduces coordination and slows the reflexes and leads to over-confidence. The facts suggest that alcohol strikes the neuronal condition of the brain that is, the theta activity of a person increases and alpha activity decreases (Theta and alpha are various frequency bands which are associated with the brain activity) [1]. There are also some conventional methods to detect the presence of alcohol in a person viz. breathalysers. Breathalysers usually estimate the alcohol concentration by estimating the volume of alcohol exhaled from our lungs. But the Breathalyzer has its own demerits such as the Breathalyzer usually detects alcohol concentration depending on the smell and the quantity of -OH ions in the exhaled breath. Hence, the breathalyzer can be easily defrauded by taking spices and breath fresheners. Also, there is another anomaly that can happen with the breathalyser i.e. as different people have a different capacity of drinking and different people feel drowsy after taking a different amount of alcohol [2]. So if a person who has consumed less amount of alcohol and is not feeling drowsy can sometimes be the victim of breathalyser's anomalies.

The EEG caps that we will be using will help us in acquiring the EEG Signal. As shown in Fig. 1, The EEG Signal that is acquired is preprocessed, where the algorithm running in the processor decomposes the EEG Signal into various brain activity frequency bands [3]. The decomposed signals are analyzed for drunken and non-drunken classification activity. The power which is generated by the algorithm drives the relay system depending upon the existence and non-existence of EEG Abnormalities.

1.1 EEG

Electroencephalography is the process of acquiring the neuron impulses that is generated by each neuron in the brain. Electroencephalography is a non-invasive technique of acquiring brain signals which are processed to obtain optimum results. This technique takes the help of a smart cap implanted with some tiny electrodes. When the cap is worn the electrodes positioned on the scalp detects the electronic signal, which is maximized with the help of filters and are acquired as the brain signals (neural oscillations) [4]. These acquired data are then decomposed into frequency bands as shown in Table 1 which. Therefore, the EEG caps are the best solution to detect if a person is drunk or normal.

Fig. 1 BCI system processing sequence

Table 1 Frequency band distribution of different waves

Frequency bands	Frequency range (Hz)
Delta (δ)	0.5–3
Theta (θ)	4–7
Alpha (α)	8–13
Beta (β)	>13

2 Related Work

In the year 2013 Ekşi [3] studied people to check if they are drunk or not. EEG signals were used to classify the drowsiness in the subject. From the data set two different data files containing 256 EEG for each 50 drunk subjects and 256 EEG for each 50 non-drunk subjects were retrieved. The preprocessing of the EEG Data Set was done by Yule-Walker Method and the both drunk and non-drunk subjects' data were reduced from 256 to 8. The Preprocessed Data is then used for training using Artificial Neural Network and. Simulation test of network 300, 900 and 1500 epochs were performed and the best result was obtained in 900 epochs with 95 % success rate using the reduced data. Also it was observed that simulation can be increased by increasing the training Dataset.

In 2012, Kumar [5] studied the how we can track people for Oscitancy. The paper proposed two modules, the sensor and microprocessor module and the alarm module. The advantages the system gave is regarding its low power consumption and small size. The paper used non-invasive way of extracting the data from the brain. Sensor and microprocessor module used the EEG and msp430 microprocessor. The sensor unit holds the preamplifier, bandpass filter and an analog-to-digital converter which was design to amplify the EEG Signal. The Microprocessor unit was basically used to process and control the ADC signal and send the processed EEG signal to the alarming module [6]. The alarming module consisted of the receiver which captures the signal transmitted by the microprocessor and an alarm system which releases the alarm tone as soon as the warning signal from the microprocessor is received. The implementation of the work is very helpful for all the highway drivers as if the oscitancy state occurs then it may lead to road mishaps.

3 Experimental Study

The experimental study consists of 4 stages as shown in Fig. 1. Data acquisition, Data preprocessing, Feature Extraction and Feature Classification. The data was retrieved from the web portal of Neurodynamics Laboratory at the State University of New York Health Center at Brooklyn [7] where it consists neural oscillation from 64 electrodes placed on the scalp sampled at 256 Hz. Here two different sets of data are used, that contains 256 EEG data for 50 drunk subjects and 256 EEG data for 50 non-drunk subjects. Among the 50 Drunk and 50 non-drunk subjects, 40 subjects from each are utilised to train the network using ANN (Artificial Neural Network), while the remaining data from the 10 drunk and 10 non-drunk subjects are used to test the network.

3.1 Feature Extraction Using DWT

The DWT is calculated passing a sequence of filters. The signal is then decomposed at the same time with the use of high-pass filter. The output gives the detail coefficient and the approximation coefficient.

$$y[n] = (x^*g)[n] = \sum_{k=-\infty}^{\infty} x[k]g[n-k] \tag{1}$$

DWT(Discrete Wavelet Transform) is used which is a classical technique for statistical data analysis, feature extraction [8] and decomposition of the waves in the form of five brain frequency levels that is Alpha Band (8–12 Hz), Beta band (12 −20 Hz), Gamma band (<20 Hz) and Delta band (0–4 Hz) [4] in Fig. 2.

In Fig. 2, the difference of the signal between the drunken and the non-drunken subjects. The signal was decomposed into alpha, beta, theta and gamma band. The green signal marks the drunken whereas the blue signal marks as the non-drunken signal of the subject. Beta band which is gradually enhanced in drunken subject and denotes excitable conditions in the subject. Similarly Alpha bands are spiked when the subject has consumed alcohol.

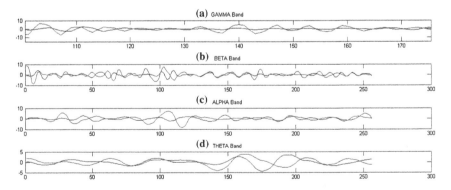

Fig. 2 Decomposition of the drunken (*green*) and not drunken (*blue*) subject into its respective 4 bands

3.2   Feature Classification Using ANN (Artificial Neural Network)

Artificial Neural Networks are the electronic models of the brain which are based on the neural schema of the brain. It is a computer-based system that is used to solve critical problems. ANN is classified as feedback and feedforward according to their architectural schemes. Also, it is classified into three groups according to the learning approaches i.e. supervised learning, unsupervised learning and reinforcement learning. Delta, Hopfield, Kahonen and Hebb are used as learning paths [5].

Here in this paper, we will be using feedforward error backpropagation artificial neural network (FEBANN) [9]. The reason that we are implementing this ANN type is because of its high speed and low training sets requirements.

The four decomposed features alpha, beta, theta and delta where considered for classification using ANN. 40 Drunk and 40 Non Drunk Subjects where used to train ANN. In the simulation 4 epochs 300, 600, 1200 and 1500 were executed. Figure 3 shows the output performance of 1200 epochs.

In Fig. 4, Performance comparison was done using four different epochs 300, 600, 1200 and 1500 where the success rate for 300 epochs was 80 %, 600 epochs was 87 %, 1200 epoch was 95 % respectively and the result is shown in Fig. 3. Which seems 1200 Epoch gave the best performance [10].

These derived results from the classification can be used in the intelligence system which can help the system to differentiate between drunk and non-drunk data which will help in preventing drunk and driving situation and reduce the chances of accidents.

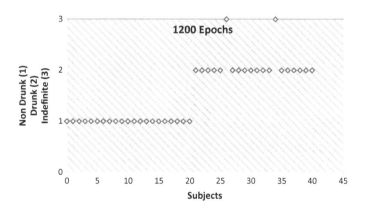

Fig. 3 Simulation results for 1200 epochs

Fig. 4 Performance
comparison of 4 epochs

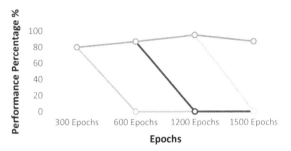

4 Proposed System for Accident Prevention

Our proposed system for accident control mainly consist of three units

1. Interface Unit
2. Operating Unit
3. Intelligence Unit

The system architecture of the accident control system consist follows the same process as discussed in Sect. 1. Additionally we implement a relay system which triggers the car fuse which is placed between the car key and the engine. The fuse is triggered on the basis of decision made by the intelligence unit which determines whether the subject is drunk or not (Fig. 5).

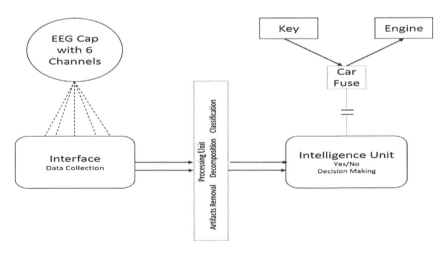

Fig. 5 Proposed system architecture for accident control

4.1   Interface Unit

The cap that was used for acquiring the driver's EEG Signal for the detection of alcohol ensure that it does not irritates the user. The cap used contained five dry electrodes embedded on the forehead of the user and one electrode was embedded behind the left ear of the user for proper acquisition of the EEG Signal. The electrodes that were attached to the scalp of the driver mainly focused on two regions that is frontal region and the central occipital region, as studied in the literature survey, the EEG signal in the central occipital region increases with the increasing amount of alcohol and the power of signal in the frontal region decreases with respect to increase in the intake of alcohol [11]. The signal that was acquired from the driver's head was transmitted wirelessly through Bluetooth to a data receiver where the data was processed in real time.

4.2   Operating Unit

The intelligence unit fetches the data that is send by the operating unit (processing unit) for the further decision making and optimum preprocessing. The Operating Unit performs the cleaning of the data with digital filters which uses to remove all kinds artifacts within it which arises from the AC (power line) interferences and the other anomalies. The system uses amplification technique which have high input impedance and high signal to noise ratio [12]. This amplification is done to avoid the mixing of noises and to keep the significant parts of the low frequency low amplitude bio signal.

4.3   Intelligence Unit

The Intelligence system is used for enabling and disabling the engine of the vehicle with regards to the presence or absence of the abnormalities in the EEG signal. Whenever some kind of abnormalities is observed in the EEG signal which may arise due to alcohol consumption of the driver in real time situation, then the relay system is activated which in turns stops the engine. The relay system simply composes of a Microprocessor, a relay of (12 V–40 A), a car fuse and wires and solder. The car fuse is simply disconnected and the terminals of the fuse is connected to the relay that is installed in the system and then the relay board is connected to the microprocessor [13]. Every time a person wears the cap the EEG is acquired and sent to the Operating unit for the basic processing. After the basic processing is completed by the operating unit, the signal is then send to the Intelligence unit, if the intelligence unit finds any kind of abnormalities in the EEG signal corresponding to the drunken state of the driver than the relay system is activated and if the car is in moving condition then it will stop or if it is in stationary condition then the vehicle will not start till the EEG abnormalities is resolved.

5 Conclusion

It was observed that the power of the EEG signal in frontal region decrease when more amount of alcohol is consumed and the power of the EEG signal in central occipital region also increases which shows that the subject is excited when he consumes alcohol. The above results gives a hypothetical proof to prevent accidents while drunk and driving. Since the driver is wearing EEG cap which transfers the recorded signals to the interface unit and the intelligence unit decide whether to allow the driver to start the car or not, it makes the high possibility of preventing the accident and saving the human life.

References

1. V. Karthick Kumar, A. Julian, Oscitancy tracking using brain-computer interface, in *2012 IEEE Students' Conference on Electrical, Electronics and Computer Science* (2012)
2. E. Malar, M. Gauthaam, D. Chakravarthy, A novel approach for the detection of drunken driving using the power spectral density analysis of EEG. Int. J. Comput. Appl. (0975–8887) **21**(7), 10–14 (2011)
3. Z. Ekşi, A. Akgül, M.R. Bozkurt, The classification of EEG signals recorded in drunk and non-drunk people. Int. J. Comput. Appl. 0975–8887 (2013)
4. S.N. Abdulkader, A. Atia, M.-S.M. Mostafa, Brain computer interfacing: applications and challenges. Egyptian Inform. J. **16** (2), 2015, 213–230. ISSN 1110-8665
5. E. Malar, M. Gauthaam, D. Chakravarthy, A novel approach for the detection of drunken driving using the power spectral density analysis of EEG. Int. J. Comput. Appl. 0975–8887 (2011)
6. B.M.R. Pre-Processing and classification of EMG signals by using modern methods, Ph.D. thesis, Sakarya University Graduate School of Applied and Natural Sciences, Sakarya, Turkey (2007)
7. http://archive.ics.uci.edu/ml/databases/eeg/eeg_full/, Accessed 5 Mar 2016
8. S.D. Puthankattil, P.K. Joseph, Classification of EEG signals in normal and depression conditions by ANN using RWE and signal entropy. J. Mech. Med. Biol. **12**(4), 1–13 (2012)
9. V.Nikulin Vadim, V.Nikulina Anna, Hidehisa yamashita, Effects of alcohol on spontaneous neuronal oscillations: a combined magneto encephalography and electroencephalography study. Prog. Neuropsychopharmacol. Biol. Psychiatry **19**, 687–693 (2005)
10. R. Lin, R. Lee, C. Tseng, H. Zhou, C. Chao, J. Jiang, A new approach for identifying sleep apnea syndrome using wavelet transform and neural networks. Biomed. Eng. Appl. Basis Commun. **18**, 138–143 (2006)
11. A. Yazdani, S.K. Setarehdan, Classification of EEG signal correlated with alcohol abusers. Control and Intelligent Processing Centre of Excellence, School of ECE, Faculty of Engineering, University of Tehran, Iran. ISBN-1-4244-0779-6/07-2007

12. P.D. Welch, The use of fast fourier transform for the estimation of power spectra: a method based on time averaging over short, modified periodogram. IEEE Trans. Audio Electroacoust. **15**, 70–73 (1967)
13. Tracking and Controlling your car remotely using Arduino and Android, http://www.instructables.com/id/Tracking-and-Controlling-your-car-remotely-using-A/, Accessed 8 Apr 2016

Robust Flare Phase Auto Landing of Aircrafts, Based on Modified ASBO

G. Parimala Gandhi, Nagaraj Ramrao and Manoj Kumar Singh

Abstract The presented research work is focused on automatic flight landing control of an aircraft for synthesis of optimal flare phase with considering the dynamic deflection angle as a control parameter. The behavior of the aircraft has considered in terms of four first-order differential equations and an explicit one step Runge Kutta has an order of 4 to 5 has applied to solve that. Computational intelligence based approach; modified adaptive social behavior optimization (*m*ASBO) has applied to estimate the optimum deflection angles on discrete time to deliver the desired flare phase performances. The proposed modification in adaptive social behavior optimization has a better balance between exploration and exploitation by providing a competitive environment for leader selection, in result, faster convergence achieved. Height based on ascent rate controlling function presented here provides an adaptive control mechanism to obtain the desired landing performances in the presence of change in starting landing altitude in compare of predefined reference altitude due to poor visibility or wind disturbance.

Keywords Aircraft landing control · Flare phase · Deflection angle · ASBO

G.P. Gandhi (✉)
Department of ECE-RRIT, Bangalore, Karnataka, India
e-mail: gparimanju@gmail.com

N. Ramrao
DhirubhaiAmbani-IICT, Gandhinagar, Gujarat, India
e-mail: nagaraj.ramrao@gmail.com

M.K. Singh
Manuro Tech Research Pvt. Ltd, Bangalore, India
e-mail: mksingh@manuroresearch.com

© Springer Nature Singapore Pte Ltd. 2017
S.C. Satapathy et al. (eds.), *Proceedings of the 5th International Conference on Frontiers in Intelligent Computing: Theory and Applications*, Advances in Intelligent Systems and Computing 516, DOI 10.1007/978-981-10-3156-4_25

1 Introduction

Among different stages of flight stage, final approach and landing is the most important and decision sensitive stage where pilot has to manage several things within short period of time and any small mistake can cause of major accident. Statistically, among total fatal accidents happen in time duration of 2003 to 2012, around 23 % occurred during landing, while average flight time taken in landing phase is around 1 % in compare to all [1]. Even though a number of rules and safety measures have been taken all over the world, accidents at the time of landing has not stopped, instead there is a slight increment observed at present [2]. There are many factors which affect the flight directly or indirectly like flight stage, atmospheric condition, human pilot cognitive strength and condition along with his flight operational knowledge, aircraft status and condition. These factors are dynamic and together make the flight activity is extremely complex. Dynamic conditions of these factors are ultimately responsible to make aircraft flight parameters to change with time. Approach and landing information with desired trajectory are available in advance and pilot try to achieve this by controlling the various state parameters like height, velocity, pitch angle, angle of attack, rate of decent etc. If there is an automatic control mechanism available which could take care of various parameters to define the proper landing, risk of accident can be reduced very much.

Solution for longitudinal motion control in flare manoeuvre has proposed in [3]. Application of fuzzy neural networks and genetic algorithm in automatic landing systems has explored in [4]. A hybrid approach based on the cerebellar model articulation controller (CMAC) and genetic algorithms (GA) has applied in [5] for the development of aircraft automatic landing control. Recurrent neural networks (RNN) with genetic algorithms (GAs) have applied for ALS in [6], where real-time recurrent learning (RTRL) is applied to train the RNN. A structure for the automatic landing system has proposed in [7] which applied the dynamic inversion concept and proportional-integral-derivative (PID) controllers in conventional and fuzzy variants. Control of aircraft in the longitudinal plane by linearization of aircraft dynamics along with considering the wind shears and errors in sensors has proposed in [8]. Automatic architecture based on neural approach has proposed in [9] for the control of aircraft lateral-directional motion during landing.

2 Aircraft Landing Problem Definition

Process of aircraft landing aircraft consists of two main stages: the glide-path phase and the flare-out phase. First, aircraft is guided by RDF equipment towards airport and later through the instrument landing system (ILS), radio contact is made with a radio beam when there is distance of a few miles from the airport. In following this beam, the pilot guides the aircraft along a glide path angle of approximately $-3°$ toward the runway. Finally, at an altitude of approximately 100 feet, the flare-out phase of the landing begins. During this final phase of the landing, due to electromagnetic disturbances, the ILS radio beam is no longer effective in guiding the aircraft, nor −3° glide path angle is desirable for safety and comfort. Hence, the pilot through visual contact with ground guide the aircraft along the desired flare-path. Certain assumptions have made in this paper to make the problem simple by assuming that the aircraft is guided to the proper location by the air traffic control and only longitudinal motion of aircraft is considered. Pitch angle control the longitudinal dynamics of the aircraft through the elevator which is a rotatable trailing edge flap traditionally located on the horizontal stabilizer. The pitch angle is the angle made by the nose of the aircraft with the earth in the vertical plane. When there is a change in elevator angle the pitching moment around the center of mass of the aircraft is altered, causing a change in the pitch angle.

3 Mathematical Modeling of Aircraft Motion

Due to landing geometry, the assumption is made that the glide path angle γ is sufficiently small. It is also assumed that during the landing the velocity V of the aircraft is constant by utilizing throttle control. Thus, the longitudinal motion of the aircraft is governed entirely by the elevator deflection $\delta_e(t)$ and this becomes the only control signal. Inclusion of these assumptions leads to the short period equations of motion of the aircraft. It is convenient to describe the behavior of the aircraft in terms of four first-order differential equations as given in (1)–(4). The variables of the aircraft motion used in this equation are the pitch angle ($\theta(t)$), pitch angle rate ($\theta'(t)$), the altitude ($h(t)$) and the altitude rate ($h'(t)$). The control variable is the elevator deflection angle $\delta_e(t)$.

$$\frac{d\left(\theta'(t)\right)}{dt} = B_1 \frac{d(\theta(t))}{dt} + B_2\,\theta(t) + B_3 \frac{d(h(t))}{dt} + C\,\delta_e(t) \tag{1}$$

$$\frac{d(\theta(t))}{dt} = \theta'(t) \tag{2}$$

$$\frac{d\left(h'(t)\right)}{dt} = B_4\,\theta(t) + B_5 \frac{d(h(t))}{dt} \tag{3}$$

$$\frac{d(h(t))}{dt} = h'(t) \tag{4}$$

where

$$B_1 = \frac{1}{T_s} - 2\zeta\,w_s \quad ; \quad B_2 = \frac{2\zeta\,w_s}{T_s} - w_s^2 - \frac{1}{T_s^2} \quad ; \quad B_3 = \frac{1}{V\,T_s^2} - \frac{2\zeta\,w_s}{V\,T_s} + \frac{w_s^2}{V}$$

$$B_4 = \frac{V}{T_s} \quad ; \quad B_5 = -\frac{1}{T_s} \quad ; \quad C = w_s^2 \, K_s \, T_s$$

where K_s, T_s, w_s, ζ are the aircraft parameters and represent short period gain, path time constant, short period resonant frequency and short period damping factor. These parameters are assumed to be time invariant and taken numerical values are:

$$K_s = -0.95\text{s}^{-1}, T_s = 2.5\text{s}, w_s = 1 \ radian/s, \zeta = 0.5$$

It can be seen that the elevator deflection angle $(\delta_e(t))$ has a direct affect on the pitch angle rate (θ'), which in turn affects the parameters like: pitch angle (θ), the altitude rate (h') and the altitude (h).

3.1 Performance Requirements and Constraints

The desired duration of the flare-out is considered as 20 s, which includes 5 s over the runway. To make the aircraft landing satisfactory, it is necessary to deliver the performance requirements and satisfy certain constraints in terms of desired response signals, desired control signals, and in terms of limits on these signals. Fundamentally, following requirements and constraints are considered to be of primary importance.

(i)   Desired altitude: an exponential trajectory followed by linear trajectory over runway considers as a safe and comfortable form of landing. The desired path is defined by (5).

$$h_d(t) = \begin{cases} 100e^{-t/5} & 0 \le t \le 15 \\ 20 - t & 15 \le t \le 20 \end{cases} \tag{5}$$

(ii) Desired rate of ascent: the magnitude of the rate of ascent other than zero is desired at touchdown to prevent aircraft from floating over the runway. A large negative value is also undesirable because of over stress applied to the landing gear. Desired rate is the time derivative of the desired altitude as given by (6).

$$h_d'(t) = \begin{cases} -20e^{-t/5} & 0 \le t \le 15 \\ -1 & 15 \le t \le 20 \end{cases} \tag{6}$$

Desired trajectory for altitude and rate of ascent in flare out phase has shown in Fig. 1, where first 15 s of trajectory comply over ground while last 5 s over the runway.

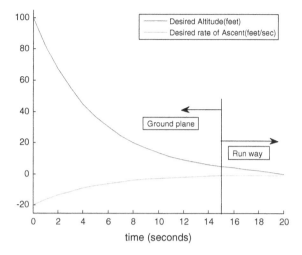

Fig. 1 Desired altitude and rate of ascent for flare out phase

(iii) Pitch angle range: pitch angle at the time of touchdown should must be greater than zero to prevent the nose wheel to touch the runway first and must have an upper limit value to prevent the tail gear touch down first. The desired pitch angle is having the range as given by (7).

$$0^0 \leq \theta(T) \leq 10^0 \tag{7}$$

(iv) Deflection angle range: the elevator, which controls the longitudinal behavior of the aircraft, is restricted to motion between mechanical stops. For this work it is assumed that defined range of deflection angle is given by (8).

$$-35^0 \leq \delta_e(T) \leq 15^0 \tag{8}$$

4 Proposed Solutions

An important and challenging issue exists in multiobjective problem is the formulation of single function which has to include not only all performance requirements but also carry constraints if there is any. In this research sum of the signal error for the aircraft altitude, aircraft rate of ascent, pitch angle at touchdown have considered for each sample of time to estimate the usefulness of selected value of deflection angle at that moment. An explicit one step Runge Kutta having order of 4 to 5 has applied to solve the differential equation. The total flare phase of 20 s has been described and for each second, optimal value of deflection angle obtained which has a minimum value of error index in terms of the objective function. Estimation of optimum deflection angles for all 20 s has done with modified adaptive social behavior optimization (mASBO) which has faster exploration by exploiting the previous global best solution in comparing with a current best solution.

4.1 Modified Adaptive Social Behavior Optimization Based Landing Control

It is true that, compared to genetic evolution, the effect of social interactions is a more effective means to have progressed in a short period of time. It is well obvious that in compared to other species, human society is more innovative and progressive. ASBO is a new meta-heuristic concept considering the model of influence based human social behavior which is adaptive and inspired by the various macro level social factors like leader, self motivation, and logical neighbors [10]. Depths of influences are self adaptive, which makes the ASBO free from any tuning constraints. There are two stages under ASBO; the first stage is based on a multicultural concept to promote exploration where different independent population evolves independently for a predefined number of iterations and in second stage, to promote exploitation, a new population created by selecting the best solutions available in the result of the first stage. In each iteration of the both stages the influence depths are calculated through self adaptive mutation strategy. The change in existing value because of influences (F) and new value is defined for each and every member of the population using (9) and (10).

$$\Delta x_{i+1} = \sum_{k=1}^{3} \left(C_K R_K [F_K - X_i] \right) \tag{9}$$

where C_k are adaptive constants ≥ 0;
$R_k \in U[0\ 1]$

$$X_{i+1} = X_i + \Delta x_i \tag{10}$$

New influence depth level p_i', σ_i' generated from each old individual p_i, σ_i by (11) and (12)

$$p_i'(j) = p_i(j) + \sigma_i(j) \times N(0, 1) \tag{11}$$

$$\sigma_i'(j) = \sigma_i(j) \times e^{\left[\tau' N(0, 1) + \tau N_j(0, 1) \right]} \tag{12}$$

$$\forall j \in \{1, 2, 3\}, \quad \text{and} \quad \tau = \left(\sqrt{2\sqrt{n}} \right)^{-1}, \quad \tau' = \left(\sqrt{2n} \right)^{-1}$$

Table 1 Aircraft initial condition

Parameters name	Initial value
Altitude [h(0)]	100 feet
Vertical velocity [h'(o)]	−20 feet/s
Pitch angle [θ(0)]	−0.0781 rad
Pitch rate [θ'(0)]	0 rad/s

where $p_i(j), p_i'(j), \sigma_i(j), \sigma_i'(j)$ denotes the 'jth' component of the vectors $X_i, X_i', \sigma_i, \sigma_i'$ respectively and N (0,1) is a random number from a Gaussian distribution. N_j (0, 1) is a new random number sampled for each new value of 'j' using Gaussian distribution and 'n' is the number of parameters which has to evolve.

In the original form of ASBO, in each iteration, there is a new leader appear which is decided by best solution in the current population. This is a deviation from events observed in practical social culture, where a previous leader is replacing only when there is fitter leader appeared at present. This approach has applied in *m*ASBO to ensure that population is guided by best leader always. The constant population size constraint is maintained by removal of the least fitter solution. This approach makes the rate of convergence faster towards an optimal solution.

ASBO and *m*ASBO have applied to estimate the optimal deflection angles for each second sample in flare phase. Totally there are 21 deflection angles required in the time duration of 0 to 20 s. The initial conditions of the different parameters have shown in Table 1. Error index value obtained from both methods have shown in Fig. 2 and obtained deflection angles with *m*ASBO have presented in Table 2. It can observe that there is a faster convergence achieved with *m*ASBO in comparing to ASBO.

It is clear from Figs. 3 and 4 that obtained an altitude trajectory and rate of ascent through estimated deflection angles from *m*ASBO is nearly close with desired one.

Fig. 2 Convergence characteristics of ASBO and *m*ASBO over flare phase

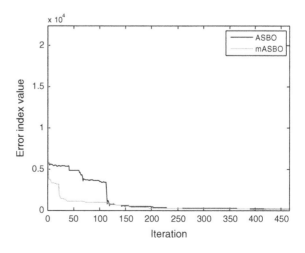

Table 2 Obtained deflection angles using *m*ASBO for reference altitude of 100 feet

Sampled time	Deflection angle (degree)
0	−0.3297
1	−3.9575
2	−0.4427
3	−0.5205
4	−3.2610
5	−0.0605
6	−1.1340
7	−0.1508
8	−0.7076
9	−1.0553
10	0.7242
11	−1.0040
12	−0.5976
13	0.5001
14	−0.0187
15	−0.3515
16	0.1604
17	−0.0227
18	−0.0439
19	−0.6451
20	0.6978

Fig. 3 Altitude trajectory obtained with achieved deflection angles

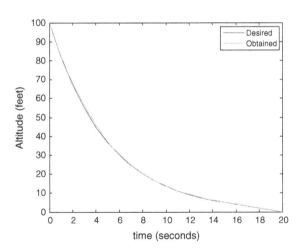

The constraint of having greater than zero pitch angle at touch down also has achieved as shown in Fig. 5, while the rate of pitch angle is in a very small range, especially at the time over the runway as shown in Fig. 6.

Fig. 4 Rate of ascent obtained with achieved deflection angles

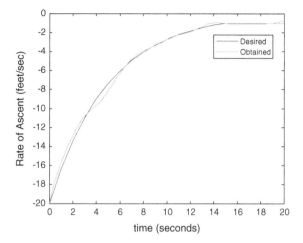

Fig. 5 Pitch angle obtained with achieved deflection angles

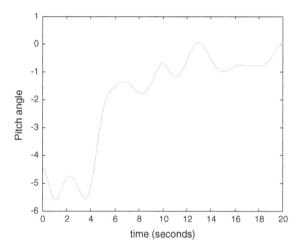

Fig. 6 Pitch angle rate obtained with achieved deflection angles

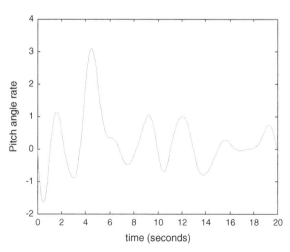

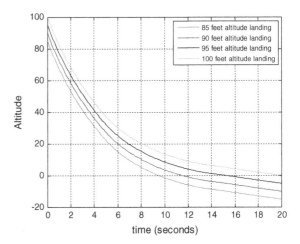

Fig. 7 Altitude trajectory with change in reference altitude of flare phase

4.2 Disturbance Rejection in Reference Altitude Variation

At the time of landing various natural disturbances like poor visibility, wind turbulence etc. may occur which force to change the starting altitude of the flare phase instead of start with predefined reference altitude. But the obtained deflection angles for reference altitude are not optimal for different altitude, in the result obtained performance characteristics does not meet the desired requirement. Moreover, there may physical constraint like limited length of the runway or fixed period of flare phase. This force to develop the adaptive mechanism which take care of control parameters in such a manner that obtained performance characteristics could very close to desired one. In the Fig. 7 it is clear that with pre-estimated deflection angles for altitude of 100 feet, when the altitude change to 85, 90, and 95 feet, obtained altitude trajectory is not acceptable at all. In all cases, there is much earlier and non-smooth landing taken places.

Hence, when there is variation in starting altitude of flare phase in compare to reference, it is not possible to utilize the same set of deflection angles for proper landing. To overcome this problem we have proposed an adaptive controlling function (13) for rate of ascent which is function of: (i) difference in starting altitude from reference altitude and (ii) predetermined reference altitude based deflection angles.

$$\varphi = -20 - \left[h_n - h_{rf}\right] \times \left[\alpha + \frac{\left(h_n - h_{rf}\right) \times \beta}{h_{rf}}\right] \tag{13}$$

where h_n, h_{rf} are present altitude and reference altitude, while α and β are empirical constants and taken as $\alpha = 7$; $\beta = 0.03$.

Performance for same altitudes (85, 90, 95 feet) with applying an adaptive rate of ascent as it given in (13) with predetermined deflection angles for reference altitude

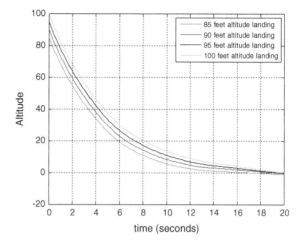

Fig. 8 Adaptive rate of ascent based altitude trajectory with different altitude

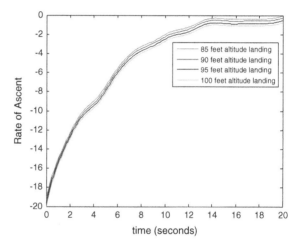

Fig. 9 Adaptive rate of ascent based rate of ascent with different altitude

(equal to 100 feet) has shown in Figs. 8, 9 and 10. It can be observed that for all cases there is very smooth landing as well as satisfied all the constraints properly.

Figure 8 has shown the altitude trajectory obtained with different altitudes and it can observe that slopes of all trajectories are nearly same at all time and close to the ideal trajectory even-though there is variation in altitude up to 10 %. In both cases (90 feet and 95 feet altitude), touch down takes place only at the 20th sec, not only that rate of ascent as it is shown in Fig. 9 is always a negative value during all time of flare phase. Consistent very small negative value of ascent rate at last five seconds ensure that touch down will smooth. Pitch angle information has shown in Fig. 10 and observed that small change takes place with time and at the time of touch down, in all cases it is greater than zero and lesser than upper limit, which ensure that rear wheels will touch the runway first and safer landing will take place.

Fig. 10 Adaptive rate of
ascent based pitch angle with
different altitude

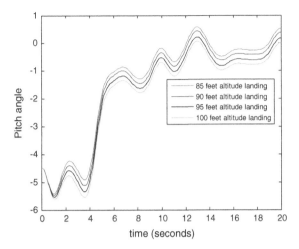

While for 85 feet altitude, touch down has occurred on 17th sec as can observe in
Fig. 8 but, negative ascent rate close to zero and positive pitch angle (≈0.5) makes
sure that landing will be proper.

5 Conclusions

This research work has proposed an automatic flight control solution for synthesis
of optimal flare out phase with smoother landing of an aircraft by considering the
dynamic deflection angle as a control parameter. A modified version of ASBO
applied to evolve the optimal value of deflection angles. Focus has also given to
develop the automatic solution to counteract against the disturbance like poor
visibility or wind disturbance, which may force to start the flare phase at different
height other than pre decided height. This counter act process should satisfy the
landing performance requirement, including specified fix time of flare phase, which
may be exist as a runway length constraint. But there is no time available to
estimate the new set of deflection angles to achieve desired performance objectives.
The proposed solution has a height range based modeling for rate of ascent to make
the operational flare phase very close to the desired one by use of predetermined
available set of deflection angles for reference altitude.

Acknowledgments This research has completed in Manuro Tech Research Pvt. Ltd., Bangalore,
India. The Authors express their thanks to associated members for their valuable suggestions to
accomplish this research.

References

1. Boeing: Statistical summary of commercial jet airplane accidents, worldwide operations, 1959–2012 (Boeing Commercial Airplanes, Seattle, WA 2013)
2. Flight Safety Foundation. Data's source: aviation safety network (2014), http://aviation-safety.net/statistics/phase/stats.php/phase=LDG
3. P. MaSłowskI. Longitudinal Motion control for flare phase of landing. Trans. Inst. Aviation **217**, 79–93 (2011)
4. J.-G. Juang K.-C. Chin, J.-Z. Chio, Intelligent automatic landing system using fuzzy neural networks and genetic algorithm, in *Proceeding of the 2004 American Control Conference* Boston, Massachusetts, 30 June–2 July 2004
5. J.-G. Juang, W.-P. Lin, Aircraft landing control based on CMAC and GA Techniques, in *Proceedings of the 17th World Congress, The International Federation of Automatic Control* Seoul, Korea, 6–11 July 2008
6. J.-G. Juang, H.-K. Chiou, L.-H. Chien, Analysis and comparison of aircraft landing control using recurrent neural networks and genetic algorithms approaches. Neurocomputing **71**(16–18), 3224–3238 (2008)
7. R. Lungu, M. Lungu, L.T. Grigorie, Automatic control of aircraft in longitudinal plane during landing. IEEE Trans. Aerospace Electron. Syst. **49**(2), 1338–1350 (2013)
8. R. Lungu, M. Lungu, Design of automatic landing systems using the H-inf control and the dynamic inversion. ASME. J. Dyn. Sys. Meas. Control. **138**(2), 024501-5 (2015). doi:10.1115/1.4032028
9. M. Lungu, R. Lungu, Automatic control of aircraft lateral-directional motion during landing using neural networks and radio-technical subsystems. Neurocomputing. **171**, 471–481 (2016)
10. M.K. Singh, A new optimization method based on adaptive social behavior: ASBO. (Springer, AISC 174, 2012), pp. 823–831

RSentiment: A Tool to Extract Meaningful Insights from Textual Reviews

Subhasree Bose, Urmi Saha, Debanjana Kar, Saptarsi Goswami, Amlan Kusum Nayak and Satyajit Chakrabarti

Abstract Every system needs continuous improvement. Feedback from different stakeholders plays a crucial role here. From literature study, the need of textual feedback analysis for an academic institute is well established. In fact, it has been perceived that often a textual feedback is more informative, more open ended and more effective in producing actionable insights to decision makers as compared to more common score based (on a scale from 1: n) feedback. However, getting this information from textual feedback is not possible through the traditional means of data analysis. Here we have conceptualized a tool, which can apply text mining techniques to elicit insights from textual data and has been published as an open source package for a broader use by practitioners. Appropriate visualization techniques are applied for intuitive understanding of the insights. For this, we have used a real dataset consisting of alumni feedback from a top engineering college in Kolkata.

Keywords Textual feedback · Sentiment analysis · Topic models

S. Bose · U. Saha (✉) · D. Kar · S. Goswami · A.K. Nayak · S. Chakrabarti
Institute of Engineering and Management, Kolkata, India
e-mail: urmisaha.88@gmail.com

S. Bose
e-mail: subhasree10.7.94@gmail.com

D. Kar
e-mail: debanjana.kar@gmail.com

S. Goswami
e-mail: saptarsi.goswami@iemcal.com

A.K. Nayak
e-mail: amlan.nayak@iemcal.com

S. Chakrabarti
e-mail: satyajit.chakrabarti@iemcal.com

© Springer Nature Singapore Pte Ltd. 2017   259
S.C. Satapathy et al. (eds.), *Proceedings of the 5th International Conference on Frontiers in Intelligent Computing: Theory and Applications*, Advances in Intelligent Systems and Computing 516, DOI 10.1007/978-981-10-3156-4_26

1 Introduction

Growth of textual data has been on the rise for last few years. Some examples of textual data are reviews or feedback, emails, chat or transcripts, tweets, blogs etc. Feedback and reviews have been effectively used for many significant insights. It is critical for any dynamic system or process. The feedback can be quantitative where users are asked to rate on a given scale of (say) 1 to 5. A specific question can be asked and the participant will be asked to choose options like Strongly Agree, Somewhat Agree, Neither Agree nor Disagree, Somewhat disagree, Strongly Disagree. On the other hand, it can be qualitative where feedback about a particular area or areas of interest may be asked for. A sentiment analysis strategy on the qualitative data can also give enough actionable intelligence. The quantitative feedback can alternatively be thought as the structured data, whereas the qualitative feedback represents unstructured data.

Undoubtedly having quantitative feedback has its own advantages. In paper [1], some of the shortcomings pointed out by authors on quantitative feedback are (i) Good survey questions are hard to write (ii) data may provide a generic picture but lacks depth. Strictly, as far as an academic sector is concerned, from our own experience it was felt that, a question on infrastructure may fetch a score of 3 out of 5, however it fails to reveal anything actionable. On the contrary, a qualitative feedback has been received as "Labs and Classrooms are okay. More flexible Library timings would help. Maybe some Cloud setup for doing complex and memory intensive/CPU intensive jobs." So it can be well understood that, the user is giving feedback on laboratories, classrooms and libraries. Now, the challenge is tough, the textual feedback is more informative, finding the aspects or features and then attaching a score is not trivial. These challenges motivate our current work. As observed in [2], a fixed questionnaire actually limits the user's capacity to give feedback, because there may be various other aspects outside the questionnaire on which user has an opinion or a feedback.

Textual feedback has been successfully used in various domains, mostly in assessing customer feedback. In [3], the authors have described a tool with interactive visualization for hotel customers. Article [4] is about tourism industry. As acknowledged in [2], feedback is extremely important for academic institutes in maintaining quality. Text mining has been applied for tracking feedback for online learning systems, to extract concepts from learning materials [5] and also for teacher evaluation. In education as well, there are various kinds of feedback, which are collected from different stakeholders like industry practitioners, faculties, students, parents, alumni. Alumni forms a very important part of any academic institute's ecosystem. In this paper, alumni feedback in terms of textual remarks has been collected on various topics like alumni interaction, placements, infrastructure, academic discipline, faculty, extracurricular activities, focus on R&D and focus on entrepreneurship. These are more insightful than mere quantitative feedback. With the help of topic modeling coupled with sentiment analysis, different finer aspects of each of the areas are identified. Finally, simple visualization techniques are used to interpret the results. However, as it is textual data, there might be semantic ambiguity as well as some spelling errors.

The unique contributions of the paper are as follows:

- The model is available as an open source package for use by all [6].
- Produces better results than reference methods.
- The data are real data and hence different engineering techniques to handle negation and the degree of sentiments are discussed.
- The process is generically built, so for any textual data, this can be extended.

The rest of the paper is organized as follows. In Sect. 2, we have discussed the related works in this area. In Sect. 3, proposed methodology is presented. In Sect. 4, different parameters of experimental setup are covered. In Sect. 5, the results are presented with necessary analysis. Section 6 contains the conclusion.

2 Related Works

In [7], authors have discussed a technique for evaluating the teachers' feedback sent over SMS. It consists of standard natural language processing (NLP) steps like Part of Speech Tagging (POS), Named Entity Recognition (NER), and Stemming etc. SMS also has the challenge of misspelled or abbreviated words. Different concepts were extracted and a sentiment analysis is performed. There is a strong need of identifying the hidden topics or sub themes in the feedback. In paper [1], authors have proposed an ontology based solution for this problem. Mathias et al. [8] have presented their findings from large corpora of feedback about a specific course on industrial design. In [2], authors have collected feedback about teachers in running text. Usual preprocessing steps are done, followed by aspect identification and then sentiment analysis is applied. The text based method appears to extract feedback about much more features which were not generally captured by a numeric score based system. In the paper [9], authors have worked with MOOC Comments. Apart from traditional text preprocessing and sentiment analysis, a correlation analysis between various sentiments is also performed. In paper [10], the authors have demonstrated how an NLP (Natural Language Processing) based system can quickly identify major areas of concerns in an e-learning system, without having the need of going through volumes of survey data. It may be noted that many of the research works as mentioned above are quite recent and are from the years 2014 and 2015, which signifies the relevance of current work.

3 Proposed Methodology

In this section, the proposed methodology in terms of the sequential steps has been elaborated.

Data Collection: Data from alumni have been collected using Google Forms. The form invited textual feedback about various dimensions like alumni interaction, placements, infrastructure, academic discipline, faculty, extracurricular, focus on R&D and focus on Entrepreneurship. These aspects are refereed as the academic dimensions in subsequent discussions.

Pre Processing: Standard techniques like whitespace, punctuation removal, spelling correction have been performed for sentiment analysis. For the sub dimension sentiment analysis, techniques like normalization, stemming, stop word removal have been performed for topic modeling.

Topic Modeling: Various sub themes, or aspects from each dimension are extracted using topic modeling. It is a standard process in text mining, where the hidden semantic structure of the text can be discovered. For a detail understanding of topic model, the paper [11] can be referred.

Sentiment Analysis: Sentiment analysis is performed at two levels. At level 1, sentiment analysis is performed directly on the academic dimensions. At level 2, it is performed at sub dimensions, as extracted or discovered through topic modeling. The sentiment analysis used a lexicon based method detailed in Sect. 4. Simple strategies have been adopted for handling negation and sarcasm.

Visualization: Simple visualizations like work clouds for identifying the topic discussed in an academic dimension and bar charts to depict the score range of sentiment are suggested.

The diagrammatic representation is shown in Fig. 1.

The entire methodology is available in a package named as 'RSentiment' [6] in the Comprehensive R Archive Network (CRAN).

4 Experimental Setup

The feedback was collected using Google Forms and at the time of the analysis the numbers of respondents were well over 60. This was conducted in an anonymous fashion for better inputs. 'R' [12] is used as the computational environment. For list of positive and negative words, works of Liu et al. [13] was used. This is an exhaustive list with 2000+ positive and 4000+ negative words. The intensity has been assumed to increase from positive to very positive or negative to very negative with the use of superlative or cooperative words. A sentence or a phrase is neutral, if it does not contain any positive or negative words. Negations are treated in a simple lexicon based manner on the basis of occurrence of certain words or punctuations. For level 2, topic modelling is done after stop word removal and stemming. The most discussed topics were identified and with each topic, we

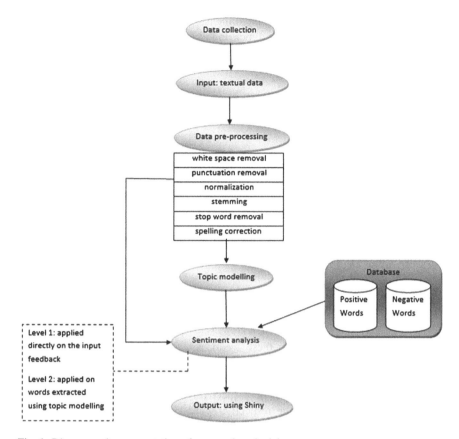

Fig. 1 Diagrammatic representation of proposed methodology

extracted the feedback focusing on these topics in that academic dimension, and applied sentiment analysis algorithm to find the sentiments of the feedback on that topic.

For preprocessing packages 'plyr' [14] and 'stringr' [15] have been used. For topic modeling "tm" [16] has been used. For visualization, "Shiny" [17] and WordCloud [18] have been used respectively.

5 Results and Discussion

The result section is organized in three sub sections. In the first section, we have chosen one academic dimension and analyzed the various feedbacks we received regarding it. We identified the total number of sentences or feedback under each category of sentiment, assigned scores to each sentence and plotted a graph and presented a word cloud to identify the most discussed topics in this area. In the second section, we chose one academic dimension and analyzed sentiment on each of the most discussed topics of that dimension to get some valuable insights on that dimension. In the last section, we have given comparison with some other open source reference methods.

5.1 Overall Sentiment

In this section, we have chosen one academic dimension say 'faculty' or 'placement' and analyzed the feedback on that dimension. The result of the sentiment analysis on dimensions 'faculty' and 'placement' are shown in Table 1.

From the above table, we can see sentiment analysis of academic dimensions 'faculty' and 'placement'. The above dimensions got, 6 into negative and 4 in the very negative category. 'Faculty' got 13 feedbacks in neutral while 'placement' got 21 in that category. There are 22 positive feedbacks in dimension 'faculty' while 'placement has got 20 positive feedbacks. In the very positive category, 'faculty' got 19 feedbacks while 'placement' got 13. We thus obtain an overall sentiment of the academic dimensions of the concerned college and to demonstrate the same we have used the following visualization techniques:

i. With the scores assigned to these feedbacks, the following plots have been generated. Figures 2a and 3a show the graphical representation of the number of feedbacks in each assigned score category of the two dimensions mentioned above—'faculty' and 'placement' respectively.

ii. Figures 2b and 3b show the respective word clouds of 'faculty' and 'placement' generated to highlight the sub-topics mentioned in the feedback text, thus gaining a proper insight of the areas which are of high concern in these academic dimensions.

An analysis of total positive opinions as a percentage of total opinions (Removing neutral) was performed. Alumni Interaction, Placements, Infrastructure, Academic Discipline, Faculty, Extracurricular, Focus on R & D, Focus on entrepreneurship gets 48 %, 77 %, 71 %, 61 %,80 %, 38 %, 50 % and 54 % respectively. This can allow the institute management an insight into the strength and weak academic dimensions very conveniently.

Table 1 Sentiment analysis result for academic dimension 'faculty' and 'placement'

Sentiment category	Number of feedback (faculty)	Number of feedback (placement)
Very negative	4	4
Negative	6	6
Neutral	13	21
Positive	22	20
Very positive	19	13

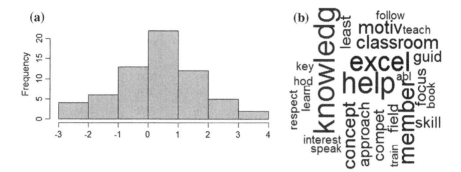

Fig. 2 a Histogram with number of feedback on dimension 'faculty' against the scores of feedback. **b** Wordcloud generated by feedback of dimension 'faculty'

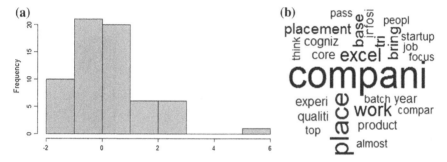

Fig. 3 a Histogram with number of feedback on dimension 'placement' against the scores of feedback. **b** Wordcloud generated by feedback of dimension 'placement'

5.2 Sentiment with Topic Modeling

In this section, we have selected one academic dimension—'Infrastructure' and divided it into the various sub-topics obtained by applying the method of topic modelling on the dimension 'Infrastructure'.

We considered the most discussed topics by observing the frequency of their occurrence in the feedback and analyzed them to find sentiment on this topics. Table 2 shows the tabular structure representing the correlation between the sub-topics and the different sentiment categories.

In the above table, we can see that lab is most discussed topic and there are 2 neutral, 1 positive, 2 very negative and 5 very positive feedback. Thus we gain a detailed insight of the academic dimensions and each of its sub-topics extracted from the feedback text by the method of topic modeling. Further analysis of extracurricular activities shows that the technical fests and cultural fests have received relatively less negative remarks compared to sports.

Table 2 Correlation between sub-topics of the dimension 'infrastructure' and the six sentiment categories

	Very negative	Negative	Neutral	Positive	Very positive
Lab	2.00	0.00	2.00	1.00	5.00
Classroom	1.00	0.00	0.00	0.00	2.00
Library	0.00	0.00	0.00	0.00	0.00
Seminar	1.00	0.00	0.00	3.00	2.00
Computer	1.00	0.00	1.00	0.00	4.00

Table 3 Comparison with QDAP

Phrases	Qdap package	Our method
Very less	0	−2
We got very less scope to interact with our seniors	0	−2
No interaction	0	−1
Has become very minimal	0	−2
Very less to none	0	−2
Absolutely nil	0	−2

5.3   Comparison with Other Methods

The results were compared with QDAP [19] and Sentiment [20]. In Table 3, we have provided with some comparisons (the sentences for which the polarity values did not match) over phrases from the alumni interaction academic dimension.

It is observed that the proposed methodology can correctly classify the polarity of the phrases. Especially with sentences which have negations, our method seems more effective. We have also compared with Sentiment [20], however it classifies a phrase as only positive or negative.

6   Conclusion

It is well established, that textual feedback is very critical for any academic institute and often produces superior insights than quantitative feedback. In this paper, a methodology has been proposed to perform sentiment analysis in conjunction with topic modeling on textual feedback. The proposed methodology is available as an open source package [6] in the Comprehensive R Archive Network (CRAN). The efficacy of this method is tested over a good amount of alumni feedback collected using Google forms. Our proposed method, firstly can identify how the institute is doing in each of the academic dimensions, secondly with the help of topic modeling, it can identify the various sub dimensions and sentiment in each of them. As an example, the tool automatically identified various areas under infrastructure, namely lab, classroom, seminar facility, library and computers. Some engineering improvements are applied to handle negation. The results in terms of visualization can be proved to be quite beneficial to authorities. Particularly, it may be noted, statements like 'no interaction', 'has become very minimal' may often be classified incorrectly by current open source tools. As an extension, we intend to test with bigger corpora of feedback and compare with more state of the art techniques.

References

1. P.K. Agrawal, A.S. Alvi, Textual feedback analysis: review, in *International Conference in Computing Communication Control and Automation (ICCUBEA)* pp. 457–460
2. A. Kumar, R. Jain, Sentiment analysis and feedback evaluation, in *2015 IEEE 3rd International Conference on MOOCs, Innovation and Technology in Education (MITE)*. (IEEE, 2015), pp. 433–436
3. Y. Wu, F. Wei, S. Liu, N. Au, W. Cui, H. Zhou, H. Qu, OpinionSeer: interactive visualization of hotel customer feedback. Visual. Comput. Graph. IEEE Trans. **16**(6), 1109–1118 (2010)
4. S.H. Liao, Y.J. Chen, M.Y. Deng, Mining customer knowledge for tourism new product development and customer relationship management. Expert Syst. Appl. **37**(6), 4212–4223 (2010)
5. C. Romero, S. Ventura, Educational data mining: a review of the state of the art. IEEE Trans. Syst. Man Cybern. Part C Appl. Rev. **40**(6), 601–618 (2010)
6. S. Bose, RSentiment: Analyse Sentiment of English Sentences. Rpackage version 1.0.4 (2016). https://CRAN.R-project.org/package=RSentiment
7. C.K. Leong, Y.H. Lee, W.K. Mak, Mining sentiments in SMS texts for teaching evaluation. Expert Syst. Appl. **39**(3), 2584–2589 (2012)
8. M. Funk, M. van Diggelen, Feeding a monster or doing good? Mining industrial design student feedback at large, in *Proceedings of the 2014 Workshop on Interaction Design in Educational Environments* (ACM, 2014), p. 59
9. B.K.P. Conrad, A. Divinsky, Mining student-generated textual data in moocs and quantifying their effects on student performance and learning outcomes, in *2014 ASEE Annual Conference*, Indianapolis, Indiana (2014)
10. W.-B. Yu, R. Luna, Exploring user feedback of a e-learning system: a text mining approach, in *Human Interface and the Management of Information. Information and Interaction for Learning, Culture, Collaboration and Business* (Springer Berlin Heidelberg, 2013), pp. 182–191
11. M.W. Hanna, Topic modeling: beyond bag-of-words, in *Proceedings of the 23rd international conference on Machine learning* (ACM, 2006), pp. 977–984
12. R Core Team, A language and environment for statistical computing. R Foundation for Statistical Computing, Vienna, Austria (2013). ISBN 3-900051-07-0, http://www.R-project.org/
13. B. Liu, M. Hu, J. Cheng, Opinion observer: analyzing and comparing opinions on the web, in *Proceedings of the 14th International World Wide Web conference (WWW-2005)*, 10–14 May 2005, Chiba, Japan
14. H. Wickham, The Split-apply-combine strategy for data analysis. J. Stat. Softw. **40**(1), 1–29 (2011), http://www.jstatsoft.org/v40/i01/
15. H. Wickham, stringr: Simple, consistent wrappers for common string operations. R package version 1.0.0 (2015), https://CRAN.R-project.org/package=stringr
16. I. Feinerer, K. Hornik, D. Meyer, Text mining infrastructure. R. J. Stat. Softw. **25**(5), 1–54 (2008), http://www.jstatsoft.org/v25/i05/

17. W. Chang, J. Cheng, J.J. Allaire, Y. Xie, J. McPherson, shiny: Web Application Framework for R. R package version 0.13.1 (2016), https://CRAN.R-project.org/package=shiny
18. Ian Fellows, wordcloud: Word Clouds. R package version 2.5 (2014), https://CRAN.R-project.org/package=wordcloud
19. T.W. Rinker, qdap: Quantitative Discourse Analysis Package. 2.2.4 (University at Buffalo. Buffalo, New York, 2013), http://github.com/trinker/qdap
20. T.P. Jurka, sentiment: Tools for sentiment analysis. R package version 0.2 (2012), https://CRAN.R-project.org/package=sentiment

A Purely Localized Random Key Sequencing Using Accelerated Hashing in Wireless Ad-Hoc Networks

Amit Kumar, Vijay K. Katiyar and Kamal Kumar

Abstract Wireless Ad hoc networks represent a form of cooperative networking through peer to peer behavior with others nodes in the networks. Hop by hop communication is default way of communication. Most of the communications are localized and interaction among local nodes requires local security provisioning. In the absence of any centralized certification authority and absence of viable localization and synchronization hardware, schematic localization and periodic refreshing proved to be a feasible solution. Several solutions have exploited GPS based localization and periodic refreshing cycles to provide a viable security solution for wireless Ad hoc networks. In this paper, we have proposed an accelerated hashing mechanism with schematic localization based on variable or multiple transmission range of few nodes. The solution has been evaluated mathematically for performance parameters like connectivity, storage overhead and computation efficiency.

Keywords Localization · Hashing · Periodic · Schematic · Transmission range

1 Introduction

Ad hoc network has replaced infrastructure based network to infrastructure less-networks. Since its invention Ad hoc networks has been applied in widely ranging fields from daily life to industries, from health services to traffic monitoring, from

A. Kumar (✉) · V.K. Katiyar
Department of Computer Engineering, M.M. Engineering College,
M.M. Universty, Mullana, Ambala, Haryana, India
e-mail: amit@mmumullana.org

V.K. Katiyar
e-mail: vkk@mmumullana.org

K. Kumar
Center for Information Technology (CIT), UPES, Dehradun, Uttarakhand, India
e-mail: kkumar@ddn.upes.ac.in

© Springer Nature Singapore Pte Ltd. 2017   269
S.C. Satapathy et al. (eds.), *Proceedings of the 5th International Conference on Frontiers in Intelligent Computing: Theory and Applications*, Advances in Intelligent Systems and Computing 516, DOI 10.1007/978-981-10-3156-4_27

agriculture to defense applications. One of the challenges issue is the security of information being received or sent. To maintain trust in Ad hoc [1] and ensure the availability of correct data and information; key management schemes and secure routing are widely used approach.

Key management is the management of key infrastructure. Key infrastructure includes keys and various procedures which may include key creation, secure bootstrapping, exchanging keys between users, managing storage for key material, usage, revocation and refreshing of keys. Secure routing utilizes underlying keying material in nodes while selecting next hop on routes towards destination.

Key Management must be efficient in terms key-distribution, key-refresh and key-revocation besides being efficient on storage requirements, connectivity and resilience to well-known attacks. Attacks in MANETs varies from eavesdropping on transmissions including traffic analysis or disclosure of message contents, to modification, fabrication, and interruption of the transmissions through node capturing, routing attacks, or flooding [2, 3]. This paper proposes localization based key management scheme which uses Generating Keys (GKs) through an accelerated hashing by the use of special kind of tree called Hash Binary Tree (HBT). The localization will help to localize the nodes and help in averting any node replication and node relocation attacks. Paper is organized into five sections. Section 2 presents related work and identifies the gaps. Section 3 presents the proposal and Sect. 4 presents a brief evaluation approach using probability. Section 5 finally concludes the paper.

2 Related Work

In [4] this paper attacks and their effects on AODV and DSR are discussed. All identified attacks are countered by Authenticated Routing for Ad hoc Networks (ARAN), using public-key cryptographic mechanisms. In [5] authors proposed a solution for AODV to counter attacks like impersonation, collusion and black hole. These solutions use malicious node detection which is based on reputation of a node Reputation is opinion of a node about another node. In [6] authors have proposed Secure Route Discovery Protocol (SRDP) for reducing communication and computation overheads. SDRP uses MAC and Digital Signatures. SDRP is unique in aggregation of signatures and multi-signatures. In [7] authors proposed a new automated approach to evaluate secure routing protocols. This new scheme has analyzed lesser number of topologies in an optimized manner by reducing the number of topologies by using equivalence classification. In [8] authors presented a review to highlight strengths and security limitations of AODV and SAODV. In [9] authors proposed a new approach named as Trust Based Secure OLSR to secure OLSR routing protocol. Trust based approach is better for Ad hoc network due to its environment which functionally depends upon the co-operations with adjoining nodes. In [10] authors proposed a scheme to counter wormhole attack on the basis of round trip time for packets. To detect wormhole in network, RTT is computed on prevailing data rates. In [11] this paper, authors proposed a KM–SR integrated scheme by using identity based cryptography (IBC) that addresses KM–SR inter-dependency cycle problem. The proposed scheme gives a feasible security solution to a wide range of MANETs In [12] authors proposed a Modified DSR (MDSR) routing protocols that detect intrusion and malicious nodes in the network and isolate them from being part of the network by adding to black list of each node. In [13] authors proposed a solution to guard against black hole attack. A modification is applied to any RREQ source. For storing more than one RREP related to a RREQ. In [14] authors proposed a scheme by considering game theoretic approach and Least Cost Total Factor (LCTF) to identify the selfish nodes and guarantee of shortest path between source and destination respectively. Nodes which do not forward RREQ packets are allowed to repeat selfish behavior until their potential misbehavior have not reached upper threshold value. In [15] authors proposed an algorithm to counter the problem of malicious node between source and destination during RREP. By using Group Diffie-Hellmen key exchange protocol, the exchange of key is established.

3 Proposal

3.1 Localization

Consider the network shown in Fig. 1. A large number of nodes are deployed randomly. For convenience deployment has been shown uniform. In general, nodes in the network transmit in a given or fixed transmission range. Occasionally, few nodes assume Anchor Nodes (ANs) roles and may transmit in multiple ranges.

The selection of ANs is based on the Eq. 1 and role is rotated such that a node will become AN next time only after each node has performed as AN after it. Each transmission range is associated with a different nonce. The selection process is governed by Eq. 1 and guarantees that next AN comes from the set of nodes waiting for their turn as ANs. Network completes multiple rounds during its lifetime.

Fig. 1 Localization by using multiple transmission range

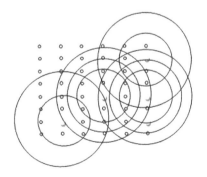

3.1.1 Anchor Selection Process

Each node generates a random number between 0 and 1 and if random number's value is smaller than threshold T(n), the node will be a new Anchor Node (AN); otherwise, it performs as normal member.

The value of T(n) is computed as follows:

$$T(n) = \begin{cases} \frac{p}{1 - p(r \bmod 1/p)} & \text{if } n \in G \\ 0 & \text{othewise,} \end{cases} \qquad (1)$$

where p is the percentage of ANs desired out of total number of nodes. Variable r is the identifier value for current round, and G is the set of nodes that have not been anchored in the last $1/p$ rounds [16].

After the election of ANs, the nodes in a network perform the operation as shown in Fig. 2. The setup phase in Fig. 2 represents the behaviors of ANs, while the steady phase represents the behavior of all the nodes.

3.2 Pre-deployment Generating Key Pre-distribution

Eachnode is pre-distributed with Generating Keys (GKs) from a pool of GKs and is denoted as G. Each node's ID is used to obtain the set of k numbers by using Pseudo Random Number Generator (PRNG) with large period (Eq. 2). This set of k numbers is equivalent to IDs of GKs to be pre-deployed in the said node and constitutes Key Pool (K). A Keyed Hashing Function (H) may be used as Generating Function for generation of K from GK (Eq. 3). The unions of all keys that can be generated from GKs constitute Key Pool (K) (Eq. 4).

$$GK_i = PRNG(ID_i), \qquad (2)$$

where GK_i is set of GKs, i.e., $GK_i = \{Gr_i^1, GK_i^2, \ldots, GK_i^k\}$

$$K_i = H(GK_i^j), \qquad (3)$$

where $j = 1...k$ and GK_i represents the GKs pre-deployed in node with ID_i.

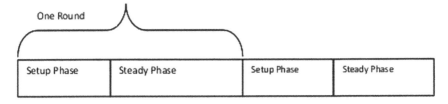

Fig. 2 Protocol operation

$$K = K_1 \cup K_2 \cup K_3 \cup \cdots \cup K_N, \tag{4}$$

where K_i denotes the key set generated at node with ID_i.

3.3 Key Establishment Phase

Key establishment phase proceeds in two steps. First step is to obtain localization information transmitted by ANs. Any node which is close to an AN may receive multiple nonce from nearest ANs and no nonce from distant AN. Each node thus receives nonce as per its location in deployment field. This is highly localized information and can be shared with only closely co-located node and contrasted from distant nodes. This may be used to localize the keying information which may be pre-distributed or post-distributed. Each node may maintain as many bit vectors as the number of ANs in its close proximity. Bit vector is most concise representation when it comes to exchange of nonce metadata and ANs related information (Table 1a, b).

Nonce Bit Vector (NBV) and Nonce Bit Vector Concise (NBVC) (Table 1) are two suggested representations and reduces overhead during exchange of NBV or NBVC as compared to that of exchanging nonce values or nonce IDs. Table 1a describes that a node received all nonce from AN_i except nonce transmitted in shortest range. Table 1b further reduces the bits in NBV by using length to dictate the length of 1 s in NBV. NBVC will save more with more transmission ranges. Each node possesses multiple NBVC. A modified NVBC may be used to store multiple bit vectors into one unified NBVC (Table 1b). After obtaining nonce, nodes perform NBVC exchange. Depending upon the desired percentage of ANs, each node may receive nonce from at the most $100*p$ ANs. Value of p determines the communication overhead of NBVC exchange. As MANET is hop-by-hop communication paradigm, so first of the security requirement is at hop level and second requirement is to sustain semantic security between source and destination.

Hop by hop security requirements are met by localized keying. Any two neighboring nodes may decide upon the common nonce set on the basis of exchanged NBVC. All common nonce may be used to obtain localized GK. The set of GKs available at any node may be computed without exchange of any metadata

Table 1 (a) Nonce Bit Vector. (b) Nonce Bit Vector Concise

(a)

AN	N_i^0	N_i^1	N_i^3	N_i^4	
ID_i	0	1	1	1	

(b)

AN	Length	AN	Length	AN	Length
ID_i	3	ID_j	2	ID_k	1

on GKs by using PRNG. Common GKs are selected for localization. Localized GK is used for generation of session key by using keyed hash function H. The generation of pairwise key for hop-to-hop communication is time stamped. Equation 5 describes the procedure and Fig. 3 describes the total process involved.

3.3.1 Hash Binary Tree

Hash Binary Tree (HBT) is way forward to reduce the number of steps in generation of a derived key of a sequence generated through hashing. Consider that root is represented as $S(0,0)$ and denotes seed. In this paper $S(0,0)$ is localized GK. Shifting $S(0,0)$ to left by one bit position and then hashing with H gives $S(1,0)$. Instead if $S(0,0)$ is shifted right by one bit position and then hashed using H gives $S(1,1)$. Left and right child respectively presents 0 and 1 in binary. The example shown is to derive any key in a Key Chain (KC) of length 8.

Leaves of HBT represent usable keys. Any one key in this sequence can be generated using three shifts and three hash operation. Extending the height of the HBT helps further increase the length of the chain and randomness in keying domain.

$$C = N_{i,j} \oplus N_{k,l} \ldots \oplus N_{m,n}, \tag{5}$$

where C represents the common nonce set received from ANs. Each component nonce $N_{i,j}$ is deciphered as nonce from AN with ID_i associated with transmission range number j.

$$S(0,0) = H\left(GK_i^k, C\right) \tag{6}$$

GK_i^k may remain same during different generations of communication between same pair of nodes. The sheer existence of localization factor (C) ensures the confidentiality and integrity of any communication.

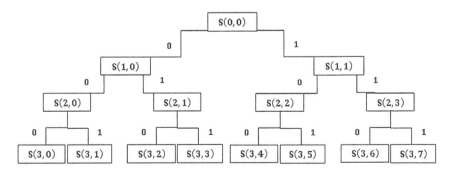

Fig. 3 Hash binary tree

3.4 Secure Communication

To achieve secure communication nodes should share a GK. Each node can generate the list of GKs pre-deployed in each node. If concerned pair of nodes shares at least one GK, common GK with lowest ID may be used for generation of localized GK and then use any random key in corresponding KC generated through the use of HBT. If a successive pair of nodes (let nodes A and B in the path Source-A-B-C-...-Destination) doesn't share any key, communication on path ceases. Node then may request previous node to establish secure communication with node's successor. This is adopted only in case when consecutive nodes fails to establish secure communication. In-fact a tunnel has been established between node's predecessor and successor nodes. This way a secure communication is still established. This way of communication is called Back-off Communication.

4 Performance Analysis

Performance analysis of the proposal is done mathematically using number theory and binary arithmetic. The parameters of concern include integrity and confidentiality, Communication overhead, computational complexity, connectivity and storage requirements.

4.1 Computational Complexity

HBT is used in this proposal for generation of random key from a KC for establishing integrity and confidentiality requirements. A normal sequential KC may be used instead. Sequential KC will require n hashing operations to arrive at nth key in KC. With increase of randomness length of key chain. i.e., n will continue to increase.HBT uses an accelerated approach. To obtain nth key in KC, a conversion of n into binary number is performed. A m bit binary number will require m shifts and m hashing operations.

This can be expressed as $2^m \leq n \leq 2^{m+1}$. This inequality implies that a most $m+1$ bits suffice to represent any number in binary format. HBT requires $m+1$ hash and shift operation to derive any key at index n such that $m+1$ bits are required to represent n. The gain in terms of reduced number of hashing operations is multifold. Graph in figure expressed the comparative analysis (Fig. 4).

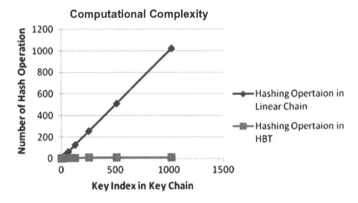

Fig. 4 Computational Efficiency of HBT based Key Chain Generation

4.2 Connectivity Ratio

To communicate any node requires at-least one GK common with other node. If any pair of nodes doesn't share any GK between them, they will not be able to communicate. As the pre-deployment key pre-distribution is based upon some pseudo random strategy, a degree of randomness is available. PRNG allows some of the keys to be repeated after a period, which may be very large. As a result it is very probabilistic scenario when any pair of node shares at least one GK. Let size of key pool K is K and each node has been allocated k keys such that $K \gg k$. Corresponding scenario is analyzed with the help of probability theory in Eqs. 7–12.

Let there be a path $p = S, A, B, C, \ldots, D$, where S and D respectively denotes source and destination and A, B and C denotes any successive node on the path. Probability to share at least one key between A and B may be denoted by eqn.

$$P(TwonodesSharesatLeastOneKey) = 1 - P(TwoNodesSharesnoKeys) \quad (7)$$

Probability of no share requires that keys are allocated with no-replacements. If any node is to be allocated k keys from a pool of K keys then the numbers of ways in which k keys can be chosen are:

$$\binom{K}{k} = \frac{K!}{k!K - k!} \quad (8)$$

Any node that don't share key with this node is may be allocated k keys from remaining key pool $K - k$ can be specified as:

$$\binom{K - k}{k} = \frac{K - k!}{k!K - 2k!} \quad (9)$$

Probability that two nodes share at least one key is as shown below:

$$P(Two nodes Shares at Least One Key) = 1 - \frac{\binom{K-k}{k}}{\binom{K}{k}} = \frac{(K-k!)^2}{K!K-2k!} \quad (10)$$

Consider when two consecutive nodes. i.e., A and B shares at least one GK but nodes B and C don't share any GK. Then Back off Communication method proposed in section is used and the probability that node A and C share at least one key is as below: Let $|A \cap B| = s$ and $|B \cap C| = 0$. Also assume that $X = (|A \cap B| = s)$ and $(|B \cap C| = 0)$ and $Y = (A \cap C \neq \emptyset)$. Thus $\sim Y = (A \cap C = \emptyset)$.

Because Y is non-empty therefore we will select the k keys for node C from a pool of $K - k$ keys (By removing k keys of B because $B \cap C = \emptyset$).

$$P(Y/X) = 1 - P(\sim Y/X) \quad (11)$$

$$P(\sim Y/X) = P(A \text{ and } C \text{ does not share})$$
$$= \frac{\binom{(K-k)-(k-s)}{k}}{\binom{K-k}{k}} \quad (12)$$

Back off communication proposed in Sect. 4.4 is also dealt with. Nodes needs to check for shared GK between its previous node and its next node. There is huge probability of obtaining the shared GK between previous and next node. Numerical value of probability computed in Eq. 11 is higher than probability value in Eq. 10. Back-off communication improves the connectivity ratio of normal random key distribution by almost 20 %. The increased degree of connectivity requires extra pair of message exchange. Two extra messages are exchanged between a node and node's previous. It is assumed that nodes on the path from source to destination are part of the data packet being routed securely.

4.3 Storage Overhead

The concept of GKs in this proposal has reduced the storage requirements with respect to proposal in [17]. The usage of GKs as key generators results not only reduced space requirements but also increased randomness in key usage from KC.

4.4 Node Capture and Node Replication Attack

Node Capture and Node Replication Attack is most severe attack in Ad hoc networks due to its deployment in hostile and unattended environments. Captured nodes may be moved somewhere else. The relocation of captured node at different location in the network offers resistance as nodes has to use their localization information before getting into secure communication. Moreover, nodes have already shared their NBVC after localization during Steady phase. This renders relocation attack useless and ineffective. Similarly, node replication will stay ineffective as replication also equivalent of relocation. Localization has improved resilience against node capture, relocation and replication attack tremendously.

5 Conclusion and Future Work

Ad hoc network opens new application areas of networked computing but also is flood gate for plethora of challenges. The information gathered by using the Ad hoc networks must be routed to a place inside or outside of Ad hoc network for further processing. The information is routed through open air and is susceptible to different types of attacks. The key management scheme proposed in this paper overcomes some of well-known attacks which includes node capture and node replacement and node replication attack. The solution presented in this paper also proposed a Back-off Communication process to overcome the issues related to connectivity of the nodes. The use of HBT highly affected the computational complexity of the key generation step in Steady Phase.

References

1. P. Khatri, S. Tapaswi, U.P. Verma, Trust evaluation in wireless ad-hoc networks using fuzzy system. Comput. Syst. Sci. Eng. **29**(1), 43–50 (2014)
2. V.C. Gungor, G.P. Hancke, Industrial wireless sensor networks: challenges, design principles, and technical approaches. IEEE Trans. Ind. Electron. **56**(10), 4258–4265 (2009)
3. D.G. Padmavathi, M. Shanmugapriya, A survey of attacks, security mechanisms and challenges in wireless sensor networks. arXiv preprint arXiv:0909.0576 (2009)
4. K. Sanzgiri, D. LaFlamme, B. Dahill, B.N. Levine, C. Shields, E.M. Belding-Royer, Authenticated routing for Ad-hoc networks. IEEE J. Sel. Areas Commun. **23**(3), 598–610 (2005)
5. I. Raza, S.A. Hussain, Identification of malicious nodes in an AODV pure Ad-hoc network through guard nodes. Comput. Commun. **31**(9), 1796–1802 (2008)
6. J. Kim, G. Tsudik, SRDP: secure route discovery for dynamic source routing in MANETs. Ad-hoc Netw. **7**(6), 1097–1109 (2009)
7. T.R. Andel, G. Back, A. Yasinsac, Automating the security analysis process of secure Ad-hoc routing protocols. Simul. Model. Pract. Theory **19**(9), 2032–2049 (2011)

8. J. Von Mulert, I. Welch, W.K. Seah, Security threats and solutions in MANETs: a case study using AODV and SAODV. J. Netw. Comput. Appl. **35**(4), 1249–1259 (2012)
9. A. Adnane, C. Bidan, R.T. de Sousa Júnior, Trust-based security for the LSR routing protocol. Comput. Commun. **36**(10), 1159–1171 (2013)
10. S. Qazi, R. Raad, Y. Mu, W. Susilo, Securing DSR against wormhole attacks in multirate Ad-hoc networks. J. Netw. Comput. Appl. **36**(2), 582–592 (2013)
11. S. Zhao, R. Kent, A. Aggarwal, A key management and secure routing integrated framework for mobile Ad-hoc networks. Ad-hoc Netw. **11**(3), 1046–1061 (2013)
12. M. Mohanapriya, I. Krishnamurthi, Modified DSR protocol for detection and removal of selective black hole attack in MANET. Comput. Electr. Eng. **40**(2), 530–538 (2014)
13. D.R. Choudhury, L. Ragha, N. Marathe, Implementing and improving the performance of AODV by receive reply method and securing it from black hole attack. Procedia Comput. Sci. **45**, 564–570 (2015)
14. D. Das, K. Majumder, A. Dasgupta, Selfish node detection and low cost data transmission in MANET using game theory. Procedia Comput. Sci. **54**, 92–101 (2015)
15. M.K. Dholey, G.P. Biswas, Proposal to provide security in MANET's DSR routing protocol. Procedia Comput. Sci. **48**, 440–446 (2015)
16. W.R. Heinzelman, A. Chandrakasan, H. Balakrishnan, Energy-efficient communication protocol for wireless microsensor networks, in *Proceedings of the 33rd Annual Hawaii International Conference on System Sciences (HICSS-33'0)* (2000), p. 223
17. L. Eschenauer, V.D. Gligor, A key-management scheme for distributed sensor networks, in *Proceedings of the 9th ACM Conference on Computer and Communications Security (CCS'02)* (ACM Press, Washington, DC, USA, 2002), pp. 41–47

GOASREP: Goal Oriented Approach for Software Requirements Elicitation and Prioritization Using Analytic Hierarchy Process

Nikita Garg, Mohd. Sadiq and Pankaj Agarwal

Abstract Software requirements elicitation is a valuable process for the identification of software requirements according to the need of different types of stakeholders. There are different methods for the elicitation of software requirements like traditional methods, group elicitation methods, goal oriented methods, etc. Among these methods, goal oriented methods have received much recognition by software requirements engineering community. On the basis of our literature review, we identify that *"goal oriented requirements elicitation processes do not support how to select and prioritize the requirements using analytic hierarchy process on the basis of the cost and effort criteria"*. Therefore, in-order to address this issue, we proposed a method, i.e. GOASREP, for the elicitation of software requirements using *"goal oriented approach"* and the prioritization of the elicited requirements using *"analytic hierarchy process"*. In the proposed method, we used function point analysis approach for the estimation of the cost of each requirement. COCOMO model has been applied to estimate the effort of each requirement. Finally, the usage of the GOASREP is explained using Institute Examination System.

Keywords Requirements elicitation · Goal oriented requirements elicitation process · Analytic hierarchy process · Function point analysis · COCOMO model

N. Garg · P. Agarwal
Department of Computer Science, Institute of Management Studies,
Ghaziabad, UP, India
e-mail: ngnguptacs@gmail.com

P. Agarwal
e-mail: pankaj7877@gmail.com

Mohd. Sadiq (✉)
Computer Engineering Section, UPFET,
Jamia Millia Islamia (A Central University), New Delhi 110025, India
e-mail: msadiq@jmi.ac.in; sadiq.jmi@gmail.com

© Springer Nature Singapore Pte Ltd. 2017
S.C. Satapathy et al. (eds.), *Proceedings of the 5th International Conference on Frontiers in Intelligent Computing: Theory and Applications*, Advances in Intelligent Systems and Computing 516, DOI 10.1007/978-981-10-3156-4_28

1 Introduction

Software requirements elicitation is one of the difficult tasks of requirements engineering (RE) process. It is the first sub-process of RE and error occurring during this process will influence the other sub-processes of RE like *"requirements modelling"*, *"requirements analysis"*, *"requirements management"*, *"requirements verification and validation"*, and *"requirements management"* [6]. There are different methods for the elicitation of software requirements like, traditional methods, group elicitation methods, cognitive methods, contextual methods, goal oriented methods, etc. Among these methods, software requirements engineering community have focused on goal oriented methods because in this method *"AND/OR graph helps requirements analyst to elicit and model the functional and non-functional requirements of software"* [6]. Based on our literature review, we identify that *"goal oriented requirements elicitation process (GOREP) do not support how to select and prioritize the requirements using analytic hierarchy process on the basis of the cost and effort criteria"*. Therefore, in-order to address this issue, we proposed a method, i.e., GOASREP, for the elicitation of software requirements using goal oriented approach and prioritization of the elicited requirements using analytic hierarchy process (AHP). In the proposed method, for the estimation of the cost of each requirement, we used function point analysis approach; and COCOMO model has been applied to estimate the effort of each requirement. Finally, the usage of the GOASREP is explained using "Institute Examination System".

This paper is organized as follows: In Sect. 2, we present the related work. Proposed method is given in Sect. 3. In Sect. 4, we present the case study; and finally, in Sect. 5 we present the conclusion and future scope.

2 Related Work

There are different methods in literature *"to elicit and prioritize the requirements"* using goal oriented methods. For example, in 2015, Sadiq and Jain [4] proposed a *"fuzzy based approach for the selection of goals/software requirements in GOR-EPs"*. In a similar study, in 2014, Sadiq and Jain [7] proposed a *"fuzzy based method for the prioritization of software requirements/goals in GOREP"*. There are different methods for the selection and prioritization of software requirements like Mathematical Programming (e.g., Integer Linear Programming), Metaheuristics approaches like Swarm Intelligence (e.g. Ant Colony Optimization), Evolutionary Approaches (e.g., Genetic Algorithm, Differential Evolution Algorithms, and Teaching Learning Based Algorithms), Analytic Hierarchy Process (AHP), TOP-SIS, and the hybridization of the Mathematical Programming and Metaheuristics approaches, and other approaches [1, 3]. In GOREP literature, less attention is given to Mathematical Programming, Metaheuristics approaches, AHP, and other approaches for the selection and prioritization of requirements. Therefore, in this paper, we proposed a method for the elicitation of software requirements using goal oriented approach; and prioritized the elicited requirements using AHP. In the proposed method, we have used cost and effort as the criteria for the pairwise comparison among requirements. Finally, we explain the proposed method with the help of a case study.

3   Proposed Algorithm

In this section we present an algorithm for the elicitation of requirements using goal oriented approach and the prioritization of the elicited requirements using AHP. Proposed method includes the following steps:

Step 1:   Elicitation of software requirements using goal oriented approach
Step 2:   Estimation of cost and effort of each software requirements
Step 3:   Prioritization of software requirements using AHP.

Step 1 Elicitation of software requirements using goal oriented approach

In goal oriented approach, *"AND/OR graph is constructed by decomposing the high level objective of stakeholders, say G, into sub-goals or requirements"*. AND/OR graph is employed to find out the different types of requirements, i.e., functional and non-functional. In AND decomposition, *"until and unless all the sub-goals would not be achieved, their parent goal will not be achieved"*. In case of OR decomposition, *"the parent goal would be achieved, if any sub-goal of the parent goal would be achieved"* [4, 7, 11].

Step 2 Estimation of cost and effort of each software requirements

Function point describes the functionality of software system. It is a well known method to estimate the size of software systems. It depends on five different measuring parameters, i.e. *"External Input (EI)"*, *"External Output (EO)"*, *"External Inquiries (EQ)"*, *"Internal Logical File (ILF)"*, and *"External Interface File (EIF)"*. In our work, we compute the FP of each requirement. The Constructive Cost Model (COCOMO), an algorithmic software cost estimation model, was developed by Barry W. Boehm. In this model, the software projects are divided into three parts, Organic, Semi-detached, and Embedded on the basis of the lines of code (LOC). In our work, we have considered, Organic mode to estimate the effort of each software requirement [10].

Step 3 Prioritization of software requirements using AHP

Analytic Hierarchy Process (AHP), introduced by Thomas L Saaty in 1970, is an *"effective tool for dealing with complex decision making problems"*. Optimization is an interdisciplinary research area and it has been applied successfully in software engineering, requirements engineering, management science, etc. Among various optimization algorithms, *"Multi-criteria Decision Making"* (MCDM) algorithms have received much attention by software engineering and operation research community [5, 8, 9]. Among various MCDM algorithms, AHP has been applied in many areas of management science and engineering, for example, in 2010, Sadiq et al. [9] applied *"AHP method for the prioritization of software requirements"*. Therefore, in the proposed method we apply AHP for the prioritization of software requirements.

4 Case Study

In this section, we present the case study of our work. In this study, we elicit the requirements of *"Institute Examination System (IES)"* using goal oriented approach; and then compute the cost and effort of each requirement. We use the cost and effort as the criteria to prioritize the requirements using AHP.

Step 1: In this step, we apply the goal oriented approach to elicit the requirements of Institute Examination System. After applying the goal oriented approach, we have identified the following functional requirements (FR):

FR 1: *Login Modules.* This is further subdivided into five sub-goals like, fr 1.1: User name; fr 1.2: Password; fr 1.3: Student; fr 1.4: Admin; fr 1.5: Faculty; fr 1.6: Submit

FR 2: *Printout of bank receipt of student's fee.* It is divided into following sub-goals: fr 2.1: Payment mode through cash, cheque, draft or by card; fr 2.2: Name of the student; fr 2.3: Course Name; fr 2.4: Account Number; fr 2.5: Registration Code; fr 2.6: Branch Name; fr 2.7: Signature; fr 2.8: Total amount; fr 2.9: Verification and validation that is fee is paid or not; fr 2.10: Verified with Stamp and Signature on it.

FR 3: *View semester result.* It has the following sub-goals: fr 3.1: Log on to University site; fr 3.2: Click on result; fr 3.3: Result of M. Tech.; fr 3.4: University Region or Noida Region; fr 3.5: Result according to semester wise; fr 3.6: Name of the student; fr 3.7: Father Name of student; fr 3.8: Roll number of student; fr 3.9: Captcha; fr 3.10: Click on submit button; fr 3.11: Print out of result

FR 4: *Generate examination seating arrangement.* After decomposing it into sub-goals we have identified following sub-goals: fr 4.1: A room has at least 60 seats; fr 4.2: Allocate one computer system to particular student with roll number and name fr 4.3: A room should be allocate for 3 h; fr 4.4: A room should be well ventilated and the atmosphere should be congenial for student; fr 4.5: There should be investigator to check the details of the student; fr 4.6: A room should have two faculty members in it; fr 4.7: In case of problem in a computer system there should be a computer engineer to test the system.

FR 5: *Online conduct of examination.* In this case, we have the following sub-goals, i.e., fr 5.1: Each student is having a registration number and password with them; fr 5.2: Instruction page is open; fr 5.3: After reading the instruction page student can click on start button; fr 5.4: A timer button is available on the right hand side; fr 5.5: An answer sheet having different color options in it such as red color is used for not reading questions and green color is used for attending the question; fr 5.6: On upper right corner photo of student should be there; fr 5.7: Scheme of Examination includes: fr 5.7.1: Aptitude; fr 5.7.2: Reasoning; fr 5.7.3: English; fr 5.7.4: Professional Knowledge; fr 5.8: Student has choice to start with which section he/she want to start either English, Aptitude, Reasoning and Professional knowledge; Fr 5.9: students have their own choice if she/he wants to switch in between; fr 5.10: Student can also check his/her answer sheet; fr 5.11: Student can also check the status of the questions that they have attempted; fr 5.12: At the end, student will have to click on submit button. In the similar way, we decompose the following FRs:

FR 6: *"Fill examination forms"*
FR 7: *"Upload any exam related activities"*
FR 8: *"Generate examination hall ticket"*
FR 9: *"Approve examination form"*
FR 10: *"On line payment of examination fee"*

Step 2: Now, we estimate the cost and effort of each requirement. To estimate the cost of R1, we first compute the FP of R1. For the computation of R1, we first identify the EI, EO, EQ, ILF, and EIF. The value of these parameters for FR 1 is given as below:

$$EI = 24, EO = 5, EI = 8, ILF = 10, \text{ and } EIF = 7.$$

The value of the total count is 54; and the value of the FP is 66. In India, the cost of one FP is 125\$ [2]. Therefore, the cost of R1 is 8250 \$. For the computation of the effort, we apply the following equation:

$$E = a(KLOC)^b \tag{1}$$

$$D = c(E)^d \tag{2}$$

where "E is the *effort*", "KLOC is the *size typically measured in thousand lines of code*", "a is the *productivity parameter* and b is an *economic or not economic of scale parameter*", "D is the *development*"; c and d are the parameters of D.

The values of a, b, c, and d are the standard values as per the basic COCOMO. We know that the LOC/FP = 128, if the software is written in C++. Therefore, in our case, for R1 the value of KLOC = 128 X 66 = 8.44 and the value of the Effort = 22.53 PM.

In similar way, we compute the cost and effort of all the remaining requirements. We summarized the values of FP, cost, and effort for the remaining requirements in Table 1.

Table 1 The values of FP, Cost, and Effort for different FRs

S. no.	FRs	FP	Cost	Effort
1	FR 1	66	8250 $	22.53 PM
2	FR 2	95	11875 $	33.06 PM
3	FR 3	110	13750 $	38.56 PM
4	FR 4	71	8875 $	24.33 PM
5	FR 5	133	16625 $	47.06 PM
6	FR 6	109	13625 $	38.19 PM
7	FR 7	43	5375 $	14.38 PM
8	FR 8	86	10750 $	29.76 PM
9	FR 9	38	4750 $	12.62 PM
10	FR 10	100	12500 $	34.89 PM

In Table 1, FP value of the FR2 is 95 and the cost to implement this requirement is 11875 $. The effort required to implement FR2 is 33.06 PM. These values would be used to find out the ranking values of the requirements in the next step.

Step 3: Now we apply AHP on the elicited requirements, i.e., FR1, FR2, ... , FR10. After applying the AHP, we have the following ranking values for 10 FRs:

FR1 = 0.295
FR2 = 0.0397
FR3 = 0.163
FR4 = 0.0186
FR5 = 0.2528
FR6 = 0.105
FR7 = 0.0099
FR8 = 0.0284
FR9 = 0.0099
FR10 = 0.0673

5 Conclusions and Future Scope

In order to strengthen the *"goal oriented requirements elicitation process" (GOREPs)*, in this paper, we proposed a method, i.e., GOASREP. It is used for the elicitation of the software requirements using goal oriented approach and the prioritization of the elicited requirements using AHP. For the computation of the cost we used the FP approach; and we have employed COCOMO model for the estimation of the effort. In our study, after eliciting the requirements of IES, we have identified 10 FRs; and among these requirements, FR1 has the highest priority, i.e., 0.295 and FR7 and FR9 have the lowest priority. Selection and prioritization of software requirements before the development of software project may enhance the software release planning, and can control the budget and scheduling. Development of the software projects based on the selected and prioritized requirements may have the low probability of being rejected. In future studies, we will apply the GOASREP on some other real life based projects like railway reservation system, and will develop the tool to automate all the steps of GOASREP.

References

1. P. Achimugu, A. Selamat, R. Ibrahim, M. Mahrin, A systematic literature review of software requirements prioritization research. Inf. Softw. Technol. **56**, 868–870 (2014)
2. B.C. Chrobot, What is the Cost of One IFPUG Method Function Point? Case Study. worldcomp-proceedings.com/proc/p 2012/SER2400.pdf
3. W. Ho, X. Xu, P.K. Dey, Multi-criteria decision making approaches for supplier evaluation and selection: a literature review. Eur. J. Oper. Res. **202**, 16–24 (2010)
4. M. Sadiq, S.K. Jain, A fuzzy based approach for the selection of goals in goal oriented requirements elicitation process. Int. J. Syst. Assur. Eng. Maintenance, Springer **6**(2), 157–164 (2015)
5. M. Sadiq, S. Ghafir, M. Shahid, An approach for eliciting software requirements and its prioritization using analytic hierarchy process, in *IEEE International Conference on Advances in Recent Technologies in Communication and Computing*, Kerala, India (2009), pp. 790–795
6. M. Sadiq, S.K. Jain, An insight into requirements engineering processes, in *3rd International Conference on Advances in Communication, Network, and Computing*, ed. by V.V. Das, J. Stephen (LNCSIT-Springer, Heidelberg, Chennai, India, 2012), pp. 313–318
7. M. Sadiq, S.K. Jain, Applying fuzzy preference relation for requirements prioritization in goal oriented requirements elicitation process. Int. J. Syst. Assur. Eng. Maintenance, Springer **5**(4), 711–723 (2014)
8. M. Sadiq, M. Shahid, Elicitation and prioritization of software requirements. Int. J. Recent Trends Eng. **2**(3), 138–142 (2009). Academy Publisher, Finland
9. M. Sadiq, J. Ahmad, M. Asim, A. Qureshi, R. Suman, More on elicitation of software requirements and prioritization using AHP, in *IEEE International Conference on Data Storage and Data Engineering*, Bangalore, India (2010), pp. 232–236
10. M. Sadiq, M. Asim, J. Ahmed, V. Kumar, S. Khan, Prediction of software project effort estimation: a case study. IACSIT Int. J. Model. Optim. Singap. **1**(1) (2011)
11. M. Sadiq, S.K. Jain, A fuzzy based approach for requirements prioritization in goal oriented requirements elicitation process, in *25th International Conference on Software Engineering and Knowledge Engineering*, Boston, USA (2013)

Confidentiality and Storage of Data in Cloud Environment

Prerna Mohit and G.P. Biswas

Abstract In this paper, we provide a secure storage of data in the cloud server using Identity-Based Encryption. Where the data owner shares the stored data with the cloud users on the basics of pay-as-you-use principle. The cloud user request for the encrypted data, stored on the cloud server's database (by applying encryption on keyword) and the server verifies the request by performing test on encrypted data. So, that illegal users or unauthorized servers cannot attack on the data.

Keywords Cloud storage · Security · IBE · Cryptography

1 Introduction

Cloud computing is one of the dream of computing, where users can remotely store and search data to enjoy on-demand services. The secure storage of data and its access are two most essential processes of cloud computing. And to evaluate the storage service of the cloud there are some based category such as: Reliability, Security, Speed, Usability. Among them security is one of the most important facture, and to provide the same, data needs to be encrypted before transferring. Similarly the Speed of uploading data file needs to be fast, while it is not necessary that the fastest service is best.

The storage of data over public cloud is one among the many services provided by cloud to its customers, for access the file from any location whenever needed (availability) without worries about backups (reliability). However, confidentiality of data is required before uploading the data to the cloud server, as data is travelling over insecure media.

P. Mohit (✉) · G.P. Biswas
Department of Computer Science and Engineering, Indian Institute of Technology (ISM), Dhanbad 826004, Jharkhand, India
e-mail: prernamohit@outlook.com

G.P. Biswas
e-mail: gpbiswas@gmail.com

© Springer Nature Singapore Pte Ltd. 2017
S.C. Satapathy et al. (eds.), *Proceedings of the 5th International Conference on Frontiers in Intelligent Computing: Theory and Applications*, Advances in Intelligent Systems and Computing 516, DOI 10.1007/978-981-10-3156-4_29

In 2000, Song et al. [1] was the first to propose the searchable encryption scheme, where song maintain the sequential search pattern over encrypted data. After that Goh [2] developed an index based searchable encryption technique using Bloom filters to achieve more efficient system. In 2005 Change et al. [3] proposes two index based scheme over set of documents. Golle et al. [4] design the concept of conjunctive searchable encryption for keywords and also explain the security model for it. In 2004 Boneh et al. [5], give a public-key based searchable encryption technique called Public Key Encryption with Keyword Search (PEKS), suitable for mail server. Where user search for particular file over the mail gateway by sending a trapdoor. Cuetmola et al. [6] review the previous scheme and present an efficient symmetric-key based searchable scheme. Kurosawa et al. [7] develop a variance of SSE, which is secure against active adversaries and universally composable (UC). There are more recent research in cloud storage is shown in [8, 9].

In this paper, we illustrate a new scheme for data storage in cloud server, by using sharing of data among users. The design of our proposal is twofold. (1) It provides secrecy by encryption of data which are stored in public cloud servers. (2) It also provides sharing of data among cloud users, so that untrusted users cannot access or search over data without the knowledge of the data owner.

The rest of the paper is organized in this way: Sect. 2 give introduction of the cloud service model and its components followed by Sect. 3, explain the proposed protocol over the cloud for data storage. Section 4 gives a security analysis of our scheme and finale the conclusion in Sect. 5.

2   Cloud Computing

Cloud computing is a internet based services, provided by cloud servers. It can provides infinite space for the storage of data, on demand. Where users have no idea about the storage and management of data. Hence, we briefly discuss the components of cloud computing.

2.1   Cloud Service Model

Depending on the requirement of users the services of cloud are classified into three types as shown in Fig. 1 and described below.

1. Infrastructure-as-a-service (IaaS)
2. Platform-as-a-service (PaaS)
3. Software as a Service (SaaS).

The term as-as-service means pay-as-you-go. In IaaS, the customer has to pay only for hardware and have to control the operating system and application by its own. Some of common providers are Amazon Web Services, Windows Azure. While

Fig. 1 Cloud service model

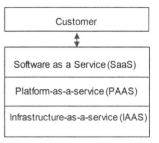

Table 1 List of symbols used

Notation	Description
ID_i	Unique identity of user i
p	Sufficient large prime number
W	Keyword
$H(.)$	One-way cryptographic hash function. $H_1 : \{0,1\}^* \longrightarrow G_1$ $H_2 : G_2 \longrightarrow \{0,1\}^n$
$X \oplus Y$	X xor with Y
$X =?Y$	Whether X equal Y or not
Z_P^*	Set of integers of order p

PaaS is one step forward to it where consumer have to pay for the platform like the operating system and have to maintain the application by its own. Finally, SaaS the customer has to pay for application and does not have to maintain anything. Cloud service provider manage the infrastructure and platforms for running the applications. Some of the most well-known SaaS solutions are Google Apps, Salesforce (Table 1).

2.2 Type of Cloud Computing

The cloud infrastructure can be mainly categorized into three types: public-, private- and hybrid- cloud [10] as shown in Fig. 2.

1. Private clouds are contained within an enterprise. The objective of private cloud is to share the information of organization among themselves.
2. Public clouds are referring to as the cloud computing for public use. It is managed and owned by the cloud service provider (CSP) such as Amazons Simple Storage Service (S3), Microsoft windows azure, cloud sigma etc. on the basis of pay-as-you-use principle. The public cloud data are broadly available and easily seen by other cloud user. Hence, its protection and security is very important.
3. Hybrid cloud are combination of public and private cloud.

Fig. 2 Public, Private and
Hybrid cloud

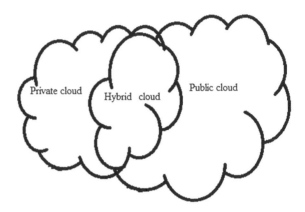

3 Proposed Cloud Scheme

In the proposed scheme, there are four entities (1) The data owner, (2) data consumer, (3) cloud server, (4) Private key Generator (PKG). PKG is responsible for generating public/private key corresponding to users. Data owner encrypts the file along with the keyword and outsource it to the cloud service provider (CSP) for storage. The data consumer request to the CSP for accessing the stored file corresponding to particular owner. And use the public key of owner to encrypt the keywords as a trapdoor and requests the cloud server for the file using keyword by sending trapdoor with its own identity. The cloud server compares the value of trapdoor with the owners encrypted keyword. And if both the value matches, then the server send the encrypted file to the owner. Where owner decrypts the file with its private key (provided by PKG) and again encrypt file with the consumer public key and send to recipient via the cloud server. Where the user can decrypt the file with its private key. The architecture of proposed protocol is shown in Fig. 3.

3.1 Protocol Description

1. The data owner A encrypts file F and keyword W using Encrypt (F, W) function.
2. The data user B send request to the cloud server regarding file of owner A without knowing file name. And the request is send in the form of Trapdoor (W') function using encryption of keyword, along with it's identity.
3. PKG generate key pair for the user B as $KeyGeneration_B(ID_B)$ function.
4. The cloud server perform Test (K, K') function on the received encrypted keywords (K, K'). If invalid terminate, otherwise
5. The server send the encrypted file C along with ID_B to the owner.
6. After receiving, owner perform Decrypt (U, V) with its own secret key.

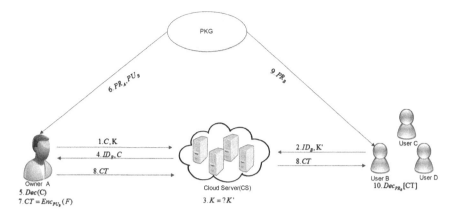

Fig. 3 Architecture of protocol

7. Re-encrypt (F) the file F with the public key of user B and send the cipher-text (CT) to B via cloud server.
8. B will use its private key to Re-decrypt (U', V') the received file F.

Function definition

- *Setup*: To set up the system, two groups G_1 and G_2 of order q, and an arbitrary generator $P \in G_1$. The bilinear map $e : G_1 \times G_1 \longrightarrow G_2$ and two cryptographic hash functions $H_1 : \{0, 1\}^* \longrightarrow G_1$ and $H_2 : G_2 \longrightarrow \{0, 1\}^n$. The system parameters are $(p, q, G_1, G_2, e, P, P_1)$, where $P_{pub} = sP$, where $s \in Z_q^*$ is master secret and only known to receiver.
- *KeyGeneration$_A$(ID$_A$)*: PKG compute $Q = H_1(ID_A)$ and secret key as $d = s \times Q$ where d is private key
- *Encrypt(F, W)*: The data owner encrypt File $C = \langle r \times P, F \oplus H_2(Q, P_{pub})^r \rangle$ and keyword $K = w \times P_{pub}$ where $w = H_1(W)$
- *Trapdoor(W)*: Data uses B compute $K' = w' \times P_{pub}$ where $w' = H_1(W')$ and P_{pub} is public key of owner.
- *KeyGeneration$_B$(ID$_B$)*: PKG generate the public, private-key for user B by computing $P'_{pub} = b \times P$; $Q' = H_1(ID_B)$; $d' = b \times Q'$. where P'_{pub}
- *Test(K, K')*: Cloud server verify both the received values of encrypted keywords from owner and user and compute. $K = K'$
- *Decrypt(U, V)*: Data owner decrypt the file as $F = V \oplus H_2(U, d)$
- *Re-encrypt(F)*: Data owner perform $CT = \langle t \times P, F \oplus H_2(Q', bP'_{pub})^t \rangle$ where t is randomly selected by owner.
- *Re-decryption(U', V')*: The user perform $F = V' \oplus H_2(U', d')$

4 Security Analysis

In this section, we give security analysis of our proposed protocol in terms of important parameters, such as: confidentiality of data and its access control for secure storage of data over cloud server.

1. **Confidentiality**: We provide confidentiality of data using identity-based encryption technology for secure data storage and sharing purposes. The data owner encipher the data before storing it to the server with its public key. The data user request for accessing data by deciphering the stored data locally by asking for its private key from PKG and again encrypt the data with the requested users public key and send it to the desired receiver where the user can decrypt the data with its private key to get the original data. Hence, in the whole process, confidentiality is maintained as encrypted data are travelling over insecure channels. In case the attacker gets access to the data he cannot obtain any information regarding the data. Hence, impossible for the cloud server or attacker to retrieve.

2. **Data Sharing**: The data sharing is performed among the users of the cloud. Whenever a particular user wants to access on uploaded data, he first sent its identity to the cloud server, based on that the server checks whether the user is a genuine cloud user or not. If it is an authenticated user then only the server further process otherwise it will terminate the process. Even though an attacker gain access to the data, confidentiality is insured. Hence, unauthorized user or servers are not able to see the data. In other words, attackers can only have access to encrypted data and don't have the private key of user to decipher data.

3. **Secure Storage**: The cloud server securely store the data into its database without knowing the content of data as the encrypted data are stored corresponding to the data owner in linked list structure so that server can easily search the encrypted file.

4. **Access control to data**: The data stored in cloud server is shared with cloud users. The user first authenticate itself to server whether s/he is genuine cloud user or not and request for particular file. After performing some operations the server send the encrypted file to data user. Hence, access of data is granted to genuine cloud user (Table 2).

Table 2 Required security

Security against	Achieve
Confidentiality	Yes
Data sharing	Yes
Secure storage	Yes
Access control	Yes

5   Conclusion

With the fast development of cloud storage services leads us to define the solution for secure data uploading. The proposed protocol is based on an identity-based encryption technology in the cloud environment for secure storage and searching of data. Where a data owner encrypts the data and stores it on cloud server for the excess of required authenticated users.

References

1. D.X. Song, D. Wagner, A. Perrig, Practical techniques for searches on encrypted data, in *IEEE Symposium on Security and Privacy. S&P 2000. Proceedings*. IEEE (2000), pp. 44–55
2. E.J. Goh, Secure Indexes. IACR Cryptology ePrint Archive, 216 (2003)
3. Y.C. Chang, M. Mitzenmacher, Privacy preserving keyword searches on remote encrypted data, *Applied Cryptography and Network Security* (Springer, Berlin, 2005)
4. P. Golle, J. Staddon, B. Waters, Secure conjunctive search over encrypted data, in *ACNS 2004, Lecture Notes in Computer Science*, vol. 3089 (Springer, 2004), p. 314
5. D. Boneh, G.D. Crescenzo, R. Ostrovsky, G. Persiano, Public key encryption with keyword search, in *Proceedings of EUROCRYP04, LNCS*, vol. 3027 (Springer, 2004)
6. R. Curtmola, J. Garay, S. Kamara, R. Ostrovsky, Searchable symmetric encryption: improved definitions and efficient constructions, in *Proceedings of the 13th ACM Conference on Computer and Communications Security* (ACM, 2006), pp. 79–88
7. K. Kurosawa, Y. Ohtaki, UC-secure searchable symmetric encryption, *Financial Cryptography and Data Security* (Springer, Berlin, 2012)
8. Z. Jianhong, D. Qiaocui, Efficient ID-based public auditing for the outsourced data in cloud storage. Inf. Sci. **343**, 1–14 (2016)
9. V. Chang, W. Gary, A model to compare cloud and non-cloud storage of big data. Future Gener. Comput. Syst. **57**, 56–76 (2016)
10. L. Yan, C. Rong, G. Zhao, Strengthen cloud computing security with federal identity management using hierarchical identity-based cryptography, in *Cloud Computing* (2009), pp. 167–177

A Method for the Selection of Agile Methods Using AHP

Bushra Sayed, Zeba Shamsi and Mohd. Sadiq

Abstract There are different types of lightweight methods for the development of software like eXtreme Programming (XP), scrum, agile modeling, etc. These methods are also referred to as agile methods. Different criteria's are involved during the selection of agile methods so we visualize the agile methods selection problem as a multi-criteria decision making problem. Selection of an appropriate agile method according to the need of the project is an important research issue. Therefore, in order to address this issue, we present a method for the selection of agile methods using Analytic Hierarchy Process (AHP). Following criteria's have been used for the selection of agile methods, i.e., *positive response in dynamic requirements* (PRDR), *incorporation of requirements changes* (IRC), *communication with the customer* (CWC), and *the size of development team* (SDT). Finally, a case study is given to explain the proposed method.

Keywords Agile methods · SDLC · Multi-criteria decision making · AHP

1 Introduction

A *"software development life cycle"* (SDLC) model is a process which is employed to describe and represent the various stages of software development, i.e., planning, analysis, design, coding, and testing. In [1] we classify the SDLC models into three

B. Sayed · Z. Shamsi
Department of Computer Science and Engineering, School of Computer Science,
AL-Falah University, Dhauj, Faridabad, Haryana, India
e-mail: bushrasayed77@gmail.com

Z. Shamsi
e-mail: zebashamsi1@gmail.com

Mohd. Sadiq (✉)
Computer Engineering Section, UPFET, Jamia Millia Islamia (A Central University),
New Delhi 110025, India
e-mail: sadiq.jmi@gmail.com; msadiq@jmi.ac.in

© Springer Nature Singapore Pte Ltd. 2017
S.C. Satapathy et al. (eds.), *Proceedings of the 5th International Conference on Frontiers in Intelligent Computing: Theory and Applications*, Advances in Intelligent Systems and Computing 516, DOI 10.1007/978-981-10-3156-4_30

297

parts: lightweight models (or agile models), heavyweight models, and hybrid models. In the past few years, among these models lightweight models have received much attention by researcher and academicians because it produces the quick software development without considering the much documentation.

The word *"agile"* can be defined as the *"ability to think and draw conclusion quickly"* [1]. In literature, different types of agile methods have been proposed like *"Extreme Programming(XP)"*, *"Dynamic System Development Methodology (DSDM)"*, *"Adaptive Software Development (ASD)"*, *"Scrum"*, *"Crystal"*, and *"Feature Driven Development (FDD)"*, etc. [1–7].

In agile software development environment, it is difficult to decide that which agile method should be used for the development of software. There are different methods for the selection of SDLC models. For example, Geambasu et al. [8] *"identify the influence factors for the choice of software development methodology"*. In 2012, Hicdurmaz [9] present a *"fuzzy criteria decision making approach to software lifecycle model selection"*. In [1] we *"proposed a method for the selection of SDLC models using AHP"*. In this method we focus on lightweight methods, heavyweight methods, and hybrid methods, without giving much consideration about the selection of agile methods. So it motivates us to extend our previous work by giving more attention on agile methods. Therefore, the objective of this paper is to propose a method for the selection of agile methods using AHP.

The remaining part of the paper is organized as follows: Sect. 2 presents an insight into analytic hierarchy process. In Sect. 3, we present the proposed method, i.e., *selection of agile methods using AHP*. Case study is given Sect. 4; and finally, Sect. 5 contains the conclusion and future work.

2 Analytic Hierarchy Process (AHP)

AHP was developed by *Thomas L. Saaty* in 1970. It is a *"multi-criteria decision making"* (MCDM) method which is used to select alternatives or requirements on the basis of the different criteria's [10]. There are different applications of AHP in literature [5, 11–13], for example, Sadiq et al. [11] proposed "an approach for eliciting software requirements and its prioritization using AHP". With the help of AHP, in 2014, Sadiq et al. [13] proposed another method *"for the selection of Software Testing Techniques"*.

In AHP, pair-wise comparison matrix is employed to give the preference of one alternative over another. For pair-wise comparison in AHP, Saaty rating scale is used, as given in Table 1.

Table 1 Rating scale proposed by Saaty

Numeric values for comparisons	Meaning
1	"Equal importance"
3	"Somewhat more importance"
5	"Much more important"
7	"Very much important"
9	"Absolutely more important"
2, 4, 6, 8	"Intermediates values (when compromise is needed)"

3 Proposed Method

In this section, we present a method for the selection of agile methods using AHP; and it includes the following steps:

Step 1: Identification of the criteria
Step 2: Draw the hierarchical structure of agile models
Step 3: Pair-wise comparison matrix
Step 4: Computation of ranking values
Step 5: Selection of an agile method

Step1: Identification of the criteria

This step is used to identify the criteria for the selection of the agile methods. We have identified the following criteria for the selection of agile methods: (i) positive response in dynamic requirements (PRDR), (ii) incorporation of requirements change (IRC), (iii) communication with the customer (CWC), and (iv) size of development team (SDT).

Step 2: Draw the hierarchical structure of agile models

In the step, the hierarchical structure of the agile method is constructed. In Fig. 1, at first level, the overall objective of the problem is given, i.e., selection of agile method. At level two, different agile methods are given; and the criteria for the selection of the agile methods are given at level three.

Step 3: Pair-wise comparison matrix

Decision matrix would be created by using AHP method. Detailed description about the construction of pair-wise comparison matrix is given in Sect. 4, i.e., case study.

Step 4: Computation of ranking values

For the computation of the ranking values we have used the following algorithm:

4.1: Construct the pair-wise comparison matrix for criteria and alternatives; and it is represented by criteria matrix (CM) and alternative matrix (AM) respectively.

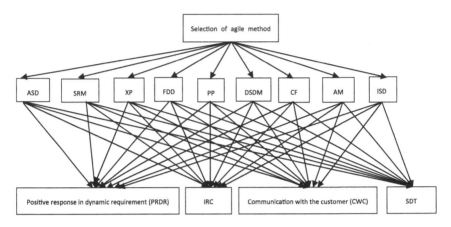

Fig. 1 Hierarchical structure for the selection of agile methods

4.2: *Normalize CM and AM.*
4.3: *Calculate the sum of each row of the normalization matrix and it is represented by weight sum vector matrix i.e., WSVM and then calculate the average of weight sum vector matrix (WSVM), it is called Criteria Weight Matrix (CWM).*
4.4: *Find out the consistency vector by using the following relationship WSVM* (1/CWM) and find out the average of the consistency vector.*

Step 5: Selection of an agile method

On the basis of the ranking values an appropriate agile model would be selected.

4 Case Study

Before starting the project, selection of an agile method is an important research issue. We know that success of any project depends upon the selection of an appropriate SDLC model. The objective of the proposed method is the selection of an agile method. In this work, we have considered the following agile methods: "*Adaptive Software Development (ASD), Scrum (SRM), Extreme Programming (XP), Functional Driven Development (FDD), Pragmatic Programming (PP), Dynamic System Development Method (DSDM), Crystal Family (CF), Agile Modeling (AM), and Internet Speed Development (ISD)*". An example of Institute Examination System is used in this study.

Step 1: Following criteria's are employed for the selection of agile methods, i.e., PRDR, IRC, CWC, and SDT.
Step 2: The hierarchical structure of the selection of agile methods is given in Fig. 1.

Step 3: We construct decision matrix by evaluating the different agile methods on the basis of the criteria, as given in step 1. Firstly, we will have to construct the initial matrix for pair-wise comparison. In initial matrix, the diagonal entries are one, as shown in Table 2.

Step 4: Now we compare different criteria's according to their importance. We first compare, PRDR with IRC. In our project, dynamic adaptation of the requirements is more important than other criteria. Therefore, we put 3 in the PRDR (row) and IRC (column); and according to the AHP method the contents of the IRC (row) and PRDR (column) would be 1/3. On the basis of the importance of one criterion over another, as per the need of the project, we complete the enterers of Table 3.

Once we have evaluated different criteria's, then the next step would be to evaluate different agile methods on the basis of different criteria's. Therefore, in Table 4, we evaluate different agile methods on the basis of PRDR.

In similar way, we evaluate the different agile methods on the basis of remaining three criteria's, i.e., IRC, SDT, and CWC.

After applying the algorithm of step 4, we identify that PRDR is an important criterion because its ranking value is 0.405.

Finally, we get the following ranking values for agile methods: Scrum = 0.2384; ASD = 0.25151596; CF = 0.3843; DSDM = 0.09416; FDD = 0.2097; ISD = 0.0786; PP = 0.0478; AM = 0.3364262.

Step 5: On the basis of our analysis we have found that Crystal Family (CF) and Agile Modeling (AM) have the highest priority because the ranking values of Crystal Family (CF) and Agile Modeling (AM) are 0.3843 and 0.3364262 respectively. Therefore, these two agile methods would be suitable for the development of IES.

Table 2 Diagonal matrix

Criteria	PRDR	IRC	SDT	CWC
PRDR	1			
IRC		1		
SDT			1	
CWC				1

Table 3 Overall performance matrix

Criteria	PRDR	IRC	SDT	CWC
PRDR	1	3	3	3
IRC	1/3	1	1	1/5
SDT	1/3	1	1	5
CWC	1/3	5	1	1

Table 4 Evaluation of different agile methods on the basis of PRDR

Agile methods	XP	SRM	ASD	CF	DSDM	FDD	ISD	PP	AM
XP	1	3	1/3	1	3	1	5	5	1/3
SRM	1/3	1	3	1	2	1	3	5	1
ASD	3	1/3	1	1	5	1	5	7	3
CF	1	1	1	1	3	1/3	3	5	1/3
DSDM	1/3	1/2	1/5	1/3	1	1/5	3	3	1/5
FDD	1	1	1	3	5	1	3	3	1
ISD	1/5	1/3	1/5	1/3	1/3	1/3	1	1	1/5
PP	1/5	1/5	1/7	1/5	1/3	1/3	1	1	1/5
AM	3	1	1/3	3	5	1	5	5	1

5 Conclusion and Future Scope

In this paper we present a method for agile methods selection using AHP. There are five steps in the proposed method, i.e., (i) identification of the criteria, (ii) draw the hierarchical structure of agile models, (iii) pair-wise comparison matrix, (iv) computation of ranking values, (v) selection of an agile method. Proposed method selects an agile method for the development of the project namely Institute Examination System (IES). As a result we identify that Crystal Family (CF) and Agile Modeling (AM) are suitable for the development of IES. In our study, we identify that *"there is a need to improve the agile methods by intertwining of decision making approaches for the selection and prioritization of requirements under fuzzy environment"*. Therefore, in future, we will propose a method to address the above issue.

References

1. M.A. Khan, A. Parveen, M. Sadiq, A method for the selection of software development life cycle models using analytic hierarchy process, *in IEEE International Conference on Issues and Challenges in Intelligent Computing Techniques* (2014), pp. 539–563
2. J. Highsmith, A. Cockburn, Agile Software Development: the People Factor. IEEE Comput. **34**, 131–133 (2001)
3. M. Sadiq, T. Hassan, An extended adaptive software development process model, in *IEEE International Conference on Issues and Challenges in Intelligent Computing Techniques* (2014), pp. 552–558
4. M. Sadiq, J. Ahmad, M. Asim, A. Qureshi, R. Suman, More on elicitation of software requirements and prioritization using AHP, in *IEEE International Conference on Data Storage and Data Engineering* (Bangalore, India 2010), pp. 232–236
5. M. Sadiq, F. Firoze, A method for the selection of software testing automation framework using analytic hierarchy process. Int. J. Comput. Appl. **102**(1) (2014)
6. K. Schwaber, M. Beedle, *Agile Software Development with Scrum, Upper saddle River* (Prentice Hall, NJ, 2002). ISBN 0130676349

7. I. Sommerville, *Software Engineering*, 5th edn. (Addison Wesley, New York, 1996)
8. C.V. Geambasu, I. Jianu, Influence factors for the choice of a software development methodology. Account. Manag. Inform. Syst. **10**(4), 479–494 (2011)
9. M. Hicdurmaz, A Fuzzy Multi Criteria Decision making Model Approach to Software Life Cycle Model Selection, in *38 IEEE EUROMICRO Conference on Software Engineering and Advanced Applications* (2012), pp. 384–391
10. T. Saaty, *The Analytic Hierarchy Process* (McGraw-Hill, New York, 1980)
11. M. Sadiq, S. Ghafir, M. Shahid, An approach for eliciting software requirements and its prioritization using analytic hierarchy process, in *IEEE International Conference on Advances in Recent Technologies in Communication and Computing* (Kerala, India 2009), pp. 790–795
12. M. Sadiq, S.K. Jain, A fuzzy based approach for requirements prioritization in goal oriented requirements elicitation process, in *25th International Conference on Software Engineering and Knowledge Engineering*, Boston, USA, June 27–June 29 (2013)
13. M. Sadiq, M.Sultana, A method for the selection of software testing techniques using analytic hierarchy process, in *International Conference on Computational Intelligence in Data Mining*, (Springer, 2014) vol 1, pp. 213–220
14. M.R.J. Qureshi, S.A. Hussain, An Adaptive Software Development Process Model. Adv. Eng. Softw. Elsevier **39**, 654–658 (2008)
15. A.M. Sen, S.K. Jain, An agile technique for agent based goal oriented requirements engineering, in *IEEE International Conference on Advance Computing and Communication* (2007), pp. 41–47

Load Balancing with Job Switching in Cloud Computing Network

Ranjan Kumar Mondal, Subhranshu Roy, Palash Samanta,
Payel Ray, Enakshmi Nandi, Biswajit Biswas, Manas Kumar Sanyal
and Debabrata Sarddar

Abstract Cloud computing, described as distributed online computing, is a kind of
Internet-based computing that provides pooled web resources and applications to
connected servers and other machines on user's demand. It is a web system for
enabling ubiquitous, on-demand access to a shared pool of configurable computing
resources which can be rapidly provisioned and released with minimal management
effort. Load balancing is an important issue in the cloud computing. Cloud computing comprises of many web resources and managing. This plays a vital role in
executing a user's request. In this present condition the load balancing algorithms
should be very efficient in allocating the user request and also ensuring the usage of
the resources in an intelligent way so that underutilization of the resources will not
occur and preemptive based resource management be there in the cloud environment. Cloud computing services different types of nodes connected to cloud to

R.K. Mondal (✉) · P. Ray · E. Nandi · D. Sarddar
Department of Computer Science and Engineering, University of Kalyani, Kalyani, India
e-mail: ranjan@klyuniv.ac.in

P. Ray
e-mail: payelray009@gmail.com

E. Nandi
e-mail: pamelaroychowdhurikalyani@gmail.com

D. Sarddar
e-mail: dsarddar1@gmail.com

S. Roy · B. Biswas · M.K. Sanyal
Department of Business Administration, University of Kalyani, Kalyani, India
e-mail: subhranshu_81@yahoo.com

B. Biswas
e-mail: biswajit.biswas0012@gmail.com

M.K. Sanyal
e-mail: manas_sanyal@rediffmail.com

P. Samanta
Department of Engineering and Technology, University of Kalyani, Kalyani, India
e-mail: samanta.palash30@gmail.com

© Springer Nature Singapore Pte Ltd. 2017   305
S.C. Satapathy et al. (eds.), *Proceedings of the 5th International Conference on Frontiers
in Intelligent Computing: Theory and Applications*, Advances in Intelligent Systems
and Computing 516, DOI 10.1007/978-981-10-3156-4_31

assist the execution of great number of tasks. As a result, to select suitable node or machine for executing a task is able to develop the performance of cloud computing system. A job switching is the switching of the job from one machine to another machine to minimize the overall completion time. In this paper, we propose a load balancing algorithm combining minimum completion time as well as load balancing strategies with job switching.

Keywords Job switching · Minimum completion time · Load balancing · Distributed system · Cloud computing

1 Introduction

Nowadays cloud computing [1] can be well-defined as an extension of online web computing. It means to offer secure, quick, convenient data storage and computing service. It conveys three types of services: Software as a Service, Platform as a Service and Infrastructure as a service.

Cloud provides online resources flexible manner that involves virtualization web services. It has several elements like clients (users) and servers. The objective of cloud system is to provide web services with lowest cost at any time. Nowadays, there are more than millions of computer machines connected to the Internet. These machines submit their request and receive the response without delay. The main objectives of cloud system are to reduce cost, enhance response time, provide better performance etc.

Cloud computing is considered as the best developing web technology relying on distributed online resources over different area to provide web services efficiently to users on their demand. Requests from online cloud users are distributed to different machines unbalancing the load assignment and this is considered as one of the disadvantages of cloud computing.

2 Load Balancing

Loads are various kinds like CPU, bandwidth, memory concern and so on. Therefore, loads need to be achieved and it could be done in two ways i.e. load balancing as well as load sharing. Many load balancing algorithms are recommended through which loads can be distributed equally with minimum completion time (MCT). Load Balancer is used here to balance the loads.

Load balancing [2] algorithm can be classified into two category i.e., one is static LB and another is dynamic LB.

Load balancing is applied to a great amount of data traffic and servers to distribute work. Advanced architectures in cloud are adopted to achieve speed and efficiency. There are several characteristics of load balancing such as: equal division of work across all the nodes, improve total performance of system and reduce response time.

As an example, if we create one application on cloud and hundreds of users are expected to access it at any time. Therefore, response time to user would be very slow and servers will become busy very quickly, resulting in slow response and unsatisfactory users. If we apply load balancing on our application, then work will be distributed at other nodes and we can get high performance and better response.

3 The Proposed Work

There are different machines (or nodes) in a cloud computing system. Each machine has different ability to perform jobs; hence, only consider the CPU remaining of the machine is not so sufficient when a machine is selected to execute a job. So, therefore, to choose an appropriate machine to execute a job is very important issue in cloud computing network.

Due to task that it has different skills for customer to pay execution. Hence it is needed to some of web resources, for instance, when implementing organism sequence assembly; it is probably have to big requirement toward CPU and memory remaining of nodes. And in order to reach the best efficient in the execution all tasks, so we will aim to tasks property to adopt a different condition decision variable in which it is accorded to resource of task requirement to set decision variable.

3.1 Problem Definition

Due to the problem of the load balancing with minimum completion time of a task. In this paper we are going to discuss proposed algorithms that have been developed under some assumptions.

3.2 Method

The development of our algorithm is offered as following:

Part 1

The progress of our proposed work is presented as following:
  Step 1: It is to compute Average Completion Time (ACT) of each node for all tasks, individually.

Step 2: *It is to find the corresponding node that has the maximum average completion time.*

Step 3: *It is to find the task (i.e. not assigned) has the minimum completion time for the task selected in Step 2. Then, this task is forwarded to the selected unassigned node from the matrix. If there are two or more unassigned minimum task then select the corresponding task having maximum task average value (threshold of task) of all nodes.*

Step 4: *Now if there is no unassigned task selected in Step 2, then all nodes should be reexamined.*

Step 5: *Then repeat Step 2 to Step 4, until all nodes have been assigned by their corresponding task.*

Part 2

Step 1: *Find the task of corresponding node that has to be switched*

Step 2: *Find the node having minimum execution time of that task if available that node otherwise select next minimum node having next minimum execution time*

Step 3: *If node available for switching that task so it takes the responsibility of the task as soon as completion its first task of that node*

After first job finish (that has minimum execution time) switches that job of remaining parts to corresponding nodes of first job and wait for execute remaining parts by of first node

Step 4: *Calculate Total Completion Time (TCT) of related executing nodes.*

4   Example

Example 1

Part 1: We assign each task to its corresponding nodes with part 1.

Tasks	Nodes			
	C_{11}	C_{12}	C_{13}	C_{14}
T_1	**18**	14	38	26
T_2	14	12	24	**18**
T_3	26	**18**	66	42
T_4	19	20	**24**	36
ACT	19.25	16	38	30.5

Part 2

Part 2 minimize more completion time with job switching as follows.

All Tasks finish their job in 18 s except Task T_4 of C_{13}. But after 18 s C_{13} node executes its corresponding task 75 % (total task is finished in 24 s so 75 % task executes in 18 s).

Now A_4 task switches its remaining task to the node that takes minimum completion time of Task T_4. That is C_{11} it takes 19 s of task T_4.

Now the calculation is execute T_4, C_{13} takes 24 where same task to execute C_{11} takes 19 s so 100 % task of T_4 meaning C_{11} takes 5 s (approximately) to execute remaining 25 % task.

Total time to execute of T_4 takes 18 + 5 = 23 s (18 s for C_{13} and 5 s for C_{11}). And C_{11} executes T_1 and T_4 in 18 + 5 = 23 s.

Tasks	Nodes			
	C_{11}	C_{12}	C_{13}	C_{14}
T_1	**18**	14	38	26
T_2	14	12	24	**18**
T_3	26	**18**	66	42
T_4	19(5)	20	24(18)	36
TCT	**23**	**18**	**18**	**18**

Example 2

Part 1: Similarly we assign each task to its corresponding nodes with part 1 like example 1.

Tasks	Nodes			
	C_{11}	C_{12}	C_{13}	C_{14}
T_1	12	13	10	**14**
T_2	**16**	24	13	25
T_3	26	31	**12**	33
T_4	17	**24**	18	31
ACT	17.25	**23**	13.25	23.75

Part 2

Part 2 minimize more completion time with job switching as follows.

Task T_3 of C_{13} node finish its job in 12 s. But after 12 s C_{12} node executes its corresponding task 50 % (total task is finished in 24 s so 50 % task executes in 12 s).

Now T_4 task switches its remaining task to C_{13}.

Now the calculation is execute T_4, C_{12} takes 24 where same task to execute C_{13} takes 18 s so 50 % task of T_4 meaning C_{12} takes 12 s then C_{13} will take 9 s.

Total time to execute of T_4 takes 12 + 9 = 21 s. And C_{13} executes T_3 and T_4 in 21 s.

Tasks	Nodes			
	C_{11}	C_{12}	C_{13}	C_{14}
T_1	12	13	10	**14**
T_2	**16**	24	13	25
T_3	26	31	**12**	33
T_4	17	**12(24)**	**9**(18)	31
TCT	16	**12**	**21**	14

5 Comparison

Example 1: Part 1 is showing minimum completion time of all nodes and part 2 is showing to balance the minimum completion time between nodes.

Part 1: The completion time (in second) of subtasks of different nodes (Fig. 1).

Part 2: The completion time (in second) of subtasks of different nodes after job switching (Fig. 2).

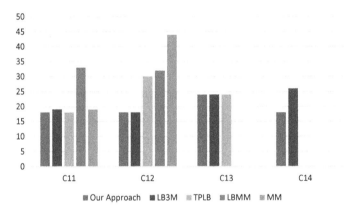

Fig. 1 The comparison of minimum completion time of subtasks of different nodes

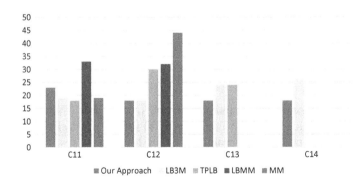

Fig. 2 The balancing of completion time of each subtask with job switching of different nodes

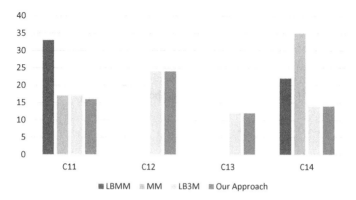

Fig. 3 The comparison of minimum completion time of all subtasks of different nodes

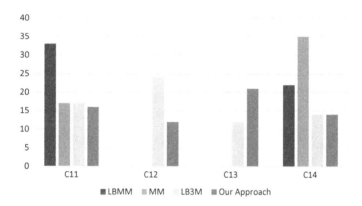

Fig. 4 The balancing of completion time of each subtask with job switching of different nodes

Example 2

Part 1: The completion time (in second) of subtasks of different nodes (Fig. 3).
Part 2: The completion time (in second) of subtasks of different nodes after job switching (Fig. 4).

6 Conclusion

In this present paper, we have proposed a load balancing with job switching scheduling algorithm, LBJS for the cloud computing network to assign subtasks of a task or tasks to available machines according to their subtask completion time. However, the load balancing of cloud network is used. Thus, the objective of loading balancing and better resources handling could be achieved. Our proposed algorithm is very helpful to balance the load in cloud computing networks. This algorithm minimizes the completion time of executing nodes in cloud computing system. So that, all nodes in the system complete their corresponding subtask nearly about their average complete time. Similarly, LBJS can get improved load balancing and performance than other load balancing algorithms, such as LBMM [3], MM [4], LB3M [5] and TPLB [6] from the above case study.

References

1. A. Sinha. Cloud computing in libraries: opportunities and challenges. Pearl: A J. Libr. Inf. Sci. 113–118 (2016)
2. R.K. Mondal, P. Ray, D. Sarddar, Load balancing. Int. J. Res. Comput. Appl. Inf. Technol. 01–21 (2016)
3. T. Kokilavani, A. DI George, Load balanced min-min algorithm for static meta-task scheduling in grid computing, in *IJCA* (2011), pp. 43–49
4. M. Maheswaran, S. Ali, J. Siegal, D. Hensgen, F. Freund, Dynamic matching and scheduling of a class of independent tasks onto heterogeneous computing systems, in *Heterogeneous Computing Workshop Proceedings Eighth* (IEEE,1999), pp. 30–44
5. L. Hung, H. Wang, C. Hu, Efficient load balancing algorithm for cloud computing network, in *ICIST*, pp. 28–30, Apr 2012
6. C. Wang, Q. Yan, P. Liao, S. Wang, Towards a load balancing in a three level cloud computing network, in *CSIT* (2010), pp. 108–113

An Efficient Detection and Mitigation of Selfish and Black Hole Attack in Cognitive Mobile Ad-hoc Network

Deepa and Jayashree Agarkhed

Abstract Cognitive radio network (CRN) is one where the licensed unused band can be redistributed among the demanding of users without an access to the licensed bands. It is essentially a programmable software radio. Such architecture suffers from different types of attacks, i.e. selfish attack and black hole attack. In this work, the problem is solved by providing an integrated solution for detecting selfish and black hole attacks and once the attack is detected that is being mitigated to all secondary user (SU), such that the SU would blacklist the attacking node. The result shows that. Such attack detection and mitigation improve the network quality significantly by allowing the node to black listed the attacking node and re-utilize of the spectrum among the among the SU nodes.

Keywords Cognitive radio network · Selfish attack · Homogeneous

1 Introduction

The wireless mesh network is literally a CRN, to connect the underneath node to the internet, generally it connects to the internet through a gateway router. This gateway is just like a router, which gets connected to the internet. The other device has got connected through the internet using a modem or using the 3G connection to the mobile, from mobile it gets connected directly through the gateway over a 3G connection, this is how the basic traditional way of getting connected to the internet. The problem here is that the number of devices which are increased in the last five years. Every device likes mobile, laptop they require an internet connection, but every device doesn't have 3G connectivity, like a laptop, it is not possible to get connected to the internet. It can get connected, laptop to the internet through the

Deepa (✉) · J. Agarkhed
Poojya Doddappa Appa College of Engineering, Kalaburagi, Karnataka 585102, India
e-mail: deepaammanna@gmail.com

J. Agarkhed
e-mail: jayashreeptl@yahoo.com

© Springer Nature Singapore Pte Ltd. 2017
S.C. Satapathy et al. (eds.), *Proceedings of the 5th International Conference on Frontiers in Intelligent Computing: Theory and Applications*, Advances in Intelligent Systems and Computing 516, DOI 10.1007/978-981-10-3156-4_32

modem. These devices are called unlicensed users. For example Wifi is the type of access point that connected to the modem and the modem is attached to the gateway, through the internet. This wifi can service up to 4–5 devices, as the number of devices increases one access point is not supported. For this multiple access point is used and these are getting connected with each other. This entire network is known as a mesh network.

Different type of underneath radio is known as a channel. For example Bluetooth is a channel, 3G is a channel, Ethernet is also known as a channel.

The phenomena of a device switching from 3G to any of the multiple radio channels are known as vertical handoff. The decision is took by the network, when the devices are going away from one base station to another base station, the current base station automatically passes our data to the next base station. Such kind of hand off is known as horizontal hand off.

In a wireless network, a fixed spectrum is allocated to the licensed users (primary user), the primary user (PU) has purchased the spectrum from the government. The base station uses a frequency of 2.4–5 GHz, which is used by unlicensed users (secondary users). Due to fixed spectrum, there is a lot of wastage of bandwidth & the unused users become overcrowded. To overcome this, the new architecture comes into existence that is Cognitive Radio Network (CRN). In CR, the Spectrum is always sensed the primary user is active or not, if the primary user is inactive then it is called a spectrum hole. In CR allows the unlicensed users (SU) to use the primary user's unused bandwidth, if the PU is active than SU gives back the service.

The other type of network is cognitive, ad-hoc network. The ad-hoc network is used in CRN to avoid the network setup problem. If consider the example of airport wifi network it has a huge area, single router cannot handle all the devices. The wifi repeaters is used. Routers are used to handle the both PU and SU. CR network is opportunistic to the SU; it switches the spectrum to the PU to the SU, when PU is inactive. These devices are moving randomly, as they move the network topology and protocol is changed. To avoid this, ad-hoc network is used in the CR network. Cognitive ad-hoc network is also having some threats and challenges that degrade the performance of network i.e. selfish attack and black hole attack.

The Proposed work is organized in 6 sections. Section 1 presents a general introduction of cognitive radio network and ad-hoc network. Section 2 presents the related work of the different types of attacks. Section 3 presents the design of the proposed system with block diagram are discussed. Section 4 presents Results and Discussion. Section 5 presents performance analysis. Section 6 concludes the work with future enhancement.

2   Related Work

The main objective of the authors [1] is to analyze the software defined radio (SDR) based on cognitive radio, threats and security issue of the main recent advancements and architectures of Software Defined radio and cognitive radio networks [2]. The primary user emulation (PUE) attacks in cognitive radio networks; this type of attack is operating in the white spaces of the digital TV band. In this attack, the SU emulates the primary user and then access the primary user signal. To avoid this attack primary user generates a pseudo-random AES-encrypted reference signal that is used as the segment sync bits. The synch bit is used to detect a malicious user [3]. Primary user emulation (PUE) attacks on CR network. They used different methods to prevent the PUE attack in CR network. The method is database assisted PUE detection approach. In this method, they have taken two databases, one for local database is integrated in each SU, the second is the global database is built up in the cognitive BS [4]. Problem of localization of the base station in wireless networks has been mainly studied in a non-adversarial setting. A number of solutions have been proposed to detect and prevent attacks on localized systems. They propose a new method to secure localization based on hidden and mobile base stations using the covert base station [5]. Investigated how to improve the security of collaborative sensing. Particularly, they develop a malicious user detection algorithm that calculates the suspicious level of consistency of secondary users based on their past reports and they calculated that how much time the node behaves badly.

From the literature survey [6–10], it can be observed that, the problems of selfish attack and black hole attacks are non deterministic hence the method for detecting these attacks is proposed here detection. The proposed method is based on the route error (RERR) and nodes of energy.

3 Proposed Work

The BS handles both PU and SU. If primary user is not accessing the channel, CRN It's allocate spectrum dynamically to the users.

Figure 1 shows the channel pre-occupation selfish attack; an attack can occur in the communication environment that is used to broadcast available channel information. This information is broadcast to the neighboring secondary legitimate user (LSU). This information includes all of the other neighboring user's channel allocation information. The selfish secondary user (SSU) will broadcast fake or available channel information. In wireless network channel is divided into two bandwidths, one is controlled bandwidth and the other one is a data bandwidth both are orthogonal to each other, such that they don't intercept with each other and control bandwidth will always be given higher priority to control channels. The base station periodically broadcasts the data on how many channels are free. Now in this attack there are 5 slots are there in that 1 slot is used. Selfish user is receiving the signal of 4 slots is free, but SSU what it does mean it showed only one free slot in only one particular user and 2 free slots for the other legitimate user. This legitimate user is not able to use this channel. In this case, SSU is not emulating any PU rather it emulates the data coming from the base station. In some case, the black hole attack is also done in order to provide the false route information.

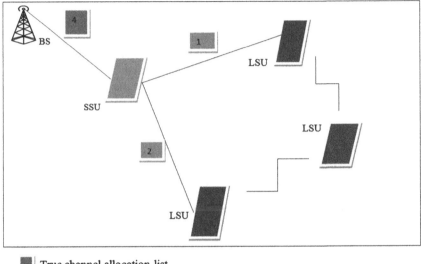

True channel allocation list

Fake channel allocation list

Fig. 1 Selfish attack

Figure 2 shows the detection of selfish attack and black hole attack. In CRN there are two frequencies. One is licensed and another one is an unlicensed frequency. The licensed frequency is operated by the BS. The licensed frequency is operating the PUS and unlicensed frequency is operating the SUS. The unlicensed frequency is also called free frequency. The distinction between the licensed and unlicensed frequency is that power. The power of unlicensed frequency i.e. 2.4 GHz is very low compare to the PU frequency. The detection mechanism is done by analyzing the signal energy and the counting maximum hop [11, 12]. The PU is emulated by the

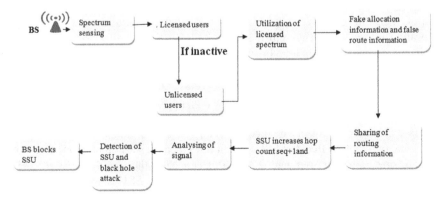

Fig. 2 Block diagram of attack detection

SSU. This SSU is changing channel allocation information and control data one thing it can't change is the energy value of the packet. In the beginning all the nodes exchange hello packet with the energy of each node because it is an ad-hoc network. SSU energy is low compared to the LSU.

The Fig. 3 shows ad-hoc network the nodes exchange the RREQ packet to check the neighbor node, but SSU increases the hop count by one and it also do a black hole attack by sending false route information as shown in Fig. 4, for this node continuously receive the RERR from the neighbor node, this means an SSU node is doing selfish attack and black hole attack.

The attacker continuously sends RERR, at that time a node request is unreachable to the destination. In ad-hoc one node sends the request to the other node to communicate with their data. The selfish node and black hole are isolated, whenever the RREQ comes to attacker node, it's going to increment the hop count; this is generated black hole attack. The attacker sends RERR and false allocation information to neighbors node. The intermediate maximum node count is set by the network. If the RERR is occurring again and again from the same node, neighbor nodes, remove that node from the queue and that node is put into the blacklist. After that all the other nodes also know about this node, i.e. This is attacker node. After that, all the nodes get that BS has energy. The BS blocks the SSU node and it blocks Mac-address of the attacker node. In this case, the entire RREQ packet coming from the attacker node it drops. Then it considered attacker node is not the part of the any path, in this way it saves its BW.

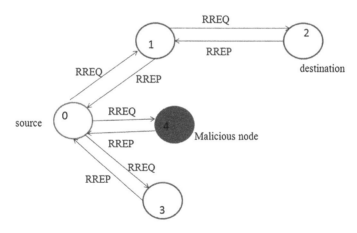

Fig. 3 Black hole attack

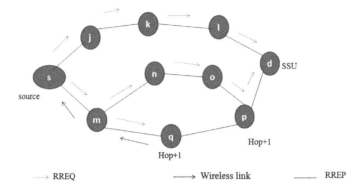

Fig. 4 Attacker increase hop count

4 Result Analysis

A simulation model based on NS2 is used, assumed that the dimension of the scenario as 1000 m × 1000 m in that 7 wireless node randomly deployed. Each or wireless node initials energy is 3.4j, 10 Mbps bandwidth and each packet size 512 kbps (Table 1).

5 Performance Analysis

- **End-End delay**

  The end to end delay is calculated using the total packet dropped by the attacking node to the simulation time. Simulation time is varying 5–20 s.

Table 1 These are the simulation parameters

Parameter	Setup
Set channel	Wireless channel
Set propagation	Two ray ground
Set network	Wireless physical
Set Mac-address	Mac/802_11
Set queue	Drop tail/priority queue
Set link layer	LL
Set antenna	Omni antenna
Set queue length	50
Set number of nodes	7
Set routing protocol	AODV
Set X-axis	1131
Set Y-axis	909
Set simulation time	20

$$\text{Packet loss} = \frac{Total\ packet\ dropped}{Simulation\ time} \tag{1}$$

- **Jitter**

  Jitter can be calculated using the nodes which drop the packet by attacking node to the simulation time. In this calculation total delay is calculate using the current latency versus previous latency.

$$\text{Total latency} = Current\ latency - Previous\ latency \tag{2}$$

$$\text{Average jitter} = \frac{Total\ latency}{Simulation\ time} \tag{3}$$

- **Throughput**

  Throughput can be calculated using the total packets are dropped by the attacking node to the simulation time. The simulation time can be varied 5–20 s. In less simulation packets dropped is less [13].

$$\text{Throughput} = \frac{Bytes\ over\ one\ second}{Simulation\ time} \tag{4}$$

Figure 5 shows per packet delay it clearly shows the amount of packet require for gathering the information about an attacking node is significantly low in between 20 and 30 s BS able to detect the attacking node and mitigate that attack. The delay can be calculated using the delay versus simulation time. If simulation time is lower than packet loss is decreased if the simulation time increases to 20 s the packet loss is more.

In Fig. 6 shows per packet jitter the two peaks which are the instance where the attacker is starting the attack; therefore it can detect the attack and mitigating the attack. Latency is calculated using a variation of simulation time. In less simulation of time it is quite high. If simulation time is high the latency is low, there is not a enough bandwidth to transmit the data because the attacker is attacking.

Figure 7 shows throughput, whenever there is been attack instance 6–40 s the throughput is going to be down, once the attack was detected and mitigated again throughput starts increasing.

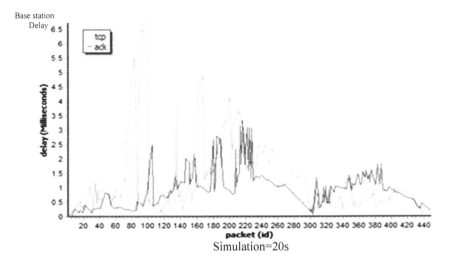

Simulation=20s

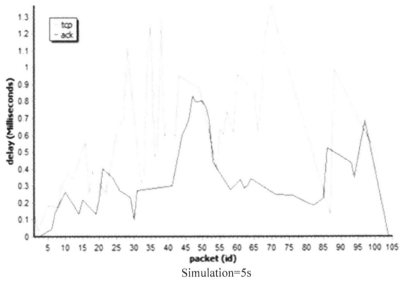

Simulation=5s

Fig. 5 Per packet. (Simulation = 20 s, Simulation = 5 s)

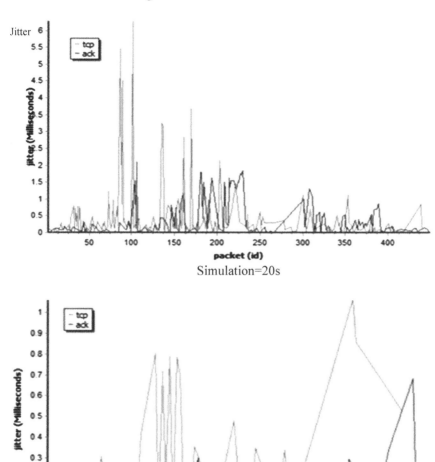

Simulation=20s

Simulation=5s

Fig. 6 Per packet. (Simulation = 20 s, Simulation = 5 s)

Throughput

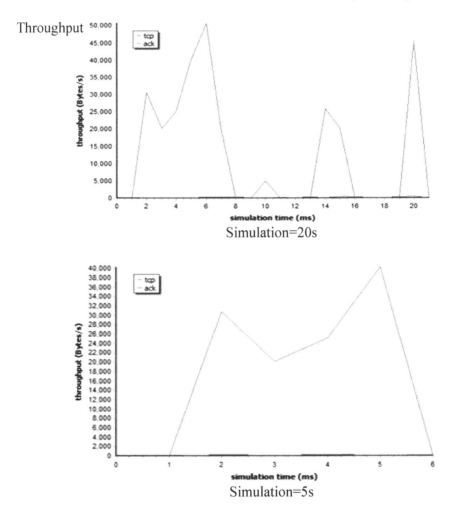

Simulation=20s

Simulation=5s

Fig. 7 Throughput. (Simulation = 20 s, Simulation = 5 s)

6 Conclusion

As the fixed spectrum is used in a wireless network, it causes wastage of bandwidth. To make efficient utilization of fixed spectrum, CR network is introduced that dynamically allocates spectrum to the users. The frequent handoff (both horizontal as well as vertical) within CR ad-hoc network is very challenging. The work provides a mechanism for detecting most challenging attacks, i.e. Selfish and black hole attack by mitigating the information from the base station to all the SU. The proposed technique also offers a blacklisting of the attacking nodes; this enables the other nodes to block the Mac-address of the attacking node. The detection mechanism is based on the threshold of the hop count. This can need to be further improved based on the attacker. The attacker could change Mac-address or additional signature of the attack, therefore using the rule based attack detection like Basien classifier or Fuzzy based detection that gives better result as far as the attacker.

References

1. M. Kaushik, A.K. Kaushik, S. Kaushik, An overview of technical aspect for mobile network technology. Int. J. Adv. Eng. Sci. Technol. (IJAEST) **1**(02), 97–103 (2012). ISSN: 2319–1120
2. A. Alahmadi, M. Abdelhakim, J. Ren, T. Li, Mitigating primary user emulation attacks in cognitive radio networks using advanced encryption standard, in *Global Communications Conference (GLOBECOM), 2013 IEEE* (IEEE, 2013), pp. 3229–3234
3. R. Chen, J.M. Park, J.H. Reed, Defense against primary user emulation attacks in cognitive radio networks. IEEE J. Sel. Areas Commun. **26**(1), 25–37 (2008)
4. S. Capkun, M. Srivastava, M. Cagalj, Securing localization with hidden and mobile base stations, in *IEEE Conference on Computer Communications (INFOCOM)* (2006)
5. W. Wang, H. Li, Y.L. Sun, Z. Han,. Attack-proof collaborative spectrum sensing in cognitive radio networks, in *43rd Annual Conference on Information Sciences and Systems, 2009. CISS 2009* (IEEE, 2009), pp. 130–134
6. Y. Tan, S. Sengupta, K.P. Subbalakshmi, Primary user emulation attack in dynamic spectrum access networks: a game-theoretic approach. IET Commun. **6**(8), 964–973 (2012)
7. C. Chen, H. Cheng, Y.D. Yao, Cooperative spectrum sensing in cognitive radio networks in the presence of the primary user emulation attack. IEEE Trans. Wirel. Commun. **10**(7), 2135–2141 (2011)
8. S. Arkoulis, L. Kazatzopoulos, C. Delakouridis, G.F. Marias Cognitive spectrum and its security issues, in *The Second International Conference on Next Generation Mobile Applications, Services and Technologies, 2008. NGMAST'08* (IEEE, 2008), pp. 565–570
9. A. Ghasemi, E.S. Sousa, Opportunistic spectrum access in fading channels through collaborative sensing. J. Commun. **2**(2), 71–82 (2007)
10. Z. Gao, H. Zhu, S. Li, S. Du, X. Li, Security and privacy of collaborative spectrum sensing in cognitive radio networks. IEEE Wirel. Commun. **19**(6), 106–112 (2012)
11. E. Romero, A. Mouradian, J. Blesa, J. Moya, A Araujo, Simulation framework for security threats in cognitive radio networks. IET Commun. **6**(8), 984–990 (2012)
12. Z. Chen, T. Cooklev, C. Chen, C. Pomalaza-Ráez, Modeling primary user emulation attacks and defenses in cognitive radio networks, in *2009 IEEE 28th International Performance Computing and Communications Conference (IPCCC)* (IEEE, 2009), pp. 208–215
13. D. Pu, Y. Shi, A.V. Ilyashenko, A.M. Wyglinski, Detecting primary user emulation attack in cognitive radio networks, in *2011 IEEE Global Telecommunications Conference (GLOBECOM 2011)* (IEEE, 2011), pp. 1–5

An Innovative Method for Load Balanced Clustering Problem for Wireless Sensor Network in Mobile Cloud Computing

Debabrata Sarddar, Enakshmi Nandi, Anil Kumar Sharma, Biswajit Biswas and Manas Kumar Sanyal

Abstract Mobile Cloud Computing is a revolutionary way where global world is progressing in massive way. Connecting wireless sensor network with Mobile Cloud computing is a novel idea in this era. In this year several research has demonstrated to integrate wireless sensor networks (WSNs) with mobile cloud computing, so that cloud computing can be exploited to process the sensory data collected by WSNs and allow these date to the mobile clients in fast, reliable and secured way. For rising lifetime of wireless sensor network, minimizing energy consumption is an important factor. In this case clustering sensor nodes is one of the effective solutions. It is required to gain some excessive load for cluster heads of cluster based WSN in case of collection of huge data, aggregation and communication of this respective data to base station. Particle Swarm Optimization or PSO is an efficient solution of for this problem.

Keywords Mobile cloud computing (MCC) · Wireless sensor network (WSN) · Clustering · Particle swarm optimization (PSO)

1 Introduction

Assembling of Data in wireless sensor networks (WSNs) as well as the data storage and processing in mobile cloud computing (MCC), WSN-MCC integration highly affected in two fields, such as academic and industry. Mobile users send requests for services to clouds through web browser then the cloud provider allocated resources according to request to establish connection. MCC will be distributed in a access outline heterogeneously with respect of Wireless Network Interfaces [1].

D. Sarddar · E. Nandi (✉)
Department of Computer Science and Engineering, University of Kalyani, Kalyani, India
e-mail: pamelaroychowdhkalani@gmail.com; enakshminanditechno14@gmail.com

A.K. Sharma · B. Biswas · M.K. Sanyal
Department of Business Administration, University of Kalyani, Kalyani, India

© Springer Nature Singapore Pte Ltd. 2017
S.C. Satapathy et al. (eds.), *Proceedings of the 5th International Conference on Frontiers in Intelligent Computing: Theory and Applications*, Advances in Intelligent Systems and Computing 516, DOI 10.1007/978-981-10-3156-4_33

In cluster based WSN all sensor nodes are arranged in group into discrete clusters with a cluster leader, known as cluster head for each [2]. Cluster based WSN has several advantages as:

[i] It can bring down energy consumption, so that cluster head per cluster required involving in routing process and data aggregation.

[ii] It can preserve communication bandwidth as the sensor nodes want to communicate with respective cluster head and the exchange of extra message among them can be ignored [3].

2 Overview of Particle Swarm Optimization or PSO

Particle swarm optimization is basically a population based stochastic optimization algorithm. The system is initialized with population of random solutions and searches for optima by upgrading generations. Here particles fly through problem space by observing the recent optimum particles [4]. The PSO algorithm is starting with a population of random candidate solutions, gestated as particles. Distinct particle assigned their velocity randomly and move through the problem place iteratively.

Algorithm of original PSO [5]

Step1: Initialize a population array of particles and selects positions and velocities on D dimensions randomly.

Step2: loop

Step3: In case of distinct particle, through D variables the required least fitness function has been assessed.

Step4: In this case, comparison between particles fitness value with $pbest_i$, if current value is better than $pbest_i$, then $spbest_i$ will be equals to the current value, and $\vec{P_i}$ will be equals to the current location $\vec{x_i}$ in D dimensional

Step5: Distinguish the particle in the neighbourhood with the best success as yet, and allotted this index with the new variable, g.

Step6: Then, alter the position and velocity of the particle on the basis of following equation

$$\vec{v_i} \leftarrow \vec{v_i} + \vec{U}(0, \varnothing_1), \otimes (\vec{p_i} - \vec{x_i}) + \vec{U}(0, \varnothing_2) \otimes (\vec{p_g} - \vec{x_i})$$
$$\vec{x_i} \leftarrow \vec{x_i} + \vec{v_i}$$

Step7: If a criterion has matched, exit loop.

Step8: **end loop**

Here $\vec{U}$ $(0, \varnothing_1)$ represents a vector of random numbers uniformly distributed in $(0, \varnothing_1)$ and "$\otimes$" is denoted by component-wise multiplication. The required equation for PSO algorithm as follows,

$$v_i^{k+1} = wv_i^k + c_1 rand_1 \times (pbest_i - x_i^k) + c2rand2 \times (gbest - x_i^k) \tag{1}$$

$$x_i^{k+1} = x_i^k + v_i^{k+1} \tag{2}$$

v_i^k particle's velocity i in kth iteration
v_i^{k+1} particle's velocity i in k + 1th iteration
w Inertia weight
c_j acceleration coefficients, j = 1, 2
x_i^k particles recent position i in kth iteration
$rand_i$ random number between 0 and 1, i = 1, 2
x_i^{k+1} particle's position i in k + 1th iteration

pbesti = *particle's best position i*

gbest = best particle's position in a population

$$\text{Fitness function} = \text{Minimize}(\text{Cost}(M) \,\forall\, M) \tag{3}$$

3 Proposed Method

TS1	TS5	TS2		TS3	TS4
Sn1		Sn3	Sn2	Sn3	Sn1

In this problem, the particles signify the task allotted for each sensor nodes (Sn) and the dimension of the particles implies number of tasks allotted to distinct Sn and the number of sensor nodes denotes the respective values allotted for each dimension of particles. Here particles signifies mapping of number of Sn with their various tasks. According to, Fig. 1. The evaluation of each particle has performed by the fitness function [6], as Eq. 3 and it has continued until the particular count of iterations and ends the process by mentioning the stop comment by the clients.

Fig. 1 Here shows 5 different tasks (TS) and 3 sensor nodes (Sn) with their storage and each task will be allotted for optimizing the load on cluster head node

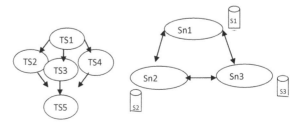

Flow Chart of PSO Algorithm
See Fig. 2.

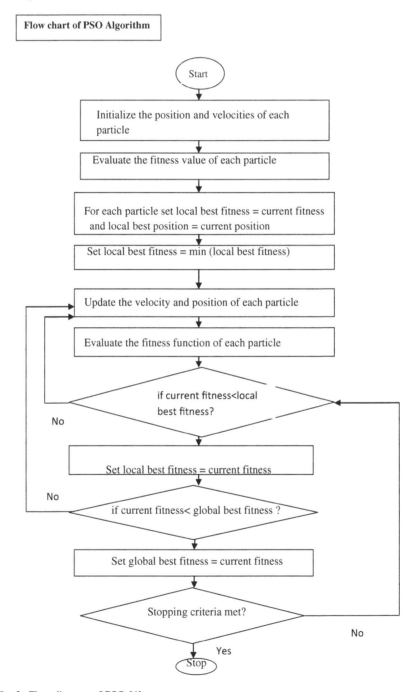

Fig. 2 Flow diagram of PSO [4]

4 Result Analysis

We calculated the distribution of different tasks onto available sensor nodes for different size of total data processed, mentioned in Fig. 1, according to this figure PSO distributes tasks to sensor nodes on the basis of the size of data. If the total size of data is small, PSO distributed tasks is proportional to all the sensor nodes (Sn1–Sn3). However, when the size of data is enhanced then more tasks were allocated to Sn1 and Sn3.

Here, we also calculate the execution time and shows the comparison of execution time with all other algorithm and result as depicted in Fig. 3, shows that our method is superior and required execution time is less (Fig. 4).

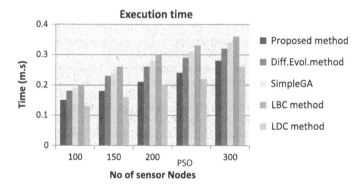

Fig. 3 Comparison between execution time of various sensor nodes

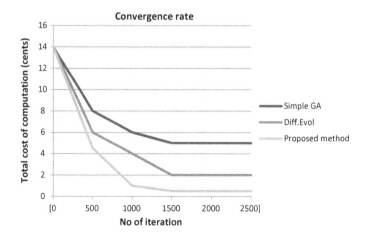

Fig. 4 Comparison between convergence rate [3]

5 Conclusion and Future Scope

In this paper, we introduce a PSO based clustering method for WSN. From the above result it signifies that our technique is much better than some other existing technique. It gives better performance in case of different factors like, energy consumption, execution time and rate of convergence. We hope that this simple method affects great impact in Mobile Clouds computing by solve problem of load balanced clustering of WSN [7] and help clients for suitable and fast communication through mobile cloud system. In near future we have planned to implement better approach to solve load balanced clustering problem for WSN in different way.

References

1. Simrat Kaur, Sarbjeet Singh, Comparative analysis of job grouping based scheduling strategies in grid computing. Int. J. Comput. Appl. **43**(15), 28–35 (2012)
2. B. Olutayo et al., A survey on clustering algorithms for wireless sensor networks, in *Proceedings of 13th IEEE International Conference on Network-Based Information Systems* (2010), pp. 358–364
3. P. Kuila, S.K. Gupta, P.K. Jana, A novel evolutionary approach for load balanced clustering problem for wireless sensor networks. Swarm Evol. Comput. **12**, 48–56 (2013)
4. S. Pandey et al., A particle swarm optimization-based heuristic for scheduling workflow applications in cloud computing environments, in *2010 24th IEEE International Conference on Advanced Information Networking and Applications (AINA)* (IEEE, 2010)
5. P. Kuila, P.K. Jana, Energy efficient clustering and routing algorithms for wireless sensor networks: particle swarm optimization approach. Eng. Appl. Artif. Intell. **33**, 127–140 (2014)
6. K.G. Srinivasa, K.R. Venugopal, L.M. Patnaik, A self-adaptive migration model genetic algorithm for data mining applications. Inf. Sci. **177**(20), 4295–4313 (2007)
7. P. Kuila, P.K. Jana, Energy efficient load-balanced clustering algorithm for wireless sensor networks. Procedia Technol. **6**, 771–777 (2012)

Dengue Fever Classification Using Gene Expression Data: A PSO Based Artificial Neural Network Approach

Sankhadeep Chatterjee, Sirshendu Hore, Nilanjan Dey, Sayan Chakraborty and Amira S. Ashour

Abstract A mosquito borne pathogen called Dengue virus (DENV) has been emerged as one of the most fatal threats in the recent time. Infections can be in two main forms, namely the DF (Dengue Fever), and DHF (Dengue Hemorrhagic Fever). An efficient detection method for both fever types turns out to be a significant task. Thus, in the present work, a novel application of Particle Swarm Optimization (PSO) trained Artificial Neural Network (ANN) has been employed to separate the patients having Dengue fevers from those who are recovering from it or do not have DF. The ANN's input weight vector are optimized using PSO to achieve the expected accuracy and to avoid premature convergence toward the local optima. Therefore, a gene expression data (GDS5093 dataset) available publicly is used. The dataset contains gene expression data for DF, DHF, convalescent and healthy control patients of total 56 subjects. Greedy forward selection method has been applied to select most promising genes to identify the DF, DHF and normal (either convalescent or healthy controlled) patients. The proposed system perfor-

S. Chatterjee (✉)
Department of Computer Science and Engineering, University of Calcutta,
Kolkata, India
e-mail: sankha3531@gmail.com

S. Hore
Department of Computer Science & Engineering, Hooghly Engineering
& Technology College Chinsurah, Hooghly, India
e-mail: shirshendu.hore@hetc.ac.in

N. Dey
Department of Information Technology, Techno India College of Technology,
Kolkata, India
e-mail: neelanjandey@gmail.com

S. Chakraborty
Department of CSE, B.C.E.T, Durgapur, West Bengal, India
e-mail: sayan.cb@gmail.com

A.S. Ashour
Department of Electronics and Electrical Communications Engineering,
Faculty of Engineering, Tanta University, Tanta, Egypt
e-mail: amirasashour@yahoo.com

© Springer Nature Singapore Pte Ltd. 2017
S.C. Satapathy et al. (eds.), *Proceedings of the 5th International Conference on Frontiers
in Intelligent Computing: Theory and Applications*, Advances in Intelligent Systems
and Computing 516, DOI 10.1007/978-981-10-3156-4_34

331

mance was compared to the multilayer perceptron feed-forward neural network (MLP-FFN) classifier. Results proved the dominance of the proposed method with achieved accuracy of 90.91 %.

Keywords Dengue fever · Dengue hemorrhagic fever · Artificial neural network · Multilayer perceptron feed-forward neural network · Particle swarm optimization

1 Introduction

Dengue virus (DENV) is found to be a life threatening mosquito borne pathogen. It affects an enormous number of people (50–100 million) every year all over the world consistent with World health Organization (WHO). Phylogenetically distinct four major serotypes can cause mild disease, such as DF, DHF, and dengue shock syndrome (DSS). Researches were interested to determine the key determinant of DF and DHF [1–4]. Other research studies were interested with the prediction models of Dengue disease outbreak [5–7]. However, DF and DHF positive patients' detection from the normal ones remains a challenging process.

Consequently, various aspects regarding Dengue Fever can be predicted and studied using the ANN (Artificial neural network), where a learning algorithm can be employed to train the network [8–10]. In this stage, the weight vectors of neural network are optimized to achieve maximum accuracy. This is accomplished by optimizing an objective function (generally minimizing error). However, researches have revealed that traditional learning algorithms may lead to a premature convergence to local optima especially with real life problems. Thus, achieving the required accuracy is challenging. Since, the ANN based on back-propagation of error strategy is generally trained with learning algorithms that are devised on the scheme of local optimization. Thus, it is highly probable that the optimization process may converge to local optima instead of better global optima value. Consequently, the NN performance to predict/classify the intended target may be poor. In order to overcome the poor performance problem, NN can be trained using meta-heuristic based optimization algorithms. This approach of improving the performance has been found to be extremely successful [11–14].

Generally, several researchers were interested with the predication and classification of the dengue fever. A prediction of defervescence of fever in dengue patients using ANN was studied in [10]. The proposed system was based on the prediction using the clinical symptoms and signs. The results established 90 % prediction accuracy for the day of defervescence in dengue patients. A combination of the self-organizing map (SOM) and multilayer feed-forward neural networks (MFNN) was employed to predict the risk in dengue patients [15]. Seven significant risk predictors were used for the classification process. The results proved 70 % predicative accuracy with 0.121 sum squared error. An extensive review was conducted on arboviral diagnosis and prognosis in [16].

The current work proposed a PSO trained Neural Network (NN-PSO) for detecting the DF, DHF, and normal patients with superior accuracy. First, Greedy forward selection algorithm [17, 18] has been applied to find the most promising features to classify the patients into three classes, namely DF, DHF, and normal. Subsequently, during the NN training phase, PSO has been employed to find the optimized weight vector for the different NN layers. Finally, the NN-PSO was tested with well-known MLP-FFN classifier trained by scaled conjugate gradient descent approach using some metrics including the precision, recall, accuracy, and F-measure.

The rest of the work has been arranged as follows. Section 2 described the proposed PSO based training phase of the NN. Section 3 represented the methodology used in the present study. Section 4 included an explanation of the experimental flow followed by the experimental results and analysis. Finally, the conclusion is included in Sect. 5.

2 Proposed Method

2.1 Artificial Neural Network Model

A three layer ANN has been employed to classify the blood samples into two aforesaid classes as illustrated in Fig. 1. The network is trained using scaled conjugate gradient descent algorithm, which is a benchmark against the back-propagation. The cross-entropy is considered to be the objective function of the minimization process. The artificial neurons are connected with each other in layer

Fig. 1 The structure of the MLP-FFNN

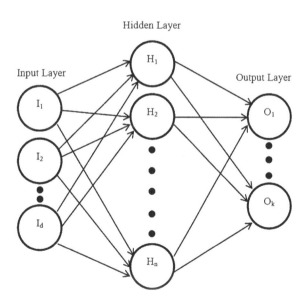

wise fashion. Each connection assigned an initial weight, which is gradually updated according to the learning algorithm. The weighted output of each neuron in the hidden/output layers is operated by an activation function that decided the actual output of the particular neuron.

2.2 Particle Swarm Optimization (PSO)

The PSO is a standard optimization algorithm [19, 20]. Typically, it is consists of particles population in a D-dimensional search space. Every particle is companioned through a fitness value to evaluate the effectiveness of a particle in achieving the objective. Randomly, the particles are located, where the swarm moves to find the optimal fitness. All particles have '*pbest*' that represents the best reachable solution by the particle, while the best realized value by any particle is signified by '*gbest*' (Global best). Randomly, a position and a velocity for each particle are initialized. The particles fitness values are calculated after every iteration, while adjusting the position/velocity of each particle depending on the succeeding expression:

$$v_i(t+1) = v_i(t) + a_1 * r_1 * (pbest_i(t) - x_i(t)) + a_2 * r_2 * (gbest_i(t) - x_i(t)) \quad (1)$$

$$x_i(t+1) = x_i(t) + v_i(t+1) \quad (2)$$

The general PSO algorithm is deployed as follows:

Algorithm 1: The PSO Algorithm

Randomly, the particles are placed at positions and zero velocities

    for $n = 1$: Size of swarm **do**

      Calculate fitness

    end for

    for $i = 1$: No. of repetitions **do**

      for $j = 1$: Size of swarm **do**

      Update pbest and gbest

      Modify position/ velocity

      Compute the new population fitness values

      end for

    end for

Initially in the present work, the particles allocated at random location in the D-dimensional space having zero velocity. Each particle position is a D-dimensional vector that considered being the input weight vector for the NN input layer. The used objective function is the Root Mean Squared Error (RMSE) that calculated as the difference between the classifier anticipated values and the discovered actual values. The classifier prediction RMSE of a regarding the computed variable v_{c_k} is determined as follows:

$$RMSE = \sqrt{\frac{\sum_{k=1}^{n}\left(v_{d_k} - v_{c_k}\right)^2}{n}} \tag{3}$$

where, v_{d_k} is the noticed value of the kth data, and v_{c_k} represents classifier predicted value. Accordingly, the PSO is used to minimize the RMSE through the NN training stage. The PSO objective is to determine the weight vector that leads to minimum RMSE as illustrated in Fig. 2.

The pre-processing step included in Fig. 2 is conducted for features selection. Subsequently, the proposed NN-PSO is used to overcome being trapped in local

Fig. 2 The flowchart of the ANN training using PSO

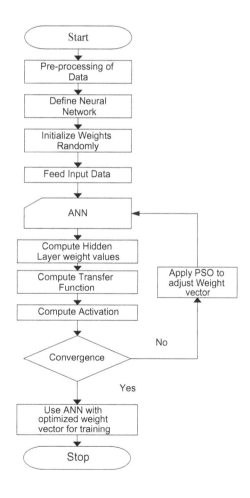

optima during classifying the patients' blood samples. The computed weight values of hidden layer are used to obtain the optimal ANN weights based on the PSO objective function to achieve the required convergence. The ANN weights are considered to be optimal values once the PSO algorithm convergence occurs. The optimal weight values are used further during the classification process.

3   Methodology

In the present work, the Gene-Expression dataset called Acute Dengue patients that obtained from Gene Expression Omnibus (GEO) is used. Through this dataset blood samples are publicly available at (http://www.ncbi.nlm.nih.gov/geo). It consists of expression level of 54,715 genes for 56 homo-sapiens subjects. This number of features makes the classification task computationally infeasible. Since, the forward greedy algorithm is computationally efficient as it deals with a sparse solution. Hence, Greedy forward selection method is used to select the significant genes as demonstrated in Fig. 3 through the pre-processing phase.

Figure 3 depicts that before classifying the patients into different classes, it is imperative to reduce the number of features by selecting the most promising features of the dataset. Greedy forward selection algorithm [21, 22] has been applied to accomplish this task. Greedy forward selection method selected 4217 genes to be considered for the experimental purpose from the 54,715 genes. However, for training phase and other NN classification phases this is considered to be unsuitable number of features. Therefore, the Greedy forward selection algorithm has been employed iteratively until a reasonable size of features set is obtained. Finally, the selected set of features contains five genes, namely CLEC5A, IRF3, MYD88, TLR7, and TLR3. The forward feature selection process is initiated by the evaluation of all feature subsets to determine the best subset. It offered the major improvement at each stage to realize sparsity. Generally, the classification process consists of three main steps, namely the training stage, the validation stage and the evaluation stage.

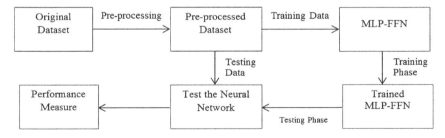

Fig. 3 Block diagram of flow the MLP-FFN classifier

4 Results and Discussion

The current work classifies the patients into three different classes concerning the Dengue Fever by using two different classifiers, namely MLP-FFN (trained with scaled conjugate gradient descent and cross entropy as objective function), and NN-PSO (Neural Network trained with Particle Swarm Optimization and RMSE as objective function). Figure 4 shows the confusion matrix of different phases involved in developing and testing the MLP-FFN based classifier.

The MLP-FFN achieves 87.5 % classification accuracy in the test phase as illustrated in Fig. 4. Furthermore, the error histogram is presented in Fig. 5 and the

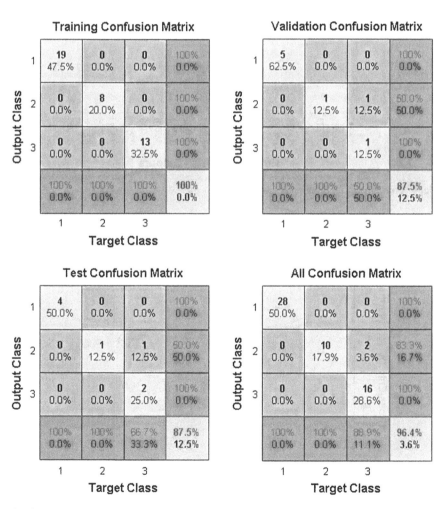

Fig. 4 Confusion matrix of different phases of MLP-FFN classifier

cross-entropy versus epochs for different phases plot is included in Fig. 6 for the MLP-FFN.

Figure 6 depicts that the paramount validation performance is 0.13182 at epoch 24. Figure 7 illustrates the ROC (receiver operating characteristic) curve.

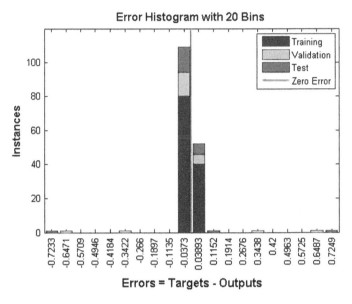

Fig. 5 Error Histogram of all phases of MLP-FFN

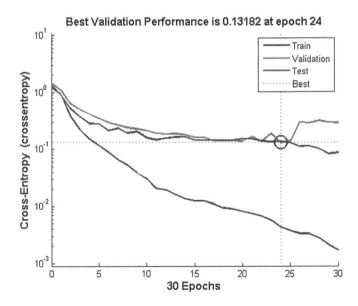

Fig. 6 A plot of cross-entropy versus epochs for all three phases of MLP-FFN

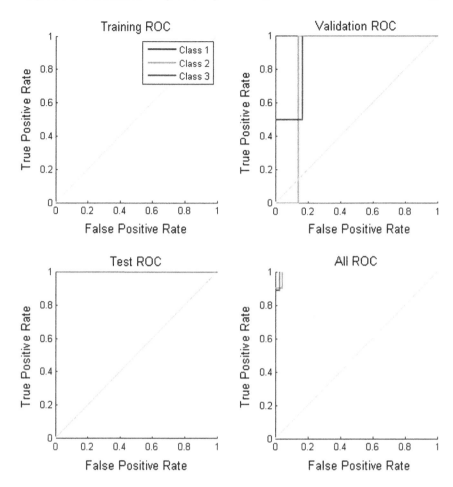

Fig. 7 Receiver operating characteristic of different phases for MLP-FFN classifier

Table 1 The NN-PSO confusion matrix of testing phase

Actual class	Predicted class		
	DF	DHF	Normal
DF	11	1	0
DHF	1	3	0
Normal	0	0	6

Table 2 Comparative study of MLP-FFN and NN-PSO

	MLP-FFN (%)	NN-PSO (%)
Accuracy	87.5	**90.91**
Precision	80	**91.67**
Recall	100	**91.67**
F-Measure	88.89	**91.67**

The error histogram and ROC plot reveals that the performance of MLP-FFN is moderate. Table 1 depicts the confusion matrix of NN-PSO model. In Table 2 a comparative study has been carried out in terms of accuracy, precision, recall and F-measure metrics.

Tables 1 and 2 establish that the NN-PSO outperforms the MLP-FFN model with reasonable improvement over its result. The proposed NN-PSO proves its ingenuity in detecting Dengue Fever with greater accuracy.

5 Conclusion

The present work proposed an efficient, robust and accurate Neural based model to predict the Dengue Fever in terms of classifying them into DF, DHF and normal (either convalescent or healthy controlled) classes; respectively. Initially, MLP-FFN based model has been employed to perform the required classification. The results proved the dominance of the NN-PSO based model compared to the MLP-FFN. It improved the classification accuracy of the patients who are suffering from Dengue fevers from those who are not or recovering from it. The NN-PSO based model achieved an accuracy of 90.91 %.

References

1. S. Ubol, P. Masrinoul, J. Chaijaruwanich, S. Kalayanarooj, T. Charoensirisuthikul et al., Differences in global gene expression in peripheral blood mononuclear cells indicate a significant role of the innate responses in progression of dengue fever but not dengue hemorrhagic fever. J. Infect. Dis. **197**, 1459–1467 (2008)
2. S.B. Halstead, Dengue. Lancet **370**, 1644–1652. 4. WHO (1997) *Haemorrhagic Fever: Diagnosis, Treatment, Prevention and Control*, 2nd edn. (2007)
3. L.L. Coffey, E. Mertens, A.C. Brehin, M.D. Fernandez-Garcia, A. Amara et al., Human genetic determinants of dengue virus susceptibility. Microbes Infect. **11**, 143–156 (2009)
4. L. Tanner, M. Schreiber, J.G.H. Low, A. Ong, T. Tolfvenstam et al., Decision tree algorithms predict the diagnosis and outcome of dengue Fever in the early phase of illness. PLoS Negl. Trop. Dis. **2**, e196 (2008)
5. A. Bakar, Z. Kefli, Predictive models for dengue outbreak using multiple rule base classifiers, in *2011 International Conference on Electrical Engineering and Informatics* (IEEE, 2011), pp. 1–6
6. M. Mousavi, A.A. Bakar, S. Zainudin, Z. Awang, Negative selection algorithm for dengue outbreak detection. Turk. J. Electr. Eng. Comput. Sci. 1–25 (2013)
7. L. Tanner, M. Schreiber, J.G.H. Low, A. Ong, T. Tolfvenstam, Y.L. Lai, L.C. Ng, Y.S. Leo, L. Thi Puong, S.G. Vasudevan, C.P. Simmons, M.L. Hibberd, E.E. Ooi, Decision tree algorithms predict the diagnosis and outcome of dengue fever in the early phase of illness. PLoS Negl. Trop. Dis. **2**(3) (2008)
8. N. Rachata, P. Charoenkwan, T. Yooyativong, Automatic prediction system of dengue haemorrhagic-fever outbreak risk by using entropy and artificial neural network, in *2008 International Symposium on Communications and Information Technologies, ISCIT* (2008), pp. 210–214

9. G. Guthrie, D.A. Stacey, D. Calvert, V. Edge, Detection of disease outbreaks in pharmaceutical sales: Neural networks and threshold algorithms, in *Proceedings of IJCNN'05 on Neural Networks*, vol. 5 (IEEE, 2005), pp. 3138–3143

10. F. Ibrahim, M.N. Taib, W.A.B.W. Abas, C.C. Guan, S. Sulaiman, A novel dengue fever (DF) and dengue haemorrhagic fever (DHF) analysis using artificial neural network (ANN). Comput. Methods Programs Biomed. **79**, 273–281 (2005)

11. S. Chatterjee, R. Chakraborty, N. Dey, S. Hore, A quality prediction method for weight lifting activity, in *Conference: Michael Faraday IET International Summit 2015 (MFIIS 2015): An IET International Conference Focused Area Power-Control-Automation*, At Kolkata, India, pp. 95–98

12. S. Chatterjee, S. Ghosh, S. Dawn, S. Hore, N. Dey, Forest type classification: a hybrid NN-GA model based approach, in *Information Systems Design and Intelligent Applications* (Springer India, 2016), pp. 227–236

13. S. Hore, S. Chatterjee, R. Shaw, N. Dey, J. Virmani, Detection of chronic kidney disease: A NN-GA based approach, in *CSI—2015; 50th Golden Jubilee Annual Convention*, At Delhi, Springer – AISC, Nov 2015

14. S. Hore, S. Chatterjee, S. Sarkar, N. Dey, A.S. Ashour, D. Bălas - Timar, V.E. Balas, Neural-based prediction of structural failure of multi storied RC buildings. Struct. Eng. Mech. (2016)

15. T. Faisal, F. Ibrahim, M.N. Taib, A noninvasive intelligent approach for predicting the risk in dengue patient. Expert Syst. Appl. **37**, 2175–2181 (2010)

16. A.S. Fathima, D. Manimegalai, N. Hundewale, A review of data mining classification techniques applied for diagnosis and prognosis of the arbovirus-dengue. IJCSI Int. J. Comput. Sci. **8** (2011)

17. N. Kausar, A. Abdullah, B.B. Samir, S. Palaniappan, B.S. AlGhamdi, N. Dey, Ensemble clustering algorithm with supervised classification of clinical data for early diagnosis of coronary artery disease. J. Med. Imag. Health Inf. **6**, 78–87 (2016)

18. N. Zemmal, N. Azizi, N. Dey, M. Sellami, Adaptive semi supervised support vector machine semi supervised learning with features cooperation for breast cancer classification. J. Med. Imag. Health Inf. **6**, 53–62 (2016)

19. E. Zitzler, L. Thiele, An evolutionary algorithm for multiobjective optimization: the strength pareto approach. Gloriastrasse 35, CH-8092 Zurich, Switzerland, Technical Report 43 (1998)

20. C.A. Coello, G.T. Pulido, Multiobjective structural optimization using a microgenetic algorithm. Struct. Multidiscip. Optim. **30**(5), 388–403 (2005)

21. T. Zhang, On the consistency of feature selection using greedy least squares regression. J. Mach. Learn. Res. **10**, 555–568 (2009)

22. S. Chatterjee, S. Sarkar, S. Hore, N. Dey, A.S. Ashour, V.E. Balas, Particle swarm optimization trained neural network for structural failure prediction of multistoried RC buildings. Neural Comput. Appl. 1–12 (2016)

Modified Non Linear Diffusion Approach for Multiplicative Noise

Vikrant Bhateja, Aditi Sharma, Abhishek Tripathi and Suresh Chandra Satapathy

Abstract Synthetic Aperture Radar (SAR) is a useful coherent imaging tool for extracting information from various fields such as astronomy and meteorology. SAR images are often corrupted by granular noise known as speckle which follows a multiplicative model. Speckle reflection in homogenous as well as heterogeneous areas obscures the contrast between the target-of-interest and its surroundings. This paper proposes a modified Non Linear Diffusion Approach for despeckling SAR images. The essence is to develop an approach that can suppress speckle and preserve the structural content as an improvement over conventional anisotropic diffusion filtering.

Keywords Non linear diffusion · Despeckling · Homogenous areas

1 Introduction

SAR is a device that actively results in radiation and also captures the backscattered signals from a small portion of the image scene. The received signal is complex, as output from the in-phase and quadrature channels can be viewed as the incoherent

V. Bhateja (✉) · A. Sharma · A. Tripathi
Department of Electronics and Communication Engineering,
Shri Ramswroop Memorial Group of Professional Colleges (SRMGPC),
Lucknow, UP, India
e-mail: bhateja.vikrant@gmail.com

A. Sharma
e-mail: aditiii065@gmail.com

A. Tripathi
e-mail: abhishek1.srmcem@gmail.com

S.C. Satapathy
Department of Computer Science and Engineering,
ANITS, Visakhapatnam, AP, India
e-mail: sureshsatapathy@gmail.com

© Springer Nature Singapore Pte Ltd. 2017   343
S.C. Satapathy et al. (eds.), *Proceedings of the 5th International Conference on Frontiers in Intelligent Computing: Theory and Applications*, Advances in Intelligent Systems and Computing 516, DOI 10.1007/978-981-10-3156-4_35

sum of several backscattered waves [1, 2]. SAR creates images of objects, such as landscapes, mining, oceanography, etc. [3–5]. These images can be either two or three dimensional representations of the object. The reflected wave from the target material consists of various independent scattering points. The interference of these de-phased but coherent waves result in the granular pattern of noise called speckle. However, speckle inherently exists in SAR images and is generally modeled as a multiplicative noise. In coherent illumination, speckle is caused by the objects with roughness of the order of a wavelength and is unable to resolve the micro-scale of the objects roughness [6]. The ability of noisy SAR images for locating and extraction of information and other fine details becomes difficult, thereby area discrimination of both textural and radiometric aspects become less efficient [7]. Therefore, for better extraction of information, speckle reduction is required to improve the quality of the image. Speckle filtering algorithms are broadly classified as on Bayesian and non-Bayesian approaches [8, 9]. The Linear Minimum Mean Squared Error (LMMSE) and Minimum Mean Square Error (MMSE) criterion minimizes the mean squared error that corresponds to shrinkage of the noisy coefficients by a factor inversely related to its Signal to Noise Ratio (SNR). However, in the lowest levels of wavelet decomposition, it does not respect the Gaussian assumption and hence is inferior to Bayesian despeckling approaches. Bayesian approaches (Lee Filter, Frost Filter and Kuan filter [9]) are based on local statistics where the centre pixel intensity is computed inside the moving window using the average intensity values. Further, the aforesaid local statistics filters [9] were extended by Lopes et al. [7]. In their work, the image is filtered by segmenting it into three regions, i.e. homogeneous, non-homogeneous and isolated points. Bayesian approaches provided satisfactory speckle reduction but smoothens the image which leads to edge distortion and blurring. The Non-Bayesian approaches employing Anisotropic Diffusion (AD) filtering [8] use the diffusion method that removes speckle by modifying it with the Partial Differential Equations (PDE) [10–15]. It involves the usage of a continuous and monotonically decreasing conduction function that classifies the image into heterogeneous and homogeneous regions while the use of gradient factor helps in discrimination of true and false edges. This helps in edge preservation and restoration of image content; however, conventional AD is computationally complex and does not prove satisfactory for images corrupted with speckle, which is a multiplicative noise [16, 17]. It is found that the conventional AD performs well for images which are corrupted by additive noise while for images with multiplicative noise; AD enhances speckle rather than reducing it. Recently, various improvements have been reported in Bayesian approaches by Bhateja et al. [18–20] and Bogdan et al. [21] but these filters display good results for other noise models like impulse or salt pepper noise and hence blur the speckle affected images. Also, recent works based on Non-Bayesian approaches are described by Bhateja et al. [13–15, 22–24] and Parrilli et al. [25]; these yielded improved filtering but resulted in higher complexity with usage of large window sizes leading to significant blurring. Therefore, the speckle filtering approach described in this paper involves modification of the conventional AD by hybridizing it with Gaussian mean filter for better speckle suppression in homogeneous areas along with subsequent feature extraction. The window sizes have been restricted to 3×3 only to minimize complexity.

2 Proposed Methodology

2.1 Background

The anisotropic diffusion algorithm modifies the image via PDE to remove noise. Filtering the image using Gaussian filter is equivalent to modifying the image according to the isotropic heat equation. The classical isotropic diffusion equation was replaced with the AD equation in continuous form as given below:

$$\frac{\partial I(x, y, t)}{\partial t} = div[g(\|\nabla I\|)\nabla I] \tag{1}$$

where: $g(\|\nabla I\|)$ is an edge stopping function and $\|\nabla I\|$ is the gradient magnitude. The diffusion equation stated in Eq. (1) has been further stabilized by Perona Malik [8]; by presenting it in its discretized form as given in Eq. (2).

$$I_s^{t+1} = I_s^t + \frac{\lambda}{|\eta_s|} \sum_{p \in \eta_s} g(\nabla I_{s,p}) \nabla I_{s,p} \tag{2}$$

where: t denotes discrete time steps, s is the pixel position and I_s^t is the discretely sampled image. The rate of diffusion is determined by λ, where $\lambda \in IR^+$, η_s represents spatial neighborhood of pixel s and $|\eta_s|$ gives the number of neighbors. The image gradient in the particular direction is approximated as:

$$\nabla I_{s,p} = I_p - I_s^+, \quad \text{where: } p \in \eta_s \tag{3}$$

Equation (3) gives the image intensity difference between pixel s and its neighboring pixel p. At constant image regions, these neighbor differences will be normally distributed and small.

2.2 Proposed Algorithm-Modified AD Filtering

In the proposed work, the modification of PMAD approach has been carried out to acquire better filtering. The operational procedure of the proposed approach begins by initializing the window (w) of size 3×3 over the noisy image. The diffusion function PMAD uses the gradient operator for the purpose of preserving edges.

PMAD uses the fundamental concept to modify the conductivity in the nonlinear diffusion equation:

$$\frac{\partial}{\partial t}I(x,y,t) = \nabla \cdot (c(x,y,t)\nabla I) \tag{4}$$

where $c(x, y, t)$ is defined as:

$$c(x,\sigma,t) = \begin{cases} \frac{1}{2}\left[1-(x/\sigma)^2\right]^2, & |x| \le \sigma \\ 0, & otherwise \end{cases} \tag{5}$$

where: $I(x, y, t)$ is an image, t is the time scale and $c(x, y, t)$ is the monotonically decreasing function of the image gradient which is defined as the conduction function.

When $c(x, y, t)$ approaches unity then the transformed pixel is estimated within a spatial sub-window of size 3×3 using the local gradient, which is calculated using nearest-neighbor differences as:

$$\begin{aligned} \nabla_N I_{i,j} &= I_{i,j} - I_{i,j} \\ \nabla_S I_{i,j} &= I_{i+1,j} - I_{i,j} \\ \nabla_E I_{i,j} &= I_{i,j+1} - I_{i,j} \\ \nabla_W I_{i,j} &= I_{i,j-1} - I_{i,j} \end{aligned} \tag{6}$$

Subscripts N, S, E, and W (North, South, East, and West) describe the direction of computation of the local gradients. The equation for the transformed pixel is given by:

$$I_t(i,j) = I(i,j) + \lambda(\nabla_E I \cdot c_E + \nabla_W I \cdot c_W + \nabla_N I \cdot c_N + \nabla_S I \cdot c_S) \tag{7}$$

While, when $c(x, y, t)$ does not approaches unity; the image is processed using the Gaussian filter which is given by:

$$G = \frac{1}{\sqrt{2\pi\sigma}}e^{-(I-\mu)^2/2\sigma^2} \tag{8}$$

Then, the transformed pixel is given by:

$$I_t = I * G \tag{9}$$

Unlike, SRAD and DPAD [9] which are approximations of Lee and Kuan filter respectively, here the non-homogeneous region is processed using Gaussian filter. The procedural steps for implementation are mentioned under Algorithm-1 in the table below.

Algorithm 1: Procedural Steps for Modified AD Filtering.

BEGIN		
Step 1	:	*Input:* Noisy image (*I*) of size MXN.
Step 2	:	*Input:* Window size (w).
Step 3	:	*Process:* Image (*I*) sequentially with window (w).
Step 4	:	*Compute:* Standard deviation (*Sd*). Gradient (G_x) and (G_y) in x and y direction: $G_i = |G_x| + |G_y|$ Conduction Coefficient (*c*): Refer Eq. (5)
Step 5	:	*If* | (*c* = 1) | Perform Eqs. (6)–(7)
		Else | Perform Eqs. (8)–(9)
Step 6	:	*Process:* The window until it reaches last pixel.
Step 7	:	*Display:* The denoised images.
END		

3 Results and Discussions

The results obtained with proposed AD filtering (as in Algorithm 1) has been demonstrated on two test images of SAR. The various parameters initially set for simulations are w = 3 × 3 and λ = 1/4. The proposed algorithm 1 has been effective to remove the speckle in homogenous areas without blurring edges. The

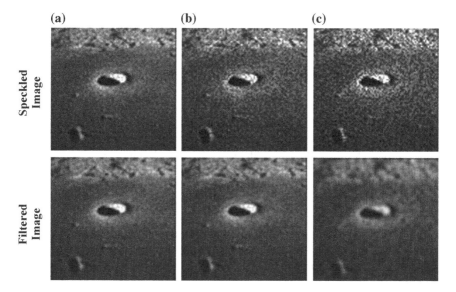

Fig. 1 Performance comparison for proposed speckle suppression methodology on test image 1 at **a** Speckle variance of 0.01, **b** 0.04 and **c** 0.06

(a)   (b)   (c)

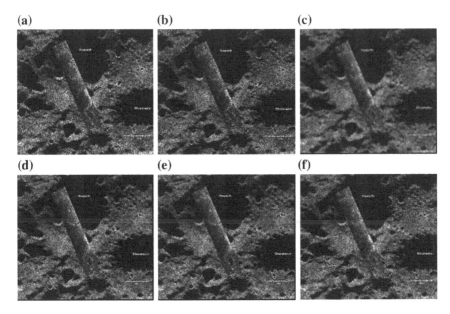

(d)   (e)   (f)

Fig. 2 Comparative analysis of results of speckle suppression for various filtering techniques on test image 2. **a** Noisy SAR image, results using **b** Lee filter [9], **c** Frost filter [9], **d** Kuan filter [9], **e** Wang et al. [26], **f** Proposed modified AD filtering

Table 1 IQA in terms of PSNR (in dB) for various speckle filtering techniques

Noise variance	Lee filter [9]	Frost filter [9]	Kuan filter [9]	Wang et al. [26]	Modified AD filtering
0.01	19.7630	20.9739	20.1494	24.2334	25.3848
0.04	18.4289	20.7853	19.2013	23.1644	24.1722
0.06	18.5117	20.6848	18.1531	20.0433	22.5192

obtained results on Test image 1 with are shown as Fig. 1. It is evident from Fig. 1 that there is smoothening in homogenous area and edges are preserved.

In order to benchmark the performance, the proposed modified AD filtering algorithm has been compared with various other techniques like: Bayesian approaches [9] and one of the recent work by Wang et al. [26] (Fig. 2).

It can be interpreted that the Bayesian approaches enhance the quality of the image at the cost of over smoothening the edges. However, the proposed work has demonstrated to overcomes the limitation of the aforesaid techniques and provide better despeckling. Tables 1 and 2 provide a comparative evaluation of Image Quality Assessment (IQA) parameters computed for validating the obtained results of speckle filtering. Higher values of PSNR and SSIM for the proposed modified AD filtering algorithm, demonstrates its efficiency in image smoothening and edge preservation of SAR images.

Table 2 IQA in terms of SSIM for various speckle filtering techniques

Noise variance	Lee filter [9]	Frost filter [9]	Kuan filter [9]	Wang et al. [26]	Modified AD filtering
0.01	0.9118	0.9278	0.9226	0.8725	0.9816
0.04	0.8863	0.9250	0.9088	0.8563	0.9760
0.06	0.8881	0.9235	0.8891	0.7943	0.9661

4 Conclusion

SAR images are contaminated during acquisition with a multiplicative noise-speckle which poses a threat in processing—in terms of edge preservation and noise suppression. The proposed work therefore formulates an effective speckle suppression approach based on modified AD filtering to tackle images with higher variance of speckle. The computed values of IQA parameters depict an improved response in terms of edge preservation and speckle suppression. Future works in this arena will include analysis of various categories for conductance functions; and selecting the suitable function for AD based speckle suppression algorithm.

References

1. Y.K. Chan, V.C. Koo, An introduction to synthetic aperture radar (SAR). J. Prog. Electromagnet. Res. B **2**, 27–60 (2008)
2. A. Lay-Ekuakille, V. Pelillo, C. Dellisanti, F. Tralli, SAR aided method for rural soil evaluation, in *SPIE 2002 Remote Sensing, Crete (Greece)* (2002), pp. 103–112
3. G. Griffo, L. Piper, A. Lay-Ekuakille, D. Pellicano, E. De Franchis, Modelling a buoy for sea pollution monitoring using fiber optics sensors, in *4th Imeko TC19 Symposium*, Lecce, Italy, June 2013, pp. 182–186
4. A. Lay Ekuakille, A.V. Scarano, Progressive deconvolution of laser radar signals, in *SPIE Remote Sensing*, Honolulu, Nov 2004, pp. 319–326
5. P. Vergallo, A. Lay-Ekuakille, Spectral analysis of wind profiler signal for environment monitoring, in *IEEE I2MTC*, Graz, Austria, May 2012, pp. 162–165
6. V. Bhateja, A. Tripathi, A. Gupta, A. Lay-Ekuakille, Speckle suppression in SAR images employing modified anisotropic diffusion filtering in wavelet domain for environment monitoring. Measurement **74**, 246–254 (2015)
7. A. Lopes, R. Touzi, E. Nezry, Adaptive speckle filter and scene heterogeneity. IEEE Trans. Geosci. Remote Sens. **28**(6), 992–1000 (1990)
8. P. Perona, J. Malik, Scale space and edge detection using anisotropic diffusion. IEEE Trans. Pattern Anal. Mach. Intell. **12**(7), 629–639 (1990)
9. F. Argenti, A. Lapini, T. Bianchi, L. Alparone, A tutorial on synthetic aperture radar images. IEEE Geosci. Remote Sens. Mag. **1**(3), 6–35 (2013)
10. S. Singh, A. Jain, V. Bhateja, A Comparative evaluation of various despeckling algorithms for medical images, in *Proceedings of (ACMICPS) CUBE International Information Technology Conference & Exhibition*, Pune, India (2012), pp. 32–37
11. A. Gupta, A. Tripathi, V. Bhateja, Despeckling of SAR images via an improved anisotropic diffusion algorithm, in *Proceedings of (Springer) International Conference on Frontiers in*

Intelligent Computing Theory and Applications (FICTA 2012), Bhubaneswar, India, AISC, Dec 2012, vol. 199, pp. 747–754

12. A. Gupta, A. Tripathi, V. Bhateja, Despeckling of SAR images in contourlet domain using a new adaptive thresholding, in *Proceedings of (IEEE) 3rd International Advance Computing Conference (IACC 2013)*, Ghaziabad (U.P.), India, Feb 2013, pp. 1257–1261

13. V. Bhateja, G. Singh, A. Srivastava, J. Singh, Despeckling of ultrasound images using non-linear conductance function, in *Proceedings of (IEEE) International Conference of Signal Processing and Integrated Networks (SPIN-2014)*, Noida (U.P.), India (2014), pp. 722–726

14. A. Srivastava, V. Bhateja, H. Tiwari, Modified anisotropic diffusion filtering algorithm for MRI, in *Proceedings of (IEEE) 2nd International Conference on Computing for Sustainable Global Development (INDIACom-2015)*, New Delhi, India (2015), pp. 1885–1890

15. V. Bhateja, A. Tripathi, A. Gupta, A. Lay-Ekuakille, Speckle suppression in SAR images employing modified anisotropic diffusion filtering in wavelet domain for environment monitoring. Elsevier Measure. J. **74**, 246–254 (2015)

16. Q. Zhang, Y. Wu, F. Wang, J. Fan, L. Zhang, L. Jiao, Anisotropic-scale-space-based salient-region detection for SAR images. IEEE Geosci. Remote Sens. Lett. **13**(3), 457–461 (2016)

17. V. Bhateja, G. Singh, A. Srivastava, A novel weighted diffusion filtering approach for speckle suppression in ultrasound images, in *Proceedings of (Springer) International Conference on Frontiers in Intelligent Computing Theory and Application (FICTA 2013)*, Bhubaneswar, India, vol. 247 (2013), pp. 459–466

18. V. Bhateja, A. Tripathi, A. Gupta, An improved local statistics filter for denoising of SAR images, in *Proceedings of (Springer) 2nd International Symposium on Intelligent Informatics (ISI'13)*, vol. 235, Mysore, India (2013), pp. 23–29

19. V. Bhateja, M. Misra, S. Urooj, A. Lay-Ekuakille, Bilateral despeckling filter in homogeneity domain for breast ultrasound images, in *Proceedings of 3rd (IEEE) International Conference on Advance in Computing, Communication and Informatics (ICACCI-2014)*, Greater Noida (U.P.), India (2015), pp. 1027–1032

20. V. Bhateja, A. Verma, K. Rastogi, C. Malhotra, A non-iterative adaptive median filter for image denoising, in *2014 International Conference on Signal Processing and Integrated Networks*, pp. 113–118, Feb 2014

21. S. Bogdan, K. Malik, B. Machala, Noise reduction in ultrasound images based on the concept of local neighborhood exploration. Adv. Intell. Syst. Comput. **313**, 103–110 (2015)

22. V. Bhateja, G. Singh, A. Srivastava, J. Singh, Speckle reduction in ultrasound images using an improved conductance function based on anisotropic diffusion, in *Proceedings of (IEEE) 2014 International Conference on Computing for Sustainable Global Development* (2014), pp. 619–624

23. A. Tripathi, V. Bhateja, A. Sharma, Kuan modified anisotropic diffusion approach for speckle filtering, in *Proceedings of (Springer) First International Conference on Intelligent Computing and Communication*, Kalyani, India (2016), pp. 1–8

24. A. Sharma, V. Bhateja, A. Tripathi, An improved Kuan algorithm for despeckling SAR images. Inf. Syst. Des. Intell. Appl. **434**, 663–672 (2016)

25. S. Parrilli, M. Poderico, C. Angelino, L. Verdoliva, A nonlocal SAR image denosing algorithm based on LLMMSE wavelet shrinkage. Pattern Recogn. Lett. 606–616 (2012)

26. Wenbo Wang, Xiaodong Zhang, Xiangli Wang, Speckle suppression method in SAR image based on curvelet domain Bivashrink model. J. Softw. **8**(4), 947–954 (2013)

L-Slotted Microstrip Fed Monopole Antenna for Triple Band WLAN and WiMAX Applications

Chandan, Toolika Srivastava and B.S. Rai

Abstract In this paper, a monopole antenna is presented for triple band WLAN and WiMAX applications. The antenna consists of four L-slots on a radiating rectangular patch and a truncated ground plane. The multiband characteristic of the antenna is achieved by a rectangular patch with four L-slots and bandwidth of the antenna is improved by cutting slots on truncated ground plane. The entire volume of the antenna is $29 \times 34 \times 0.8$ mm^3 which is very compact with operating bands of (2.276–2.58 GHz)/2.411 GHz, (3.585–3.623 GHz)/3.609 GHz and (5.508–5.765)/5.56 GHz which covers operating bands for WLAN as per IEEE 802.11 a/b/g/n standards with 12 %, 1 % and 4.5 % impedance bandwidth respectively.

Keywords VSWR · A monopole antenna · Slots · Defected ground plane · Dual-band

1 Introduction

Microstrip patch antenna is very famous among researchers despite some drawbacks like lower bandwidth, lower gain, and extraneous radiation at junctions. The reasons for its popularity are: it is very low cost, easy to fabricate in bulk also and conformal to the surface because of its planar dimension. The drawbacks of lower

Chandan (✉) · B.S. Rai
Department of Electronics & Communication Engineering,
Madan Mohan Malaviya University of Technology,
Gorakhpur 273010, UP, India
e-mail: chandanhcst@gmail.com

B.S. Rai
e-mail: bsr_54@yahoo.co.in

T. Srivastava
Department of Electronics & Communication Engineering,
Kanpur Institute of Technology, Kanpur 208001, UP, India
e-mail: toolikaec1049@gmail.com

© Springer Nature Singapore Pte Ltd. 2017   351
S.C. Satapathy et al. (eds.), *Proceedings of the 5th International Conference on Frontiers in Intelligent Computing: Theory and Applications*, Advances in Intelligent Systems and Computing 516, DOI 10.1007/978-981-10-3156-4_36

bandwidth and lower gain can be overcome by using different techniques like cutting different types of slots, using defected ground structure and also increasing substrate height or using metamaterials. Like any resonant antenna, conventional patch antenna also has dimension λ/2 which can be reduced by using different methods, some of them are mentioned above. All these properties make patch antenna very useful for the present time where the optimum response is required in a compact size and low cost. Besides it, multiband antennas are another requirement for wireless applications like WLAN and WiMAX. WLAN is one of the very popular application in present time which is used in almost each field for example, at home, in schools, in office and also in military purpose. According to IEEE standards 802.11 a/b/g/n operating bands for WLAN is 2.4 GHz (2.40–2.484 GHz), 5.2 GHz (5.15–5.35 GHz), 5.8 GHz (3.585–5.825 GHz) and operating bands for the WiMAX are 2.4 GHz (2.5–2.8 GHz), 3.5 GHz (3.2–3.8 GHz) and 5.5 GHz (5.2–5.8 GHz). So there is need of antenna which can be operated on multiple frequencies. An antenna which can be operate for more than one application is very valuable as customers want many applications from one device and using different antennas for different application may increase the size and cost of the device.

In literature, popular designs suitable for WLAN and WiMAX have been reported such as an antenna with meandering split-ring slot [1], an antenna using defect in the ground plane [2], both for triple band applications. An NRI-TL metamaterial loaded broadband monopole antenna [3] and a single-cell metamaterial loaded monopole antenna with for triple band application [4] but in these the size of the antenna is relatively larger. A planar antenna [5], a multiband planar antenna [6] and printed antennas [7, 8], are presented for triple band applications and having very compact structure but complex design. ACS-fed antenna open-ended slots [9], triple-frequency monopole antenna with defected ground structure [10] and compact multi-resonator-loaded planar antenna [11] have also been reported for multiband applications but in these the size of the antenna is relatively larger as compared to proposed antenna. In [12], an antenna is designed using connected U-slots and truncated ground plane for four band application. Another antenna [13] is designed for WLAN and WiMAX application which cover only two bands.

In this paper, an L-slotted rectangular antenna is presented with defected ground structure to cover all WLAN 2.4/5.2/5.8 GHz and WiMAX 2.4/3.5/5.5 GHz frequency bands. Multiple frequency bands are obtained by cutting L-slots in rectangular patch and frequency bandwidth is increased by using defected ground structure. The maximum % impedance bandwidth covered by the antenna is 12 %. The antenna designed is a low profile compact antenna with good radiation pattern and good gain.

2 Antenna Geometry

The geometry of antenna proposed is shown in Figs. 1 and 2. In the top view of the antenna (Fig. 1), there is a rectangular patch in which four L-shaped slots are cut. The dimension of the rectangular patch is $W_p \times L_p$ (14 × 17 mm^2) and width of L-slots are 1 mm. The dimension of antenna is calculated by the help of formula given below [14]:

$$W_p = \frac{c}{2f_r} \sqrt{\frac{2}{\epsilon_r + 1}}$$

$$\varepsilon_{reff} = \frac{\epsilon_{reff} + 1}{2} + \frac{\epsilon_{reff} - 1}{2} \left[1 + 12 \frac{h}{wp} \right]^{-1/2}$$

$$L_{eff} = \frac{c}{2f_r \sqrt{\epsilon_{reff}}}$$

$$\Delta L = 0.412 \, h \frac{\left[\epsilon_{reff} + 0.3 \right] \left[\frac{wp}{h} + 0.264 \right]}{\left[\epsilon_{reff} - 0.258 \right] \left[\frac{wp}{h} + 0.8 \right]}$$

$$L_p = L_{eff} - 2\Delta L$$

where c (= 3 × 10^8 m/s) is the speed of light.

Bottom view of antenna in Fig. 2 shows a truncated ground plane with some defects which is done by cutting rectangular slots. The overall dimension of the ground plane is $W_g \times L_g$ (27 × 11 mm^2) and dimension of rectangular slots is $L_s \times W_s$ (12.55 × 0.5 mm^2). Antenna is fabricated on the substrate FR4_epoxy of thickness h (0.8 mm) and fed by 50-Ω microstrip line of dimension $W_f \times L_f$ (1.9 × 15 mm^2). The various parameters of antenna are optimized using parametric analysis. All the dimension of optimized design is given in Table 1.

Fig. 1 *Top view* of the proposed antenna

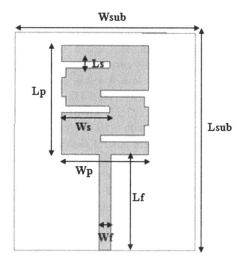

Fig. 2 *Bottom view* of the
proposed antenna

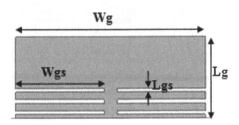

Table 1 All the dimensions
are in mm. Substrate used:
FR4_epoxy, Dielectric
constant $\varepsilon_r = 4.4$, loss tangent
$\delta = 0.02$

Parameters	Dimension	Parameters	Dimensions
W_{sub}	29	W_g	27
L_{sub}	34	L_g	11
W_p	14	W_s	7.7
L_p	17	L_s	1
h	0.8	W_f	1.9
L_f	15	W_{gs}	12.55
L_{gs}	0.5		

3 Results and Discussions

The antenna is designed and simulated using HFSS v16 software and characteristic
of the antenna is studied and analyzed in terms of an s11 parameter, current distribution, and radiation pattern. In this section, there is a discussion of simulated
results. Return loss curve of the final design is given in Fig. 3; the antenna covers
three bands with four resonant points. Frequency bands obtained by the antenna
proposed are (2.276–2.58 GHz)/2.411 GHz, (3.585–3.623 GHz)/3.609 GHz and
(5.508–5.765)/5.56/5.737 GHz. This covers WLAN 2.4/5.7 GHz and WiMAX
2.4/3.5/5.6 GHz with −10 dB impedance bandwidth of 12 %, 1 % and 4.5 % for
band 1, band 2 and band 3 respectively.

To analyze the effect of various parameters on the results of antenna parametric
analysis is done. A parametric analysis is given in Fig. 4, in which the effect of
variation in the width of patch Wp is used as variable. At $W_p = 14$ mm, there is
three bands with good return loss are achieved. But as we increase or decrease the
value of Wp, shift in band and less number of bands is achieved. Figure 5 shows

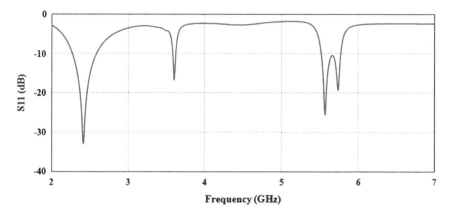

Fig. 3 Return loss of the proposed antenna

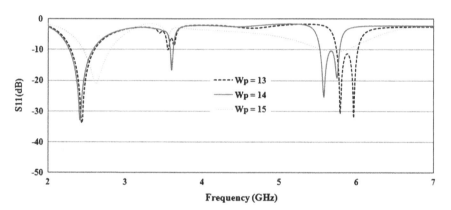

Fig. 4 Variation in width of the patch Wp

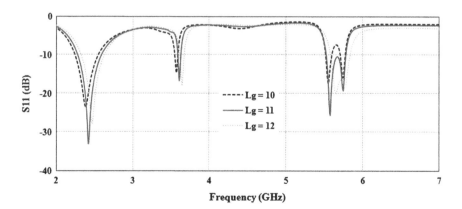

Fig. 5 Variation in length of ground plane Lg

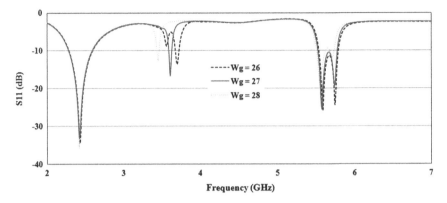

Fig. 6 Variation in width of ground plane Wg

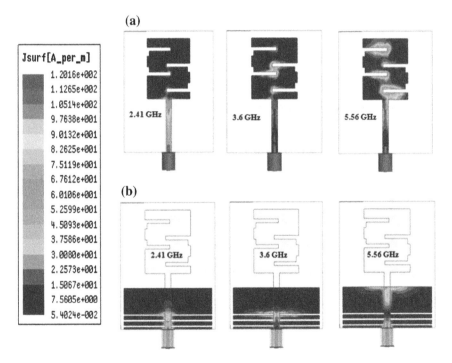

Fig. 7 Surface current distributions of proposed antenna at different frequencies. **a** *Front view*. **b** *Rear view*

the variation in length of ground plane Lg, by which it can clearly observe that when $L_g = 11$ mm, a better return loss is obtained and also there is a small shift in frequency when the length of ground is varying. In Fig. 6 effect of the width of the ground plane on the s11 parameter is presented. At $W_g = 28$ mm, return loss is

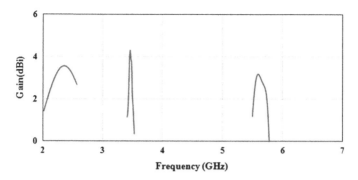

Fig. 8 Gain of the proposed antenna

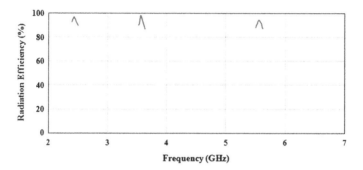

Fig. 9 Radiation efficiency of the proposed antenna

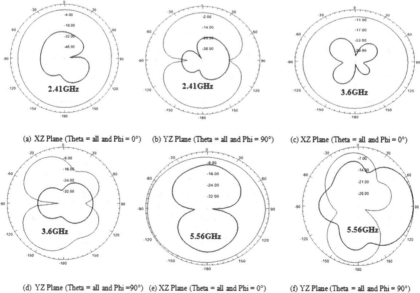

Fig. 10 Radiation pattern at different frequencies

quite good for first but for other two bands it is not good, but at $W_g = 27$ mm better results are achieved for all the three bands and at $W_g = 26$ mm, shift is obtained in middle band.

Figure 7 represents surface current distribution on the patch which shows the variation in field distribution on various frequencies. More radiation is observed near the junction at the microstrip feed line. The gain of the antenna is shown in Fig. 8 in gain vs. frequency curve. Maximum gain obtained is 3.56 dBi for band1 (2.276–2.58 GHz)/2.411 GHz, 4.32 dBi for band2 (3.585–3.623 GHz)/3.609 GHz and 3.17 dBi for band3 (5.508–5.765)/5.56/5.737 GHz. Figure 9 shows the curve of radiation efficiency with respect to frequency. Radiation efficiency of 96.06 %, 98.34 % and 94.47 % have been observed for the lower, middle and higher band. Two-dimensional radiation patterns of proposed antenna at different frequencies are given in Fig. 10. It is observed that the antenna has good omnidirectional radiation pattern in H-plane and bidirectional radiation pattern in E-plane.

4 Conclusions

An L-slotted rectangular patch antenna is successfully designed and various characteristics of the antenna are successfully studied and analyzed. Antenna proposed can be operated for (2.276–2.58 GHz)/2.411 GHz, (3.585–3.623 GHz)/3.609 GHz and (5.508–5.765)/5.56 GHz frequency bands with a gain of 3.56, 4.32 and 3.17 dBi for respective bands. The antenna is best suited for WLAN and WiMAX applications with radiation efficiency of 96.06 %, 98.34 % and 94.47 % for the respective band1, band2 and band3. Good omnidirectional radiation pattern in H-plane and bidirectional radiation pattern in E-plane is observed.

References

1. P.A. Liu, Y.L. Zou, B.R. Xie, X.L. Liu, B.H. Sun, Compact CPW-fed tri-band printed antenna with meandering split-ring slot. IEEE Antenna Wirel. Propag. Lett. **11**, 1242–1244 (2012)
2. J. Pei, A. Wang, S. Gao, W. Leng, Miniaturized triple-band antenna with a defected ground plane for WLAN/WiMAX applications. IEEE Antennas Wirel. Propag. Lett. **10**, 298–302 (2011)
3. M.A. Antoniades, G.V. Eleftheriades, A broadband dual-mode monopole antenna using NRI-TL metamaterial loading. IEEE Antennas Wirel. Propag. Lett. **8**, 258–261 (2009)
4. J. Zhu, M.A. Antoniades, G.V. Eleftheri-ades, A compact triband monopole antenna with single-cell metamaterial loading. IEEE Trans. Antennas Propag. **58**(4), 1031–1038 (2010)
5. H. Chen, X. Yang, Y.-Z. Yin, S.-T. Fan, J.-J. Wu, Triband planar monopole antenna with compact radiator for WLAN/WiMAX applications. IEEE Antenna Wirel. Propag. Lett. **12**, 1440–1443 (2013)
6. H.-Q. Zhai, Z.-H. Ma, Y. Han, C.-H. Liang, A compact printed antenna for triple-band WLAN/WiMAX applications. IEEE Antennas Wirel. Propag. Lett. **12**, 65–68 (2013)

7. S. Verma, P. Kumar, Compact triple-band antenna for WiMAX and WLAN applications. Electron. Lett. **50**(7), 484–486 (2014)
8. A. Mehdipour, A.R. Sebak, C.W. Trueman, T.A. Denidni, Compact multiband planar antenna for 2.4/3.5/5.2/5.8-GHz wireless applications. IEEE Antennas Wirel. Propag. Lett. **11**, 144–147 (2012)
9. X. Li, X.-W. Shi, W. Hu, P. Fei, J.-F. Yu, Compact tri-band ACS-fed monopole antenna employing open-ended slots for wireless communication. IEEE Antenna Wirel. Propag. Lett. **12**, 388–391 (2013)
10. W.-C. Liu, C.-M. Wu, Y. Dai, Design of triple-frequency microstrip-fed monopole antenna using defected ground structure. IEEE Trans. Antennas Propag. **59**(7), 2457–2463 (2011)
11. W. Hu, Y.-Z. Yin, X. Yang, P. Fei, Compact multiresonator-loaded planar antenna for multiband operation. IEEE Trans. Antennas Propag. **61**(5), 2838–2841 (2013)
12. Chandan, Toolika Srivastava, B.S. Rai, Multiband monopole U-slot patch antenna with truncated ground plane, Microwave Opt. Technol. Lett. (MOTL) **58**(8), 1949–1952 (2016)
13. Chandan, B.S. Rai, Dual-band monopole patch antenna using mirostrip fed for WiMAX and WLAN applications, in *Springer Proceedings of Third International Conference*, India 2016, vol. 2 (2016), pp. 533–539
14. C.A. Balanis, *Antenna Theory: Analysis and Design*, 3rd edn. (Wiley, New York, 2005)

Performance Analysis of Fully Depleted SOI Tapered Body Reduced Source (FD-SOI TBRS) MOSFET for Low Power Digital Applications

Vimal Kumar Mishra and R.K. Chauhan

Abstract The fully depleted silicon-on-insulator metal oxide semiconductor field effect transistor (FD- SOI MOSFET) have been considered a promising candidate to extend scaling of planar CMOS technology beyond 100 nm. This technology has been used to reduce leakage current, parasitic capacitances, and fabrication complexity as compared to planar CMOS technology at 50 nm gate length. This paper presents the performance analysis of proposed Tapered Body Reduced Source (FD-SOI TBRS) MOSFET. The proposed structure consumes less chip area and better electrical performance as compared to conventional FD-SOI MOSFET. The proposed structure exhibits higher I_{on} to I_{off} ratio when compared with conventional FD-SOI MOSFET. The structures were designed and simulated using the Cogenda device simulator.

Keywords TBRS · FD-SOI MOSFET · Area · I_{on} to I_{off} ratio

1 Introduction

Silicon technologies are increasing popularity day by day. The main drawback of these technologies is the effect of reducing the dimension of devices. This can be overcome by scaling down of devices to get precise and better output. Due to a reduction in the channel length, the short-channel effects and leakage current become an important issue that degrades the device performance. Fully depleted silicon-on-insulator MOSFET are a promising candidate to reduce short channel

V.K. Mishra (✉) · R.K. Chauhan
Department of Electronics and Communication Engineering, Madan Mohan Malaviya
University of Technology, Gorakhpur, India
e-mail: vimal.mishra34@gmail.com

R.K. Chauhan
e-mail: rkchauhan27@gmail.com

© Springer Nature Singapore Pte Ltd. 2017
S.C. Satapathy et al. (eds.), *Proceedings of the 5th International Conference on Frontiers in Intelligent Computing: Theory and Applications*, Advances in Intelligent Systems and Computing 516, DOI 10.1007/978-981-10-3156-4_37

effects at 50 nm gate length. The silicon area consumed by MOSFET structure in any CMOS technology is another important concern for most of the researchers. The economics related to the layout of semiconductor devices on a single substrate is an important issue, as more efficient layouts lead to a greater number of devices fabricated on a given chip area.

Another main area concerns are the leakage current, low parasitic capacitances, low power dissipation, low subthreshold slope, and DIBL. These problems were removed by using SOI MOS technology. The SOI MOS has divided in the PD-SOI MOSFET and FD-SOI MOSFET. In the present scenario, the FD-SOI MOS devices are mostly used in the low power dissipation. Cheng K. G, and Khakifirooz A. has been detailed studied on Fully depleted SOI (FDSOI) technology [1].

Most of the researchers work to improve the electrical performance of FD-SOI MOS devices. Tseng et al. discuss the effect of floating body in FD-SOI MOSFET [2]. Ohtou et al. and Chouksey et al. has been studied to improve the electrical performance of FD-SOI MOSFET at a very thin BOX thickness [3, 4]. Zhang et al. proposed the threshold voltage model (LVT) of FD-SOI MOS device for low power digital applications [5]. Wen-Kuan et al. studied the impact of junction doping on FD–SOI device performance variability and reliability [6]. Agarwal et al. discuss the analytical surface potential model for study the impact of the electric field in FD-SOI MOSFET [7]. Rathnamala et al. study the effect of RDF on the FD-SOI MOS devices. [8]. Mishra et al. discuss the FD-SOI MOSFET based inverter circuit and improve the transient delay in inverter circuit [9].

In continuation of all these work, the tapered body reduced source (TBRS FD-SOI) MOS devices are proposed for less area consumed and high electrical performance.

2 Device Structure and Specifications

The schematic structure of conventional FD-SOI MOSFET is shown in Fig. 1a. The structure has channel length 50 nm. The Si thickness of conventional FD-SOI MOSFET is 10 nm. The doping concentration at the source and drain region is 10^{20} cm^{-3} and the doping concentration of p-substrate is 10^{16} cm^{-3}. The oxide thickness, kept at 2 nm whereas the material used in gate oxide is HfO_2. The work function of the gate material is 5.0 eV.

The schematic structure of proposed TBRS FD-SOI MOSFET is shown in Fig. 1b. The proposed device structure has channel length and doping profile at source, drain and substrate region are taken as same as the conventional FD-SOI MOSFET. The difference in proposed structure is the tapering body and reducing of source till the depletion width of the channel.

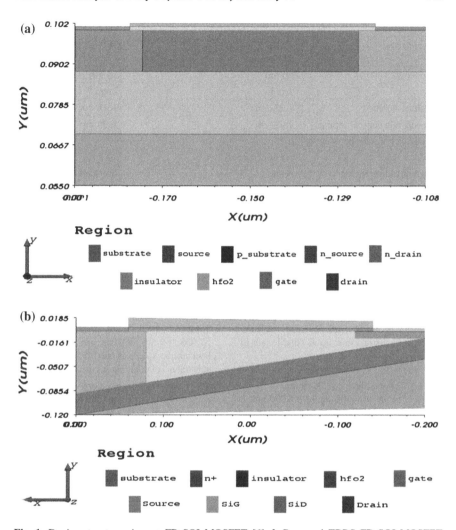

Fig. 1 Device structure view: **a** FD-SOI MOSFET [6]. **b** Proposed TBRS FD-SOI MOSFET structure

Table 1 shows the comparison of various device parameters of the proposed structure with FD-SOI MOSFET [6]. Table 2 compares the area consumed in proposed TBRS FD-SOI MOSFET with the FD-SOI MOSFET. From the table, it can infer that proposed structure consumed approx 68 % less area than the FD-SOI MOSFET [6].

Table 1 Parameters of TBRS FD-SOI MOSFET and FD-SOI MOSFET device [6]

Parameters	TBRS FD-SOI n-MOSFET (Proposed)	FDSOI MOSFET [6]
Gate length	50 nm	50 nm
Tox	1 nm	2 nm
Substrate doping	1e16 cm^{-3}	1e16 cm^{-3}
Source/Drain region doping	1e20 cm^{-3}	1e20 cm^{-3}
Work function of gate material	5.0	–
Insulator thickness (BOX)	5 nm (Tapered)	20 nm
Reduced source region thickness	5 nm	Constant Source = 10 nm
Tapered drain region thickness	10 nm	Constant Drain = 10 nm
Silicon substrate	Tapered cover remaining source area	10 nm

Table 2 Comparison of area consumed in TBRS FD-SOI MOSFET with FD-SOI MOSFET [6]

Plane	FD-SOI MOSFET [6]	Proposed TBRS FD-SOI MOSFET
X-plane	Total length in x-plane = 100 nm	Total length in x-plane = 100 nm
Y-plane	10 nm Si + 20 nm SiO$_2$ + 10 nm substrate thickness, Total length in y-plane = 40 nm	3 nm for source and 10 nm for drain in NMOS, +5 nm SiO$_2$, the area saved beneath the source is used for the substrate, Total length in y-plane = 13 nm
Total Area	Total area covered along x × y = 100 × 40 = 4000 nm^2	The Total area covered along x × y = 100 × 13 = 1300 nm^2

3   Results and Discussions

The optimized Tapered Body Reduced Source MOS device simulations were performed by using Cogenda Device Simulator [10].

Figure 2 shows the input characteristics of TBRS SOI n-MOSFET and FD-SOI MOSFET. From the figure, it can be inferred that the off-state current in proposed structure is lower than the conventional FD-SOI MOSFET. The I_{off} current obtained from the simulation of the proposed structure is 0.6 pA and Ion current as 9 mA. The ratio of I_{on}/I_{off} is 10^{10}. The proposed TBRS FD-SOI MOSFET shows a sub-threshold slope of 60 mV/decade at room temperature. In conventional FD-SOI MOSFET, the I_{off} current is 60 pA and Ion current as 7 mA hence the I_{on}/I_{off} ratio is therefore 10^8. Also, the proposed exhibits a higher I_{on} current as compared to FD-SOI MOSFET at 50 nm channel length.

As per the International Technology Roadmap for Semiconductors (ITRS), the thickness of the gate oxide decreases the metal gate play an important role in the high-performance operation of the device [11]. The high gate material work function (Φm) is important for low off-state current. Figure 2 shows that the input

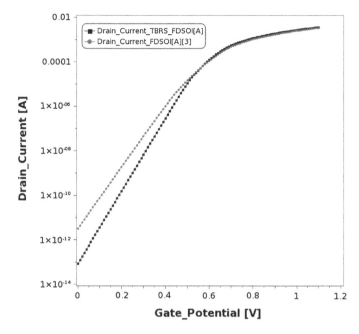

Fig. 2 The comparison of the input characteristic curve of the proposed structure with that of conventional FD-SOI MOSFET structure [6] which shows that the I_{on} to I_{off} ratio is higher in the proposed structure

characteristic curves for varying gate work function from 4.5 to 4.9 eV. To get low off-state current, the channel doping can be increased, but the random dopant fluctuation (RDF) occurs on the channel, so work function increases are preferred rather than the substrate doping increases. However, with decreased in channel doping threshold voltage roll-off for NMOS gets worse. From Fig. 3 the saturation current (Idsat) increases with the gate work function, decrease but the off-state current is to highly reduces as compared to the saturation current. Hence, the I_{on} to I_{off} ratio is increasing as the work function increases. Thus, to fabricate proposed TBRS FDSOI MOSFET with a suitable threshold voltage and drive currents, the gate electrode should have a suitable work function of 5.0 eV.

For high fan out of any digital circuit required high drive current. The high drive current has been observed by the drain current analysis in the MOSFET, which can be shown by the output characteristics of the device. Figure 4 shows the output characteristics of proposed MOSFET for varying the gate voltage. It has been seen that the kink effect, which appears in the conventional bulk MOSFET, reduces in the proposed structure. This is because the body-effect reduces when the tapered insulator region of 5 nm at the bottom of silicon is used for reducing the leakage current from the body. From the figure, it can be inferred that the highest drive current of the proposed structure is 35 mA was found at Vgs = 1.1 V and drive current are reduced as the gate voltage reducers. The large driving current is desired

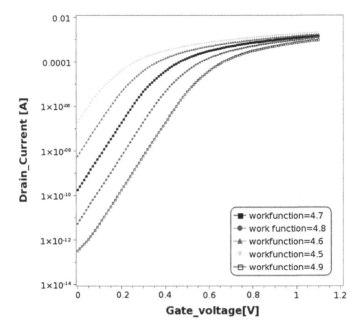

Fig. 3 Input characteristic curve of TBRS FD-SOI n-MOSFET for varying workfunction

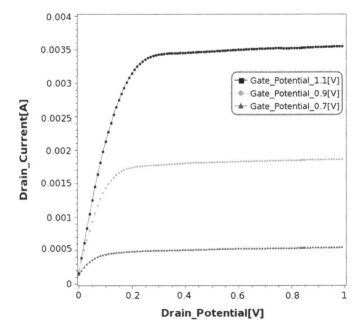

Fig. 4 Output characteristic curve of TBRS FD-SOI n-MOSFET for varying gate potential

Table 3 Performance comparison of various SOI MOSFET devices reported in the literature with the proposed work

References	Channel length (nm)	I_{on}	I_{off}	Vth	Tox	I_{on}/I_{off}
FD-SOI [6]	50	0.1 mA	1pA	0.3 V	1 nm	10^8
FD-SOI [9]	50	1 mA	2pA	0.17 V	2 nm	10^9
SOI FinFET [12]	50	1 mA	10pA	0.27 V	2 nm	10^8
FD-SOI [13]	45	31uA	28pA	0.3 V	1.3 nm	10^6
Proposed work	50	3 mA	80fA	0.3 V	1 nm	10^{10}

for improved transient analysis of any digital circuits. Because of the low I_{off} current in proposed structure, the power dissipation will be reduced in the proposed device. Hence, the performance of the digital circuit is improved when using the proposed TBRS FD-SOI MOSFET (Table 3).

4 Conclusions

In this work the TBRS FD-SOI MOS structure was designed and simulated using the Cogenda device simulator. The proposed structure of SOI n-MOSFET exhibits lower sub-threshold slope, low I_{off} current, and high I_{on} to I_{off} ratio when compares with conventional FD-SOI MOSFET. Proposed structure consumes 50 % less area as compared to conventional FD-SOI MOSFET. High metal gate work function of 5 eV is used to get low off-state current and high I_{on} to I_{off} ratio. Further, the high work function is more compatible with high-κ gate dielectrics (HfO_2). The proposed structure is compatible with the CMOS inverter circuit and other digital circuits.

Acknowledgments This work is supported by the AICTE under research, promotion scheme (RPS-60).

References

1. K.G. Cheng, A. Khakifirooz, Fully depleted SOI (FDSOI) technology. Sci. China Inf. Sci. **59**, 1–15 (2016)
2. Y.-C. Tseng, W.M. Huang, D.C. Diaz, J.M. Ford, J.C.S. Woo, AC floating-body effects in submicron fully depleted (FD) SOI nMOSFET's and the impact on analog applications. IEEE Electron Device Lett. **19**(9), 351–353 (1998)
3. Siddharth Chouksey, Jerry G. Fossum, Shishir Agrawal, Insights on design and scalability of Thin-BOX FD/SOI CMOS. IEEE Trans. Electron Devices **57**(9), 2073–2079 (2010)
4. Tetsu Ohtou, Kouki Yokoyama, Ken Shimizu, Toshiharu Nagumo, Toshiro Hiramoto, Threshold-voltage control of AC performance degradation-free FD SOI MOSFET with

extremely thin BOX using variable body-factor scheme. IEEE Trans. Electron Devices **54**(2), 301–308 (2007)

5. Guohe Zhang, Zhibiao Shao, Kai Zhou, Threshold voltage model of short-channel FD-SOI MOSFETs with vertical gaussian profile. IEEE Trans. Electron Devices **55**(3), 803–809 (2008)

6. Wen-Kuan Yeh, Cheng-Li Lin, Tung-Huan Chou, Wu Kehuey, Jiann-Shiun Yuan, The impact of junction doping distribution on device performance variability and reliability for fully depleted silicon on insulator with thin box layer MOSFETs. IEEE Trans. Nanotechnol. **14**(2), 330–337 (2015)

7. P. Agarwal, G. Saraswat, M. Jagadesh Kumar, compact surface potential model for FD-SOI MOSFET considering substrate depletion region. IEEE Trans. Electron Devices, **55**(3), 789–795 (2008)

8. Rathnamala Rao, Nandita DasGupta, Amitava DasGupta, Study of random dopant fluctuation effects in FD-SOI MOSFET using analytical threshold voltage model. IEEE Trans. Device Mater. Reliab. **10**(2), 247–253 (2010)

9. V.K. Mishra, R.K. Chauhan, Performance analysis of fully depleted ultra thin body [FD UTB SOI MOSFET] based CMOS inverter circuit for low power digital applications. Adv. Intell. Syst. Comput. (2016), pp. 375–382

10. Visual TCAD User Guide, version: 1.8.2-8, Cogenda device simulator, (March 2015)

11. International Technology Roadmap for Semiconductors, (SIA), edition of ITRS, (2013). http://itrs2.net

12. M. Saremi, A. A. khusa, Saeed Mohamadi, Ground plane fin-shaped field effect transistor (GP-FinFET): a FinFET for low leakage power circuits. Microelectron. Eng. **95**, 74–82 (2012)

13. D. Tang, Y.H. Li, G.H. Zhang et al., Single event upset sensitivity of 45 nm FDSOI and SOI FinFET SRAM. Sci. China Tech. Sci. **56**, 780–785 (2013)

A Study and Analysis of Different Brain Tumor Segmentation Techniques

K. Mahantesh, B.V. Sandesh Kumar and V.N. Manjunath Aradhya

Abstract In this paper, a thorough study and quantitative analysis of different brain tumor segmentation techniques will be addressed. Several significant algorithms are proposed in literature for partitioning MRI (Magnetic Resonance Imaging) brain image into considerable multiple disjoint regions indicating tumor tissues and normal brain tissues. But, benchmarking brain tumor segmentation (BTS) techniques is found very less. In this regard, we study, explore and create benchmark for most popular and widely accepted segmentation techniques such as histogram thresholding, adaptive k-means clustering, region based active contour SVM and PCA based K-NN classifier. A detailed quantitative evaluation of aforesaid techniques on the MRI images from the standard datasets as well as collection of our own datasets is presented. An analysis of results reported will have an excellent impact on current and future research efforts in brain tumor segmentation.

Keywords Brain tumor segmentation · Benchmark · k-means · Histogram thresholding · SVM · PCA, K-NN

1 Introduction

Image segmentation is considered to be a very significant stage in medical image processing and pattern recognition, especially characterizing image structure well suitable for clinical diagnosis and analysis of medical images. Segmentation is

K. Mahantesh (✉) · B.V. Sandesh Kumar
Department of ECE, Sri Jagadguru Balagangadhara Institute of Technology,
Bangalore, India
e-mail: mahantesh.sjbit@gmail.com

B.V. Sandesh Kumar
e-mail: bvsk39@gmail.com

V.N. Manjunath Aradhya
Department of MCA, Sri Jayachamarajendra College of Engineering, Mysore, India
e-mail: aradhya.mysore@gmail.com

© Springer Nature Singapore Pte Ltd. 2017
S.C. Satapathy et al. (eds.), *Proceedings of the 5th International Conference on Frontiers in Intelligent Computing: Theory and Applications*, Advances in Intelligent Systems and Computing 516, DOI 10.1007/978-981-10-3156-4_38

defined as a process of partitioning an image into its essential regions, and the level of partitions is classified depending upon the problems being solved. MRI (Magnetic Resonance Imaging) is one of the most admired medical imaging techniques, particularly in diagnosing and predicting brain tumors. Since MRI images are advantageous in showing variety of tissues with high resolution and excellent contrast, partitioning some explicit cells and tissues (most of the times tumor) from the rest of the image (healthy cells and tissues) is considered to be as exigent task. Segmentation assists and guides general practitioner to detect lesion/tumor more accurately. Manually locating and segmenting tumors involves manual variations and subjective opinions misleading the presence or absence of tumors and are usually expensive, time consuming and is monotonous task. This strongly suggests that there is a definite need of automated brain tumor segmentation technique [1].

Being motivated from the above challenges, several state- of-the-art BTS techniques have been explored and survey on few crucial algorithms are discussed in Sect. 2. Section 3 gives a summary of four most popular image segmentation techniques considered for analysis and evaluation. Dataset creation and evaluation metrics used to measure the performances of each of these techniques are explained in Sect. 4. Finally, conclusion and some future research avenues are discussed in Sect. 5.

2   Related Works

Since brain tumor is one of the serious and life threatening disease, automating its detection is a challenging task. Brain tumor segmentation and classification is a crucial process in detecting the tumors at initial stages and has evolved as a major clinical requirement for brain tumor diagnosis and radiotherapy planning [2]. Hybrid self-organizing map (SOM) with fuzzy K means (FKM) offers successful classification of tumor from healthy tissues in MRI brain image [3]. Multi-level wavelet features are used with morphological filtering and compared the ROI areas with neighboring pixels, PNN is used to increase the classification accuracy [4]. Integration on k-means and Fuzzy C-Means followed by thresholding and level-set methods provided accurate tumor information and also observed minimal computation time for k-means clustering [5]. Statistical features proved to be highly efficient with reduced dimension and lesser computational cost in discriminating tumor tissues in comparison with Gabor-wavelet based features [1]. Features extracted by DWT given to two tier classifier i.e. SOM neural network and knn-classifiers to produce deterministic results [6].

Enhanced Possibility Fuzzy C-Means (EPFCM) combines region based (i.e. contour based methods) and FCM methods to improve the detection [7]. Generative probabilistic model for segmentation of brain lesions in multi-dimensional images that generalizes the EM segmenter [8], Data-driven algorithm for 3D multi-modal MRI brain glioma tumor and edema segmentation in different modalities [9]. Multi-resolution fractal model extracts spatially varying multi fractal features [10]. With this outbreak of novel segmentation and classification approaches, it is very difficult to evaluate and/or benchmark the benefits of a particular algorithm over alternate methods. Hence, there is a vital need for effective analysis, integration and testing of these ideas with a standard metrics with its manually segmented/ground-truth images.

3 Proposed Methodology

In the literature, it is evident that several algorithms proposed have been experimented on very few images/image dataset and performance evaluation of these algorithms with the standard metrics is still concealed. The reason might be due to the unavailability of the standard evaluation metrics and standard set of images and its ground-truth. Also, openly accessible implementations of state-of-the-art segmentation techniques are rarely to be found in order to experiment. In this paper, we have considered very popular segmentation techniques such as histogram thresholding, adaptive k-means clustering, Contour based SVM and PCA and K-NN classifier based techniques to study, evaluate and benchmark their performances. Brief explanation of aforesaid techniques can be found in the following subsections.

3.1 Histogram Thresholding

In histogram thresholding firstly the input image is divided into two equal parts, histogram is plotted for both the parts. Both histograms are compared to calculate the threshold point then based on thresholding tumor is localised in brain, Image is filtered to remove the unwanted distortion during segmentation [10].

3.2 Adaptive K-Means Clustering

In adaptive k-means algorithm, the value of "k" is chosen randomly and they are made as initial centroids [6]; here measurement of distance between pixel and cluster is very important. Consider two elements S_1 and S_2, distance between elements S_1 and S_2 can be calculated using, $\sqrt{(S_{11} - S_{21})^2 + (S_{12} - S_{22})^2 + \cdots + (S_{1n} - S_{2n})^2}$. Where, $S_1 = (S_{11}, S_{12}, \ldots, S_{1n})$ and $S_2 = (S_{21}, S_{22}, \ldots, S_{2n})$.

The distance between two clusters should be d_{min}, K-means algorithm will be carried out as follows.

Step-1: If the gap between cluster and element (pixel) is '0', put that particular element (pixel) to that cluster and continue process with next element(pixel).

Step-2: When the gap between pixel and cluster is less than required d_{min}, put this particular pixel to nearby cluster.

Step-3: If d_{min} is less than the gap between pixel and closest cluster then Combine cluster C_2 in cluster C_1 and vanish the cluster C_2 and delete all the pixels from cluster C_2. After that new pixel is added into empty cluster, hence new cluster is formed and finally recalculate the distance between all clusters of an image data, process continues, till no pixel switches clusters.

3.3 Region Based Active Contours and Support Vector Machines

Active contour is the process which looks onto nearby edges and localizes them more accurately [11]. Active contour can be defined as $\vec{Z}(s) = (\vec{A}(s), \vec{B}(s))$. Where $A(s)$ and $B(s)$ are along x, y coordinates through the contour, where s is the normalized index, energy function used to describe active contour is made up of two components, internal energy and external energy. Curve compactness is obtained by the internal forces; curve around the object's borders is obtained by External forces. Adding of bending energy and elastic energy gives internal energy as shown below:

$$E_{int} = E_{elastic} + E_{bend} = \alpha(s) \, |dv/ds|^2 + \beta \, |dv/ds|^2 \qquad (1)$$

Continuity is specified by adjustable constant denoted by α and contour curving is denoted by β. Support Vector Machine is supervised classification algorithm, SVM finds hyper plane which optimally separates two classes, SVM is a best segregator of two classes here In this algorithm, each data point is plotted in n-dimensional space with the features of considered coordinate.

3.4 PCA + K Nearest Neighbor Classifier

Principal component methodology is an important and powerful boon for feature extraction; it is used to reduce the feature space dimension hence PCA gives more accurate and efficient result [3]. Principal Component Analysis method utilizes the eigenvectors from a correlation matrix for converting or transforming to lower dimensional feature space. PCA transforms the image into principal components i.e.

$$Z = W^T X. \qquad (2)$$

where, Z is feature vector matrix of order MXN and $X = (X_1, X_2, \dots, X_i, \dots, X_n)$ of order NXN and $W^T = NXM$ is a transformation matrix of W. In this, eigenvectors are placed in columns, Eigen values computed as $\lambda_{ei} = S_{ei}$. Where, ei, λ and S are eigen vectors, eigen values and scatter matrix respectively. Scatter matrix is represented as,

$$S = \sum_{i=1}^{n}(X_i - \mu)(X_i - \mu)^T \tag{3}$$

where μ is mean of the data X. To increase the value of determinant of scatter matrix W_opt is considered,

$$W_{opt} = arg^{max-w}\left|W^T SW\right| = [W_1, W_2, \ldots, W_m] \tag{4}$$

K-Nearest Neighbor is an instance method of learning, In classification process, k is a constant defined by user and the test point which is un-labelled is classified by the process of allocating the label of the most frequently appearing considered k training samples close to the test point, Euclidean distance is used as a distance metric here. The classification of any object is based on neighbors majority belongingness and that object is considered to be of the class which is very common among its K-Nearest Neighbor, Training phase of K-NN is super fast, here in local sub region training samples of K-NN is stored, which is essential when testing process of K-NN classifier is carried out.

4 Experimental Results and Performance Analysis

In this section, firstly, we explained about collection of MRI images and creating dataset along with a usual experimental setup. Later, experimental results on each techniques and performance analysis using some standard metrics are demonstrated.

4.1 Dataset Collection and Experimental Set up

Brain Tumor Segmentation (BraiTS—241) dataset is first of its kind created for brain tumor detection and evaluation, it comprises of 241 MRI images with its ground truth images. Images in dataset are collected from Google images and stored in ".JPEG" format with different resolution, high intensity variations, varying contrast and multiple tumors making it even more challenging. Figure 1, shows some of the example images of MRI images (1st row) and it ground-truth generated with the help of an expert (2nd row). BraiTS—241 dataset will be made available on-line very shortly for various research community in order to evaluate and benchmark their algorithms.

The proposed method is implemented in MATLAB 14.0 on an Intel core2duo 2.20 GHz with 3GB RAM. As an experimental setup, all 241 images in BraiTS-241 dataset are considered for segmentation using above discussed techniques. Further, quantitative analysis is carried out by comparing resultant segmented images with the ground truth images by means of six different metrics to depict the rate of similarity between them.

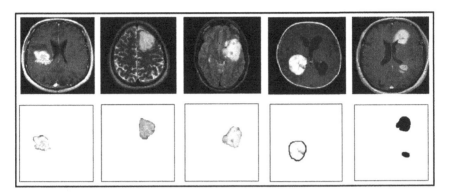

Fig. 1 Example images. *1st row* Original MRI image. *2nd row* Ground truth/manually segmented images showing tumor

4.2 Experimental Results

Figure 2 shows the resultant segmented images of the above discussed algorithms. Histogram thresholding produces most favorable peaks and valleys and makes us very easy in discriminating tumor (peaks) and non-tumor (valleys) tissues in MRI image. The results of Histogram thresholding for MRI images considered and its associated results are shown in Fig. 2a. Random selection of k-clusters has yielded promising results for successfully segmenting the tumors and the results of adaptive k-means clustering for MRI images is shown in Fig. 2c along with its constructed ground truth images and segmented output using adaptive k-means algorithm.

Active localization of nearby edges and describing the contours with its energy information has given satisfactory results in defining the contours of tumor. The results of Region based active contours and SVM for MRI images considered is shown in Fig. 2d. PCA has once again proved to be efficient in extracting discriminant features and KNN to identify the minimum distance between the neighboring pixels to identify between tumor and non-tumor regions. The main disadvantage of this technique is that it fails to remove skull part and few crucial healthy part of the brain. The results of PCA+KNN Classifier for MRI images considered is as shown in Fig. 2b.

4.3 Performance Analysis

In order to evaluate the performances of each of the segmentation technique, resultant image obtained is compared with its associated ground-truth using six different quantitative metrics such as MSE, PSNR, RMSE, Jaccard, Dice, and Correlation. The comparative study of segmentation methods discussed is noted in Table 1 and the values shown in table are the average metric values obtained considering all 241 images in dataset.

(a)   (b)

(c)   (d)

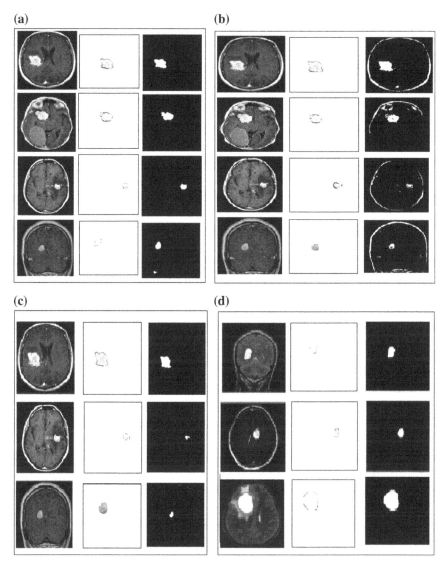

Fig. 2 Example images of MRI, ground truth and resultant segmented images using **a** Histogram Thresholding **b** PCA + KNN respectively **c** Adaptive k-means **d** SVM respectively

Greater values of Jaccard and Dice reveals exceedingly closest match with the ground truth. Minimal values of MSE and RMSE obtained are evident for a lesser amount of errors between ground truth and segmented images. Higher values of correlation put forward high statistical dependency between any two images. High PSNR value reveals the quality of segmentation with respect to the homogeneity of tumor sections in ground-truth and segmented images obtained with different methods.

Table 1 Performance analysis

Methods	MSE	PSNR (dB)	RMSE	Jaccard	Dice	Corr
Histogram Thresholding [10]	0.0241	60.9376	0.1553	0.2662	0.2243	0.3338
Adaptive k-means [6]	0.0126	64.306	0.1576	0.698	0.6130	0.6018
Contour-based SVM [11]	0.0523	61.1074	0.2289	0.1519	0.1986	0.2742
PCA based KNN [3]	0.0752	59.366	0.2743	0.1137	0.2043	0.2655

Comparative study using six different quantitative metrics proves that, adaptive k-means clustering segments out the tumor section more efficiently in comparison with the other three most admired techniques for brain tumor detection.

5 Conclusion and Future Scope

The proposed work mainly concentrates on quantitative evaluation of some popular brain tumor detection techniques and a small effort in generating a benchmark dataset for segmentation and analysis of brain tumors. A study and thorough analysis on BraiTS—241dataset concludes that adaptive k-means technique has outperformed histogram thresholding, Contour based SVM, and PCA based KNN techniques in detecting the tumor. It further suggests that the adaptive k- means algorithm is a promising technique for medical imaging applications and in future, it can be combined with different classifiers.

References

1. N. Nabizadeh, M. Kubat, Brain tumors detection and segmentation in MR images: Gabor wavelet vs. statistical features. Comput. Electr. Eng. **45**, 286–301 (2015)
2. N. Gordillo, E. Montseny, P. Sobrevilla, State of the art survey on MRI brain tumor segmentation. Magn. Reson. Imaging **31**, 1426–1438 (2013)
3. G. Vishnuvarthananan, M. Pallikonda Rajasekaranb, P. Subbarajc, A. Vishnuvarthanan, An unsupervised learning method with a clustering approach for tumor identification and tissue segmentation in magnetic resonance brain images, Appli. Soft Comput. **38**, 190–212 (2016). Elsevier
4. M. Havaei, A. Davy, D. Warde-Farley, A. Biard, Brain tumor segmentation with deep neural networks. Med. Image Anal. (2016)
5. Eman Abdel-Maksoud, Mohammed Elmogy, Rashid Al-Awadi, Brain tumor segmentation based on a hybrid clustering technique. Egypt. Inf. J. **16**, 71–81 (2015)
6. V. Anitha, S. Murugavalli, Brain tumour classification using two-tier classifier with adaptive segmentation technique. Inst. Eng. Technol. ISSN **1751–9632**, 1–9 (2015)
7. A. Rajendran, R. Dhanasekaran, Fuzzy clustering and deformable model for tumor segmentation on MRI brain image: a combined approach, in *International Conference on Communication Technology and System Design 2011, Procedia Engineering*, vol. 30, (2012), pp. 327–333

8. I. Njeh, L. Sallemi, I. Ayed, 3D multimodal MRI brain glioma tumor and edema segmentation: a graph cut distribution matching approach. Comput. Med. Imaging Graph. **40**, 108–119 (2015)
9. B.H. Menze, K. Van Leemput, A generative probabilistic model and discriminative extensions for brain lesion segmentation with application to tumor and stroke. IEEE Trans. Med. Imaging **35**, 933–946 (2016)
10. A. Islam, S.M.S. Reza, K.M. Iftekharuddin, Multifractal texture estimation for detection and segmentation of brain tumors. IEEE Trans. Biomed. Eng. **60**(11) (2013)
11. J. Sachdeva, V. Kumar, I. Gupta, N. Khandelwal, C.K. Ahuja, A novel content-based active contour model for brain tumor segmentation. Magn. Reson. Imaging **30**(2012), 694–715 (2012). Elsevier

A Novel Representation for Classification of User Sentiments

Steven Lawrence Fernandes, S.N. Bharath Bhushan, Vinlal Vinod and M.J. Adarsh

Abstract In this paper we present a term sequence preserving representation for text documents called as label matrix for classification of opinions. This is an efficient yet effective representation in polarity of opinions with the use of only three parts of speech feature set viz verb, adverb and adjective. To draw out the efficiency of our proposed technique we led experimentation on one publically accessible extremity audit dataset furthermore our own made motion picture survey and item survey datasets. We have explored quantitative comparative analysis between existing classifiers and proposed method.

Keywords Opinion analysis · Representation · Sentiment classification

1 Introduction

The World Wide Web (WWW) is a widely distributed and dynamic information gallery. It contains huge amount of subjective textual information. In recent years we have seen many discussion groups and review sites such as Amazon.com, Internet Movie Database (IMDB) [1], and Discussion forums etc. This contains overall opinion towards the products or services, which make it difficult to track and understand customer opinions. Feature characterization has a few applications, for instance in business to comprehend the voice of client as communicated in regular

S.L. Fernandes (✉) · S.N. Bharath Bhushan · V. Vinod
Sahyadri College of Engineering and Management, Mangalore, Karnataka, India
e-mail: steven.ec@sahyadri.edu.in

S.N. Bharath Bhushan
e-mail: sn.bharath@gmail.com

V. Vinod
e-mail: vinodvinlal@gmail.com

M.J. Adarsh
JNN College of Engineering, Shimoga 577 201, India
e-mail: mjadi14@gmail.com

© Springer Nature Singapore Pte Ltd. 2017
S.C. Satapathy et al. (eds.), *Proceedings of the 5th International Conference on Frontiers in Intelligent Computing: Theory and Applications*, Advances in Intelligent Systems and Computing 516, DOI 10.1007/978-981-10-3156-4_39

correspondences. In politics, to understand the opinion of the voter about the candidate. To purchase the products in online shopping or in advertisement. In education for e-learning, research furthermore in Blog investigation the capacity to handle ambiguities is one of the critical markers of vigorous conclusion characterization strategy.

There are several challenging aspects in characterized classification. The first is to figure out if an archive or partition is subjective. Second test is that the trouble lies in the wealth of human dialect utilized i.e. individuals don't generally express sentiment same way [1]. With a specific end goal to land at sensible conclusions, examination of the notion connection needs to get it. Be that as it may, "The motion picture was incredible" is altogether different from "The motion picture was not extraordinary". In the more casual manner like twitter or other websites the more probable individuals are to consolidate distinctive conclusions in the same sentences which is simple for a human to see be that as it may, more troublesome for a PC to parse.

In this paper, we explore a label matrix representation to arrange the user comments as a positive or the negative with the utilization of just three sections of discourse (i.e., verb, adverb, and adjective) feature set. We have conducted experimentation on movie review and product review datasets and also publically available polarity movie review dataset. We examine the quantitative comparative analysis between existing classifiers and proposed method.

For the comfort, remaining section of the paper is sorted out as follows: Sect. 2 presents Literature Survey; Sect. 3 presents working principle of proposed method to classify the opinions with the use of Label Matrix Representation. Section 4 contains the details of different datasets and experimentation. Last section includes conclusion on this work.

2 Literature Survey

As of late numerous analysts, have proposed numerous machine learning ways to deal with order the extremity of suppositions. In [2], creators have inspected the adequacy of applying machine learning procedures like Support Vector Machines and Naïve Bayes, to the opinion order issue. In [3], author presents an unsupervised learning method known as semantic introduction for characterizing a survey as positive or negative. In [4], they proposed a novel machine learning strategy that applies content classification systems by considering of subjective parts of the user comments. Separating subjective segments can be executed utilizing proficient strategies for discovering least cuts in diagrams. In [5] they address the rating - deduction issue: as opposed to simply figure out if an audit is "thumbs up" or "thumbs down". In [6], writer exhibits that utilizing emoticons decreases the user comment size which intern reduces the memory for processing with point and time for feeling characterization. In [7], addresses the undertaking of characterizing blog entries to positive and negative my making use of SVM. In [8] they proposed machine learning approach in view of string/character segmentation for identification of Chinese reviews. In [9] they introduced a condition of workmanship survey by mining the user review from online client criticism. In [10], they proposed NN based dictionary construction which consolidates the benefits of machine learning and data recovery methods. In [11], they have presented a novel methodology for natural ordering the assumptions of Twitter messages by discarding the input ordering. In [12], they proposed a machine learning algorithm to deal with fuse scattered data into a label assessment order framework. In [13], they proposed a multi-information based methodology utilizing Self Organizing Maps (SOM) and motion picture learning keeping in mind the end goal to model conclusion over a multi-dimensional estimation space. In [14], they introduce machine learning procedure to deal with ordering the client surveys. In [15], they give a general overview about estimation investigation identified with user reviews related to day to day things. In [16], they presents a review that covering the strategies and techniques in feeling investigation and difficulties to show up in the field. In [17], they explain a methodology with estimation investigation which utilizes SVM to compare differing user review data. In [18], they introduce a study on part of user review in opinion examination and proposed a sentence label feeling recognizable proof utilizing word label segmentation technique. In [19], they proposed semi directed learning technique to group items or things that includes conclusion in terms of positive and negative which will be termed as conclusion mining. In [20], the present the construction effective representation model which can be extended for sentiment classification. In [21], they concentrate on utilizing Twitter, the most well known miniaturized scale blogging stage, for the errand of slant examination. In [22], they exhibit a basic model variation where a SVM is worked over Naive Bayer's log-consider proportions highlight values and get great results.

3 Proposed Method

Classifying the polarity of the opinions based on only three parts of speech features viz verb, adverb and adjective play an important role in opinions. In the proposed method, since only few terms are considered the user comments reduces the feature space which intern reduces the processing classification time. These features reasonably taking less time for classification of opinions compared to using all eight parts of features. In feature extraction stage, features namely verb, adverb, adjective are extracted from the database. Once the feature extraction stage is accomplished, these extracted features are subjected for classification. Knowing the sequence of term, one can understand the content and flow of the review. The existing models will contain only set of terms without preserving the term sequence, hence they does not preserve the contextual information of the terms present in reviews. In this article, a new approach is proposed for effective representation of the user reviews called as "label matrix", which is adequately protects the grouping of terms in a content reports. Our aim in this work is to examine whether this method classifies the polarity of reviews same as that in the topic categorization. A label Matrix is a binary matrix, where contents are either 0 or 1.

3.1 Representation Stage

Consider polarity of classes C_i, such that class C_i ($i = 1$ and 2;) there are m number of training reviews D_i^j, ($j = 1,..,m$). A review D_{ij} is pre-processed to eliminate all stopwords (terms which do not preserve any domain knowledge) resulting in t_j terms. The pre-processed reviews of each class are pooled together to construct a dictionary which will be considered as knowledge base for classification of unknown reviews. M_{ij} is a label matrix of dimension $i \times t$, where i is the dimension of classes and t is the quantity of terms in query review after pre-proceesing. A new method for voting is proposed for voting of unknown sample to classify as positive or negative samples. It is defined as if the term t_j is available in the learning base of the class C_i, then review related to the sentence of C_i and the section t_j is set to 1, else it is set to 0.

$$M_{ij} = \begin{cases} 1 & \text{if } t_i \in C_i \\ 0 & \text{otherwise.} \end{cases} \tag{1}$$

3.2 Classification Stage

This section of will address development of classification algorithm for the proposed lable matrix representation of the user query data. After constructing the label matrix the row sum gives the voting for corresponding class. Hence first level of classification will be achieved by considering voting rule defined in the above mentioned equation. If two classes such as positive and negative classes have same voting value, then searching of longest matching sequence in the label matrix for the unknown samples among the classes which involved in conflict will be performed. Now the test review is assigned with the class label with which has longest matching sequence of terms present in the unknown review. The above explanations of data representation and classification is pictorially represented in the Fig. 1.

4   Experimentation

To demonstrate the viability of the proposed methodology, we have conducted experiential trials utilizing four classifiers viz., Centroid, Nearest Neighbor, Naive Bayes, Voting classifier with label matrix representation and also voting classifier

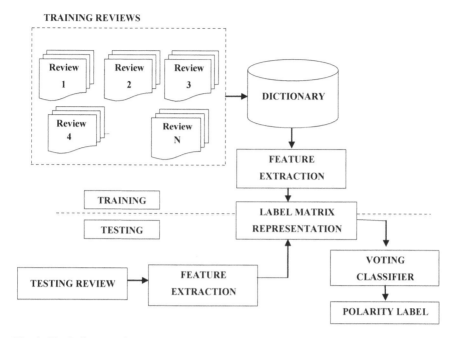

Fig. 1 Block diagram of classification of opinions using Status Matrix

without voting matrix representation. Each classifier is empirically evaluated by publically available movie review datasets and also our datasets viz., movie review datasets and product review datasets. Given a reviews dataset we split the dataset into two sets i.e., 60 % of dataset is used to train the system and 40 % of dataset is used to evaluate the system. We also explored the quantitative comparative analysis between all four classifiers and Status Matrix representation. To find efficiency and robustness of our proposed work we have applied Precision, Recall and F- Measure.

4.1 Data Sets

Any system has to be tested to find its effectiveness and efficiency for which it has been designed. In this angle, following are the details of the datasets used for evaluation of the proposed algorithms.

4.1.1 Data Set 1: Movie Reviews

This dataset contains two sets of movie reviews along with the other important details of the movie such as duration, director name and other details of the movie are also manually downloaded from the website. The reviews are in .txt format and which are collected from IMDB review site.

4.1.2 Data Set 2: Product Reviews

This Product review contains reviews of four different products such as Mobile, Home appliances, Watch and Camera, which are collected from different review sites. Ex: Amazon, Epinions etc. All surveys of this item are in .txt format.

4.1.3 Data Set 3: Polarity Review Movie Dataset

This dataset is created by Pang et al. in [4]. All surveys of this dataset were composed before 2002, with a top of 20 reviews for every author (312 creators' aggregate) per classification. The dataset is collected from the url present in the above mentioned paper. The reviews are in .txt format.

4.2 Quantitative Comparative Analysis

In this section, we show the quantitative comparison between existing four classifiers and voting classifiers with label matrix representation. Comparing to all classifiers, voting classifier with label matrix representation gives best result. This Quantitative comparative analysis is as shown in Table 1.

Table 1 Quantitative comparative analysis between existing method and proposed method

Data sets	Classifiers (Maximum F-Measure)				
	Centroid	Nearest neighbor	Naïve Bayes	Voting without abel matrix (Proposed)	Voting with label matrix (Proposed)
Movie review 1	0.7720	0.7670	0.7370	0.7741	0.8645
Movie review 2	0.7023	0.7126	0.7318	0.6017	0.8956
Movie reviews	0.7371	0.7398	0.7344	0.6877	0.8800
Mobile review	0.8376	0.6790	0.7240	0.7141	0.9578
Home appliances	0.6700	0.7023	0.7500	0.7023	0.8498
Watch review	0.7281	0.7510	0.7453	0.7689	0.8745
Camera review	0.6638	0.6741	0.7782	0.6810	0.8856
Product reviews	0.7248	0.7016	0.7493	0.7165	0.8919
Polarity review movie dataset	0.7175	0.7072	0.7407	0.6789	0.8578

5 Conclusion

This paper, we have made an effort to handle the extremity of sentiment classification by utilizing label matrix representation with the utilization of just three sections of discourse list of capabilities. We have empirically checked the effectiveness of the system by conducting an extensive experiment with the use of movie review, product review and also publically available polarity review movie datasets. We have explored quantitative comparative analysis between existing classifiers and proposed method. From this analysis we have concluded that, the voting classifier with label matrix representation will give best result in less time and also preserves the sequence of terms in testing review.

References

1. B. Liu, Sentiment Analysis and Opinion Mining, Morgan & Claypool Publishers, (May 2012)
2. B. Pang, L. Lee, S. Vaithyanathan, Thumbs up? sentiment classification using machine learning techniques, in *Proceedings of the Conference on Empirical Methods in Natural Language Processing (EMNLP)*, (2002), pp. 79–86
3. P.D. Turney, Thumbs up or thumbs down? semantic orientation applied to unsupervised classification of reviews, in *Proceedings of the ACL*, (2002), pp. 417–424
4. B. Pang, L, Lee, A sentimental education: sentiment analysis using subjectivity summarization based on minimum cuts, in *Proceedings of the ACL*, (2004), pp. 271–278
5. B. Pang, L. Lee, Seeing stars: exploiting class relationships for sentiment categorization with respect to rating scale in *Proceedings of ACL*, (2005)
6. G. Mishne, experiments with mood classification in blog posts, in *1st Workshop on Stylistic Analysis of Text for Information Access*, (2005)
7. J. Read. Using emoticons to reduce dependency in machine learning techniques for sentiment classification, in *Proceedings of ACL*, (2005)
8. C. Zhang, W. Zuo, T. Peng, F. He, Sentiment classification for Chinese reviews using machine learning methods based on string kernal, in *Third International Conference on Convergence and Hybrid Information Technology(ICCIT)*, (IEEE, 2008)
9. T. Bhuiyan, Y. Xu, A. Josang, State-of-the-art review on opinion mining from online customers' feedback, in *Proceedings of the 9th Asia-Pacific Complex Systems Conference*, (2009)
10. L.-S. Chen, H.-J. Chiu, Developing a neural network based index for sentiment classification, in *Proceedings of the International Multi Conference of Engineers and Computer Scientists (IMECS)*, vol. I, (2009)
11. A. Go, R. Bhavani, L. Huang, Twitter sentiment classification using distant supervision. Processing, (2009), pp. 1–6
12. S. Li, S.Y.M. Lee, Y. Chen, C.-R. Huang, G. Zhou, Sentiment classification and polarity shifting, in *Proceedings of the 23rd International Conference on Computational Linguistics*, (2010), pp. 635–643
13. W.B. Claster, D.Q. Hung, S. Shanmuganathan, Unsupervised artifical neural nets for modeling sentiment, in *Second International Conference on Computational Intelligence, Communication systems and Networks*, (IEEE, 2010)
14. J. Zizka, V. Rukavitsyn, Automatic categorization of reviews and opinions of internet e-shopping customers, in *Annual Conference on Innovations in Business and Management*, (2011)
15. N. Jebaseeli, E. Kirubakaran, A survey on sentiment analysis of (Product) reviews. Int. J. Comput. Appl. **47** (2012)

16. G. Vinodhini, R.M. Chandrasekaran, Sentiment analysis and opinion mining: a survey. Int. J. Adv. Res. Comput. Sci. Softw. Eng. **2** (2012)
17. T. Mullen, N. Collier, Sentiment analysis using support vector machines with diverse information sources, in *Proceedings of EMNLP*, **4**, 412–418 (2004)
18. M. Wiegand, A. Balahur, B. Roth, D. Klakow, A. Montoyo, A survey on the role of negation in sentiment analysis, in *Proceedings of the Workshop on Negation and Speculation in Natural Language Processing*, (2010), pp. 60–68
19. Z. Zhai, B. Liu, H. Xu, P. Jia, Clustering product features for opinion mining, in *Proceedings of the 4th ACM International Conference on Web Search and Data Mining*, (2011), pp. 347–354
20. A. Danti, S.N.B. Bhushan, Document vector space representation model for automatic text classification, in *Proceedings of International Conference on Multimedia Processing, Communication and Information Technology*, (Shimoga, 2013), pp. 338–344
21. A. Pak, P. Paroubek, Twitter as a corpus for sentiment analysis and opinion mining, in *Proceedings of LREC*, (2010)
22. S. Wang, C.D. Manning, Baselines and bigrams: simple, good sentiment and topic classification, in *Proceedings of the 50th Annual Meeting of the Association for Computational Linguistics: Short Papers,* vol. 2, (2012), pp. 90–94

A 3D Approach for Palm Leaf Character Recognition Using Histogram Computation and Distance Profile Features

Panyam Narahari Sastry, T.R. Vijaya Lakshmi, N.V. Koteswara Rao and Krishnan RamaKrishnan

Abstract Handwritten character recognition has been a well-known area of research for last five decades. This is an important application of pattern recognition in image processing. Generally 2D scanning is used and the text is captured in the form of an image. In this work instead of regular scanning method, the X, Y co-ordinates are measured using measuroscope at every pixel point. Further a 3D feature, depth of indentation, 'Z', which is proportional to the pressure applied by the scriber at that point, is measured using a dial gauge indicator. In the present work the profile based features extracted for palm leaf character recognition are 'histogram' and 'distance' profiles. The recognition accuracy obtained using the Z-dimension, a 3D feature, is very high and the best result obtained is 92.8 % using histogram profile algorithm.

Keywords Palm leaf character recognition · Histogram profile · Distance profile · k-NN classifier · 3D feature

1 Introduction

Technological developments gave the printed character recognition a new dimension called Optical Character Recognition (OCR). In the initial stages, most of the work was contributed towards the printed English characters, due to more number

P.N. Sastry (✉) · N.V. Koteswara Rao
CBIT, Hyderabad, India
e-mail: ananditahari@yahoo.com

N.V. Koteswara Rao
e-mail: nvkoteswararao@gmail.com

T.R. Vijaya Lakshmi
MGIT, Hyderabad, India
e-mail: vijaya.chintala@gmail.com

K. RamaKrishnan
ADRIN, Indian Institute of Space Science and Technology, Trivandrum, India
e-mail: drrkdrrk@gmail.com

© Springer Nature Singapore Pte Ltd. 2017
S.C. Satapathy et al. (eds.), *Proceedings of the 5th International Conference on Frontiers in Intelligent Computing: Theory and Applications*, Advances in Intelligent Systems and Computing 516, DOI 10.1007/978-981-10-3156-4_40

of English speakers compared to any of the Indian languages. However, handwritten character recognition became important in due course of time for the application of automatic mail sorting (zip code identification). Further, signature verification of forensic departments and authenticity of a document written by a specific scriber increased the importance of handwritten character recognition.

2 Previous Work

Sastry and Krishnan developed database and test characters for Palm Leaf Character Recognition (PLCR) pertaining to Telugu (a south Indian language) [1–5]. They developed many models using 2D correlation, PCA and Radon transform [1–3] in the area of Palm Leaf Character Recognition. Based on the measure of similarity all the Telugu characters were divided into 3 Co-ordinate planes which are XZ, XY and YZ. The best recognition accuracy was reported in the YZ plane for all the methods and found to be 90 %.

Patvardhan et al. [6, 7] presented a denoising approach using discrete curve-let transform and binarization technique using wavelets.

Manjunath Aradhya and others [8] proposed character identification using combination of FT and PCA. The documents were scanned on a HP 2400 scan jet scanner and subsequently skew corrected. Aradhya et al. [9] presented text recognition from videos and images using Gabor filter and wavelet transforms. Aradhya et al. also presented recognition of numerals and multilingual text using wavelet transforms and wavelet entropy, respectively.

Vijaya Lakshmi et al. [10–12] worked on isolated Telugu handwritten characters using zoning techniques and hybrid classification approaches. In [12] they reported that the recognition accuracy can be improved by classifying the characters using two classifiers viz., k-NN and SVM. For various feature extraction methods they reported improved recognition accuracy using two stage classification approach.

3 Data Acquisition

In general for any character recognition system the documents are scanned and stored in computer for further analysis. When the documents are scanned noise and skew problems were inevitable. In the present work the co-ordinates of X, Y and Z were measured in a laboratory set up using measuroscope and dial gauge indicator with plunger assembly [1, 3]. After selecting the basic Telugu isolated alphabet both the co-ordinates (X and Y) are measured using a measuroscope. Probe attached to the moving axis of this machine is used to measure the X and Y co-ordinates.

The selected points for measurements are chosen at the starting point, bends, intersections, turns and the end points of the character as shown in Fig. 1. The left most pixel in the character is considered as origin 'O' and the remaining pixel co-ordinates are measured with reference to origin 'O' [1]. The number of pixel points for

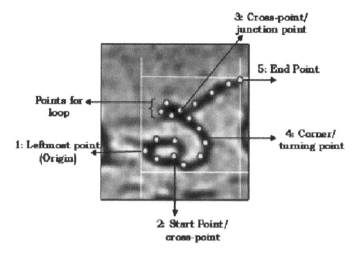

Fig. 1 Selection of points along the contour

different characters vary from 13 to 30 depending on the shape of the Telugu character [1]. The depth of indentation (Z) is proportional to the scriber's stylus stress applied on the palm leaf [1–3]. A needle made up of Teflon is attached to the dial indicator plunger to measure the depth of a character at a selected point. The distance of the bottom of the pixel point is measured and recorded as $Dist_1$. Then the distance is measured on the top of the palm script and recorded as $Dist_2$. The difference between $Dist_1$ and $Dist_2$ gives the depth of indentation for a selected pixel. This method is followed to obtain different alphabets of Telugu on the palm leaf and the measurements of X and Y are recorded. This is the 3D feature which differentiates highly similar characters and gives the best results in PLCR. The depth of indentation varies from 10 to 150 microns which is obtained utilizing dial indicator and a plunger assembly. For every Telugu character, three co-ordinates (X, Y, Z) are found to test and train the images. Using the X and Y co-ordinates, a pattern is developed and termed as the image of XY plane. Using the values of X, Y, Z images are developed for the XZ, XY and YZ co-ordinate system. The number of training samples developed in each plane of projection is 112 from 28 different classes i.e., 4 samples/class. Whereas the number of samples used for testing are 28 from different classes. A total of 420 character patterns are developed in all three planes. The training and testing samples are mutually disjoint.

4 Proposed Recognition Model

The proposed recognition model consists of various steps such as preprocessing, feature extraction and Classification. For every projection plane all these steps are performed and their recognition accuracies are reported in Sect. 5.

4.1   Preprocessing

The character images are binarized using Otsu's thresholding technique. The minimum boundary rectangle method is used for normalizing all the images to a size of 50×50. The data acquisition method does not involve the scanning of palm leaf, thereby avoiding the problems of skew and noise.

4.2   Feature Extraction

The two features extracted from the characters and used for recognition are Histogram profile and Distance profile.

4.2.1   Histogram Profile

The histograms projected in specified directions are created. The directions considered are vertical, horizontal and two diagonals left and right. The vertical (V_q) and horizontal (H_p) histograms are computed using Eq. (1).

$$V_q = \sum_{p=1}^{N} I(p,q),$$
$$H_p = \sum_{q=1}^{N} I(p,q) \tag{1}$$

for $q = 1{:}N$ and $p = 1{:}N$ respectively, and $I(p, q)$ is the character image of size $N \times N$.

The left diagonal (LD_j) and right diagonal (RD_j) histograms are computed using Eq. (2).

$$LD_j = \sum_{p} \sum_{q} Diag((I(p,q),j)),$$
$$RD_j = \sum_{p} \sum_{q} Diag(flip(I(p,q),j)) \tag{2}$$

for $j = -(N{-}1){:}(N{-}1)$

For a sample character 'va' the four histograms are shown in Fig. 2.

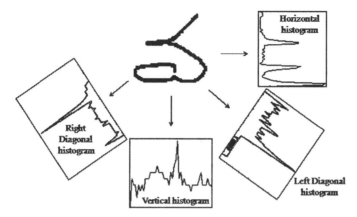

Fig. 2 Histogram profile for a Telugu character 'va'

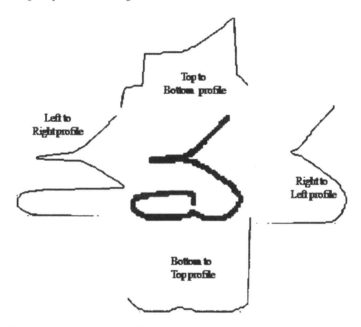

Fig. 3 Distance profile features for a Telugu character 'va'

4.2.2 Distance Profile

The distances of a character image are measured from a boundary box in specified directions. The directions considered to compute distance are top to bottom, bottom to top, left to right and right to left. The size of the feature vector for an N × N image is 4N.

For a sample character 'va' the four distance profiles are shown in Fig. 3.

5 Experimental Results and Discussions

The similar Telugu characters are classified into 6 different groups based on the correlation coefficient [1, 3]. The correlation coefficient is more than 0.75, for the characters in the same group, as reported in the literature [1, 3]. There are many characters in Telugu which are highly similar and hence their patterns are always confusing for recognition. Table 1 shows the three co-ordinates of a character "Ba".

A few similar Telugu characters "Ae", "Na" and "Pa" from the same group having high similarity as shown in Fig. 4.

The corresponding YZ images are shown in Fig. 5. The problem of recognizing Telugu characters is resolved to a maximum extent with the method proposed, if we consider these patterns which are highly uncorrelated, for any 2 different Telugu palm leaf characters. Further these patterns are highly repetitive in nature for any palm leaf character.

The corresponding XZ images are shown in Fig. 6. With these patterns the recognition rate could be improved to a greater extent. It is very clear from these patterns that these patterns are completely different to each other; thereby recognition accuracy would naturally increase for the proposed approach.

Table 1 Three co-ordinates of a character "Ba"

S.no.	X in mm	Y in mm	Z in μm
1	0.213	1.387	45
2	0.243	1.057	53
3	1.234	1.711	69
4	1.123	1.311	75
5	0.669	0.777	76
6	1.145	0.567	85
7	1.549	0.936	89
8	1.678	0.379	82
9	2.098	0.841	74
10	2.256	1.435	49
11	1.324	2.333	99

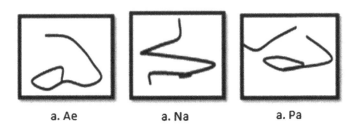

a. Ae   a. Na   a. Pa

Fig. 4 A few similar Telugu characters in XY plane

a. Ae   a. Na   a. Pa

Fig. 5 A few similar Telugu characters in YZ plane

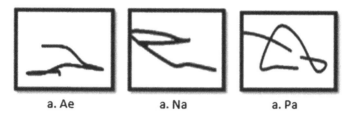

a. Ae   a. Na   a. Pa

Fig. 6 A few similar Telugu characters in XZ plane

Table 2 Proposed methods compared with the existing methods on the same dataset

Plane of projection	% Recognition accuracy				
	Existing methods			Proposed methods	
	PCA	2D Correlation	Radon	Histogram profile	Distance Profile
XY	40	54	76	51	35.7
YZ	40	90	89	92.8	89.2
XZ	37	70	80	89.2	85.7

The proposed recognition model is compared with the published methods on the same dataset which is in Table 2. All the proposed and existing methods used k-NN classifier for character classification. Sastry et al. [1–3] contributed work on palm leaf character recognition with Radon transform, PCA and 2D correlation approaches.

Sastry et al. published that the maximum Recognition Accuracy (RA) as 90 % in YZ Co-ordinate system using 2D correlation approach [1]. This is a spatial approach by which the proposed model is compared.

Sastry and Krishnan reported the maximum RA as 89 % in YZ Co-ordinate system using Radon transform [2]. The palm leaf character recognition was experimented by considering the image intensities along a radial line [2]. This is a transform domain approach by which the comparison is made with the proposed model.

Sastry et al. [3] published that the recognition accuracy using Principal Component Analysis (PCA) to be less than 50 % for XZ, XY and YZ planes as shown in Table 2.

In the proposed method YZ plane of projection gave the best results which is in line with the existing methods. The reason is due to a large variation in the Y direction between any two similar Telugu palm leaf characters which is an inherent characteristic of the script. Also the 3D feature i.e., the depth of indentation (Z dimension) is a special feature which gave the best results. The RA is high for YZ compared to XY and XZ planes of projection. The best RA obtained is 92.8 % in the YZ plane.

The recognition rate along XZ projection plane is found to be 89.2 % using Histogram profile approach which is higher than the recognition accuracy of XY plane of projection. The reason is due to the 3D feature i.e., 'Z' dimension.

6 Conclusions and Future Work

There is no standard database for Indian languages including Telugu as reported in the literature. Since there are many Telugu characters due to its inherent curvilinear shape and more number of modifiers, achieving high recognition accuracy is a challenging task. The recognition accuracy obtained is 51 and 35.7 % using histogram computation and distance profile methods respectively in XY plane, which is very low. The scriber uses a stylus for scribing Telugu characters on the palm leaf, which gives various depth values along the contour of the character. These values become the special features in palm leaf character recognition. Since many Telugu characters are highly similar, hence the depth information of the pixels along the contour (Z measured in microns) helped to improve the recognition accuracy from 35.7 % to 92.8 % using the proposed methods. Automatic scanning of the palm leaf characters for data acquisition and decrease of human interface can be developed which would decrease the time of data acquisition.

Acknowledgements The proposed work is sponsored by AICTE under Research Promotion Scheme and so the authors thank AICTE for motivating to do this work which is, useful to the society.

References

1. P.N. Sastry, R. Krishnan, B.V.S. Ram, Classification and identification of Telugu handwritten characters extracted from palm leaves using decision tree approach. J. Appl. Eng. Sci. **5**(3), 22–32 (2010)
2. P.N. Sastry, R. Krishnan, Isolated Telugu palm leaf character recognition using radon transform, a novel approach, in *World Congress on Information and Communication Technologies (WICT)*, (2012), pp. 795–802
3. P.N. Sastry, R. Krishnan, B.V.S. Ram, Telugu character recognition on palm leaves- a three dimensional approach. Technol. Spectr. **2**(3), 19–26 (2008)
4. P.N. Sastry, T.R. Vijaya Lakshmi, R. Krishnan, N. Rao, Analysis of Telugu palm leaf characters using multi-level recognition approach. J. Appl. Eng. Sci. **10**(20), 9258–9264 (2015)

5. T.R. Vijaya Lakshmi, P.N. Sastry, R. Krishnan, N.V.K. Rao, T.V. Rajinikanth, Analysis of Telugu palm leaf character recognition using 3D feature, in *International Conference on Computational Intelligence and Networks (CINE)*, (2015). pp. 36–41

6. C. Patvardhan, A.K. Verma, C.V. Lakshmi, Document image binarization using wavelets for OCR applications, in *Proceedings of the Eighth Indian Conference on Computer Vision, Graphics and Image Processing*, (ACM, 2012), pp. 60:1–60:8

7. C. Patvardhan, A.K. Verma, C.V. Lakshmi, Document image denoising and binarization using curvelet transform for OCR applications, in *Engineering (NUiCONE), 2012 Nirma University International Conference on*, (Dec 2012), pp. 1–6

8. V.M. Aradhya, G.H. Kumar, S. Noushath, Multilingual OCR system for south Indian scripts and English documents: an approach based on Fourier transform and principal component analysis. Eng. Appl. Artif. Intell. **21**(4), 658–668 (2008)

9. V.M. Aradhya, M. Pavithra, A comprehensive of transforms, Gabor filter and k-means clustering for text detection in images and video. Appl. Comput. Inf. (2014)

10. P.N. Sastry, T.R. Vijaya Lakshmi, N.V.K. Rao, T.V. Rajinikanth, A. Wahab, Telugu handwritten character recognition using Zoning features, in *International Conference on IT Convergence and Security (ICITCS)*, (Beijing, 2014), pp. 1–4

11. T.R. Vijaya Lakshmi, P.N. Sastry, T.V. Rajinikanth, Recognition of isolated Telugu handwritten characters using 2D FFT, in *Proceedings of the 2nd International Conference on Advanced Computing Methodologies*, ser. ICACM'13, (2013), pp. 372–376

12. T.R. Vijaya Lakshmi, P.N. Sastry, T.V. Rajinikanth, Hybrid approach for Telugu handwritten character recognition using k-NN and SVM classifiers. Int. Rev. Comput. Softw. **10**(9), 923–929 (2015)

Feature Optimization to Recognize Telugu Handwritten Characters by Implementing DE and PSO Techniques

T.R. Vijaya Lakshmi, Panyam Narahari Sastry and T.V. Rajinikanth

Abstract Recognizing Indian handwritten text is relatively complex compared to recognized foreign language such as English. In this work optimization techniques are presented to recognize Telugu handwritten characters. By extracting cell-based directional features from these characters, optimum features are selected by implementing optimization algorithms such as differential evolution and particle swarm optimization. An improvement of 3.5 % recognition accuracy is achieved using differential evolution algorithm. The optimization techniques are compared with the existing hybrid approach of Telugu script.

Keywords Telugu handwritten characters · Cell-based directional distribution · Optimization techniques · Differential evolution · Particle swarm optimization

1 Introduction

Character recognition (CR) is used as an umbrella term in various application domains that cover all types of machine recognition of characters [1, 2]. Character recognition is an art of detecting, segmenting and identifying characters from image. More precisely character recognition is the process of detecting and recognizing characters from input image and converts it into ASCII or other equivalent machine editable form. Conversion of handwritten characters is important for

T.R. Vijaya Lakshmi (✉)
MGIT, Hyderabad, India
e-mail: vijaya.chintala@gmail.com

P.N. Sastry
CBIT, Hyderabad, India
e-mail: ananditahari@gmail.com

T.V. Rajinikanth
SNIST, Hyderabad, India
e-mail: rajinitv@gmail.com

© Springer Nature Singapore Pte Ltd. 2017   397
S.C. Satapathy et al. (eds.), *Proceedings of the 5th International Conference on Frontiers in Intelligent Computing: Theory and Applications*, Advances in Intelligent Systems and Computing 516, DOI 10.1007/978-981-10-3156-4_41

making several important documents related to our history, such as manuscripts, into machine editable form so that it can be easily accessed and preserved.

In this paper feature optimization is carried out by implementing optimization techniques such as differential evolution and particle swarm optimization from the literature. Cell-based directional distribution features are extracted from Telugu characters and these features are used for feature selection step. The organization of this paper is as follows: The related work is discussed in Sect. 2. The feature selection techniques are described in Sect. 3. The experimental results are discussed and compared in Sect. 4. Finally, the conclusions are quoted in Sect. 5.

2 Related Work

Pal et al. [3] developed an approach for Devanagari script containing basic, modified and compound characters. For classification all the characters are separated using the zonal information. Structural feature based binary tree classifier is used to recognize the basic and modified characters and to recognize compound characters a hybrid approach combination of structural characters and run based template is employed.

Sastry et al. developed database for Palm Leaf Character Recognition pertaining to Telugu (a south Indian language) [4, 5]. They extracted depth information from palm manuscripts. This additional feature proved to yield improved recognition accuracy.

Vijaya Lakshmi et al. [6–8] worked on isolated Telugu handwritten characters using zoning techniques and hybrid classification approaches. In [6] they reported the local feature extraction approaches yield better recognition of characters, compared to global feature extraction approaches. In [8] they reported that the recognition accuracy can be improved by classifying the characters using two classifiers viz., k-NN and SVM. For various feature extraction methods they reported improved recognition accuracy using two stage classification approach.

Manjunath Aradhya et al. [9, 10] worked on recognizing multilingual south-Indian scripts using Fourier transform and PCA. In [11] they worked on recognizing handwritten digit recognition using radon transform.

3 Proposed Methodology

In the proposed approach 50 basic isolated handwritten characters of Telugu script are considered for the experiments. The characters are collected from 360 different scribers of different profiles written in an isolated manner with different scale, translation and rotation. These documents are scanned at 300 dpi on flatbed scanner. The spacing between samples is enough to segment them by taking the horizontal and the vertical profiles. The character samples are converted into binary images using Otsu's thresholding technique. All the images are normalized to a size of 50×50. A total of 18,000 (50 × 360) characters are considered in this work. These characters are preprocessed to remove noise and make the images invariant to scale, translation and rotation.

Features are extracted from the preprocessed images by dividing an image into cells. Superimposing 8 directional masks [8] on the each cell the directional information is extracted. For a cell size of 10 × 10, the number of features extracted for a character in 8 directions is 200. For k = 1, with k-NN classifier, the Euclidean distance is calculated to find the similarity between the test image and all the training images. The test image which has the minimum distance with a particular training image is considered to be matched with that particular training image. All the experiments are cross validated as there is no proper testing set. In V-fold cross validation, each fold contains characters written by N number of scribers (for V = 8, N is 45). For a 50 class problem each fold contain 50 × N character samples. To test a tth subset the remaining (V-1) subsets are used for training the classifier. Hence the training and the test sets are disjoint. The average of recognition accuracy (RA) obtained with all the V folds is considered as the RA of the model.

Dimensionality reduction or feature selection is carried out in this work by implementing two optimizing techniques viz., Differential Evolution (DE) [12] and Particle Swarm Optimization (PSO) [13, 14]. These are discussed in the following subsections.

3.1 Differential Evolution

The steps involved in DE are population initialization, mutation, crossover and selection.

3.1.1 Population Initialization

The population $P_{X,g}$ is initialized randomly for NP patterns with D dimensions. The size of the initial population is $NP \times D$. Let $\vec{X}_{j,g}$ be the jth pattern in gth generation. The lower limit of the search space is set to '1' and its upper limit is set to 'total number of features'. Fitness is evaluated for each pattern in the population. In the current work fitness computed is the classification error using k-NN classifier (better fitness for low error rate). The block diagram of DE is shown in Fig. 1.

3.1.2 Mutation

Let the mutant population be $P_{Y,g}$. The mutant pattern $\vec{Y}_{j,g}$ is generated by taking the weighted difference between two random patterns and then adding with a third random pattern from the population, as depicted in Eq. (1).

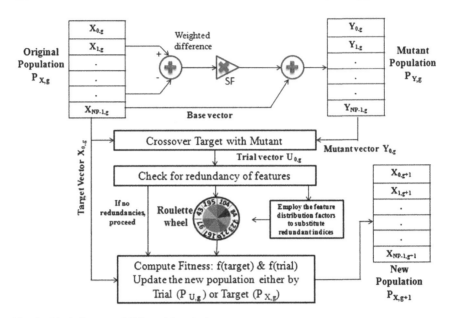

Fig. 1 Block diagram of differential evolution

$$\vec{Y}_{j,g} = \vec{X}_{s_0,g} + SF.(\vec{X}_{s_1,g} - \vec{X}_{s_2,g}) \tag{1}$$

where s_0, s_1 and s_2 are three random numbers generated. SF ϵ (0,1) is the scale factor which is inversely proportional to the maximum of the random numbers s_1 and s_2. This allows the second term in Eq. (1) to oscillate within limits without crossing the optimal solutions. In other words it controls the population evolution rate.

3.1.3 Crossover

DE employs uniform crossover in order to build the trial pattern/vector $\vec{U}_{j,g}$. It crosses each original pattern with a mutant pattern as given in Eq. (2).

$$\vec{U}_{j,g} = \begin{cases} \vec{Y}_{j,g}, & \text{if } s(0, 1) \leq C_r \\ \vec{X}_{j,g}, & \text{otherwise} \end{cases} \tag{2}$$

where $C_r \epsilon$ (0,1) is the crossover rate and s is a random number. The crossover rate controls the fraction of values copied from both the patterns/vectors. In the current work C_r is set to 0.5.

3.1.4   Selection

Roulette wheel is employed to check for redundancy of features. The feature distribution factor FD_i computed, using Eq. (3), is fed to the roulette wheel to decide which feature to choose, to replace the duplicated features.

$$FD_{i,g} = \frac{PD_i}{PD_i + ND_i} \tag{3}$$

where PD_i and ND_i denotes positive and negative distributions of the feature f_i, respectively. Positive distribution is the number of times the feature f_i contributed in forming good subsets (low fitness). Negative distribution is the number of times the feature f_i is used in less competitive subsets (high fitness).

The minimum the fitness error the better is the pattern/individual. Based on the fitness evaluation, DE replaces the original pattern with the trial pattern. In other words it selects the better pattern for the next generation.

All the above steps are repeated until the maximum generations count is reached to find an optimum solution.

3.2   Particle Swarm Optimization

To compare the performance of Differential evolution technique, PSO developed by Eberhart and Kennedy in 1995 is implemented. Let the initial population be $P_{X,g}$ generated randomly for NP particles with D dimensions. Let the initial position and velocity vectors of the jth particle in the swarm be $\vec{X}_{j,g}$ and $\vec{V}_{j,g}$, respectively, at gth generation. The flowchart of PSO is shown in Fig. 2. Fitness computed for all the particles is the classification error using k-NN classifier.

For every generation, the velocities of the particles are updated using Eq. (4). The positions of the particles are updated using Eq. (5). The particle's best value is denoted by $\vec{X}_{best}$ and the swarm's best value is denoted by $\vec{G}_{best}$.

$$\vec{V}_{j,g+1} = \omega \times \vec{V}_{j,g} + c_1 \times rand1() \times (\vec{X}_{best} - \vec{X}_{j,g}) + c_2 \times rand2() \times (\vec{G}_{best} - \vec{X}_{j,g}) \tag{4}$$

$$\vec{X}_{j,g+1} = \vec{X}_{j,g} + \vec{V}_{j,g+1} \tag{5}$$

where $rand1()$ and $rand2()$ are two numbers generated over the range [0 1], c_1 and c_2 are cognitive and social acceleration constants respectively, and ω is a linearly time varying weight given by

$$\omega = (\omega_1 - \omega_2) \times \frac{MAXGEN - g}{MAXGEN} \tag{6}$$

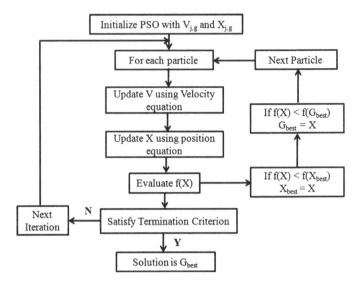

Fig. 2 PSO flowchart

where ω_1 and ω_2 are the inertia weights, g is the current generation and *MAXGEN* is the maximum number of generations. In the current work c_1 and c_2 are set to 2.

The positions and velocities of the particles are updated until the maximum number of generations count is reached to find the optimum subset of features.

4 Experimental Results

The Telugu characters collected from 360 different scribers are used for training and testing in this work. The number of classes considered are 50. In total 18,000 characters are used for both testing and training. The directional information in each cell of a character image are extracted as features from this dataset. For classification k-NN classifier is employed. The recognition accuracy using these features extracted is found to be 85.6 %.

The optimization techniques used in the current work namely, Differential Evolution and Particle Swarm Optimization are allowed to start with the same initial population. The population size is set to 40 and the number of generations is set to 50. However, the feature subset size is varied from 5 to 100, in steps of 5. For each subset, the experiment is made to run 20 times and the average recognition accuracies are shown in Fig. 3.

The best recognition accuracy achieved using PSO is 86.7 % for an optimum feature subset size of 90, as shown in Fig. 3, which is in the same band using the original feature set. It is observed from Fig. 3 that even for smaller subset sizes DE performed well compared to PSO. The highest recognition accuracy obtained is 89.1 % using

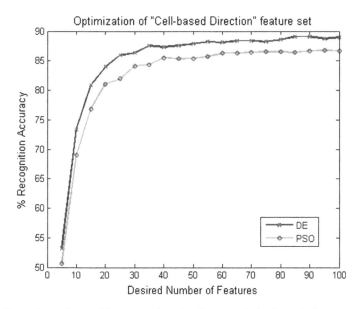

Fig. 3 Comparison of recognition accuracies using feature optimization techniques

Table 1 Comparison with the existing systems on the same dataset

Description	Single-stage classification [8]	Hybrid classification [8]	Optimization techniques	
			DE	PSO
No. of Features	200	200	85	90
Memory requirement	One unit	One unit	Reduced to half unit	
% Recognition accuracy	85.60 %	89.30 %	89.10 %	86.70 %
No. of times the system is tested	1	2	1	1
Classifiers used	k-NN	k-NN and SVM	k-NN	k-NN

DE algorithm for an optimum subset size of 85 features. This shows an improvement of 3.5 % in recognition accuracy with 85 optimum features compared to using 200 features considered earlier. Hence even with 50 % of the feature subset reduction, the recognition accuracy has increased by 3.5 %. So, more than 50 % of the redundant data is reduced using these optimization techniques.

Vijaya Lakshmi et al. [8] worked on the same dataset and employed the same feature set (cell-based directional information) to recognize Telugu handwritten characters. They presented a hybrid classification approach to improve the recognition rate. The characters misclassified using k-NN classifier in the first stage are again classified using SVM classifier in the second stage. They reported a recognition rate using this hybrid approach for cell-based directional features as 89.3 %. The results obtained with this hybrid method are compared with the optimization techniques presented in the current study and are tabulated in Table 1. Compared to the hybrid approach, with the proposed optimization techniques the computational complexity is reduced.

5 Conclusion

In this work experiments are conducted on Telugu handwritten characters by extracting cell-based directional distribution features. Differential evolution and particle swarm optimization techniques are successfully implemented on these cell-based directional distribution features extracted. By varying the desired number of features, the classification accuracies are obtained and compared using differential evolution and particle swarm optimization techniques. There is an improvement of 3.5 % in recognition accuracy using differential evolution technique when compared to single stage classification system. Even with 50 % of the feature subset reduction the recognition accuracy has increased by 3.5 %. So, more than 50 % of the redundant data is reduced using these optimization techniques.

References

1. N. Arica, F.T. Yarman-Vural, An overview of character recognition focused on off-line handwriting. IEEE Trans. Syst. Man Cybern. Part C Appl. Rev. **31**(2), 216–233 (2001)
2. R. Plamondon, S.N. Srihari, Online and off-line handwriting recognition: a comprehensive survey. IEEE Trans. Pattern Anal. Mach. Intell. **22**(1), 63–84 (2000)
3. U. Pal, B.B. Chaudhuri, Indian script character recognition: a survey. Pattern Recognit. **37**(9), 1887–1899 (2004)
4. P.N. Sastry, T.R. Vijaya Lakshmi, R. Krishnan, N. Rao, Analysis of Telugu palm leaf characters using multi-level recognition approach. J. Appl. Eng. Sci. **10**(20), 9258–9264 (2015)
5. T.R. Vijaya Lakshmi, P.N. Sastry, R. Krishnan, N.V.K. Rao, T.V. Rajinikanth, Analysis of Telugu palm leaf character recognition using 3D feature, in *International Conference on Computational Intelligence and Networks (CINE)*, (2015), pp. 36–41
6. P.N. Sastry, T.R. Vijaya Lakshmi, N.V.K. Rao, T.V. Rajinikanth, A. Wahab, Telugu handwritten character recognition using Zoning features, in *International Conference on IT Convergence and Security (ICITCS)*, (Beijing, 2014), pp. 1–4
7. T.R. Vijaya Lakshmi, P.N. Sastry, T.V. Rajinikanth, Recognition of isolated Telugu handwritten characters using 2D FFT, in *Proceedings of the 2nd International conference on Advanced Computing Methodologies*, ser. ICACM'13, (2013), pp. 372–376
8. T.R. Vijaya Lakshmi, P.N. Sastry, T.V. Rajinikanth, Hybrid approach for Telugu handwritten character recognition using k-NN and SVM classifiers. Int. Rev. Comput. Softw. **10**(9), 923–929 (2015)
9. V.M. Aradhya, G.H. Kumar, S. Noushath, Multilingual OCR system for South Indian scripts and English documents: an approach based on Fourier transform and principal component analysis. Eng. Appl. Artif. Intell. **21**(4), 658–668 (2008)
10. V.M. Aradhya, G.H. Kumar, Principal component analysis and generalized regression neural networks for efficient character recognition, in *First International Conference on Emerging Trends in Engineering and Technology*, (2008), pp. 1170–1174

11. V.M. Aradhya, G.H. Kumar, S. Noushath, Robust unconstrained handwritten digit recognition using radon transform, in *International Conference on Signal Processing, Communications and Networking*, (2007), pp. 626–629
12. R.N. Khushaba, A. Al-Ani, A. Al-Jumaily, Feature subset selection using differential evolution and a statistical repair mechanism. Expert Syst. Appl. **38**(9), 515–526 (2011)
13. S. Mirjalili, S.Z.M. Hashim, H.M. Sardroudi, Training feed forward neural networks using hybrid particle swarm optimization and gravitational search algorithm. Appl. Math. Comput. **218**(22), 125–137 (2012)
14. S. Mirjalili, S. Hashim, A new hybrid PSOGSA algorithm for function optimization, in *International Conference on Computer and Information Application (ICCIA)*, (Dec 2010), pp. 374–377

Writer Specific Parameters for Online Signature Verification

K.S. Manjunatha, S. Manjunath and D.S. Guru

Abstract In this work, we present a new model which is capable of handling variations in signatures for better representation by employing structure preserving feature selection method and representing the selected data in the form of interval representation scheme. The proposed model represents each writer with writer dependent dimension and authentication threshold. Decisions on the number of features to be used for each writer and the similarity threshold for deciding the authenticity of a given signature are arrived based on minimum equal error rate (EER) criteria. Based on the symbolic representation, a method of verification is proposed. The proposed model is tested for its effectiveness on benchmarking MCYT (DB1) and MCYT (DB2) datasets consisting of signatures of 100 and 330 writers respectively. The obtained results indicate the effectiveness of the proposed model.

Keywords Writer specific features · Writer dependent feature dimension · Symbolic representation · Online signature verification

1 Introduction

Authentication based on the signature biometric trait has been well accepted in financial and administrative organization as a legal means of verifying the identity of an individual. Depending on the way the data is captured, signature verification

K.S. Manjunatha (✉) · D.S. Guru
Department of Studies in Computer Science, Manasagangothri, University of Mysore, Mysore 570006, Karnataka, India
e-mail: kowshik.manjunath@gmail.com

D.S. Guru
e-mail: dsg@compsci.uni-mysore.ac.in

S. Manjunath
Department of Computer Science, Central University of Kerala, Kasaragod 671316, India
e-mail: manju_uom@yahoo.co.in

© Springer Nature Singapore Pte Ltd. 2017   407
S.C. Satapathy et al. (eds.), *Proceedings of the 5th International Conference on Frontiers in Intelligent Computing: Theory and Applications*, Advances in Intelligent Systems and Computing 516, DOI 10.1007/978-981-10-3156-4_42

methods can be either offline or online [1]. Authentication based on online signatures is found to be more reliable than offline signatures due to the availability of additional dynamic features [2].

Online signature verification approaches are categorized into two types' namely parametric and functional approaches. In case of functional approaches, a signature is represented by means of discrete time functions such as pressure, velocity, positional co-ordinates etc. Authenticity is decided by comparing the time functions of a test and a signature which generally gives less error rate when compared to parametric approaches [3]. In case of parametric approaches, a signature is stored in the knowledgebase by means of a fixed dimensional feature vector which takes less time for comparison and also less storage. During verification, the authenticity of a test signature is decided by means of a suitable classifier. Verification based on the classifiers such as nearest neighbor [4], Bayesian [2], neural network [5], support vector machine [6]; hidden markov model [7], decision tree [4] and fusion based approach [8] can be traced in the literature.

In most of the above cited works, what features to be used? how many features to be used? and what decision threshold must be used for a specific writer? are decided independent of a writer. But signature verification is a writer intrinsic activity where the decision on all these depends on the signatures of an individual writer. The same strategy has been adapted by human experts while manually verifying a signature. This demands the design of a verification model based on writer dependent parameters which include a different set of features and decision threshold for different writers. Writer dependent threshold has been well exploited in the works of [1, 8, 9] which has shown improved performance when compared to a common threshold. In [10], the author has proposed genetic algorithm based approach for selecting personalized features. But genetic algorithm does not guarantee optimal feature subset and requires more number of parameters to be set. In our earlier work [11], we have proposed a model making use of writer dependent features but the feature dimension and threshold used is same for all writers. The study revealed clearly that consistency of signing is not uniform across all writers i.e., intra-class variation is more and storing all template signature in the database is not required. To preserve intra-class variations and to reduce the number of template signature to be stored in the database, the selected writer dependent features are represented in the form of a single interval valued symbolic representation. With this backdrop, in this work, we propose a model in which writer dependent features, writer dependent feature dimension and writer dependent decision threshold are estimated automatically.

The difference between our earlier and the work proposed herein are as follows

- Present work automates the process of identifying writer dependent parameters with respect to features, its dimensions and the threshold. Whereas in our previous work we have worked only on user dependent features with a common feature dimensions and a common threshold.

- Even though we have used same symbolic representation scheme in this work, we have also made a study with writer dependent parameters using conventional representation scheme.

The overall organization of the paper is as follows: In Sect. 2, different stages of the proposed model are discussed. In Sect. 3, we present the details of experimentation with the obtained results. Conclusions are drawn in Sect. 4.

2 Proposed Model

The proposed model consists of the following major stages and we discuss each of these stages in the subsequent subsection.

1. Selection of features for each writer based on feature's relevancy
2. Symbolic representation of selected features to assimilate intra-class variations
3. Fixation of writer dependent parameters
4. Signature verification

2.1 Feature Selection Based on Cluster Preserving Capability

Generally, signatures of a writer contain some unique characteristics which are difficult for an imposter to forge and also discriminates each writer from other writers. Further, signatures of a writer create some natural clusters due to intra-class variations. Selecting features which can preserve the structure of the clusters provides better representation of the signature. Hence to select relevant features for a writer we specifically adapt the feature selection method proposed by [12] which is suitable for multi-cluster data. A score is computed for each feature which indicates its relevancy to the cluster. Based on the computed score, most relevant features (top score) for each writer are selected. The detail description of feature selection is as follows.

Let N be the number of writers and let each writer contributes n number of signatures for training purpose. Let P be the number of features which characterize a signature. The resulting dimension of the data matrix say X of each writer will be $n \times P$. The feature selection algorithm computes a score called Multi Cluster Feature Selection (MCFS) score for all P features and select only d features with top score. The score denotes the contribution of a feature towards preserving the cluster structure of a writer. The steps in selecting features for writer W_i are

1. A weighted graph with n vertices is constructed where the vertex denotes a training signature and its neighbors. Edge denotes the relationship between a vertex and its neighbor.

2. For the weighted graph, a weight matrix M is created by means of Heat Kernel weighting scheme. From the weighted matrix M, a diagonal matrix D is computed by summing each row or each column of M. A graph Laplacian L is computed from the weight matrix M and diagonal matrix D as follows

$$L = D - M \tag{1}$$

3. Determine a set of Eigen vectors for L by solving the Eigen problem of the form $Ly = \lambda Dy$. Let $y = (y_1, y_2, y_3, \ldots y_t)$ be the set of t eigenvectors corresponding to t smallest eigenvalues. Each y_i, $1 \le I \le t$ represents the distribution of points in Ith projection.

4. A coefficient vector a_I of dimension P is determined to approximate each y_I using least regression analysis and by minimizing the fitting error as defined in Eq. (2) a subset of features which are most relevant are obtained

$$\min_{a_I} \left\| y_I - X^T a_I \right\|^2 + \beta |a_I| \tag{2}$$

where X is the original data matrix and β is a parameter which controls the sparseness of a_I

5. Equation (2) can be reformulated as an optimization problem as

$$\min_{a_I} \left\| y_I - X^T a_I \right\| \text{ such that } |a_I| \le \gamma \tag{3}$$

where γ is a scalar

6. Each a_I contains P values each corresponds to an approximation coefficient for each feature. Choose the maximum value in each column of a_I to compute the score for each features i.e.

$$MCFS(J) = \max_I |a_{I,J}|, \ 1 \le I \le t, \ 1 \le J \le P \tag{4}$$

Here $a_{I,J}$ is the Jth element of a_I and $MCFS(J)$ denotes the score of the Jth feature. After computing the MCFS score for all P features for all writers, the features are sorted based on their MCFS score. Finally, for each writer we recommend selecting d features with highest $MCFS$ score.

Table 1 EER of the proposed model based on symbolic representation with writer dependent parameters

Testing	Approaches	MCYT (DB1)			MCYT (DB2)		
		Min	Max	Average	Min	Max	Average
Skilled_05	Symbolic	5.61	6.86	6.30	6.61	7.51	7.00
	Conventional	8.39	9.41	8.52	8.83	9.69	9.07
Skilled_20	Symbolic	0.38	1.14	0.76	0.28	1.48	0.74
	Conventional	0.68	2.30	1.59	1.34	1.63	1.32
Random_05	Symbolic	2.56	3.36	2.94	3.28	3.84	3.53
	Conventional	4.33	4.80	4.52	4.60	4.95	4.76
Random_20	Symbolic	0.01	0.47	0.23	0.14	0.61	0.35
	Conventional	0.20	1.08	0.56	0.49	0.77	0.61

2.2 Symbolic Representation

In this work we recommend to use symbolic representation which is capable of preserving the intra-class variation [9]. Once the features for each writer are selected, training signatures of each writer are stored in the knowledgebase in the form of interval valued symbolic feature vector. The symbolic feature vector for a writer W_i is created as follows.

Let $\{S_1, S_2, S_3, \ldots, S_n\}$ be n training signatures of a writer W_i, $i = 1, 2, \ldots, N$ where N is the number of writers. Let $\{f_{i1}, f_{i2}, \ldots, f_{id}\}$ be the feature vector representing the writer W_i where d is the number of features selected. The interval-valued feature vector representing W_i is obtained by computing the mean and standard deviation of each of the d features selected. Let $Mean\,(f_p^i)$ and $Std\,(f_p^i)$ be the mean and standard deviation of pth feature due to all n samples of W_i, $p = 1, 2, \ldots, d$ i.e.

$$Mean\,(f_p^i) = \frac{1}{n}\sum_{s=1}^{n} f_{sp} \text{ and } \sqrt{\left(\frac{1}{n}\sum_{s=1}^{n}(f_{sp} - Mean\,(f_p^i))\right)} \qquad (5)$$

Now every feature of a writer W_i is represented in the form of an interval valued. For example, pth feature of the writer W_i is represented as $\left[f_{ip}^-, f_{ip}^+\right]$ where f_{ip}^- and f_{ip}^+ denotes the lower and upper limits the pth feature of the writer W_i respectively which are computed as follows

$$f_{ip}^- = Mean\,(f_p^i) - Std\,(f_p^i) \text{ and } f_{ip}^+ = Mean\,(f_p^i) + Std\,(f_p^i) \qquad (6)$$

Similarly all the d features are represented in the form of an interval which results in the creation of an interval valued symbolic feature vector RF_I for the writer W_i as

$$RF_i = \left\{ \left[f_{i1}^-, f_{i1}^+ \right], \left[f_{i2}^-, f_{i2}^+ \right], \ldots, \left[f_{id}^-, f_{id}^+ \right] \right\} \ i = 1, 2, \ldots, N \qquad (7)$$

This symbolic feature vector is stored in the knowledgebase as a representative of the writer W_i. In symbolic representation, it is sufficient to store only one symbolic feature vector characterizing a writer. In conventional representation, we need to store all n training signatures of each writer. It is interesting to note that in conventional representation every feature is of crisp type as against symbolic representation where each feature is of interval valued type.

2.3 Writer Dependent Feature Dimension and Threshold Fixation

In this section, we discuss the procedure for fixing up of writer dependent feature dimension and similarity threshold. For each writer, we vary d from 5 to 75 in step of 5 and for each d the similarity threshold is varied from 0.1 to 1.0 in step of 0.1. We estimate the false acceptance rate (FAR) and false rejection rate (FRR) for each threshold. These two errors respectively denote the percentage of forgery samples accepted as genuine and percentage of genuine signatures rejected as forgeries. Finally, using receiver operating curve (ROC), EER is estimated for each d. The corresponding d and threshold which result in lowest EER is decided to be the suitable feature dimension and threshold as the suitable parameter for the respective writer. It should also be noted that for some writers the parameter d may be same but the features will be different. In symbolic representation, threshold denotes the percentage of test signature that should lie within the corresponding reference signature of the claimed writer for treating the test signature as a genuine. For instance, threshold equal to 0.5 indicates that 50 % of features of test signature must be within the interval of the corresponding feature of a reference signature. In case of conventional representation, threshold is the normalized distance estimated among the genuine signatures of the corresponding writer which is scaled to the range 0 to 1. The threshold selected for each writer varies from other writers. In the database, we store selected features indices and the threshold of each writer which will be used during the verification stage.

2.4 Signature Verification

Given a test signature F_q represented in the form of P dimensional feature vector say $F_q = \{f_{q1}, f_{q2}, \ldots, f_{qP}\}$, its authenticity is determined as follows. We compare only d features of a test signature with the corresponding d interval valued features of a reference signature. The indices of the d features to be compared are available in the knowledgebase. To keep track of number of features of test signature that lies within the corresponding interval valued feature of a reference signature we use a counter (A_{cp}). If a feature of a test signature lies within the corresponding interval-valued feature of a reference signature, (A_{cp}) is incremented by one. If the value of (A_{cp}) is greater than a predefined threshold, then the test signature is accepted as a genuine. In conventional representation, the verification is done by means of nearest neighbor classifier where the distance of the test signature to all n training signature is estimated which results in n different distances. If the minimum of the distances is less than the threshold, the test signature is accepted as a genuine or else rejected as a forgery.

3 Experimentation and Results

We conducted experiments on two variants of MCYT online signature datasets which are denoted as DB1 and DB2 [13]. DB1 consist of signatures of first 100 writers while DB2 consist of signatures of all 330 writers. Both DB1 and DB2 consist of 25 genuine and 25 skilled forgeries for each writer. For experimentation we worked on the 100 global parametric features the details of which are available in the works of [8].

3.1 Experimental Setup

For experimentation, the dataset is divided into training and testing set. Depending on the number of samples used for training and testing, we conducted experiments on 4 different categories which are denoted as Skilled_05, Skilled_20, Random_05 and Random_20 as done in the works of [8, 9]. We conducted 20 different trials and the average of all the 20 trials is considered as the EER of the respective writer. Finally, EER of all writers is averaged to estimate the EER of the system. The reason for conducting multiple trials is to ensure that the system is not biased towards a particular set of training and testing samples. The same approach has been followed in [14].

3.2 Results and Discussion

The minimum, average and maximum EER obtained with the proposed model based on symbolic representation and conventional representation is shown in Table 1.

From Table 1, it is observed that the EER obtained with writer dependent parameters and symbolic representation is much lower than the EER obtained with

Table 2 Details of writer dependent parameters for first 5 writers of MCYT-330

Writer-Id	d	Threshold	EER
1	53	0.35	3.81
2	49	0.41	10.97
3	25	0.39	11.22
4	62	0.36	5.21
5	62	0.32	2.80

writer dependent parameters and conventional representation. In Table 2, the various writer dependent parameters along with the EER obtained are shown for first 5 writers of MCYT-330 as examples.

3.3 Time Complexity Analysis of Verification

In this section we discuss the computational complexity of our model. The two core stages in our work are enrollment and signature verification. Enrollment is an offline process which is done during training stage itself and hence we have considered only verification stage. The actual time is the verification stage which is an online process which involves following steps: (a) fetching the indices of d features from the knowledgebase which takes d unit of time as it is basically a searching operation. (b) computation of features for a test signature. As the features selected vary from a writer to a writer, the time taken for computation of features depends on the selected feature. In general, if T_f is time taken to compute a feature then the total time taken is $d \times T_f$ (c) comparison of every feature of the test signature with corresponding interval valued feature of a reference signature which involves two comparisons. For d features, the total comparison involved is $2 \times d$. Hence the time complexity T_C of the verification stage is given by

$$T_C = O\left(d + d \times T_f + 2 \times d\right) = O\left(3d + d \times T_f\right) \tag{8}$$

However for details on the time complexity of the feature selection, readers are directed to refer the work of [12].

3.4 Comparative Study

In order to have fair comparisons, we compared the EER of various online signature verification models based on a common dataset as comparison of different approaches validated on different datasets is difficult. Table 3 shows the EER of our model with other existing models on MCYT (100) dataset. Even though, we worked on both DB1 and DB2, for comparative study, we considered the result

Table 3 EER of various online signature verification models on MCYT (DB1)

Model		Skilled_05	Skilled_20	Random_05	Random_20
Proposed model	Symbolic	6.30	0.76	2.94	0.23
	Conventional	8.52	1.59	4.52	0.56
[6] a. Parzen Window Classifier (PWC)		9.70	5.20	3.40	1.40
b. Nearest Neighbor descriptor (NND)		12.20	6.30	6.90	2.10
c. Mixture of Gaussian Description(MOGD_3)		8.90	7.30	5.40	4.30
d. Mixture of Gaussian Description (MOGD_2)		8.10	7.00	5.40	4.30
e. Support Vector Descriptor (SVD)		8.90	5.40	3.80	1.60
[9]		5.84	3.80	1.85	1.65
[15]		15.40	4.20	3.60	1.20
[14]		4.02	2.72	1.15	0.35

obtained on DB1 as DB1 is publicly available for research purpose. From Table 3, it can be observed that proposed models based on symbolic representation outperform all the existing models in case of skilled_20 and Random_20. Even in case of Skilled_05 and Random_05 the EER of the proposed model based on symbolic representation is lower than all other models except [9, 14]. The reason for higher EER in case of Skilled_05 and Random_05 is due to availability of less number of training samples for capturing writer dependent characteristics.

The EER of the proposed model based on conventional representation also results in lowest EER compared to other models in case of Skilled_20. In case of Random_20, the obtained EER is lower than all other models except [14]. In case of Skilled_05, obtained EER is lower than all other models except [9, 14]. Whereas, in case of Random_05 the proposed model outperforms all models except models presented in [9, 15, 14].

4   Conclusion

In this work, an approach for online signature verification based on the usage of writer dependent features dimension and similarity threshold has been proposed. Relevant features for each writer are selected by means of a feature section method which is suitable for multi-cluster data and suitably represented to assimilate intra-class variations. The efficacy of the model is well established with extensive experimentations on benchmarking dataset. The obtained result clearly indicates the effectiveness of the proposed approach. A method for signature based on the representation selected is proposed. A study on different classifiers along with their fusion suitable for various users in achieving lower EER is our future target.

Acknowledgments We thank Dr. J.F Aguilar, Biometric Research Lab-AVTS, Spain for providing MCYT online signature dataset and Prof. Anil K Jain for his support in getting the dataset from Aguilar. We also thank Deng Cai, Zhejiang University, China for sharing his work on feature selection. First and second authors acknowledge the support rendered by UGC, under faculty improvement programme and startup grant respectively.

References

1. A.K. Jain, F.D. Griess, S.D. Connel, On-line signature verification. Pattern Recogn. **35**, 2963–2972 (2002)
2. A. Kholmatov, B. Yanikoglu, Identity authentication using improved online signature verification method. Pattern Recogn. Lett. **26**, 2400–2408 (2005)
3. A. Rashidi, A. Falla, F. Towhidkhah, Feature extraction based DCT on dynamic signature verification. Sci. Iranica **19**, 1810–1819 (2012)
4. R. Doroz, P. Porwik, T. Orczyk, Dynamic signature verification method based on association of features with similarity measures. Neurocomputing **171**, 921–931 (2016)
5. M.M.M. Fahmy, Online handwritten signature verification system based on DWT features extraction and neural network classification. Ain Shams Eng. J. **1**, 59–70 (2010)
6. L. Nanni, Experimental comparison of one-class classifiers for on-line signature verification. Neurocomputing **69**, 869–873 (2006)
7. E.A. Rua, J.L.A. Castro, Online signature verification based on generative models. IEEE Trans. SMC—Part B: Cybern. **42**, 1231–1242 (2012)
8. J.F. Aguilar, L. Nanni, J.L. Penalba, J.O. Garcia, D. Maltoni, An on-line signature verification system based on fusion of local and global information. AVBPA, LNCS **3546**, 523–532 (2005)
9. D.S. Guru, H.N. Prakash, Online signature verification and recognition: An approach based on Symbolic representation. IEEE Trans. Pattern Anal. Mach. Intell. **31**, 1059–1073 (2009)
10. W.S. Wijesoma, Selecting optimal personalized features for online signature verification using GA in *IEEE International Conference on SMC* (2000), pp. 2740–2745
11. D.S. Guru, K.S. Manjunatha, S. Manjunath, User dependent features in online signature verification, in *First International Conference on Multimedia Processing, Communication and Computing Applications (ICMCCA)*, LNEE, vol. 213, (2013), pp. 229–240
12. D. Cai, C. Zhang, X. He, Unsupervised feature selection for multi-cluster data, in *Proceeding of the 16th ACM SIGKDD international conference on Knowledge discovery and data mining* (2010), pp. 333–342
13. O.J. Garcia, J.F. Aguilar, D Simon, MCYT baseline corpus: a bimodal database, in *IEEE Proceedings on Vision, Image and Signal Processing*, vol. 150 (2003), pp. 3113–3123
14. N. Sae-Bae, N. Memon, Online signature verification on mobile devices. IEEE Trans. Inf. Forensics Secur. **9**, 933–947 (2014)
15. D.S. Guru, H.N. Prakash, S. Manjunath, Online signature verification: an approach based on cluster representation of global features, in *Seventh International Conference on Advances in Pattern Recognition* (2009), pp. 209–212

Word-Level Script Identification from Scene Images

O.K. Fasil, S. Manjunath and V.N. Manjunath Aradhya

Abstract Script identification on camera based bus sign boards are presented in this work. The text localization is achieved by a series of morphological operations. Number of texture features, such as gabor features, log-gabor features and wavelet features are extracted from the segmented text images to classify the text images into three scripts, English, Kannada and Malayalam. Three different classifiers are evaluated and results are reported in this work.

Keywords Bus sign board recognition · Script identification · Texture features

1 Introduction

As many part of India are treated as a tourist hub, people across the globe visit India. During their visit it is very difficult to get information from localities as they know only local languages. For example during transportation the sign boards in buses are written in local languages. Similarly, in hotels, shops, sign boards, local languages are written. And there will both handwritten and printed. In such cases it is very difficult to read and understand the information from them. In such situations, one should identify the script in which the text is written, before the recognition of text. Developing a common OCR to recognize text in different scripts is difficult and it is not available. One option is to identify the script and use appropriate OCR [1]. Since text appears in bus sign boards and shops are at word level, one should identify the script

O.K. Fasil (✉) · S. Manjunath
Department of Computer Science, Central University of Kerala,
Kasaragod 671316, Kerala, India
e-mail: fasilok92@gmail.com

S. Manjunath
e-mail: manju_uom@yahoo.co.in

V.M. Aradhya
Department of MCA, S.J. College of Engineering, Mysuru 570006, India
e-mail: aradhya.mysore@gmail.com

© Springer Nature Singapore Pte Ltd. 2017   417
S.C. Satapathy et al. (eds.), *Proceedings of the 5th International Conference on Frontiers in Intelligent Computing: Theory and Applications*, Advances in Intelligent Systems and Computing 516, DOI 10.1007/978-981-10-3156-4_43

in sign boards at word level. This will rise additional challenges such as the information available from a word is minimal. The font size variations, different orientation, complex background and reflections in images also make the script identification challenging in sign boards at word level. In this work we propose a model for script identification on camera based bus sign board image. Various texture features, such as gabor, log-gabor and wavelets are extracted for the purpose of script identification. Features are combined and the effect of dimensionality reduction technique (PCA) is also carried out. We evaluated the performance of KNN And SVM classifiers in this work.

The paper is organised as follows. A brief review of script identification is presented in Sect. 2. In Sect. 3, the proposed script identification method is described. In Sect. 4 experimental analysis is presented and the work is concluded in Sect. 5.

2 Related Work

Script identification is a challenging issue and well addressed for scanned and printed document images. Recently script identification for camera based images has attained greater attention. According to Ghosh et al. [1], based on the nature of approach and features used, script identification methods can be classified into two classes: (1) structure-based methods [2–5] (2) visual appearance-based methods [6–13]. These two classes may further classified in to page-wise, paragraph-wise, textline-wise and word-wise, based on which level we are applying script identification.

In structure-based methods, script classes are differ from each other in their stroke structure and connections, and the writing styles associated with the character sets they use. One method under this category is connected component extraction and analysis from the document image to explore the morphological characteristics of the script. A Gradient-Angular-Features (GAF) is proposed by Shivakumara et al. in [2] for video script identification. A two-stage approach for word-wise script identification is proposed in [3], at the first stage a 64-dimensional chain-code histogram features are extracted and in the second stage a 400 dimensional gradient features are extracted. A study of script identification in video is proposed in [4, 5]. Number of low-level features such as mean and standard deviation of the edge pixels, density of edge pixels, energy of edge pixels, horizontal projection and cartesian moments of the edge pixels are extracted in [14] to discriminate Latin and Ideographic scripts. Similarly some structure based features like structural shape, profile, component overlapping information, topological properties, water reservoir concept etc. are used to distinguish English and Thai scripts in [15].

Another classification of script identification is visual appearance-based methods, in which the script is recognized on the basis of its visual appearance without really analysing the character patterns in the document. Accordingly, several features that describe the visual appearance of a script region have been proposed and used for script identification by many researchers [1]. The usefulness of traditional script identification is explored in [6]. Three feature extraction techniques, namely Zernike moments, gabor and gradient features are extracted and SVM classifier is used to classify scripts. In [7] Gabor filters have been applied to the problem of script identification in printed documents. Similarly Gabor filter analysis of textures on word-level is presented in [8]. Authors presented a classifier system of four classifiers for script identification purpose. Tan [9] also used Gabor filter analysis for distinguishing six languages (Chinese, English, Greek, Russian, Persian, and Malayalam). These all works in literature shows that the ability of Gabor filter to capture the visual appearance of various scripts in images. Busch et al. [10] have presented the combination of two popular texture features, wavelet and Gabor to identify the scripts. An attempt to identify scripts in variety of bilingual dictionaries has been carried out in [11]. 16 channel gabor features are extracted and a comparison of the performance of three classifiers is done by authors. Some other studies [12, 13] also showing the ability of the texture features like wavelets, gabor and discrete cosine transform to distinguish the script in complex images.

Overall, from the literature survey it is understood that, visual appearance based are simple and yet effective. In this work, to identify the scripts on bus sign boards we proposed a model for script identification for specific application. In the following section we present our model.

3   Proposed Model

A block diagram of proposed script identification model is shown in Fig. 1. the proposed model contain three stages after image acquisition, viz., preprocessing, feature extraction and learning. As images are acquired in real time, preprocessing is necessary. From preprocessed image, suitable features are extracted and supervised learning is used to learn scripts. For classification, we have considered bi-level script identification and we have used KNN and SVM classifiers.

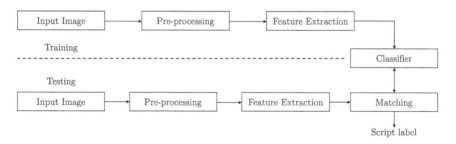

Fig. 1 Block diagram of proposed script identification model

Fig. 2 **a–c** Original text images, **d–f** Binarized text images

3.1 Pre-processing

The text in images are detected and localized based on a morphology based text localization method. Text localization is achieved by a series of morphological operations. In localization, first normalized input image is converted into gray scale image to reduce the computational complexity. Then converted image is filtered by a median filter to remove any noise present in image. Edges of filtered image is then detected using canny edge detector. After edge detection process, a series of morphological operations are applied to edge detected image to find out all individual characters. And all candidate individual characters are extracted by connected component analysis. All false alarms are removed by some heuristics, based on region properties of components. To connecting all individual characters a dilation operation is applied with a line structuring element. And again connected component analysis is applied to extract words from candidate regions by removing further false positives.

For the script identification, the input text images are first converted into grayscale images and k-means clustering is used to binarize text image as did in [6]. The sample binarized images are shown in Fig. 2.

3.2 Feature Extraction

Since texture features are good to capture information from complex backgrounds, we extracted different texture features, such as Gabor feature, Log-gabor and wavelets. The following subsections provide a brief overview of above mentioned features.

3.2.1 Gabor Filter

Gabor filters are band pass filter, and those are complex sinusoidal gratings modulated by 2-D Gaussian functions in the space domain, and shifted Gaussian functions in the frequency domain. They can be configured to have various shapes, bandwidths, orientations and center frequencies. The computational model for 2D isotropic Gabor filters are:

$$h_e(x, y) = g(x, y) \cos[2\pi f(x \cos \theta + y \sin \theta)] \tag{1}$$

$$h_o(x, y) = g(x, y) \sin[2\pi f(x \cos \theta + y \sin \theta)] \tag{2}$$

where h_e and h_o are the even-symmetric Gabor filter, and $g(x, y)$ is an isotropic Gaussian function. The spatial frequency responses of the Gabor functions are:

$$H_e(u, v) = \frac{[H_1(u, v) + H_2(u, v)]}{2} \tag{3}$$

$$H_o(u, v) = \frac{[H_1(u, v) + H_2(u, v)]}{2j} \tag{4}$$

where $j = \sqrt{-1}$

$$H_1(u, v) = \exp\{-2\pi^2\sigma^2[(u - f \cos \theta)^2 + (v - f \sin \theta)^2]\} \tag{5}$$

$$H_2(u, v) = \exp\{-2\pi^2\sigma^2[(u + f \cos \theta)^2 + (v - f \sin \theta)^2]\} \tag{6}$$

f, θ and σ are the spatial frequency, orientation and space constant of the Gabor envelope. In this work, the image is resized to 64×64, and four values of spatial frequency are selected: $0.04, 0.08, 0.16$ and 0.32. The combination of these four frequencies with four selected values of $\theta(10°, 45°, 90°, 135°)$ give a total of 16 Gabor responses and first and second order moments of each response is computed. As a result 32 dimensional feature vector's are obtained.

3.2.2 Log-Gabor Filter

One of the disadvantage with Gabor filter is, the maximum obtainable bandwidth is only 1 octave which limits the feature size that can be captured. One another option is Log-Gabor filter proposed by Field [1987]. Which allow larger bandwidths from 1 to 3 octaves which make the features more effective, reliable and informative [16]. On a linear scale, the transfer function Log-Gabor filter is expressed as:

$$\Phi_{(r_o,\theta_o)} = exp\left\{-\frac{(log(\frac{r}{r_o}))^2}{2(log(\frac{\sigma_r}{r_o}))^2}\right\} exp\left\{-\frac{(\theta - \theta_o)^2}{2\sigma_\theta^2}\right\} \tag{7}$$

where r_o is the central radial frequency, θ_o is the orientation of the filter, σ_θ and σ_r represent the angular and radial bandwidths, respectively. 16 dimensional features have been extracted using first and second order moments from 8 responses of Log-Gabor filter.

3.2.3 Wavelets

The discrete wavelet transform (DWT) employs two sets of functions, called scaling functions and wavelet functions, which are associated with low pass and high-pass filters, respectively. The decomposition of the signal into different frequency bands is simply obtained by successive high-pass and low-pass filtering of the time domain signal. Outputs of this filters are as follows:

$$a_{j+1}[p] = \sum_{n=-\infty}^{+\infty} l[n - 2p]a_j[n]. \tag{8}$$

$$d_{j+1}[p] = \sum_{n=-\infty}^{+\infty} h[n - 2p]a_j[n]. \tag{9}$$

Elements a_j are used for next step (scale) of the transform and elements d_j, called wavelet coefficients, determine output of the transform. $l[n]$ and $h[n]$ are coefficients of low and high-pas filters respectively One can assume that on scale $j + 1$ there is only half from number of a and d elements on scale j. This causes that DWT can be done until only two a_j elements remain in the analyzed signal These elements are called scaling function coefficients. A feature representation with reduced dimension of wavelet feature space is computed by finding the energy of each coefficients. First DWT is applied to image and energy of each coefficient is calculated. We have used Daubechies 1 wavelet family and up to level 4.

Also, we have tried to study the effect of combining all three different features which results in 60 dimensional vector. As the feature dimension is large, we have applied PCA to reduce dimension and used for classification.

3.3 Classification

The performance of the proposed system is tested with two classifiers, k-NN classifier and SVM classifier [17]. The number of training samples needed for each classifier for a better classification is also tested.

In k-NN classifier classification is done by assigning the label which is most frequent among the k training samples nearest to that unlabelled point measured by Euclidean distance and in case of SVM we have used two kernel functions linear and gaussian radial basis function (rbf).

4 Experimentation

As, there is no standard data set for bus sign board, we have created our own data set. We have collected around 1000 bus sign board images in out door environment from that we have considered 600 images. Total 600 word images of three scripts (200 word images of each script) i.e., English, Kannada and Malayalam are used for experimentation. Experimentation is carried out in Bi-script basis, i.e., English versus Kannada, English versus Malayalam and Malayalam versus Kannada.

We have carried out the experimentation by taking different percentage of training data (i.e.,10 %, 20 %, ..., 90 %). The images for training and testing is selected randomly and the experimentation is carried out 10 times and minimum F-measure, maximum F-measure, mean F-measure, standard deviation of F-measure and time is stored for each percentage of training data. Additionally for k-NN classifier the experimentation is carried out for different k values and best k is selected. For SVM, linear and rbf kernels are considered for experimentation. The experimentation have been conducted on a computer with 64 bit Windows operating system, having intel core i5 processor and 4GB RAM capacity using Matlab (version 2014).

For the purpose of study we consider the result of 90 % training data. The result of best feature for each classifier is shown in Table 1. We applied PCA on combined features which having a dimension of 60 and result of various reduced dimension is studied. A comparison of the classification result of combined texture features with PCA and with out PCA is presented in Table 2. Study shows that the reduced features using PCA is better than using all features.

Some of the scripts are miss-classified due to complex background of text images. Compared to all other combination miss-classification is more in the case of English v/s Malayalam combination. Compared to English and Malayalam, Kannada scripts has a sharp ending it is not in the case of English and Malayalam. Even though English and Malayalam are different scripts, after normalization some of the Malayalam text also look like English. Other reasons for miss-classification includes large font variations, poor normalization method and image acquisition issues such as, different orientation, reflection, shadow etc.

5 Conclusion

A study of script identification in three languages, English, Kannada, Malayalam from complex scene images is presented in this work. For the purpose of script identification first we have segmented word images from scene image using a series of morphological operations. Then for the purpose of script identification number of texture features such as gabor features, log-gabor features and wavelet features, the combination of these three texture features are also used for script identification. Also we have studied the effect of PCA. Study shows that the ability of the texture features to capture the script information from text images even in complex backgrounds.

Table 1 Analysis of texture features

Best classifier	English versus Kannada				English versus Malayalam				Kannada versus Malayalam			
	Feature	Min	Max	Std	Feature	Min	Max	Std	Feature	Min	Max	Std
KNN	Gabor	0.907	0.974	0.025	Gabor	0.728	0.899	0.054	Log-Gabor	0.87	0.972	0.036
SVM-L	Combined features	0.772	0.935	0.046	Combined features	0.614	0.705	0.027	Combined features	0.782	0.871	0.025
SVM-R	Gabor	0.8	0.93	0.053	Wavelet	0.695	0.855	0.047	Gabor	0.77	0.947	0.048

Table 2 Comparison of maximum F-measure values of combined features with and without using PCA

Script	KNN		SVM-L		SVM-R	
	With out PCA	With PCA	With out PCA	With PCA	With out PCA	With PCA
English versus Kannada	0.856	0.926 (18)	0.935	1.0 (21, 24, 30, 39, 42, 45)	0.719	0.921 (15)
English versus Malayalam	0.889	0.926 (18)	0.705	0.925 (45)	0.704	0.824 (21)
Kannada versus Malayalam	0.875	0.975 (24)	0.871	1.0 (24, 42)	0.871	0.777 (18)

References

1. D. Ghosh, T. Dube, A.P. Shivaprasad, Script recognition—A review, in*Proceedings of IEEE Ttansactios on PAMI* (2010), pp. 2142–2161
2. P. Shivakumara, N. Sharma, U. Pal, M. Blumenstein, C.L. Tan, Gradient-angular-features for word-wise video script identification, in *Proceedings of 22nd International Conference on Pattern Recognition* (2014)
3. S. Chanda, S. Pal, K. Franke, U. Pal, Two-stage approach for word-wise script identification, in *Proceedings of ICDR* (2009), pp. 926–936
4. D. Zhao, P. Shivakumara, S. Lu, C.L. Tan, New spatial-gradient-features for video script identification, in *Proceedings of DAS* (2012), pp. 38–42
5. T.Q. Phan, P. Shivakumara, Z. Ding, S. Lu, C.L. Tan, Video script identification based on text lines, in *Proceedings of ICDAR* (2011), pp. 1240–1244
6. N. Sharma, S. Chanda, U. Pal, M. Blumenstein, Word-wise script identification from video frames, in *Proceedings of Document Analysis and Recognition(ICDAR)*, August 2013, pp. 867, 871, 25–28
7. W.M. Pan, C.Y. Suen, T.D. Bui, Script identification using steerable Gabor filters. Proc. ICDR **29**(6), 1153–1160 (1981)
8. S. Jaeger, H. Ma, D. Doermann, Identifying script on word-level with informational confidence, in *Proceedings of ICDR* (2005), pp. 416–420
9. T.N. Tan, Rotation invariant texture features and their use in automatic script identification. Proc. IEEE Trans. PAMI **20**(7), 751–756 (1998)
10. A. Busch, W.W. Boles, S. Sridharan, Texture for script identification. Proc. IEEE Trans. PAMI **27**(11), 1720–1732 (2005)
11. H. Ma, D. Doermann, Word level script identification for scanned document images, in *Document Recognition and Retrieval XI*, January 2004
12. P.B. Pati, A.G. Ramakrishnan, Word level multi-script identification. Pattern Recogn. Lett. **2008**, 1218–1229 (2005)
13. P.S. Hiremath, S. Shivashankar, Wavelet based co-occurrence histogram features for texture classification with an application to script identification in a document image. Pattern Recogn. Lett. **2008**(29), 1182–1189 (2008)

14. J. Gllavata, B. Freisleben, Script recognition in images with complex backgrounds, in *IEEE International Symposium on Signal Processing and Information Technology* (2005)
15. S. Chanda, O.R. Terrades, U. Pal, SVM based scheme for Thai and English script identification, in *Proceedings of ICDR* (2007), pp. 551–555
16. D.J. Gopal, G. Saurabh, S. Jayanthi, A generalised framework for script identification. Int. J. Doc. Anal. Recogn. **10**, 55–68 (2007)
17. R.O. Duda, P.E. Hart, D.G. Stork, Pattern classification, 2nd edn

Recognizing Human Faces with Tilt

H.S. Jagadeesh, K. Suresh Babu and K.B. Raja

Abstract Issues related to realtime face recognition are perpetual even with many existing approaches. Generalizing these issues is tedious over different applications. In this paper, the real time issues such as tilt or rotation variation and few samples problem for face recognition are addressed and proposed an efficient method. In preprocessing, an edge detection method using Robert`s operator is utilized to identify facial borders for cropping purpose. The query images are axially tilted for different degrees of rotation. Both database and test images are segmented into one hundred fragments of 5 * 5 size each. Four different matrix characteristics are derived for each divided part of the image. Corresponding attributes are added to yield features related to final matrix. Final one hundred facial attributes are obtained by fusing diagonal features with one hundred features of matrix. Euclidean distance between the final attributes of gallery and query images is computed. The results on Yale dataset has superior performance compared to the existing different approaches and it is convincing over the dataset created.

Keywords Cropping · Matrix norm · Tilt invariance

1 Introduction

Application based face recognition system design narrow downs the constraints. Additionally controlled environment system design also reduces the number of issues to be considered. Different publically available databases considers many issues to reduce the burden on researchers. The different parameters such as pose, occlusion, [1], illumination, and expression [2] are involved in the process of

H.S. Jagadeesh (✉)
ECE Department, Rayalaseema University, Kurnool 518007, India
e-mail: jagadeesh.11@gmail.com

K.S. Babu · K.B. Raja
Department of Electronics and Communication Engineering, UVCE,
Bengaluru 560001, India

© Springer Nature Singapore Pte Ltd. 2017
S.C. Satapathy et al. (eds.), *Proceedings of the 5th International Conference on Frontiers in Intelligent Computing: Theory and Applications*, Advances in Intelligent Systems and Computing 516, DOI 10.1007/978-981-10-3156-4_44

generating databases. Many databases [3] focus on these parameters individually or in combined form. It is decided by the researcher to select these databases appropriate for the objectives set and use the same to test the performance of system designed. Some external factors such as image axis tilt, limited samples problem is not considered during the formation of database.

Many techniques are proposed and deployed in the arena of solving face recognition problem. Principal component analysis [4], scale-invariant feature transformation [5], Directional local binary pattern [6] are repeatedly used in literature as different methods. Feature extraction [7] process in face recognition has vital role to deduce input images into fewer coefficients or attributes and influences on the accuracy. The performance of any algorithm proposed, depends directly on number of images used in database step. Increasing face recognition accuracy is difficult with single image used for database, when the availability of samples is less. It is addressed as One Sample per Person Problem [OSPP] [8]. OSPP is advantageous in terms of faster processing time and less memory requirements.

Organization: Organization of rest of the paper is as follows; related work review of different researchers is in Sect. 2; Proposed TI FR model is in Sect. 3 and Sect. 4 depicts the proposed algorithm. Performance is discussed in Sect. 5 and Concluded in Sect. 6.

2 Related Work Review

Wang and Yang [9] proposed a neural network based method with single training image per person for face recognition problem. Auto-associative memory using multiple values neural network is proposed using activation function and modified rule of evolution. Results are effective on AR and FERET face databases over other approaches. Jingang and Chun [10] addressed small-sample size problem in face recognition by using coupled mappings. The low resolution probe image and high resolution gallery images are projected on a unified latent subspace. Good similarities between the images are obtained with local optimizations. Recognition performance on SC, Multi-PIE and FERET face databases is promising. Chnagxing et al. [11] developed a face recognition framework over ±90° of pose variation. Pose variation problem is converted into a partial recognition problem. Partial faces are synthesized using patches of face and the same is used for classification. Performance of the proposed framework is effective over pose variations on Multi-PIE, FERET, and CMU-PIE databases.

3 Proposed Work Description

The complete description of different steps involved in the proposed Tilt Invariant Face Recognition (TIFR) model is shown in Fig. 1 and its explanation is as follows.

3.1 Databases

The Yale [12] and Created face databases are used to test the performance of the work proposed with additional challenges such as image tilt and less number of samples. The created database contains fifty one persons with ten images per person i.e. there are 510 images in total with each image of size 352 × 288. The images are captured with different poses, expressions and ranges, while the lighting conditions are relatively constant. Logitech c170 camera with 5 mega pixel resolution is used to acquire all the images. The image acquisition setup consists of a rectangular box made up of wood. Inside the box a square block is constructed which is made up of glass; four Compact Fluorescent Lamps (CFL) are placed in between glass and the wood to provide required illumination. The main objective behind this setup is to overcome illumination problems occurring when images are captured during different time of the day. This setup will provide constant illumination anytime and anywhere. The illumination is varied by using control switches of CFL.

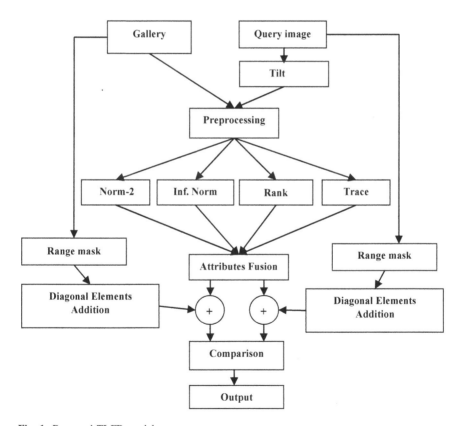

Fig. 1 Proposed TI FR model

3.2 Pre-processing

Original input images from database are preprocessed to refine and remove any unwanted trivial information content e.g. rest of face. The different steps in pre-processing are; (i) Any format to gray level format (optional), (ii) Face boundary identification using 2 * 2 Robert's mask [13], (iii) Extracting facial part through cropping, and (iv) uniform sizing of images. Any input image format is converted to intensity or gray format, which is useful in decreasing computation time. The edge operator extracts gradient maximum by using 2 * 2 Robert's mask for each gray level values.

Entire image from previous step is split into two equal parts with respect to central column. For each half part, every row is scanned for edges, and the scanning is stopped whenever an edge is detected. The corresponding coordinates of edges in each row is noted. The other part of the image is scanned similarly from opposite direction to note the coordinates of edges. Based on the result of coordinates scanned, the coordinates near to the corners are used to crop the facial region. As a last part of preprocessing, all images of previous step are resized to 50 * 50 uniformly. The query images are tilted or rotated for $\pm 1°$, $\pm 2°$, $\pm 3°$, $\pm 4°$ and $\pm 5°$ separately to observe the effectiveness of the method proposed.

3.3 Extracting Attributes of Face

The conversion of a preprocessed image into fewer symbolic values representing entire part of face involves dimension reduction and good compatibility. The 50 * 50 image is partitioned into one hundred pieces with 5 * 5 sizes each. The different matrix features such as trace, rank, infinite norm and 2-norm of the matrix are computed individually on each of 5 * 5 segment. All these features are respectively averaged algebraically to obtain 100 distinct attributes of each face image. The algebraic operator norm of a matrix is the highest value obtained by adding all absolute elements of rows or columns [14].

$$M = \begin{pmatrix} 10 & -22 & 13 \\ 45 & 57 & 16 \\ 74 & 83 & 19 \end{pmatrix} \tag{1}$$

Consider a matrix M in Eq. 1, whose different matrix features are computed as follows; (i) Infinite norm [15] is the highest value among different rows obtained by summing the absolute value of elements and it is given by $\|M\|_\infty$ = Maximum $[(10 + |-22| + 13), \quad (45 + 57 + 16), \quad (74 + 83 + 19)]$ = Max. $[45, \quad 118, \quad 176]$ = 176. $\|M\|_2$ is the highest singular value extracted by applying singular value decomposition and it is 135.25 for the matrix M. Adding all the principal diagonal elements of a matrix yield trace [16] and for matrix M, trace is given by

$(10 + 57 + 19) = 86$. Rank [17] is the total number of rows or columns of a matrix which are linearly independent. The rank of each 5 * 5 segment is computed in our work. Respective matrix attributes of trace, rank, infinite norm and 2-norm for each 5 * 5 segment matrix are fused by averaging to obtain one hundred attributes related to each image matrix.

Further, the attributes are also extracted using range of gray level computation within each 3 * 3 overlapping mask at every pixel location without preprocessing. Range is the subtraction of maximum to minimum intensity value in a 3 * 3 segment. The both diagonal elements such as secondary and principal, are combined to get other set of attributes. These 100 diagonal features are fused with 100 attributes of different matrix properties to generate final attributes.

3.4 Attributes Comparison

The final attributes of every database image are matched with the attributes of query image. Performance results are derived based on Euclidean Distance (ED) in matching process. The general formula used to compute ED is in Eq. 2 between vectors P and Q. The parameter i is varied for all the attributes of each image.

$$D\,(Pi, Qi) = \sqrt{((p1 - q1)^2 + (p2 - q2)^2 + \cdots \ldots} \tag{2}$$

4 Algorithm

Problem definition: Humans are identified using Tilt Invariant Face Recognition system (TI FR). Proposed algorithm is depicted in Table 1 and the objectives are;

(i) Recognition Rate (RR) improvement.
(ii) False Rejection Rate (FRR) and False Acceptance Rate (FAR) are to be reduced.

5 Performance Analysis

The two datasets such as Yale, and own created face database are used to test proposed algorithm. The challenges such as number of gallery or database images and tilt in query images are considered.

As in small sample size problem the gallery images count is less and is limited to one or two. The query images are tilted for each of $\pm 1°$, $\pm 2°$, $\pm 3°$, $\pm 4°$ and $\pm 5°$. Positive tilt is in anticlockwise and clockwise is negative tilt.

Table 1 Proposed algorithm of TIFR model

Input: Face images of gallery and query
Output: Recognition/Refutation of a person
1. The four different Preprocessing steps are as follows
(i) Any format to gray format conversion (optional), (ii) 2 * 2 Robert`s mask is used to identify edges, (iii) Cropping of facial part, (iv) Uniform resizing to 50 * 50
2. Only query Images are tilted for $\pm1°$, $\pm2°$, $\pm3°$, $\pm4°$ and $\pm5°$
3. Both gallery and query images of 50 * 50 size are segmented into 100 parts of 5 * 5 equal size
4. For each 5 * 5 segment, trace, matrix infinite norm, 2-norm and rank are calculated
5. Each of 50 * 50 image produces 100 feature values by averaging all four matrix attributes of step 4 to get 100 matrix attributes
6. Range mask of 3 * 3 size is applied on each pixel location and resized to 100 * 100
7. Both diagonal elements of step 6 output matrix are summed and fused with the matrix attributes of step 5 to lead final attributes
8. Euclidean distance based comparison is used for gallery and probe image attributes
9. The person is declared as recognized, for an image with lowest ED value

(i) *Yale database.* The thirteen subject gallery is used to store as known persons. FRR is computed for 10th image of 13 persons gallery and the 5th image of out of gallery persons is used to calculate FAR. It is observed that for $\pm1°$, $-2°$, $\pm3°$, and $\pm4°$ tilt, the results are robust and rotation invariant. Table 2 justifies the same, and a maximum of 53.8 % RR is obtained in one gallery image category. For double gallery images, 92.3 is the highest % RR and it is rotation invariant for $0°$, $1°$, $2°$, $3°$ and $4°$ tilt angles. The highest % RR of the work proposed is compared with other [18, 19, 20] approaches is in Table 3 and superior performance is obtained over methods listed.

(ii) *Created database.* The proposed work uses images of 50 persons to evaluate the performance. Thirty three persons each with two images are utilized for gallery purpose and FRR is computed using rest of any one image. For the remaining 17 persons FAR is computed. Highest %RR is obtained without tilting probe image is 93.93 and it is 90.9 for both $\pm1°$ angle tilt. Hence the proposed approach exhibits rotation invariant property $\pm1°$ as shown in Table 4. Further the algorithm is tested by varying the Persons Inside the Gallery (PIG) and Persons Outside the Gallery (POG) ratio. Table 5 depicts the highest % RR values for the PIG: POG ratio variation. For the cases of PIG: POG ratios 20:30, 25:25, 30:20, the highest % RR is 90, 92, 93.33 respectively increasing, but it is 91.42 for 30:20 PIG: POG ratio.

Table 2 Performance of one and two gallery images on Yale database

Threshold	One gallery image ±1°, −2°, ± 3°, ± 4° Tilt			Two gallery images 0°, 1°, 2°, 3°, 4° Tilt		
	% FAR	% FRR	% RR	% FAR	% FRR	% RR
0.0	0	100	0	0	100	0
0.1	0	100	0	0	100	0
0.2	0	100	0	0	69.2	30.7
0.3	0	61.5	30.7	0	30.7	69.2
0.4	0	46.1	46.1	0	7.6	92.3
0.5	0	38.4	46.1	0	0	92.3
0.6	50	23	53.8	100	0	92.3

Table 3 Highest % RR comparison on Yale database

Method	Highest % RR
ALBP + BCD [18]	71.9
WM (2D) 2PCA [19]	80.77
GABOR + DTW [20]	90.67
Proposed TIFR model	92.3

Table 4 Performance for two gallery images on created database

Threshold	0° tilt			± 1° tilt		
	% FAR	% FRR	% RR	% FAR	% FRR	% RR
0.0	0	100	0	0	100	0
0.1	100	0	93.93	100	0	90.9

Reasons for the results improvement are; the robustness of an algorithm is improved by fusing trace, rank, infinite norm and 2-norm attributes for various tilt angles. The fusion of diagonal attributes with features generated from four matrix attributes contributes for improvement in the accuracy of recognition.

Table 5 Highest % RR for different PIG: POG ratios on created database

PIG: POG ratio	20:30	25:25	30:20	35:15	33:17
Highest % RR	90	92	93.33	91.42	99.93

6 Conclusion

It is an elusive task to recognize human faces with fewer number of data samples. A simple, yet efficient face recognition method using fewer database images with image rotation is proposed in this paper. A edge detection based facial cropping is used in preprocessing. Tilting of query images is performed for different angles individually. The averaged or fused attributes of matrix and diagonal elements are final features considered for next step. The results of matching or non matching between gallery and query image attributes is produced using Euclidian distance metric. The results on Yale is better over other approaches and it is substantial on Created database.

References

1. S. Evangelos, G. Hatice, C. Andrea, Automatic analysis of facial affect: a survey of registration, representation, and recognition. IEEE Trans. Pattern Anal. Mach. Intell. **37**(6), 1113–1133 (2015)
2. K. Jyoti, R. Rajesha, K.M. Pooja, Facial expression recognition—a survey. In: 2nd International Symposium on Computer Vision and the Internet, Procedia Computer Science, vol. 58 (Elsevier, 2015), pp. 486–491
3. G. Castaneda, T.M. Khoshgoftaar, A survey of 2D face databases, in *16th IEEE International Conference on Information Reuse and Integration* (2015), pp. 219–224
4. E. Gonzalez-Sosa, R. Vera-Rodriguez, J. Fierrez, P. Tome, J. Ortega-Garcia, Pose variability compensation using projective transformation for forensic face recognition, in *IEEE International Conference of the Biometrics Special Interest Group (BIOSIG)* (2015), pp. 1–5
5. Y. Gao, H.J. Lee, Cross-pose face recognition based on multiple virtual views and alignment error. Int. J. Pattern Recogn. Lett. **65**, 170–176 (2015)
6. L. Shen, J. He, Face recognition with directional local binary patterns, in *6th Chinese Conference on Biometric Recognition, Lecture Notes in Computer Science*, vol. 7098 (Springer, Berlin, 2011), pp. 10–16
7. S. Ding, H. Zhu, W. Jia, C. Su, A survey on feature extraction for pattern recognition. J. Artif. Intell. Rev. **37**(3), 169–180 (2012)
8. X. Tan, S. Chen, Z.-H. Zhou, F. Zhang, Face recognition from a single image per person—a survey. J. Pattern Recogn. **39** (9), 1725–1745 (2006)
9. C. Wang, Y. Yang, Robust face recognition from single training image per person via auto-associative memory neural network, in *IEEE International Conference on Electrical and Control Engineering (ICECE)* (2011), pp. 4947–4950
10. S. Jingang, Q. Chun, From local geometry to global structure -learning latent subspace for low resolution face image recognition. IEEE Signal Process. Lett. **22**(5), 554–558 (2015)
11. D. Changxing, X. Chang, T. Dacheng, Multi-task pose-invariant face recognition. IEEE Trans. Image Process. **24**(3), 980–993 (2015)
12. http://vision.ucsd.edu/datasets/yale_face_dataset_original/yalefaces.zip
13. R.C. Gonzalez, R.E. Woods, S.L. Eddins, *Digital image processing using MATLAB* (Pearson Education India, 2004)
14. C. Ding, D. Zhou, X. He, H. Zha, R1-PCA-Rotational invariant L1-norm principal component analysis for robust subspace factorization, in *23rd International Conference on Machine Learning* (Pittsburgh, 2006), pp. 1–8

15. L. Guangcan, L. Zhouchen, Y. Shuicheng, S. Ju, Y. Yong, Ma. Yi, Robust recovery of subspace structures by low-rank representation. IEEE Trans. Pattern Anal. Mach. Intell. **35**(1), 171–184 (2013)
16. M. Jaggi, M. Sulovsky, A simple algorithm for nuclear norm regularized problems, in *27th International Conference on Machine Learning*, Israel, pp. 1–8 (2010)
17. Z. Ding, S. Suh, J.-J. Han, C. Choi, Y. Fu, Discriminative low-rank metric learning for face recognition, in *11th IEEE International Conference and Workshops on Automatic Face and Gesture Recognition (FG)*, vol. 1 (2015), pp. 1–6
18. R. Shyam, Y.N. Singh, Face recognition using augmented local binary pattern and bray curtis dissimilarity metric, in *2nd IEEE International Conference on Signal Processing and Integrated Networks* (2015),pp. 779–784
19. E. Zhang, Y. Li, F. Zhang, A single training sample face recognition algorithm based on sample extension, in *6th IEEE International Conference on Advanced Computational Intelligence* (2013), pp. 324–327
20. S. Venkatramaphanikumar, V.K. Prasad, Gabor based face recognition with dynamic time warping, in *6th IEEE International Conference on Contemporary Computing* (2013), pp. 349–353

Cloud Computing Technology as an Auto Filtration System for Cost Reduction While Roaming

S. Biswas, A. Mukherjee, M. Roy Chowdhury and A.B. Bhattacharya

Abstract Cloud computing environment appears to be indispensable in ensuring economical telecommunication. When a person changes current state (national and international), the mobile service providers change their respective charges adding incoming call charges as well. Further, from the telecom service providers end, no filtering of unnecessary calls can be done. In this paper, a cloud-based system has been proposed indicating a reduction of the communication cost between the users when one of the two users is in roaming. By making use of the technology, a robust system has been developed that automatically identifies a roaming mobile number and blocks any unknown number but notifies at either end. The paper provides an analysis of service roaming considering the ground realities of the international mobile roaming of both industry and market.

Keywords Cloud computing · Auto filtration · Communication cost · Cloud routing · Roaming system

S. Biswas · A. Mukherjee
Department of Computer Science and Engineering,
Netaji Subhas Engineering College, Kolkata, India
e-mail: subashiscse@gmail.com

A. Mukherjee
e-mail: abhishekmukherjeecse@gmail.com

M. Roy Chowdhury
Department of Computer Science and Engineering,
Techno India University, Kolkata, India
e-mail: meghdut.tig@gmail.com

A.B. Bhattacharya (✉)
Department of Electronic and Communication Engineering,
Techno India University, Saltlake, West Bengal, India
e-mail: bhattacharyaasitbaran@gmail.com

© Springer Nature Singapore Pte Ltd. 2017
S.C. Satapathy et al. (eds.), *Proceedings of the 5th International Conference on Frontiers in Intelligent Computing: Theory and Applications*, Advances in Intelligent Systems and Computing 516, DOI 10.1007/978-981-10-3156-4_45

1 Introduction

Cloud computing may be regarded as one of the most booming technology among the professional of information technology. Cloud plays a vital role in the smart economy, and the possible regulatory changes required in implementing better applications by using the potential of cloud computing [1]. In practice, the host operator passes the call via 'international transit' to the end of the home operator. The operator connects one to his friend's operator and establishes the call [2]. Now-a-days, web-based marketing largely contributes to the income for any organization. It is assumed, as an example, that an organization allocated at Bangalore in India has many customers with its own call centre and a caller calls a customer 'A' from the call centre. It is practically impossible for this caller to know whether that particular customer is in India or abroad. If the customer at the time of calling is in UK, then he is in roaming and so after speaking to the caller from India for just a few minutes, he will be charged heavily. As features of web-bed marketing, a customer also receive bulk SMS and automated calls. In this paper, a cloud model is proposed which contains the towers of various telecom service providers in different countries like India, USA, UK, Australia, Canada, Hong Kong and others. The towers of telecom service providers in a particular country are shown collectively as a specific cloud structure inside the massive cloud. Let a person 'X' stays in India and uses Vodafone India connection. For any purpose, he goes to UK. Now, as soon as he enters UK, he is no longer able to use his Vodafone India connection as he is at that moment using the telecom service of the telecom service provider, say named Orange, staying in UK. There is a professional tie-up between Vodafone India and Orange regarding the telecom service which is to be used by a person from just after arriving in UK. Originally, the person 'X' was using a certain mobile number of Vodafone India but after entering UK, he is now using a different number (a number of "Orange"). Consequently, if 'X' receives any call from India and speaks to the caller for a specific period of time, he faces high communication charge. As an emerging new technology cloud computing has a significant role about resources, economics and the environment. Many of these questions are closely related to geographical considerations in association with the data centres. With a view to interpret and analyze evidence within this environment, computer forensics finds a broader range of technical knowledge across multiple hardware platforms and operating systems. The purpose of the paper is to explore the interrelationships between the geography of cloud computing, its users, its providers as well as governments. In fact, cloud computing permits a set of heuristics for preventing overload in the system effectively while saving energy used.

2 Materials and Methods

In order to ensure economical telecommunication between two persons, communicating from two different countries, the proposed cloud computing model consists of a cutting-edge system which detects each and every call going abroad, i.e. determines whether a particular mobile number is in roaming or not. This system is named Roaming Identifier. It helps in finding out which mobile number is using the telecom services of which telecom service provider at present. To ensure easy detection of roaming mobile number, this system employs a concept called flag. As soon as a roaming mobile number is called, the status of flag becomes 1. This is the indication of a roaming mobile number. Let the person 'X' stays in USA and gets a call from India. This system if implemented, he will be able to recognize whether the call is coming from a call centre or it is from any of his known persons. Therefore, he can instantly decide whether to receive this call or not.

The paper has emphasized concepts and approaches towards effective privacy management for mobile platforms based on the prevailing key players in the mobile market, viz., Apple, RIM, Microsoft and Google. As this work has been initiated by Google the work has mainly concentrated on Android-based concept towards customizable privacy management approaches. Current mobile applications such as the introductory example 'Hike' requires highly sensitive personal data related to position, contact data, etc. For obvious reason, the user does not want to share this data with everybody in every context. Without any private data at all it becomes often impossible for an application to provide any reasonable service. The architecture of a cloud infrastructure for mobile data collection is shown in Fig. 1. The system allows the user create mobile Apps for data gathering using simple interfaces; no programming skill is needed. The data is stored in a scalable cloud

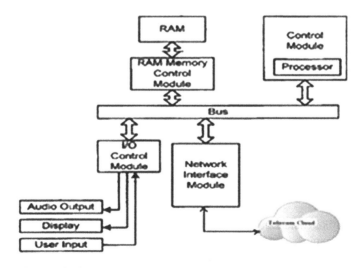

Fig. 1 Architecture for future mobile technology

infrastructure and the system is based on free software [3, 4]. The present method has designed an architecture based on a cloud-based platform to get closer to the reality of location-based applications. The demand for location-based services is increasing everyday and number of users is increasing by time. Having a stable, flexible, secure and scalable application is required these days. Interconnection is a fundamental pre-requisite for International Mobile Roaming (IMR). The "person-to-person" nature of traditional IMR services such as voice and SMS meant that all or most communications occur between one or more end-users across operator networks for which a managed model is very effective and essential.

The empirical results can help to the mobile companies for establishing concrete standards to measure the operating performance of roaming business. The impact of regulatory pressure is to reduce international mobile roaming (IMR) charges, especially in Europe, and the potential competition from other technology. In tabular form customer registration details, like date of issue, each unique customer code, status of each customer, registration no of each customer, mobile no of each customer, the particular date on which a particular mobile number is issued are available [2].

3   Results and Discussion

Roaming for voice, the ability for a user to make and receive calls when visiting another country, is taken for granted by GSM users today. The Telecommunications Industry Association (TIA) has initiated project to develop standards to enable PBX systems with wireless capability [5]. It becomes necessary to roam into foreign networks in order to access such services. This approach consists of two tokens: Passport (identification token) and Visa (authorization token) to provide a flexible authentication method for foreign network to authenticate mobile users [3]. Besides this, telecom industries are growing rapidly due to huge demand of users. Telecom infrastructures shared mutually to reduce establishment cost. Now it is time to get support by the users from telecom. Popularity of roaming service is considerable and this is necessary to connect any time for business or personal issues. It is, in fact, a trend to get a person any time anywhere and from that point of view, it is the advantage of the present technology. It is the time of cloud where the main agenda is available. Through this concept it may be possible to reduce the complexity of roaming. Huge revenue of roaming service has been produced by utilizing roaming partner's network. In this growing age communication cost can be minimized for international roaming while other communication cost becomes significantly high. Cloud technology is not only cheaper but at the same time offers infinite storage capacity along with backup and recovery.

A cloud-based architecture is to store and retrieve on need a user's important mobile data like contact details, mobile log file, call history, etc., through internet. In this cloud architecture, a robust relation between a mobile manufacturer and a mobile service provider has been established. With the help of this cloud architecture, if a users' mobile or SIM gets accidentally damaged or lost, all of their 'lost' data can be retrieved into the replaced devices or SIM without any copy and paste process. As for example, let a customer stays in India and using the service of a specific mobile service provider going to use a mobile of a specific mobile manufacturer. For some purposes, he goes to USA and accordingly starts a new SIM for communication and other utilities. Then, for getting all the data of the previous SIM (mobile log, contact details, call history etc.) in the new SIM, he does not need to start the tedious copy and paste activity. Instead, all the required data can be simply downloaded from the cloud into the new SIM by using the customer's Identification number.

Algorithm 1

```
active_roming_service()
begin
If (change network = = 1)
begin
mno = getting from sim card;
sent request to customer for activation();
If (rflag = = 1)
begin
$z = set   nroaming = 1   where   mobile_no = mno   into   cus-
tomer_register_details table;
add signal(1);
close connection();
end
Else
If(mno_request for i roaming = = 1)
begin
$zz = set   iroaming = 1   where   mobile   no = mno   into   cus-
tomer_register_details table;
add signal(1);
close connection();
end end end
```

Algorithm 2

```
national system()
begin
If(addsignal = = 1)
begin
mno = getforiegnnetwork();
check sameaction ();
If(cloudserver = = 1)
begin
q = set nroaming = 1 where mobno = mno into cservice table;
active system ();
```

```
end
Else
begin
alert("Server not available");
end end end
```

Algorithm 3

```
active system ()
begin
s = select n roaming where mobno = mno from cservice table;
while(s is active)
sign = row[n roaming];
If(signal = = 1)
begin
f = incoming call();
chk = retrieve data from clearlist table where mob no = mno &&
save no = f;
i = 0;
while(chk is active)
i ++;
If(i = = 1)
begin
call process();
end
Else
begin
caller_alert = 'The number is in roaming';
receiver_alert = 'f is trying to call you but it is not your-
known number;
send sms (f, caller);
send sms(mno, receiver);
caller tune(f, caller);
dropcall ();
end end end
```

Usually, the call centres promote their products by calling the common people which is somewhere unnecessary. In fact, the users get disturbed by frequent calls from the call centres while roaming. If a person's SIM is under international roaming the cost of incoming calls would count, and person gets really harassed by receiving an unnecessary/unwanted call as the call centres are using local numbers. Through DND system people can get relief from getting promotional bulk messages but one cannot use the call filtration process which can help in tracing the local numbers [6, 7]. While in international roaming, if someone lost his mobile or if it is destroyed one cannot trace the SIM as the so and so country does not have the person's details. And that person's family will not be able to get in touch with him for a pretty long time until he gets a new SIM card which is indeed takes quite a long process. Another significant part in international roaming is incoming cost, that is, the call charge of an incoming call of per second is equivalent to per minute. According to the existing system, there is a database named Customer Registration Details. One may add two new columns to this database named n_roaming and i_roaming. This database is maintained at the telecom service provider's end. It has thus been proposed a new cloud-based system which can be connected to all the telecom service providers in the cloud. Mobile service providers can maintain a cloud for their mutual sharing for a specific zone but the work proposes a global zone where all the small cloud will come under one single cloud for sharing the data. When any one calls a number, the signal first hits the BTS and BTS then communicates with the nearest Base Station Controller (BSC), where lots of databases are working to route the signal. Several Service Providers are maintaining the same data in several locations which is data redundancy. It can be connected through cloud implementation where the BSC communicate with Master Switching Centre (MSC) for next level routing. In this MSC, several queries can be worked for routing the call.

For developing one prototype open source MySQL database is implemented. Real cloud environment with Xen platform has been used while for virtualization VM Ware has been used. Also 4 hexacore processors with individual 4 GB memory have been used for application testing with a view to share the same customer's credentials with other service providers. According to Fig. 2a, one can connect cloud systems into BSC or MSC zone wise. Later, these small clouds will be under a global zone where the globe service providers across will be able to share the data with respect to their needs (Fig. 2b).

A new SIM can be issued from a remote location by updating the c-code column with a new SIM number through a global proposed cloud. All previous work done to design table for cloud environment is for a single zone and not for global usage. This system consists of several databases, viz., service provider_master, cloud service, service and clear list. The service provider_master table contains 3 columns —id, service provider and active. The active column in the table indicates whether telecom service is ON or not. If service is ON, then the value in this column is 1. All the telecom service providers are listed in the table with their respective ids.

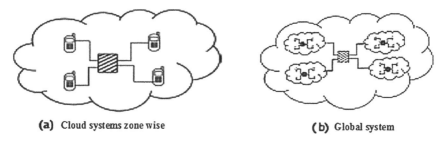

(a) Cloud systems zone wise   **(b)** Global system

Fig. 2 a Proposed connection of cloud systems zone wise. b Proposed global system

The table shows specified two new columns n_roaming and i_roaming. For a particular mobile number, if national roaming is activated, the value of the column n_roaming is 1, otherwise the value is 0. When a particular mobile number needs to use international roaming facility, it will send a request to the telecom service provider, on positively responding to the request, the value in the field i_roaming for the particular number becomes 1, otherwise the value for any mobile number in this column remains 0. If a roaming number gets a call from a mobile number not listed in the Clear List, notifications will be sent at both ends—the caller's end and the receiver's end. At the caller's end an SMS will be received mentioning that the number to which the call is done is in roaming. At the same time, at the receiver's end, it will be indicated that this number is an unknown number and consequently the call may be disconnected. But in case, if the number from which the call is done is a landline number, then the Caller Tune function of the proposed system will be activated. The caller will be indicated by a caller tune that the number to which the call is done is in roaming, specifically by a sentence like 'THIS NUMBER IS IN ROAMING'. This caller tune will also hear if the call is done from a mobile number in addition to the SMS. At the same time, the receiver whose number is in roaming will get a notification that the number from which the call is being done is an unknown number and it is certainly to be disconnected. There a simulator is used to test the latency time because a new field for checking the roaming status is added. Figure 3 clearly shows that it is not taking any extra time as the cloud is an extension of parallel computing and millions of instruction per second can be processed through this.

A 3rd party bulk SMS (transaction) has been used to incorporate this within the system. For sending the SMS, curl_init() and curl_setopt ($CH, CURLOPT_URL, $LOGINURL) functions are used. This filtration system is suitable to be used under BSC or MSC to filter the call. It is more effective to filter through mobile apps. It was further tried to use this through mobile apps on android platform, e.g., eclipse juno 4.2, Apk 5.0.1, android lollipop and also API-21 have been used to develop the prototype and open source database MySQL as storage. Connection method Driver Manager.get connection has been used for android platform. But system prototype for BSC and MSC should be fruitful for better result.

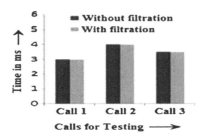

Fig. 3 Time to process call

4 Conclusions

In recent years, newer and smarter technologies are being introduced in the market with time. On using cloud technology, considerable technological cost reduction can be ensured. A mobile number, when in roaming, generates exorbitant communication charges and often experiences disconnection. Only incoming SMS is trouble-free for such mobile users since it is of no charge. The robust cloud-based system presented in this paper will enable elimination of unnecessary calls and consequently monthly communication expense can be greatly reduced. This system will eliminate all unwanted calls through its filtration mechanism [3]. A secure roaming protocol for mobile networks can be developed by the suggested technology which can minimize the overall temperature effect caused by individual systems. The reduction of heat consumption in the atmosphere by implementing the cloud computing method will reduce the pollution effect and thus will automatically cut down the global warming, even may be by a considerable amount [8]. It may be pointed out that by applying the proposed technology the system sends only filtered call to the receiver which does not depend by any means on manual receiving without filtration, since only after receiving the call one can recognize whether the call is unwanted or not.

Acknowledgments Thanks are due to Mr. Sourav Banerjee, System Head, Aircel and Mr. Ayan Chakraborty, System Engineer, Aircel for their supervision during our prototype design and architecture development.

References

1. F. Chauvel, H. Song, N. Ferry, F. Fleurey, Evaluating robustness of cloud-based systems. J. Cloud Comp. **4**, 1–17 (2015)
2. S. Biswas, T. Roy, A.K. Manna, A. Roy, Cloud computing implementation architecture: data processing mechanism for telecom medium. Int. J. Eng. Res. Tech. **3**, 1095–1102 (2014)
3. J.S. Ward, A. Barker, Cloud cover: monitoring large-scale clouds with Varanus. J. Cloud Comp. **4**, 1–28 (2015)
4. J.S. Park, H.J. Lee, M. Kim, Technical standardization status and the advanced strategies of the next generation mobile communications. Mobile Telecommunication Research Division, ETRI Conference (2006)
5. http://www.ex4u.org/yes_two_sim_FAQ.php
6. V. Sekhri, I.A. Ames, R. Gupta, A.K. Somani, CompuP2P: light-weight architecture for Internet computing, in *Broadband Networks* (BroadNets, 2005)
7. J.S. Hansen, R. Lachaize, Using idle disks in a cluster as a high-performance storage system, in *Cluster Computing: Proceedings IEEE International Conference* (IEEE Press, New York, 2002), pp. 415–424
8. A.B. Bhattacharya, B. Raha, S. Biswas, *Planetary radio signal reception & cloud computing: a remote sensing perspective* (Lap Lambert Academic Publishing, Germany, 2015), pp. 295–303

Design of Visual Cryptography Scheme Using C(m, w) Combinations

Anindya Kr. Biswas, S. Mukhopadhyay and G.P. Biswas

Abstract VCS is a perfect secure technique that allows easy concealment of images without any cryptographic computation, however, the encrypted image can be recovered by human visual system. The scheme is proposed by Naor and Shamir for binary images, called *k-out-of-n* VCS, where $k \leq n$ participants can recover images. Subsequently, a number of efficient VCS models are proposed that enhanced different VCS features. This work proposes a new model for $k = 2, 3$ that generates shares using *C(m, w)*, where a bit string of length m with $w < m$ number of 1's is taken for constructing a share. On analysis, it has been found that our scheme realizes *2-out-of-n* VCS more efficiently than the scheme proposed by Naor et al. Also our design for *3-out-of-n* VCS shows improvement as well as supports access structure similar to Ateniese et al. On simulation, it has been found that our VCS performs satisfactorily.

Keywords Visual cryptography scheme (VCS) · Access structure · Basis matrices · Image contrast and resolution

1 Introduction

The visual cryptographic scheme (VCS) is a new cryptographic paradigm, in which a secret binary image is concealed simply by encoding into multiple invisible shares. And during decoding, a set of predefined shares are stacked together for

A.Kr. Biswas (✉)
Department of Information Technology, MAKAU (WBUT), Kolkata, West Bengal, India
e-mail: anindya.kr.bws@gmail.com

S. Mukhopadhyay · G.P. Biswas
Department of Computer Science and Engineering, Indian School of Mines, Dhanbad, Jharkhand, India
e-mail: msushanta2001@gmail.com

G.P. Biswas
e-mail: gpbiswas@gmail.com

© Springer Nature Singapore Pte Ltd. 2017
S.C. Satapathy et al. (eds.), *Proceedings of the 5th International Conference on Frontiers in Intelligent Computing: Theory and Applications*, Advances in Intelligent Systems and Computing 516, DOI 10.1007/978-981-10-3156-4_46

reconstruction of the input image through human vision system. It does not involve any cryptographic knowledge and computation as required in any other security design and applications. Its implementation is easy, however the security level is perfect as the scheme is not breakable even with brute-force attack and infinite computation without having adequate shares. In 1995, Naor and Shamir introduced VCS in a seminal paper [1], where without through powerful computers, a technique similar to the security checking by human beings is proposed. It is known as *k-out-of-n* VCS and the secret image is reconstructed if at least any *k* shares out of $n \geq k$, are combined together.

In general, three major approaches for concealing image and/or data are followed and they are *encryption, information hiding* and *secret sharing*. In encryption, the messages are encrypted using known symmetric or public-key methods, where the corresponding private keys are kept secret [2, 3]. In information hiding, the watermarking techniques are used, in which the secret data are hidden into cover images such that the visibility of latter ones are not disturbed [4]. In secret sharing, Naor and Shamir proposed a cryptography scheme for sharing of secret among any of $k \leq n$ *for n* participants [1]. In recovery, any *k* or more can combine their shares to reconstruct the secret whereas any $k - 1$ or fewer get no information. The basic VCS model has been extended by different researchers including Ateniese et al. [5], Tzeng and Hu [6] and so on. In [5], authors have analyzed the structure of VCS and provided an efficient technique for realizing different VCSs. It is shown that this construction is better with the pixel expansion than the one proposed in [1]. Also the graph based access structure for VCS is described in [5], in which a qualified set contains an edge of a given graph corresponding to the participant of the scheme.

The VCS presented in [6] has in some cases better pixel expansion than previous results. Also the authors proposed an improved VCS based on the fact that human visual system cares about the contrast only and thus, it is not necessary that the recovered image must be always darker than background as in those works followed by Naor and Shamir's VCS model. There are other works on VCS that contribute to the improvement of the contrast of the recovered image. Hofmeister et al. in [7] proposed a VCS for *k-out-of-n* threshold access structure that achieves the best contrast by solving a simple linear problem. Droste proposed a very general method to construct an extended VCS for an arbitrary access structure, however, the scheme is not necessarily monotonic [8]. The access structure of negative images is developed by Kim et al. [9]. The VCSs for color images are proposed in [10–12]. The current work describes and analyzes three existing VCS models and explains their tangible features. In addition, a new design paradigm for $k = 2$ and 3 VCS is proposed that generates and implements participants' shares using combination rule $C(m, w)$, where *w* is number of 1 *s* in a share having size *m* known as pixel expansion value.

The rest of the paper is organized as follows. Three important VCSs including their models are presented in Sect. 2. In Sect. 3, the proposed VCS designs and their implementation are described. A simulation result carried out on a sample binary image is given in this section. Finally, the concluding remarks are given in Sect. 4.

2 Preliminaries of Three VCS Models for Binary Images

There are several important VCS models available in the literature, however, three of them including the scheme proposed by Naor et al. [1] are presented in this section.

2.1 VCS Model Proposed by Naor and Shamir [1]

In [1], an input binary image is encrypted using VCS and distributed among a set of $n > 1$ participants $P = \{1, 2 \ldots n\}$. For this, two basis matrices of order $n \times m$ with 0 and 1 pixels are taken and each input pixel after VCS encryption appears in n shares as a collection of m white and black pixels. A *k-out-of-n* VCS consists of two collections of $n \times m$ Boolean matrices (C_i for $i = 1$ or 2) generated from basis matrices and one from each is chosen in generating the shares. In VCS, the following three conditions are to be satisfied, where $H(V)$ is the Hamming-weight of m-vector V obtained by logical *OR*-ing of k rows of a matrix selected from C_i:

1. For $S_0 \in C_0$, any k of n rows satisfies $H(V) \leq d - \beta$, where $1 \leq d \leq m$ and $\beta = \alpha(m) \times m$, called image contrast ($\alpha(m)$ is the relative difference and $\alpha(m) = (H(V) \text{ for } S_1 - H(V) \text{ for } S_0)/m)$.
2. For $S_1 \in C_1$, any k of n rows satisfies $H(V) \geq d$.
3. For any subset $\{i_1, i_2 \ldots i_q\}$ with $q < k$, two collections of $q \times m$ matrices obtained by limiting each matrix to rows $i_1, i_2 \ldots i_q$ are indistinguishable.

The conditions 1 and 2 are called *image contrast* while the condition 3 defines *VCS security*. In fact, no one by stacking fewer than k shares and even with endless computation get any benefit in deciding whether the shared pixel was black or white.

2.2 VCS Model Proposed by Ateniese et al. [5]

An optimal VCS model proposed in [5] is briefly described. For given qualified (Γ_{Qual}) and forbidden (Γ_{Forb}) sets, a ($\Gamma_{Qual}, \Gamma_{Forb}, m$)-VCS with a set of threshold $\{(X, t_X)\}_{X \in \Gamma_{Qual}}$ is realized using two $n \times m$ basis matrices B_0 and B_1 if two conditions given below are satisfied:

1. *Construction condition:* Any subset $X = \{i_1, i_2, \ldots, i_p\} \in \Gamma_{Qual}$ can recover the secret image by combining their shares. That is, for any matrix $M \in C_0, H(V)$ of $i_1, i_2, \ldots, i_p$ satisfies $H(V) \leq t_X - \beta$, whereas for $M \in C_1, H(V) \geq t_X$, where t_X is the threshold used to identify the reconstructed pixels as black or white.

2. *Security condition:* Any subset $X = \{i_1, i_2, \ldots, i_p\} \in \Gamma_{Forb}$ has no information of secret. That is, two matrices of size $p \times m$ obtained by limiting each $n \times m$ matrix, are indistinguishable.

This model incorporates and satisfies all the characteristics defined in VCS proposed in [1], and provides a systematic formal presentation. It's a generalization of VCS as each set $X \in \Gamma_{Qual}$ is possibly associated with a different threshold t_X. Also the access structure of [5] is not as strong as presented in [1].

2.3 VCS Model Proposed by Tzeng and Hu [6]

An efficient VCS with different features is proposed in [6], where a new access structure for VCS is provided. An access structure $\Gamma = (P, Q, F)$ with $Q \cap F = \varnothing$ is presented in [6], where Q and F are respectively qualified and forbidden sets of participants P. Here Q is monotonically increasing if $X \in Q$ implies for all $X' \supseteq X$, $X' \in Q$ and F is monotonically decreasing if $X \in F$ implies for all $X' \subseteq X$, $X' \in F$. Two collections C_0 and C_1 of $n \times m$ Boolean matrices constitute a visual cryptography scheme if there exist value $\alpha(m) > 0$ and a set $\{(X, t_X)\}$ $X \in Q$ satisfying:

1. Any set $X = \{i_1, i_2 \ldots i_q\} \in Q$ can recover the secret image by combining their shares. That is, for any $M \in C_0$, $w(M, X) = t_X$ and any $M' \in C_1$, $w(M', X) \geq t_X + \alpha(m) \times m$ or $M' \in C_1$, $w(M', X) \leq t_X - \alpha(m) \times m$, where $w(M, X) = H(V)$.
2. Any set $X = \{i_1, i_2 \ldots i_q\} \in F$ contains no information of secret. That is, two matrices of size $q \times m$ obtained by limiting each $n \times m$ matrix such that

   (a) If X is not a qualified set in Q, are indistinguishable.
   (b) If X is a qualified set in Q, the collections obtained by *OR*-ing all rows of each $q \times m$ matrix are indistinguishable.

Here, the scheme changes image contrast as the revealed images become darker or lighter than the background if *condition*-1 is satisfied. Note that the revealed images in [5] are always darker than background. VCS model in [6] is non-monotonic in nature.

3 Proposed VCS Model

A new VCS design for hiding image visibility is proposed. It considers $C(m, w)$ combinations to generate binary strings of length m with w number of 1's for $1 \leq w < m$, and uses them to represent participants' shares. These strings are suitable as each share in VCS is encoded using an m-length equal weight binary string. Two basis matrices used for designing *2-out-of*-3 VCS in [1] as

$$B_0 = \begin{pmatrix} 1 & 0 & 0 \\ 1 & 0 & 0 \\ 1 & 0 & 0 \end{pmatrix} \quad B_1 = \begin{pmatrix} 1 & 0 & 0 \\ 0 & 1 & 0 \\ 0 & 0 & 0 \end{pmatrix}$$

Three shares are generated for three participants, where each input pixel is expanded by three subpixels ($m = 3$). Since each row of the matrices has weight $w = 1$, the basis matrices can be simply generated using $C(3, 1)$. For smaller n, a generalization for *2-out-of-n* VCS is proposed, where the following basis matrices are used [1]:

$$B_0 = \begin{pmatrix} 100 & \cdots & 0 \\ 100 & & 0 \\ \vdots & \ddots & \vdots \\ 100 & \cdots & 0 \end{pmatrix} \quad B_1 = \begin{pmatrix} 100 & \cdots & 0 \\ 010 & & 0 \\ \vdots & \ddots & \vdots \\ 000 & \cdots & 1 \end{pmatrix}$$

The design of *2-out-of-n* VCS in [1] is not efficient for two reasons—(i) a unique encryption pattern (using binary string of weight 1) for generating shares is followed and (ii) size of basis matrices become large for large n. This paper proposes an efficient design for *2-out-of-n* VCS, which is described below in Sect. 3.1.

3.1 Design of 2-out-of-n Threshold VCS

The design of *2-out-of-n* VCS in [1] that uses n-length binary strings of $w = 1$, can be generalized as

$$B_0 = (n \, repetitions \, of \, a \, string \in C(n, 1)) B_1 = (all \, strings \in C(n, 1))$$

Since $C(n, 1) = n$, this scheme has a limitation that the number of participants can never be increased more than n. However, if w is increased ($n/2 \geq w > 1$), the value of $C(n, w)$ becomes greater than n. Thus, the following three improvements over the scheme in [1] can be made if $C(m, w)$ is used:

1. Increase number of participants without increasing the size of basis matrices
2. Support multiple implementations for a given threshold VCS (m remains smaller, enhances image contrast)
3. Contrast of the revealed images can be increased (β increases for some of the qualified subsets)

For instance, the design of *2-out-of-4* VCS in addition to using $C(4, 1) = 4$, can also be done by using $C(4, 2) = 6$ and thus, six participants instead of four could be involved without increasing matrix-size. After reconstruction, the image contrast becomes $\beta = 1$ or 2 whereas $\beta = 1$ in [1]. The proposed VCS for 2-*out-of-n* VCS can be generalized as

1. Adjust pixel expansion parameter m and string-weight w such that the number of combinations in $C(m, w)$ is equal to n, i.e. $C(m, w) = n$, where $1 \leq w < m$.
2. Out of multiple designs, select one having lesser m and higher β values.

An alternative design for 2-*out-of*-3 VCS can be done by considering $m = 3$ and $w = 2$, where the following basis matrices are taken:

$$B_0 = (3 \, repetitions \, of \, a \, string \in C(3,2))B_1 = (all \, strings \in C(3,2))$$

That is, $B_0 = \begin{pmatrix} 1 & 1 & 0 \\ 1 & 1 & 0 \\ 1 & 1 & 0 \end{pmatrix} \quad B_1 = \begin{pmatrix} 1 & 1 & 0 \\ 0 & 1 & 1 \\ 1 & 0 & 1 \end{pmatrix}$

Its simulation results are shown in Fig. 1.

For $m = 4$ and $w = 2$, we have $C(4, 2) = 6$ combinations as 1100, 0110, 0011, 1001, 1010 and *0101*. One way of designing the basis matrices is

$$B_0 = \begin{pmatrix} 1 & 1 & 0 & 0 \\ 1 & 1 & 0 & 0 \\ 1 & 1 & 0 & 0 \\ 1 & 1 & 0 & 0 \\ 1 & 1 & 0 & 0 \\ 1 & 1 & 0 & 0 \end{pmatrix} \quad B_1 = \begin{pmatrix} 1 & 1 & 0 & 0 \\ 0 & 1 & 1 & 0 \\ 0 & 0 & 1 & 1 \\ 0 & 1 & 0 & 1 \\ 1 & 0 & 1 & 0 \\ 1 & 0 & 0 & 1 \end{pmatrix}$$

Or, $B_0 = (6 \, repetitions \, of \, a \, string \in C(4,2)) \, B_1 = (all \, 6 \, strings \in C(4,2))$

Here, $t_X = 3$ or 4 for $X \in Q$, relative difference and contrast are $\alpha(m) = 1/4$ or $1/2$ and $\beta = 1$ or 2, respectively. Thus, the qualified subset $\{1, 3\}$ with $t_{\{1, 3\}} = 4$ reconstructs images with better contrast than the subset $\{1, 2\}$ having $t_{\{1, 3\}} = 3$. Also our scheme requires 6×4 basis matrices than 6×6 as in [1]. Although the design has similarity to the scheme in [5], the approach is different.

Our scheme for 2-*out-of-n* VCS is better than the scheme presented in [1] as the values of n and m could be reduced. Also it has been found that the proposed

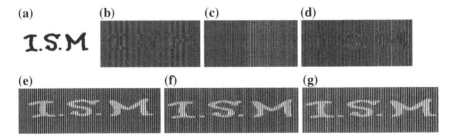

Fig. 1 Results for 2-*out-of-3* VCS **a** Input image **b** Share s_1 **c** Share s_2 **d** Share s_3 **e** Combination of s_1 and s_2 **f** Combination of s_2 and s_3 **g** Combination of s_3 and s_1

Table 1 Comparison of proposed *2-out-of-n* threshold VCS with [1]

m	Values of r	Number of participants, $n = \binom{m}{r}$	Basis matrix-size in proposed scheme	Basis matrix-size in [1]	Improvement in bits over [1]
4	1, 3	4	4 × 4	4 × 4	Nil
	2	5–6	5 × 4 – 6 × 4	5 × 5 – 6 × 6	10–24
5	1, 4	5	5 × 5	5 × 5	Nil
	2, 3	6–10	6 × 5 – 10 × 5	6 × 6 – 10 × 10	12–100
6	1, 5	6	6 × 6	6 × 6	Nil
	2, 4	7–15	7 × 6 – 15 × 6	7 × 7 – 15 × 15	14–270
	3	7–20	7 × 6 – 20 × 6	7 × 6 – 20 × 20	14–560
...	...	...	...	...	...
10	1, 9	10	10 × 10	10 × 10	Nil
	2, 8	11–45	11 × 10 – 45 × 10	11 × 11 – 45 × 45	22–3150
	3, 7	11–120	11 × 10 – 120 × 10	11 × 11 – 120 × 120	22–26400
	4, 6	11–210	11 × 10 – 210 × 10	11 × 11 – 210 × 210	22–84000
	5	11–252	11 × 10 – 252 × 10	11 × 11 – 252 × 252	22–121968

scheme requires $2n(n\text{-}m)$ bits lesser than in [1]. A comparison of the proposed *2-out-of-n* VCS with the scheme presented in [1] for $m = 4, 5, 6$ *and* 10 is given in Table 1, where $1 \leq w \leq m\text{-}1$.

3.2 Design **3-out-of-n** *Threshold VCS*

In this section, VCS design for *3-out-of-n* using $C(m, w)$ is discussed and compared with Naor and Shamir's scheme [1]. The design in [1] is briefly presented. Consider two binary matrices C and I of order $n \times (n\text{-}2)$ and $n \times n$ respectively, where C contains all 1's and I is an identity matrix. Their concatenation $B\|I$ becomes a matrix of order $n \times 2n - 2$. The basis matrices for *3-out-of-n* are $B_0 = \left(\overline{C\|I} \right)$ and $B_1 = (C\|I)$, where $\left(\overline{C\|I} \right)$ is the complementation of $C\|I$. For illustration, the basis matrices of *3-out-of-4* VCS are as follows:

$$B_0 = \begin{pmatrix} 0 & 0 & 0 & 1 & 1 & 1 \\ 0 & 0 & 1 & 0 & 1 & 1 \\ 0 & 0 & 1 & 1 & 0 & 1 \\ 0 & 0 & 1 & 1 & 1 & 0 \end{pmatrix} \quad B_1 = \begin{pmatrix} 1 & 1 & 1 & 0 & 0 & 0 \\ 1 & 1 & 0 & 1 & 0 & 0 \\ 1 & 1 & 0 & 0 & 1 & 0 \\ 1 & 1 & 0 & 0 & 0 & 1 \end{pmatrix}$$

Note that the size of the basis matrices for *3-out-of-n* VCS is increased rapidly, which would be infeasible for practical applications. Our design procedure based on $C(n, w)$ is better than it. Before presenting the proposed scheme, an illustration is given. The design of 3-*out-of*-3 VCS could be done using $(0)\|C(3, 2)$ and $C(4, 2)$ for B_0 and B_1 respectively, where (0) is a zero column matrix i.e.,

$$B_0 = \begin{pmatrix} 0 \\ 0 \\ 0 \end{pmatrix} \|C(3,2) = \begin{pmatrix} 0 & 1 & 1 & 0 \\ 0 & 0 & 1 & 1 \\ 0 & 1 & 0 & 1 \end{pmatrix} \quad B_1 = C(4,2) = \begin{pmatrix} 1 & 0 & 1 & 0 \\ 1 & 0 & 0 & 1 \\ 1 & 1 & 0 & 0 \end{pmatrix}$$

It is identical to the design presented in [1]. Although similar design for any 3-*out-of-n* ($n > 3$) using $C(n, w)$ exist, it supports the access structure (not strong) presented in [5]. For instance, the design of 3-*out-of-4* VCS using $(0)\| C(4, 3)$ and $C(5, 3)$ basis matrices B_0 and B_1 respectively, would be feasible. Here B_0 is fixed, however, B_1 has multiple options as out of $C(5, 3) = 10$, four are taken. A simple design procedure for B_1 is—(i) one of five columns contains all $1s$ and the remaining four columns must contain exactly two $0s$. The design is shown below:

$$B_0 = \begin{pmatrix} 0 & 1 & 1 & 1 & 0 \\ 0 & 0 & 1 & 1 & 1 \\ 0 & 1 & 0 & 1 & 1 \\ 0 & 1 & 1 & 0 & 1 \end{pmatrix} \quad B_1 = \begin{pmatrix} 1 & 1 & 1 & 0 & 0 \\ 1 & 1 & 0 & 0 & 1 \\ 1 & 0 & 1 & 1 & 0 \\ 1 & 0 & 0 & 1 & 1 \end{pmatrix}$$

Here, each of the basis matrices has size 4×5, which requires lesser pixel expansion ($m = 5$) than 4×6 ($m = 6$) as in [1]. The qualified set is $Q = \{\{3$-*out-of*-4$\}, \{1, 4\}, \{2, 3\}\}$. Some other designs for 3-*out-of-n* and comparison with [1] are shown in Table 2.

Table 2 Comparison of proposed *3-out-of-n* threshold VCS with [1]

3-out-of-n VCS	Proposed scheme			Matrix size in [1]	Improvement in bits
	B_0Matrix	B_1Matrix	Matrix Size		
3-out-of-3	$(0) \| C(3, 2)$	$C(4, 2)$	3×4	3×4	Nil
3-out-of-4	$(0) \| C(4, 3)$	$C(5, 3)$	4×5	4×6	4
3-out-of-5	$(0) \| C(5, 4)$	$C(6, 4)$	5×6	5×8	10
3-out-of-6	$(0) \| C(6, 5)$	$C(7, 5)$	6×7	6×10	18
...	...	...	...	...	...
3-out-of-10	$(0) \| C(10, 9)$	$C(11, 9)$	10×11	10×18	70

4 Conclusions

A new VCS design based on *C(m, w)* framework for $k = 2, 3$ is presented. Our scheme for *2-out-of-n* is far better than the VCS presented by Naor and Shamir, however, *3-out-of-n* VCS although better than [1], supports access structure similar to one presented in [5]. A comparative study with [1] and a sample simulation are provided that are found to be satisfactory.

References

1. M. Naor, A. Shamir, Visual cryptography, in *Proceedings of Eurocrypt 94*, LNCS 950 (Springer, 1994), pp. 1–12
2. J. Daemen, V. Rijmen, Rijndael: the advanced encryption standard. Fr. Dobb's J. (2001)
3. M. Bellare, A. Boldyreva, S. Micali, Public-key encryption in multiuser setting: security proofs and improvements, in *Advances in Cryptology-Eurocrypt-2000* (2000)
4. Cox et al., *Digital Water Marking* (Morgan Kaufmann Publishers, 2000)
5. G. Ateniese, C. Blundo, A. De Santis, D.R. Stinson, Visual cryptography for general access structure. J. Inf. Comput. **129**(2), 86–106 (1996)
6. W.G. Tzeng, C.M. Hu, A new approach for visual cryptography. J Des. Codes Crypt. **27**, 707–727 (2002)
7. Hofmeister et al., Contrast-Optimal *k-out-of-n* Secret Sharing Schemes in Visual cryptograph COCOON 97, LNCS, vol. 1276 (1997)
8. S. Droste, New results on visual cryptography, in *Proceedings of Advances in Cryptology-CRYPTO 96*, LNCS, vol. 1109 (1996), pp. 401–415
9. K. Kim et al., Human-machine identification using visual cryptography, in *Proceedings of IEEE International Workshop on Intelligent Signal Processing and Communication Systems* (1998), pp. 178–182
10. D. Naccache, Colorful cryptography—a purely physical secret-sharing scheme based on chromatic filters, in *Coding and Information Integrity, French-Israeli Workshop* (1994)
11. V. Rijmen, B. Preneel, Efficient colour visual encryption or shared colors of Benetton, in *UROCRYPT 96* (1996)
12. Y.C. Hou, Visual cryptography for color images. Pattern Recognit. **36**, 1619–1629 (2003)

A Secured Digital Signature Using Conjugacy and DLP on Non-commutative Group over Finite Field

L. Narendra Mohan and G.S.G.N. Anjaneyulu

Abstract In the present paper, we propose a secured scheme of digital signature connecting both conjugacy problem and discrete logarithm problem based on non-commutative group generated over a finite field. For this, we define a non-commutative group over matrices with the elements of finite field such that conjugacy and discrete logarithm problems can be executed together proficiently. By doing so, we can formulate the signature structures using conjugacy and discrete logarithm through non commutative group. In some domains, the above combination reduces to completely in discrete logarithm problem. This digital signature scheme more elemental over $F^{*}q(x) = G L_n (F_q)$. Here the security of the signature protocol depending on complexity of the problems associated with conjugacy and discrete logarithm. The security analysis and intermission of proposed protocol of digital signature is presented with the aid of order of complexity, existential forgery and signature repudiation.

Keywords Digital signature · Public key cryptography · Conjugacy problem · Discrete logarithm problem · Quasideterminant and non commutative ring

1 Introduction

A digital signature scheme is a scientific approach for demonstrating the legitimacy of an advanced message or record. A substantial digital signature delivers for a beneficiary reason to have confidence that those message might have been made by an known sender, such and such the sender cannot rebuff hosting sent those message (authentication and non-repudiation). Also note that those messages might have been not modified, when the message being transmitted(integrity). In the

L. Narendra Mohan · G.S.G.N. Anjaneyulu (✉)
Department of Mathematics, SAS, VIT University, Vellore 632014, Tamil Nadu, India
e-mail: anjaneyulu.gsgn@vit.ac.in

L. Narendra Mohan
e-mail: mohannarendra3@gmail.com

© Springer Nature Singapore Pte Ltd. 2017
S.C. Satapathy et al. (eds.), *Proceedings of the 5th International Conference on Frontiers in Intelligent Computing: Theory and Applications*, Advances in Intelligent Systems and Computing 516, DOI 10.1007/978-981-10-3156-4_47

present scenario, Digital signatures are essential and regularly utilized to pro-gramming distribution, economic transactions, what's more to other situations the place, where it is imperative should recognize falsification or damaging. Diffie and Hellman principally presented the basic notion about advanced signature scheme along with the first introduced the concept of Digital signature scheme using cryptography. This piece of information of a "digital signature" at first disclosed in Diffie and Hellman's inspiring paper, "New Directions in Cryptography" [1]. They suggested that each client must distribute an "open key"(used for validating/Confirming signatures), and at the same time keeping mystery "secret key" (used for producing signatures). At the inception, in their protocol, Entity A's signature for a message M is a worth which relies on upon M and on A's private key, so that any individual may be ensured those one can check the legitimacy of A's signature with A's open key. Be that as it may, in the meantime knowing A's open key is important to support and one to authenticate A's signatures, it doesn't concede one to just distort A's signatures.

In this proposed article, we make some components over non-commutative group/ring/field. Here we utilize digital signature scheme in cyclic group F_q^*, the place F_q will be the finite field for q components. The strength and security of this scheme in light of the challenge depending on the strain of calculating discrete logarithms in the cyclic group F_q^*. Here discrete logarithm issue is a hard-hitting experience. Let $F_q^* = GL_1(F_q)$, particular case could wonder if the group $GL_2(F_q)$ for two-by-two invertible matrices or all the more for the most part of the group $GL_n(F_q)$, which conceals natural typical form, might be a chance to utilize in Digital Signature and also there may be some preference previously, utilizing them.

Let us detect a non-singular matrix $X \in GL_n(F_q)$. Realizing X and a power X^a, it is very hard to compute a. The initial side of the point will be that having X, one can calculate $\det(X) \in F_q^*$ (the determinant of X), furthermore $\det(X^a) = (\det(X))^a$. In this way, such DL problem in matrix groups tempers to the DL problem done in F_q^*. Person could halt away from this struggle by picking a matrix X such that $\det(X)$ as unity, at the same time then by calculating characteristic values from X and from X^a and utilizing the reality that those last are those previous in the power of a, par-ticular case diminishes once more those DL problem of the particular case in some development about F_*, So there will be no improvement in view of the DL problem in the pool of non-singular matrices over a finite field, what's new additional for the most part through a finite commutative ring. In this problem, including non-commutative (semi) groups in digital signatures suggested a stage involving braid groups and the same notion was proposed in [2]. Also another platform using matrix algebra was conferred in [3]. It employs conjugation and exponential powers together to its security. A stage to this protocol utilizing braid groups might have been initially proposed in [4] also an alternate one utilizing an F_q-algebra to [5].

We might provide for a digital signature scheme of these two platforms in the area described in the Sect. 2.3, by dropping the issue to the DL problem over certain finite field. Those semi group generated on matrices in a commutative ring might have been acknowledged in [5] to an verification protocol, be that its security is in light of the strain of the conjugacy search problem, but not depending on the discrete logarithm tool. Certainly, the creators consider about matrices through some way or another convoluted ring, to be specific those ring of polynomials for k variables would create the conjugacy search problem infeasible.

1.1 Phases of Digital Signature Protocol

Another fundamental public key cryptographic scheme is a Digital signature, whose concept was first designed by Diffie–Hellman (DH76-17). The ability to build a digital signature scheme is a great lead of public cryptography over symmetric key cryptography. The present article is based on digital signature. So we discuss different primal stages of digital signature scheme. A digital signature algorithm can be designated as follows. A digital signature is an authentication mechanism that facilitates the creator to attach a code that acts as a Digital signature. The signature is molded by taking the hash of the message and encrypting the message with the creators' private key. The Digital signature guarantees that the source and integrity of the message [6, 7].

A **Digital signature** is a structure having five components (F, C, I, L, O), whether the following conditions are satisfied.

1. F is a finite set of manageable keys.
2. C is a finite set of manageable signatures.
3. I, the key space and is a finite set of manageable keys.
4. For each $k \in I$, there is a signing algorithm $\text{sig}_k \in L$ and an associated verification algorithm $\text{Ver}_k \in O$. Each $\text{Sig}_k: F \to C$ and $\text{Ver}_k : FXC$ {True, False} are functions such that the accompanying condition is fulfilled for each message $x \in F$ and for each signature $y \in C$. Ver (x, y) = {True if y = sig(x) and False if $y \neq \text{sig}(x)$}

For every $k \in I$, the functions Sig_k and Ver_k should be polynomial time functions. Ver_k will be a public function and Sig_k will be a secret. Thereafter, a signature must be not forgeable. This means that, it must be infeasible to calculate a signature of a message, with respect to a public key without the information of the corresponding secret key [8–12].

2 Computational Primitives for Proposed Signature Scheme

2.1 Discrete Logarithm Problem

Most widely used computational problem in security protocols is discrete logarithm problem. More details can be seen [13, 14].

Definition If G be a finite cyclic group of order n and g be a generator of G and y $\in$ G then discrete logarithm of y to the base h, denoted by $\log_h^y$ is the unique integer x,

0 $\leq$ x $\leq$ n-1, such that y = g ^{x}. This is called as DLP [15, 16].

2.2 *Conjugacy Problem*

The Conjugacy problem in non-commutative group G contains two components u, v in G are mutually conjugate each other, composed u $\sim$ v where v = a^{-1}u a for some element a $\in$ G. Here the elements a or its inverse is known as a conjugator and the couple (u, v) is understood to be conjugate. More details can be seen [13, 14]. Obviously '$\sim$' is an equivalence relation. The conjugacy relation has the following properties

  (i)   u = e^{-1}u e for e $\in$ G and u $\in$ G (Reflexive).
  (ii) v = a^{-1}u a $\rightarrow$ u = $(a^{-1})^{-1}$v a $\rightarrow$ u $\sim$ v (Symmetric).
  (iii) v = a^{-1}u a, w = b^{-1}u b then w = $(ab)^{-1}$ u ab $\rightarrow$ w $\sim$ u (Transitive).

So that conjugacy relation is clearly an equivalence relation and the following are the types of conjugacy problems, which have been used in cryptography [17, 18].

  (a) **The conjugacy decision problem (CDP):** The CDP inquire to find out, whether u, v are used for a particular occasion (u, v) $\in$ G x G
  (b) **The conjugator search problem (CSP):** The CSP solicits to discover a $\in$ G, satisfying v = a^{-1}u a for a specified case (u, v) $\in$ G x G such that u $\sim$ v.

2.3 *Matrices—Quasideterminents*

Let L be a square matrix of order n, with elements of a non commutative ring R, we reminder L^{ij} the matrix attain from square matrix L by removing the ith row and the jth column. We like wise reminder by the ith row of matrix with jth place excepted, and with the jth column of L by the ith place excepted. Here every position (i, j), the quasideterminant [19] of d is defined by $|L|_{ij} : = d_{ij} - r_i^j(L_{ij})^{-1}c_i^j$ We have $|L|_{ij} \in$ R and, obviously, this quasideterminant occurs if the (n $-$ 1)-by-(n $-$ 1) matrix L_{ij} is invertible. Thus, for a matrix of order n, there will be a n^2 quasideterminants.

Example L = $\begin{bmatrix} c_{11} & c_{12} \\ c_{21} & c_{22} \end{bmatrix}$ for n = 2

$|L|_{11} = c_{11} - c_{12}c_{22}^{-1} c_{21}, |L|_{12} = c_{12} - c_{11}c_{21}^{-1} c_{22},$
$|L|_{21} = c_{21} - c_{22}c_{12}^{-1} c_{11}, |L|_{22} = c_{22} - c_{21}c_{11}^{-1}c_{12}.$

Remark Where in the abelian case, a quasideterminant is not equivalent to a det, but rather additionally the proportion of two dets, to be specific,

$$|L|_{ij} = (-1)^{i+j} \frac{\det(D)}{\det(D_{ij})}$$

Here with signature perspective, we just need to ensure that there is no scheme decreasing the DL problem in the group of matrices with non-commutative elements to the Discrete Logarithm problem in the ring of coefficients. More details can be seen [20].

3 Proposed Digital Signature

Here we elucidate a digital signature scheme based on group of matrices over a finite field. This digital signature scheme has the following stages.

3.1 Initial Setup

We select a nonsingular matrix $X \in G L_n(F_q)$, and calculate power X^a. The Next step $\det(X) \in F_q^*$ (the determinant of X), furthermore $\det(X^a) = (\det(X))^a$. Here Alice needs to produce a signature for a message M. At the last step, the recipient Bob should confirm the signature, which is valid and also demonstrate the legitimacy to reveal the message [21].

Construct a square matrix L of order n, with elements over some finite field F with non commutative property. We note L^{ij} is the sub matrix derived commencing L by neglecting those ith row and the jth column. We furthermore reminder through the ith row of L with jth position excepted, and through the jth column of L with the ith position excepted [22].

3.2 Key Generation

We now combine the discrete logarithm problem and the conjugacy search problem collectively to generate the public key. So that extracting the private key from public key is not feasible as the discrete logarithm problem and the conjugacy search problem are intractable over the fundamental work structure. Suppose G may be non-commutative group generated over field elements of F and K_1, K_2 are two subgroups of G such that each component of K_1 commutes with each component K_2. We define G, K_1, K_2, also a component $X \in G$ of various higher powers of n, would be public information [23].

A picks on arbitrary a secret integer $a \in \{2, 3, \ldots, n - 1\}$ and a secret matrix $U \in K_1$ ($UX \neq XU$), $V \in K_2$; She computes a public key $\alpha = U^{-1}X^aU$.

Let us suppose that $K = K_1 = K_2 = \left\{ \begin{bmatrix} x & y \\ y & x \end{bmatrix} \in GL_2 (F_q), \ x, \ y \ \in \ F_q, \ x^2 - y^2 \neq 0 \right\}$,

which is commutative subgroup of $GL_2(F_q)$.

3.3   Signature Generation

Let a group $G = GL_2 (F_q)$ be the fundamental work infrastructure and $X \in GL_2(F_q)$ of order n, Alice performs the following simultaneously. Alice selects a secret integer 'a' $\in \{2, 3, \ldots, n - 1\}$ as her private key and a secret matrix $U \in K$ and $\alpha = U^{-1}X^aU$ as her public key. Alice calculates as follows

$\beta = V^{-1}X^aU$ and also computes
$\gamma = (V^{-1}X^aV)$
$\delta = U^{-1}X^aV$

Here Alice forms a code (β, γ, δ, M) as her signature and send it to Bob as her signature on the message M for confirmation and acceptance.

3.4   Verification

After receiving the signature (β, γ, δ, M) from Alice and to verify this signature as authenticated, Bob will do as follows.

First he computes $\theta = \alpha \beta^{-1}$. If $\delta = \theta\gamma$ then he accepts the signature, otherwise it will be rejected automatically.

3.5   Confirmation Theorem

Theorem If $\theta = \alpha \beta^{-1}$ then $\delta = \theta\gamma$
Completeness: A's signature is (β, γ, δ, M). If Alice monitors signature verification, and then Bob always approves it, as an authenticated signature and the message.

Proof We know that $\alpha = U^{-1}X^aU$ is Alice's public key and Bob accepts the signature

(β, γ, δ, M) and are parameters of the signature
Then we prove the signature as follows
Since $\theta = \alpha \beta^{-1}$

$\beta = V^{-1}X^aU$
$\gamma = (V^{-1}X^aV)$ and $\delta = U^{-1}X^aV$
$\theta = U^{-1}X^aUU^{-1}X^{-a}V = U^{-1}X^aX^{-a}V$
$\theta = U^{-1}V$

Then conformation will be done as follows in final

$$\theta\gamma = U^{-1}VV^{-1}X^aV$$
$$= U^{-1}X^aV = \delta.$$

Here Bob computes the message, when Alice follows his algorithm. Then B accepts signature algorithm.

Hence completes the proof.

4 Security Analysis

We explain the security for the digital signature scheme of matrices over non commutative group over a finite field is given below. Here n be the index of X and $\alpha = U^{-1}X^a U$, we have to calculate the secret key '**a**' and exchanged key $U^{-1}V^{-1}X^{ab}UV$ is difficult [24–27]

(i) **Existential Forgery**: No hash function will be utilized within the elgamal signature system, at that point the existential falsification may believable under direct attack. Without hash function, existential forgery is not applicable in this signature.

(ii) **Signature Repudiation**: Assume Alice intentionally refuse her signature on some valid information. Substantial signature might be changed by eve and furthermore she could sign the message M, with the forged signature $(\beta_e, \gamma_d, \delta_e, M_d)$ instead of original. That moment conformation technique and verification will be failed as

$$\theta_e\gamma_e = (U^{-1}V)_e(V^{-1}X^aV)_e \neq \delta.$$

Then this identifies non repudiation property.

(iii) **Total Break**: The security of the private key in this digital signature is additional, as we develop the secret signature key on non-commutative structure, where the problem is not tractable. These days' signatures need aid of utilizing this property. Here the problem is dependent upon unmanageability from claiming conjugacy and discrete logarithm problem.

(iv) **Selective Forgery**: Even an assailant has the ability to make a signature by selecting identified with specific message M, this contradicts with the equation $\delta_e \neq \theta_e\gamma_d$. The Making the signatures like this is not beneficial for the attacker under the selective forgery.

5 Conclusions

In this article, we effectively utilized conjugacy problem and discrete logarithm together to enhance security of signature scheme. We have designed digital signature scheme over non-commutative matrix group with elements of field. The key thought behind our plan is that we constructed non-commutative group of matrices over the field. We make them as the underlying field structure for constructing signature. We demonstrated the strength and soundness for signature scheme by proving confirmation theorem. We enlightened the security analysis of the signature scheme by proving, it is secure against data forgery, signature repudiation and existential forgery. This is secure against total break as public key and private keys are connected with conjugacy problem and discrete logarithm.

References

1. W. Diffie, M. Hellman, New directions in cryptography. IEEE Trans. Inf. Theory **IT-2.2**(6), 644–654 (1976)
2. J. Kang, J.W. Han, J.H. Cheon, S.J. Lee, K.H. Ko, S.J. Lee, C. Park, New public key crypto system using braid groups. Lect. Notes Comput. Sci. **1880**, 166–183 (2000)
3. D. Moon, K.C. Ha, S. Cho, Y.-O. Kim, Key exchange protocol using matrix algebras and its analysis. J. Korean Math. Soc.
4. A. Raulynaitis, E. Sakalauskas, P. Tvarijonas, key agreement protocol using conjugacy and discrete logarithm problems in group representation level. Informatica **18**(1), 115–124 (2007)
5. Los Alamitos, Network and System Security. IEEE Computer Society, CA, USA (2009), pp. 443–446
6. Text Book: Cryptography and network security by MR. AtulKahate
7. http://www.iacr.org (International Association for Cryptographic Research—website)
8. D. Poulakis, A Variant of Digital Signature Algorithm Designs, Codes and Cryptography, vol. 51(1) (2009), pp. 99–104
9. N.M.F. Tahat, E.S. Ismail, R.R. Ahmad, A new digital signature scheme based on factoring and Discrete logarithms. J. Math. Stat. **4**(4), 222–225 (2008)
10. C.Y. Yang, M.S. Hwang, S.F. Tzeng, A new digital signature scheme based on factoring and discrete logarithm. IJCM **81**(1), 9–14 (2004)
11. Z. Shao, Security of a new digital signature scheme based on factoring and discrete logarithms. IJCM **82**(10), 1215–1219 (2005)
12. Z. Shao, Signature schemes based on factoring and discrete logarithms, in *Computers and Digital Techniques, IEEE Proceedings-*, vol. 145. IET (2002), pp. 33–36
13. G.S.G.N. Anjaneyulu, P.V. Reddy, U.M. Reddy, Secured digital signature scheme using polynomials over non-commutative division semi ring. IJCSNS **8**(8), (2008)
14. G.S.G.N. Anjaneyulu, U.M. Reddy Secured directed digital signature over non-commutative division semirings and allocation of experimental registration number. IJCSI **9**(5), 3 (2012)
15. A.J. Menezes, Y.-H. Wu, The discrete logarithm problem in G Ln (Fq); ARS Combinatorica **47**, 23–32 (1997)
16. S. Wei, A new digital signature scheme based on factoring and discrete logarithms, in *Progress on Cryptography* (2004), pp. 107–111
17. S. Alam, A. Jamil, A. Saldhi, M. Ahamad, Digital image authentication and encryption using digital signature, in *ICACEA* (2015), pp. 332–336

18. N.A. Moldovyan, D.N. Moldovyan, A new hard problem over non commutative finite groups for cryptographic protocols. Lect. Notes Comput. Sci. **6258**, 183–194 (2010)
19. D. Boneh, A. Joux, P.Q. Nguyen, Why textbook Elgamal and RSA encryption are insecure, in *Lecture Notes in Computer Science,* vol. 1976 (2000), pp. 30–44
20. V. Retakh, S. Gelfand, I. Gelfand, R. Wilson, Quasideterminants. Adv. Math. **193**, 56–141 (2005)
21. M. Eftekhari, A Diffie-Hellman key exchange protocol using matrices over non abelian ring. http://arXiv1209.6144v1[cs.CR] (2012)
22. B. Leclerc, V. Retakh, J.-Y. Thibon, A. Lascoux, D. Krob, I. Gelfand, Non commutative symmetric functions. Adv. Math. **112**(2), 218–348 (1995)
23. V. Shpilrain, D. Grigoriev, Authentication from matrix conjugation, groups, complexity, cryptology, vol. 1, pp. 199–205 (2009)
24. B. Lynn, D. Boneh, H. Shacham, Short signatures from the Weil pairing, in *Proceedings of Asia Crypt 2001*, LNCS, vol. 2248 (Springer, 2001), pp. 533–551
25. D. Pointcheval, T. Okamoto, The gap-problems: a new class of problems for the security of cryptographic schemes, in *Proceedings of PKC 2001*, LNCS, vol. 1992 (Springer, 2001), pp. 104–118
26. A. Lincoln, Electronic signature laws and the need for uniformity in the global market, 8 J. Small & Emerging Bus. L. 67 (2004)
27. T.J. Smedinghoff, R.H. Bro, Moving with change: electronic signature legislation as a vehicle for advancing e-commerce. 17 J. Marshall J. Computer & Info. L. 723, 199

A Novel Edge Based Chaotic Steganography Method Using Neural Network

Shahzad Alam, Tanvir Ahmad and M.N. Doja

Abstract This paper provides a method of hiding sensitive information in digital image. In this paper, we introduce a chaotic edge based steganography techniques based on artificial neural network. First we find the edges of image using artificial neural network which is given by Jinan Gu et al. Secondly, the key based chaotic scheme is used to disperse the bits of the secret message randomly into edge pixel of the image to produce the stego image that take advantage of edge detection techniques. Finally the experiment results show the higher value of PSNR that indicate that there is no difference between the original and stego image. Therefore the proposed algorithms are dependent on the key which make it robust and can protect the secret data from stealing. The experimental results show the satisfactory performance of proposed method based on edge detection techniques.

Keywords Edge detection · Steganography · Chaotic · Stego image · Artificial neural network

1 Introduction

The revolution brought by recent advancements in information hiding techniques to the modern era is immense. The ease of reproduction and editing in digital media has come with two important domain steganography and watermarking. The watermarking is basically for maintaining the authentication of digital media like image, audio and video. The steganography is the method to send the secret

S. Alam (✉) · T. Ahmad · M.N. Doja
Department of Computer Engineering, Faculty of Engineering and Technology,
Jamia Millia Islamia, New Delhi, India
e-mail: shahzad5alam@gmail.com

T. Ahmad
e-mail: tahmad2@jmi.ac.in

M.N. Doja
e-mail: ndoja@yahoo.com

© Springer Nature Singapore Pte Ltd. 2017
S.C. Satapathy et al. (eds.), *Proceedings of the 5th International Conference on Frontiers in Intelligent Computing: Theory and Applications*, Advances in Intelligent Systems and Computing 516, DOI 10.1007/978-981-10-3156-4_48

information in digital media to the intended user without the undetectable of the presence of the secret message. There are various techniques to conceal the message into digital media for secure transmission [1]. These techniques are mainly based on spatial domain and transform domain. In spatial domain techniques methods, a secret message is hidden LSB bits of the pixel [2]. The changes made in LSB of pixel is imperceptible to the human eye thus making it best place for hiding the secret data without any perceptual changes in the cover media [3]. While embedding the secret data in higher bits of lsb leads more visual distortion to the cover media. There are many analytical techniques which reveal existence of the secret data by detecting the histogram, statistical difference of the original and the stego media that contain the secret message. Thus steganography technique may be taken into account two basic measures. First avoid perceptible parts when hiding the data into the cover media. Second embed large amount of information in the cover media. The demand of secure and fast image-based communication has attracted growing attention of researchers worldwide. Edges are basic notable local changes of intensity in an image. Edge detection of image is the simplest task in image analysis. The edges of image are the place where we can hide the message easily without any perceptible distortion. The embedding secret message in the edges of cover image leads high imperceptibility [4, 5]. The edges in a digital image are the point at which brightness changes abruptly or precipitate changes of discontinuities in an image. There are many techniques to find the edges of an image. There are various edge detection technique [6, 7] such as Roberts, Sobel, Prewitt, Robinson, Kirsh, LoG, Marr-Hildreth and Canny edge detection algorithms [8] etc. to find out the edges. The canny algorithm produces good result. Wen-Jan Chen et al. [9] uses canny and fuzzy logic [10] as hybrid to find out the edges of image for steganography. In this paper we used the artificial neural network proposed by Jinan Gu et al. [11] to find the edges of the image and then apply the steganography techniques on it as described it in subsequent section.

Here is the structure of the remaining paper is arranged as follows: Sect. 2 introduce the artificial neural network and proposed method for steganography, the Sect. 3 explain the simulation of embedding, Sect. 4 explains the analysis and conclusions of the work are provided in Sect. 2.

2 Proposed Algorithm

In this section, a novel steganography scheme is presented. The embedding procedure contains two phases. First phase is to training the artificial neural network and second phase is embedding process illustrated in figure. The artificial neural network, 1D chaotic logistic map are implemented in the proposed design methodology.

2.1 Training of Neural Network

In this phase we have taken 3 × 3 size window of image to select the pixels which are served as input to the artificial neural network as shown in Fig. 1a, b. These nine pixel values will input the artificial neural network. The window would move throughout the image to prepare the input dataset.

We used learning sample of edge image produced by canny algorithm [12] as it is high precision than other classic edge detection algorithm so it is opted as the desired learning sample. Again the 3 × 3 fixed size windows is traversed through the edge image, obtained by canny algorithm, of the same image that was initially fed as input. The center pixel of the window is considered as the output for each iterating window.

The learning method used is back-propagation consisting of two parts i.e. forward propagation of information and back-propagation of error. In the forward propagation, the input information is transferred to output layer by layer. If the output is not up to the desired level, the error will be measured. The error will be back-propagated to the previous layer of the ANN. According to the error measured value, the weights and threshold of every layer will be re adjusted until the desired goal is achieved.

2.2 Chaotic 1D Logistic Map

In this paper we used one-dimensional Logistic map proposed by Robert May [13] to produce the random number because it is one of the simplest nonlinear chaotic discrete systems that exhibit chaotic behavior. It has following equation.

$$T_{n+1} = \alpha T_n (1 - T_n) \tag{1}$$

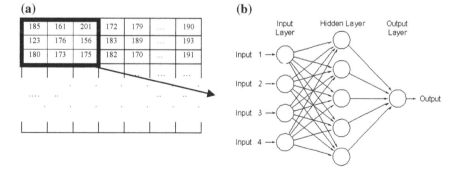

Fig. 1 **a** 3 × 3 size window input to ANN **b** structure of ANN

where α is the system parameter and T_0 is initial condition and n is the number of iterations for system. It is found that map is chaotic for $3.57 < α < 4$ and $T_{n+1} \in (0, 1)$ for all n.

2.3 Embedding Process

In the embedding algorithm, Firstly, we generate edge image of source image by using trained artificial neural network (ANN). And obtain the image that indicate the non-edge pixel and edge pixel. Finally we embed the message into pixel which is selected randomly using Random-SelectPixel(RSP) algorithm. At the receiver side reverse procedure is applied to obtain the secret message. The flow chart and block diagram describes the process of proposed embedding algorithm as shown in Figs. 2a and 3a respectively. The steps of proposed algorithm are as follows:

Phase 1: Firstly we trained the ANN. And apply the ANN on source image to generate the edge image.

Phase 2: In this phase the secret message bits is hidden in edge pixel only. The selection of the pixel is chosen according to Random-SelectPixel(RSP) algorithm based on Fisher—Yates shuffle techniques [14]. The pixel P_i is chosen randomly to scattered the data. Stego key has been used as the seed value to permute the set of pixel. The selected pixel of source image is modified. The two bit of LSB of red and green component of selected pixel is replaced by the secret message and LSB of blue component is replaced by one if pixel is edge pixel. If selected pixel is non-edge pixel then on LSB of blue component is replaced with 0 bit. This process

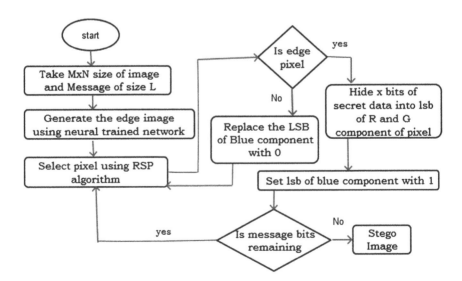

Fig. 2 **a** Flow chart of the embedding algorithm

Fig. 3 **a** Block diagram of embedding procedure

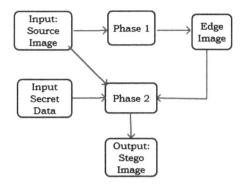

is repeated until all the bits of secret data are processed. The steps of Random-SelectPixel are as follows:

Algorithm: *Random-SelectPixel* ()
Input : Size of Image
Output : Target Pixel

A.1. An array of size M * N taken. M*N is the size of image
A.2. Initialize the array from 1 to M * N. take i = M * N
A.3. Generate the random no.(r) using 1D chaotic logistic map form 1 to i
A.4. Select the pixel by A[r] for processing the pixel
A.5. Replace the content of an array by the last pixel i.e. A[r] = A[N]
A.6. Decrease the value of i by one
A.7. Repeat the process until all the secret bits is hidden

Embedding of secret data procedure is depicted in following Fig. 1.

3 Experimental Results and Analysis

In experimental result we used the 256 × 256 grayscale images, "Lena", "Building", "Camera man" and colored pepper image of size 512 × 512, ostrich image of size 321 × 481 to demonstrate the performance of proposed method as Figs. 4, 5 and 6. We have taken Initial parameter value of $\alpha = 3.68$ and key $T_n = 0.58374587635$. The tested images and its corresponding edge image are shown in figure. We used colored, grayscale, variable size of images to test the proposed scheme. In the experiment we have hide the different size of secret data in images. We have shown all the results of experiment which is carried out in tabular form. We used two performance evaluation criteria peak to signal Noise Ratio (PSNR) and Mean square Error (MSE). Tables 1, 2 and 3 shows a comparison of the MSE and PSNR with the increase of secret data in image.

In this procedure, embedding the secret bits of message in edge pixel of original image leads to changes in edge pixels of stego image as shown in the Tables 1, 2 and 3.

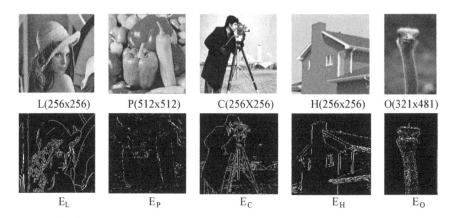

L(256x256)   P(512x512)   C(256X256)   H(256x256)   O(321x481)

E_L E_P E_C E_H E_O

Fig. 4 L: Lena, P: Pepper, C: Camera man, H: Building, O: Ostrich images and corresponding edge images obtained during phase 1

Table 1 Result of Lena image L_1 of size 256×256. M: message bits, d_L = difference in edge pixel $((L_1 - E_{Li})$

Image L_1	Total pixels	Edge pixels	Edge pixels (%)	M (bits)	Stego image edge pixels	d_L	Diff. (%)	MSE	PSNR
L_2	65536	5168	7.88	2048	5164	4	0.0773	0.0290	**63.5058**
L_3	65536	5168	7.88	4096	5161	7	0.135	0.0566	**60.6058**
L_4	65536	5168	7.88	8192	5157	11	0.212	0.1131	**57.5972**
L_5	65536	5168	7.88	16384	5152	16	0.309	0.2283	**54.5464**

Table 2 Result of Pepper image P_1 of size 512×512. d_P = difference in edge pixel $(P_1 - E_{Pi})$

Image P_1	Total pixels	Edge pixels	Edge pixels (%)	M (bits)	Stego image edge pixels	d_P	Diff. (%)	MSE	PSNR
P_2	262144	10140	3.86	4096	10142	2	0.044	0.0223	**64.6407**
P_3	262144	10140	3.86	8192	10133	7	0.069	0.0447	**61.6303**
P_4	262144	10140	3.86	16384	10136	4	0.039	0.0881	**58.6794**
P_5	262144	10140	3.86	32768	10141	1	0.009	0.1767	**55.6578**

At the extracting process we received edge image of stego images which have pixel mismatch with original image as shown in figure. The edge pixel of building image H_1 have 3212 pixel and stego image have 3205 pixel as shown in Fig. 5. The total numbers of mismatch pixel are 7 after embedding the secret data in source image. As a result, it makes it impossible to find the correct edge pixel. Hence retrieving is not possible from the stego image if edge pixel is differing. Therefore we used one

Table 3 Result of Building image H_1 of size 256×256 d_B = difference in edge pixel ($H_1 - E_{Bi}$)

Image H_1	Total pixels	Edge pixels	Edge pixels (%)	M (bits)	Stego image edge pixels	d_B	Diff. (%)	MSE	PSNR
H_2	65536	3212	4.90	1024	3205	7	0.217	0.0192	**65.2896**
H_3	65536	3212	4.90	2048	3205	7	0.217	0.0392	**62.2013**
H_4	65536	3212	4.90	4096	3206	6	0.186	0.0786	**59.1792**
H_5	65536	3212	4.90	8192	3203	9	0.280	0.1541	**56.2541**

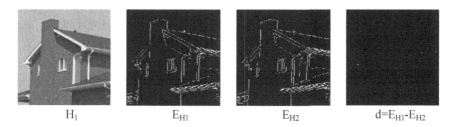

H_1 E_{H1} E_{H2} $d=E_{H1}-E_{H2}$

Fig. 5 H_1 original image, E_{H1} edge pixel of H_1, E_{H2} edge pixel of stego image and d contain mismatch pixel

component of pixel as an indicator that keep information about the edge pixel and non-edge pixel as discussed in algorithm. We compare the quality of stego image to that of original image as seen by human visual system. The PSNR measures value are higher than 50 dB which make evident the satisfactory performance of proposed method based on edge detection techniques.

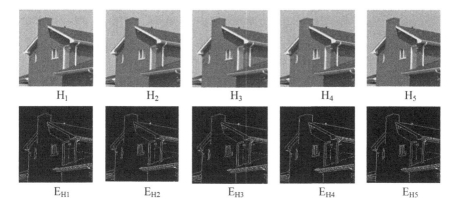

H_1 H_2 H_3 H_4 H_5

E_{H1} E_{H2} E_{H3} E_{H4} E_{H5}

Fig. 6 H_1 original image, H_2 to H_5 are stego image with different secret message bit E_{H1} to E_{H5} are the edge images respectively

4 Conclusion

In this paper a chaotic edge detection based steganography has been proposed. In the experiment, conducted on variable size of image a performance evaluation has been done on the basis of perceptibility i.e. PSNR and MSR. The standard value of PSNR should be greater than 40 dB. In our experiment PSNR values lies between 55 dB and 65 dB. The result from the experiment proved that the edge pixel difference is very less as we embed the secret message only into edge pixel. Further it become very difficult for the human visual system to recognize the changes in the original image. Initial key having large value introduces security level in the systems. The retrieval of information is not possible without having any knowledge of secret key. Hence, the presented work demonstrated successful embedding and retrieval of secret message.

References

1. S. Alam, A. Jamil, A. Saldhi, M. Ahmad, Digital image authentication and encryption using digital signature, in *International Conference on Advances in Computer Engineering and Applications (ICACEA)*, vol. 19–20 (March 2015), pp. 332–336. ISBN:978-1-4673-6911-4/15
2. Z. Li, X. Chen, X. Pan, X. Zeng, Lossless data hiding scheme based on adjacent pixel difference, in *Proceedings of the International Conference on Computer Engineering and Technology*, (2009), pp. 588–592
3. S. Islam, P. Gupta, Revisiting least two significant bits steganography, in *Proceedings of the 8th International Conference on Intelligent Information Processing (ICIIP)*, Seoul, Republic of Korea, 1–3 April 2013, pp. 90–93
4. S. Alam, V. Kumar, W.A. Siddiqui, M. Ahmad, Key-dependent image steganography using edge detection, in *International Conference on Advanced Computing & Communication Technologies*, (2014), pp. 85–88
5. S. Alam, S.M. Zakariya, M.Q. Rafiq, Analysis of modified LSB approaches of hiding information in digital images, in *International Conference on Computational Intelligence and Communication Systems*, (IEEE, Hybrid Intelligent Systems, 2013), pp. 280–285. ISBN 978-1-4799-7633-1/14/2014
6. S. Islam, M.R. Modi, P. Gupta, Edge-based image steganography. EURASIP J. Inf. Secur. **2014**, 8, (2014). http://jis.eurasipjournals.com/content/2014/1/8
7. M.R. Modi, S. Islam, P. Gupta, Edge based steganography on colored images, in *9th International Conference on Intelligent Computing (ICIC)*. Lecture Notes in Computer Science, vol. 7995 (Springer, Berlin, 2013), pp. 593–600
8. J.F. Canny, Finding edges and lines in images, Master's thesis, MIT. AI Lab. TR-720 (1983)
9. W.J. Chen, C.C. Chang, T.H.N. Le, High payload steganography mechanism using hybrid edge detector. Expert Syst. Appl. **37**(4), 3292–3301 (2010)
10. Y.H. Kuo, C.S. Lee, C.C. Liu, A new fuzzy edge detection method for image enhancement, in *Proceedings of the 6th IEEE International Conference on Fuzzy Systems*, vol. 2, (1997), pp. 1069–1074
11. J. Gu, Y. Pan, H. Wang, Research on the improvement of image edge detection algorithm based on artificial neural network. Optik **126**, 2974–2978 (2015)

12. J. Canny, A computational approach to edge detection. IEEE Trans. Pattern Anal. Mach. Intell. PAMI **8**(6), 679–698 (1986)
13. R.M. May, Simple mathematical model with very complicated dynamics. Nature **261**, 459–467 (1967)
14. S. Alam, S.M. Zakariya, N. Akhtar, Analysis of modified Triple-A steganography technique using fisher yates algorithm, in *International Conference on Hybrid Intelligent Systems*, (2014). ISBN 978-1-4799-7633-1/14/2014

A Latest Comprehensive Study on Structured Threat Information Expression (STIX) and Trusted Automated Exchange of Indicator Information (TAXII)

M. Apoorva, Rajesh Eswarawaka and P. Vijay Bhaskar Reddy

Abstract One of the important challenges in threat intelligence is to use them efficiently which can be obtained by both external and internal sources. The need for organization to have cyber threat intelligence is growing and a basic component for any such capacity is sharing threat intelligence between trusted partners, which will help us to target and compute the large cyber security information. This paper briefly explains the way of sharing the threat information which is both human and machine readable using Structured Threat Information Expression (STIX) and Trusted automated exchange of indicator information.

Keywords Threat intelligence · Cyber threat · Cyber threat sharing

1 Introduction

Cyber Security is a problem domain which is more varied and complicated and tends to prolong higher. As the confidence on the technology tend to increase at the same time the threat environment also increases and expands effectively [1]. In conventional methods of cyber security it deals with the inward focus of understanding the vulnerability and weaknesses which are very important but incomplete. It is required to balance both outward focus and inward focus to have a useful defense beside the effective Threats, wherein outward focus deals with the

M. Apoorva (✉) · R. Eswarawaka · P.V.B. Reddy
Department of Information Science and Engineering, Dayananda Sagar College
of Engineering, REVA University, Bangalore, India
e-mail: apoorva.m.06@gmail.com

R. Eswarawaka
e-mail: rajesheminent@gmail.com

P.V.B. Reddy
e-mail: Bhskar.dwh@gmail.com

© Springer Nature Singapore Pte Ltd. 2017   477
S.C. Satapathy et al. (eds.), *Proceedings of the 5th International Conference on Frontiers
in Intelligent Computing: Theory and Applications*, Advances in Intelligent Systems
and Computing 516, DOI 10.1007/978-981-10-3156-4_49

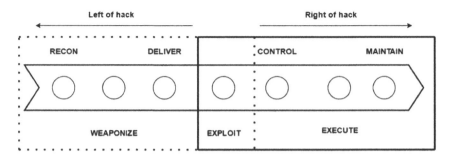

Fig. 1 Cyber kill chain

understanding of attacker's behavior and potential. By having proper understanding on the outward and inward focus helps to have complete understanding on Threats which in turn helps to take defensive decisions.

The cyber kill chain is effectively used to understand the outline of the attackers and the way to defend against the attacks. The following Fig. 1 shows the Cyber kill chain.

The cyber kill chain consists of the following phases Recon, weaponize, deliver, exploit, control, maintain and execute which are the phases of advanced cyber attack. It also shows the right of hack and left of hack where the left of hack is the beginning stage of hacking where hacker still doesn't have access to data but in the right of hack the hacker will have complete access to the data. The following table gives the details about each phase in detail (Table 1).

By having the knowledge about the cyber kill chain it will be helpful for defenders to protect themselves from the attacks, and also threat information sharing helps defenders have a worthy view on attacker's purpose and their strategy which in turn helps the defenders responds for the attack and think about their action and built the suitable protection.

One of the important tasks of sharing threat information is to have the threat information in the structured format so that it is both human and machine readable and without losing any information which can be done by using Structured Threat Information Expression (STIX) [2] and sharing of this actionable information can be done by using Trusted Automated Exchange of Indicator Information (TAXII) [3].

Table 1 Cyber kill chain phases

Cyber kill chain phase	Description
Recon	The attackers analyze and explore the destination
Weaponize	At the target system the set of attack tools are executed
Deliver	At the target system the set of attack tools are delivered
Exploit	At the target system the set of initial attack tools is executed
Maintain	The attacking actions are being directed to the target system
Control	The attackers start performing his goal requirements

2 Structured Threat Information Expressions (STIX)

STIX is a language which is growing in association with all involved parties, to built structured language which gives the complete information about cyber threat information. STIX uses the following use cases [4]:

- Analyzing cyber threats
- Specifying indicator patterns for cyber threat
- Managing cyber threat response activities
- Sharing cyber threat information

STIX gives a simple mechanism for using the structured cyber threat information with help of use cases which gives flexibility, consistency and efficiency. The below diagram shows the STIX architecture which provides the complete set of threat information and includes eight core constructs (Fig. 2):

- Cyber observable
- Indicator
- Incident
- Campaign
- Tactics, Techniques, Procedure
- Exploit Target
- Course of action
- Threat Actor.

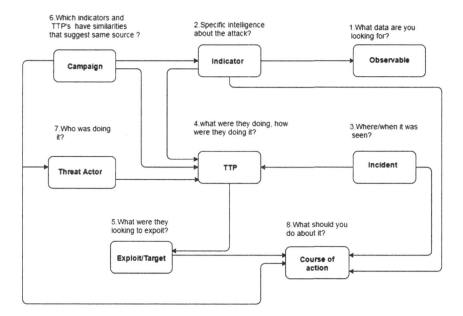

Fig. 2 STIX architecture

2.1 Cyber Observable

It is the important construct in the STIX architecture which is used to detect the events and gives the information about the file name, file size etc. Cybox is used to represent observables [5].

2.2 Indicator

It gives the specific signs unique to the malicious activity and gives information about the IP address and domain name etc.

2.3 Incident

It gives the information about the region where the malicious activity has taken place.

2.4 Campaign

It gives the information about the details of motivation and reason for the malicious activity happened.

2.5 Tactics, Techniques, Procedure

It gives the information about particular tools and methods being used in the malicious activity.

2.6 Exploit Target

It gives the information about the flaws and weakness of the software from which malicious activity would have happened.

2.7   Course of Action

It gives the information about the action which has to be performed in order to overcome the threat attack.

2.8   Threat Actor

It gives the information about who is involved and their motivation of threat attacks.

3   Trusted Automated Exchange of Indicator Information (TAXII)

TAXII is a set services and exchange of messages when built helps to sharing of actionable cyber threat information [6]. TAXII shares the structured threat information expression (STIX) between the organization in secure and automated way which helps the organization defend them by the threat attack. The use cases in TAXII are:

- Public Alerts or warning
- Private alerts and reports
- Push and pull content dissemination
- Set-up and Management of data sharing (Fig. 3).

The above gives the detailed figure of TAXII representation which describes the components of the TAXII such as the TAXII specifications and TAXII toolkit [7]. The TAXII specification explains in detail about the specifications and the use of

Fig. 3 TAXII representation

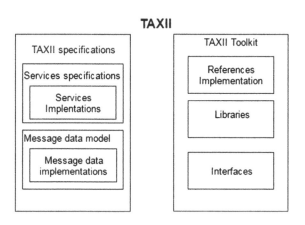

TAXII use cases. The TAXII toolkit explains the set of tools used and reference implementation, libraries and interfaces used to built TAXII.

TAXII provides many sharing models such as source-subscriber, peer-peer, hub-spoke, and push-pull sharing [8]. This model provides the way for sharing of cyber threat information. Also these models provide four services such as discovery service, collection management service, inbox service and poll service.

4 Conclusion

There is a demand for new and more collective approaches for cyber security defense. Cyber threat intelligence and cyber threat sharing are the popular unique approach between the attackers and defender. This cyber threat intelligence should be standardized so that it would be helpful for the defenders to make use of it and this is done by Structured Threat Information Expression (STIX), it provides the complete description about the cyber threat information. Trusted Automated Exchange of Indicator Information (TAXII) is a most favorable method used for exchange of threat information which is represented in Structured Threat Information Expression (STIX) which provides organization to share threat information securely.

References

1. Getting ahead of advanced threats: achieving intelligence driven information security, recommendation from global 1000 Executives, Security for business innovation council
2. Structured threat information expression (STIX). https://stix.mitre.org/
3. trusted automated exchange of indicator information, (TAXII) https://taxii.mitre.org
4. The MITRE Corp, STIX–Structured Threat Information Expression, The MITRE Corp, 2014
5. Cyber observable expression (cybox). http://cybox.mitre.org/
6. The MITRE Corp, TAXII Overview 1.0, The MITRE Corp, 2014
7. The MITRE Corp, The TAXII Services specification 1.1, The MITRE Corp, 2014
8. Cyber information sharing models an overview, MITRE Corporation, May 2012

Preventive Maintenance Approach for Storage and Retrieval of Sensitive Data

V. Madhusudhan and V. Suma

Abstract Securing user data is one of the challenges in all applications. User's sensitive data present in the storage server is expected to be highly available, secured and easily accessible from anywhere according to the demand in time. This paper provides a preventive maintenance approach to access the data, though storage server containing the sensitive data fails. This approach uses a simultaneous copying technique to save the data on both storage server and backup server during the upload process. Thus, when server fails, the data needs of the user can be served by the backup server. Data file is broken into data blocks and these blocks are encrypted and stored in the storage server instead of directly uploading the sensitive data file. Thus, when an intruder gains access to the storage server and tries to access the data, retrieval of the data file is not possible since mapping of files on the data blocks is random and encrypted. This supports preventive maintenance of data which not only secures data but also reduces risk and cost of recovery.

Keywords Data deduplication · Preventive maintenance · Sensitive data

1 Introduction

Data is a prime raw fact of any transaction in all applications. Securing the data from unauthorized access is one of the challenges in both online and offline transactions. It is more critical in online applications since it is prone to different categories of threats, from uploading to data retrieval. Thus, sensitive data needs to be protected against threats to safeguard the privacy.

V. Madhusudhan (✉) · V. Suma
Dayananda Sagar College of Engineering, Bangalore, India
e-mail: vmadhusudhan91@gmail.com

V. Suma
e-mail: sumavdsce@gmail.com

© Springer Nature Singapore Pte Ltd. 2017
S.C. Satapathy et al. (eds.), *Proceedings of the 5th International Conference on Frontiers in Intelligent Computing: Theory and Applications*, Advances in Intelligent Systems and Computing 516, DOI 10.1007/978-981-10-3156-4_50

Sensitive data can be Personal, Organizational or Classified data. Personal data is the identity of the person such as medical data, biometric information, personally identifiable financial data and distinctive identifiers like UID number or passport. Exposure of these data not only leads to identity theft but also revealing private data that the individual would favor remaining private.

Organizational data is business sensitive data includes anything that poses risk to the company such as acquisition plans, trade secrets, monetary information, supplier and customer data. With the ever increasing amount of data generated by businesses, protecting company data from unauthorized access have become a major part to company security.

Classified Sensitive Data is the data that is related to government body, and which is restricted from the access of general public. These data can be classified according to levels of sensitivity a stop secret, secret, confidential and restricted. Data is usually classified to safeguard security. Once the danger possessed by revealing of the data has passed or reduced, classified data could also be unclassified and, if possible can be made public.

These sensitive data are stored in the storage server in mass which requires high security. Data threats can be at client, network while uploading the data or at storage server. Research has been going on in these areas to protect sensitive information and avoid misuse for their own benefits leaving the owner and data at risk.

2 Related Work

The paper by He et al. [1] proposed that, of the two types of deduplication Block level data deduplication and file level data deduplication, the block level data deduplication is better in terms of efficiency. File-level technology provides compression ratio around 5:1, while the block-level technique provides compression ratio ranging from 20:1 to 50:1. With the rapid growth in the volume of the data centers, with the increase in amount of data deduplication the amount of data can be greatly reduced while optimizing storage system, which decrease heat emissions and dropping energy utilization. It also lowers the number of disks utilized to cut down disk energy utilization costs.

The paper by Wu et al. [2] proposed a framework for sharing of sensitive data in a secured way on the ig data platform. It presents a proxy re-encryption algorithm, which is based on a cipher-text transformation an the user protection method based on virtual machine monitor. The framework protect the security of user's sensitive data and shares these data securly. and also the data owners will have the complete control of their data for data security.

The paper by Chen and Zhang [3] proposed based on encryption (CP-ABE) an cryptographic access control scheme know as CS-CACS which is, and is implemented based on HDFS workstation. It combines the proxy reencryption and lazy re-encryption, which enables cloud servers do the re-encryption computing while revoking user's permission, data owners are benefited by reduction in the cost of computation. This scheme has properties of accountability of user secret key and user access permission confidentiality. Performance analysis showed that it is more efficient and secure when cloud storage is accessed by more users.

The paper by Feng et al. [4], proposes a scheme for secure and efficient revocation to overcome the challenge of providing data confidentiality unauthorized users and cloud servers. it put forth's a modified algorithm for fine grained access control method. This scheme offers least overhead to both the cloud servers and the data owners. And also data owners can send key updates to cloud server without disclosure of data content. More importantly, this scheme preserves the vital aspect that the revocation will not affect the user's, whose attribute set could be a super set of the revoked user.

The paper by Su et al. [5], proposes Data Division and Out-of-Order key stream generation encryption method to protect data in distributed storage environment. This scheme is verified to the correctness and effectiveness through simulation and synthesis, which is done on reconfigurable hardware devices. The results were encouraging as they considered implementing this scheme operations using FPGAs, and this task is pushed to the bottomlayer of data processing. The operations including data division, encryption, decryption and reassembly are apparent to users and upper layer applications programs.

The paper by Vijayarani and Tamilarasi [6], They have analyzed performances of two masking techniques namely bit transformation and data transformation used for protection of sensitive data. Their analysis consisted of four steps, (i) Identifying the sensitive data in a large data set, (ii) Then modify the sensitive data using data transformation and bit transformation, (iii) Do performance analysis, (iv) Lastly find the best technique. The results when considering the privacy protection, statistical performance and clustering accuracy, the data transformation as performed better than the bit transformation.

The paper by Bhanu and Hingwe [7], proposed a framework which supports database encryption, query encryption, and also range query encryption over an encrypted database. This scheme focuses on storing sensitive information in a secured database without any leak. Sensitive data is secured using double layred encryption and non-sensitive data using single layered encryption. For single encryption order preserving encryption (OPE) is used, but this OPE as the disadvantage of revealing information, therefore for sensitive data protection, a double layered encryption by means of format preserving encryption and order preserving encryption is used.

The paper by Bertino et al. [8], They have presented an privacy preservation and data leak detection solution to overcome the sensitive data leak caused by human errors. For detection of errors a distinctive set of sensitive data digests is used. The advantage of these techniques is that it allows the data owner to carefully delegate the detection process to a semi honest provider devoiding of revealing the sensitive data to the provider. The evaluation outcome showed that this technique helps precise detection using a small number of fake alarms under different data-leak scenarios.

The paper by Dai et al. [9], proposes an algorithm called Protector that focuses on generation of new replicas in order to replace replicas that are lost to transient or permanent failures, by means of network-wide statistical prediction. This algorithm significantly increases prediction accuracy by doing predictions across aggregate replica groups in the place of single nodes. The estimate of the number of "live replicas" shows effective data replication policies. They showed that the probability of permanent failures is given by the data on node down times, the estimate provided by this algorithm is more accurate. Guided by models or traces, they illustrated two methods to find the failure probability function. They carried out simulations based on real and synthetic traces, which demonstrated that protector closely estimated the performance of a exact "oracle" failure detector, timeout-based detectors were outperformed considerably by means of a wide range of parameters.

The paper by Liu et al. [10], proposed an application aware de-duplication in the personal computing environment for cloud backup to enhance deduplication efficiency. The deduplication policy in AA-Dedupe to reduce computational overhead with slight loss in deduplication efficiency is accomplished by utilizing file semantics. The AA-Dedupe consists of data structure, such as application-aware index structure that can considerably reduce the disk index lookup bottleneck and optimize lookup performance by splitting a central index into several independent small indices. In this prototype, AA-Dedupe increases source-deduplication method by a factor of 2–7, and also backup window size is condensed by 10–32 %, power-efficiency is enhance by a factor of 3–4, then cost of cloud for cloud backup service is saved by 12–29 %.

The paper by Sun et al. [11], proposed a technique for the cloud storage privacy protection that comprises of bit split and bit combination. Bit Split is carried out on data and then it is reassembled to create many part-files and later they are uploaded to multiple cloud storage servers. Later while downloading, the original file is reconstructed by merging part-files from different cloud servers. Experiment shows that this technique is helpful in protecting the confidentiality of the user's data, and has a 20 times performance enhancement compared to traditional encryption and decryption techniques.

3 Preventive Data Maintenance Approach for Data Security

Data on server is very important asset to an organization. Security and safety of such data is an important aspect of Software Engineering. This is part of Software Maintenance which has different dimensions, one of which is preventive maintenance.

Preventive maintenance approach is securing the data which are to be stored in the Storage server. There are chances to lose data in case of server failures due to natural disaster or manmade one. To make data highly available, this paper proposes a preventive maintenance approach along with data deduplication technique.

3.1 Data Deduplication

Data Deduplication is a technique of reducing storage space. In this technique only one unique instance of data is stored, eliminating the redundant or repetitive data. Thus the main theme of data deduplication is to optimize the storage space, it also reduces the amount of data that as to be sent over the network, thus reducing the usage of bandwidth, which otherwise would have been more. Deduplication is done in two ways depending on data granularity.

1. File level Deduplication:
   If two files are identical, only one file is stored and subsequent file points to the already existing file. It is not efficient because slight modification in one file results in storage of new file.
2. Block level Deduplication:
   The file is divided into blocks, and hash code is generated for each block using MD5 cryptographic hashing algorithm. Only unique blocks are stored, repetitive blocks points to its unique block instances already existing. When a new file arrives for storage, it takes into account of the blocks already existing in the repository and the same process is repeated storing only the unique blocks, eliminating the repetitive blocks.
   While using MD5 algorithm the possibility of having a hash collision is very negligible. Even after generating 26 trillion hash values, the possibility of getting the next hash value same as the one in previous 26 trillion values, is one out of one trillion(1/1trillion) (Fig. 1).

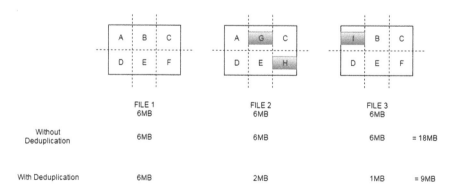

Fig. 1 Data deduplication

4 Implementation

The approach has two stages. (1) Upload Phase (2) Download Phase.

1. Upload Phase:

In this process, the sensitive data file which is to be uploaded is split into multiple blocks in the block generation step. Using MD5 hashing algorithm, hash code is generated for each block in calculating cryptographic hash step. In the deduplication step, it identifies the duplicate blocks of the file to be uploaded, which are already present in the storage server. This is done with the help of repository index (Logical block addresses) which as the hash code information of the blocks which are already existing in the storage server. For the duplicate blocks, it adds the pointer to the hash index of already existing block, and for the unique blocks, for which earlier hash code entries are not present in the repository, new entry of hash code is added. Only the unique data blocks are encrypted using RSA algorithm and uploaded to both the main storage server and the backup server (Fig. 2).

2. Download Phase:

In this process, a check is made for the availability of the server. If so, file is downloaded from the server. When the main storage server is unavailable, connection to the backup server for retrieval of data is done. Once connected to the storage server, the block information of the sensitive data file is got from repository

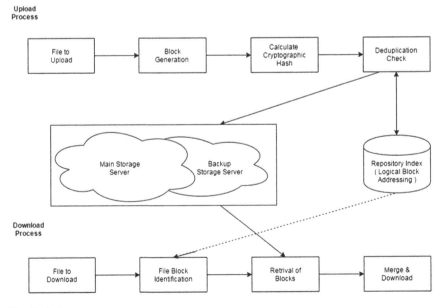

Fig. 2 Architecture

index (Logical block address). From this block information, that particular blocks are retrieved from the server and decrypted. All the retrieved blocks are merged into one file and provided to the user.

5 Conclusion

This paper focuses on secured storage of user sensitive data in the storage server using preventive maintenance approach. It provides high availability sensitive data in time. Security is provided by splitting the data file into blocks, then encrypting and uploading only the unique data blocks to the server. When an intruder gains access to the storage server, right data is not accessible as it is in random blocks and encrypted. High availability is provided by simultaneously uploading the encrypted unique data blocks to both main storage server and backup server so that even if the main server crashes, backup server can be used to retrieve sensitive data. Thus, data loss and maintenance cost is reduced. This paper limits its work with variable length data deduplication.

References

1. Q. He, X. Zhang, Z. Li, Data deduplication techniques, in *IEEE, International Conference on Future Information Technology and Management Engineering*, (2010)
2. H. Wu, Z. Xue, W. Zhou, H. He, X. Dong, R. Li, Secure sensitive data sharing on a big data platform, Tsinghua Sci. Technol. **20**(1), 72–80 (2015). ISSN1007-021408/11
3. P.S. Chen, R. Zhang, A dynamic cryptographic access control scheme in cloud storage services, Scientific Research Fund of Zhejiang Provincial Education Department, Zhejiang Provincial Key Specific Project of Science & Technology, (2010)
4. D. Feng, M. Zhang, C. Hong, Z. Lv, A secure and efficient revocation scheme for fine-grained access control in cloud storage, in *IEEE, 4th International Conference on Cloud Computing Technology and Science*, (2012)
5. Z. Su, W.-S. Ku, J. Feng, Y. Chen, D-DOG: securing sensitive data in distributed storage space by data division and out-of-order keystream generations, (IEEE ICC, 2010)
6. S. Vijayarani, A. Tamilarasi, An efficient masking technique for sensitive data protection, in *IEEE-International Conference on Recent Trends in Information Technology, ICRTIT*, (2011)
7. S.M.S. Bhanu, K.K. Hingwe, Two layered protection for sensitive data in cloud, (IEEE, 2014)
8. E. Bertino, X. Shu, D. Yao, Privacy-preserving detection of sensitive data exposure. IEEE Trans. Inf. Forensics Secur. **10**(5), 2015
9. Y. Dai, B.Y. Zhao, W. Chen, J. Tian, Z. Yang, Probabilistic failure detection for efficient distributed storage maintenance, in *IEEE, Symposium on Reliable Distributed Systems*, (2008)
10. F. Liu, L. Tian, N. Xiao, H. Jian, F. Yinjin, AA-dedupe: an application-aware source deduplication approach for cloud backup services in the personal computing environment, in *IEEE International Conference on Cluster Computing*, (2011)
11. X. Sun, T. Xu, W. Zhang, Data privacy protection using multiple cloud storages, in *IEEE, International Conference on Mechatronic Sciences, Electric Engineering and Computer*, (2013)

Join Operations to Enhance Performance in Hadoop MapReduce Environment

Pavan Kumar Pagadala, M. Vikram, Rajesh Eswarawaka
and P. Srinivasa Reddy

Abstract Analyzing large data sets is gaining more importance because of its wide variety of applications in parallel and distributed environment. Hadoop environment gives more flexibility to programmers in parallel computing. One of the advantages of Hadoop is query evaluation over large datasets. Join operations in query evaluation plays a major role over the large data. This paper Ferret outs the earlier solutions, prolongs them and recommends a new approach for the implementation of joins in Hadoop.

Keywords Hadoop · Query evaluation · Distributed environment · Performance analysis · Optimization

1 Introduction

The trendy programming model which supports distributed computing on myriad and large data sets is Map/Reduce framework over wide spread of machines It facilitates an effective process to execute distributed applications without having insightful parallel programming knowledge. Map and/or Reduce tasks are executed by each participating node which involve in the process of reading as well as writing huge data sets.

P.K. Pagadala (✉) · P.S. Reddy
Bharat Institute of Engineering & Technology, JNTUH, Hyderabad, India
e-mail: pavankumarpagadala@gmail.com

P.S. Reddy
e-mail: psrinu570@gmail.com

M. Vikram
Sri Venkateswara College of Engineering, JNTUA, Anantapur, India
e-mail: vikram.maditham@gmail.com

R. Eswarawaka
Dayananda Sagar University, Bangalore, India
e-mail: rajesheminent@gmail.com

© Springer Nature Singapore Pte Ltd. 2017   491
S.C. Satapathy et al. (eds.), *Proceedings of the 5th International Conference on Frontiers
in Intelligent Computing: Theory and Applications*, Advances in Intelligent Systems
and Computing 516, DOI 10.1007/978-981-10-3156-4_51

Typical relational database operations like projection, aggregation, and selection etc. can be performed in Map/Reduce. However, binary relational operators such as set operations, joins and cartesian products are difficult to be implemented with Map/Reduce. Homogeneous data streams can be easily processed with Map/Reduce. But for handling heterogeneous multiple input data streams, Map/Reduce does not provide the direct support. In support of the execution of the fundamental database algorithms, set up of the Map/Reduce framework is tedious task. Especially, the join algorithms seem to be more expensive in terms of input/output cost the utilization of memory is also most challenging. The current work in this paper makes comprehensive comparison and analysis between two fundamental join algorithms the - reduce and map join. Results are relied upon to give an insight of how Map/Reduce is a good fit for evaluating Joins.

The remaining paper contains the following sections:
Section 1 gives the introduction to the Map Reduce framework and the advantages of the query evaluation techniques in distributed environment. In Sect. 2, the literature survey describes the existing drawbacks of using the basic Join algorithms. Section 3 explains the methodology, Sect. 4 results and comparison and Sect. 5 draws the conclusion.

2 Related Work

Large volumes of data is intended to be processed by the Database. In general, for implementing the fundamental database algorithms, deploying of more cost effective join algorithms with the Map/Reduce framework is more complicated task as well as difficult task. In Sect. 2.1 the existing map-side join concept and in Sect. 2.1

2.1 Join at Map-Side

Depending on the volume of the data sets, Map side join, should not be used broadly where the data sets are limited in terms of the memory size as specified by Pigul [1]. In Map side join, the reduce phase is eliminated because it is a one step join i.e. it eliminates the transmission of data between the two phases over myriad network. There will be no transference of data on network because map join uses only the map phase. The map algorithm concludes a pair of data will be previously divided by the similar practitioner into matching number of splits.

The data will be read and stacked in the existing hash table before applying join operation on the two sets by the same hash table at map phase. Like skew data case, Algorithm 2.1 introduced in [1, 2] fails because of less memory i.e. all the records get buffered with the same keys present in memory.

Algorithm 2.1

```
1. The first data set to be partitioned
2. The second dataset to be partitioned
3.  Perform Join operation on the two datasets
          a) Call init() method //Initialization at map phase
   read the required partition of the output file from the
first job;
   add that first job to the hashmap.(Key, list(V)) H;
          b) Call map (K:null, V from B)//Map method is
                      called
Check the if condition of(K in H) then for r in LV do
          for l in Hget.(K) do
          Call emit (null, tuple(r,1));//emit is called
```

2.2   Join at Reduce-Side

Tagging of the data files is the predominant operation of the reduce-side join, introduced in [1]. Reduce join is a default join which performs the join operation in two
steps. From mappers, more amount of data is transferred to reducers over network.
From different reducers present in the reduce phase, the (key, value) pairs are obtained.
By applying join operation, (key, value) pairs with the similar key are joined.

As mentioned in [1], with the help of reduce side join algorithm, reduce phase joins
the data by direct join and map phase preprocesses the data. More time is consumed in
reduce-side join because it involves transmission of the data from one phase to
another over the networks as an additional phase, which is described by Ming Hao
[3]. Without restrictions on the data, the most widely used join is algorithm 2.2.

Algorithm 2.2

```
1. Call Map (K: null, V from A or B)//A and B two datasets
2. Tag = bit from of A or B;//Assigning the values
3. Call the method Emit.(Key, pair(V,Tag));
4. Call Reduce (K1: join key, LV: list of V with key K1)
5. Create buffers Buf.(A) and Buf.(B) for A and B;
6.  for x in LV do//x loop starts
       begin
          add x.v to Buf(A) or Buf(B) by x.Tag;
       end
7. for a in Buf(A) do
     begin1 //Outer loop a starts
        for b in Buf(B) do
           begin2//Inner loop b starts
              emit (null, tuple(a.V,b.V));
           end2
     end1
```

Map and reduce phases are present in algorithm 2.2. Map phase reads data from a pair of different origins and attaches labels which identifies the origin of a (key, value). The hash partitioner standard key can be utilized because the tagging does not effect the key which is specified in Konstantina Palla [4]. Nested-loop algorithms joins the data with same key and distinct tags in reduce phase. The main concern with this process is the reducer must contain enough memory to store all the records with a similar key as described in Pigul [1].

2.3 Hadoop

Hadoop is a software framework which allows parallel processing on large clusters of nodes targeted on large sets of data in a distributed environment. Hadoop is designed to achieve scalability up to millions of computers. Each node will have its own computational and storage units. Hadoop follows client-server architecture as described by Azza Abouzeid et al. [5].

In a cluster, One node acts as master where as others will be worker nodes. Name node and job tracker are given to the master node. The worker nodes will be assigned jobs by the job tracker after dividing task into sub tasks. The files in the distributed file system are pointed by meta data containing the name node. The worker nodes will have task tracker and data node. The task tracker executes the jobs and the data node will receive the data directly from distributed file system as described in [5].

Hadoop consists of distributed file system in which data is maintained in blocks of nodes in a cluster which are arranged as racks. The data is replicated among different racks so as to attain reliability. Hadoop is fault tolerant so that if one node fails the jobs will be assigned to another node.

2.3.1 Key-Value Pair Orientation

The Map/Reduce structure comprises of two main phases for processing the data viz., the map and the reduce phases. Each Map/Reduce job assigned to a mapper and/or reducer are executed by many DataNodes described in [6].

Each mapper and reducer fixtures editable concerns like the map and reduce. The (key, value) pairs are exclusively operated by map/reduce functions as mentioned in [4] (Fig. 1).

2.3.2 The Map Phase

Each device in this phase scans the input record by record during the map phase. Then it converts them into the (key, value) pairs and propagates these (key, value)

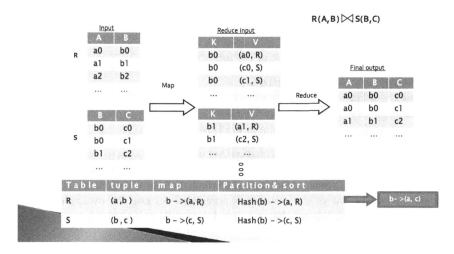

Fig. 1 Key-value pair orientation in Map Reduce

pairs to the user defined map concern. The latter changes the input pair and outputs anew (key, value). The output pairs of the map function are partitioned, sorted and grouped by their keys for further propagation to the next phases as mentioned in [4]. Grouping can be achieved by partitioning the key space with the help of partitioner using a hash function over the keys. The whole number of portions are equal to the number of reduces used for the job. At last, these midway pairs are stored locally on the mappers as specified in [4].

2.3.3 The Reduce Phase

The intermediate data is retrieved from the mappers by the DataNodes that implement the reducers during the reduce phase as specified in [4]. In particular, All values with the similar elements are injected to the similar reducer function as every reducer bring backs a interrelated portion from every mapper node. Taking the same key into account, (key, list(values)) are constructed by merging the fetched map output pairs. The recently organized pairs of (key, list) are propagated to the user defined reduce concern as described in [4]. The final output on receiving an iterator returns a new (key, value) pair by combining those values together. The final output files are written by the output pairs and are finally stored in the HDFS.

3 Methodology

This paper proposes improvements of existing one by using two basic Joins viz. map-side join and reduce-side join.

3.1 Map Side Join

This algorithm sets the map task in the following steps.

- Reading the related sets in HDFS.
- Building a Hash map in the memory.
- Reading every line in the split table.
- Searching for the output results in the Hash map.

Map side partition join is the enhanced version of existing join, which is studied in algorithm 3.1. In this algorithm, a merge join is applied if the same ordering is followed by the data sets, in adding up to their portion. The main benefit of this approach is overflow of the memory is controlled because the evaluation of the succeeding set is on-demand, but it is not done totally.

Algorithm 3.1

```
1. First dataset to be partitioned
2. Second dataset to be partitioned
3. Perform Join between two datasets
        a) init()              // Initialization of  map phase
            read the needed partition of the output file from
the
                first job;
            read the initial lines with the same key K2  from
the
                split partition and add it to the buffer Buf;

        b)Call map(K:null, V from B)
            while (k >k2) do
               begin
                     read tuple T from split Partition with key
k2;
                  end
            while( k == k2) do
                 begin
                    add tuple T to Buf;
                 end
            read T from splitPartition with key k2;
            if ( k == k2) then
            for x in Buf do
             begin
       Call  emit(null, tuple(r,V));
               end
```

3.2   Reduce Side Join

This algorithm performs reduce-side join in the following steps:

- Mapping both sets of tuples
- Bringing together tuples which are sharing the same key in execution framework.
- Actual join is performed in reducer.

Algorithm 3.2

```
Call Map (K:null, V from A or BL)
Tag = bit from A or B;//Assigning datasets to tag
Call emit (pair(Key,Tag), pair(V,Tag));
Call Partitioner(K:key, V:value, P:the number of reducers)
      return hash_f(K.Key) mod P;
Call Reduce (K': join key, LV: list of V' with key K1)
  create buffers Buf(A) for R;
for t in LV with x.Tag corresponds to R do
 begin
         add t.v to Buf(A);
 end
for l in LV with b.Tag corresponds to L do
begin1
        for r in Br do
           begin2
               call emit (null, tuple(a.V,b.V));
           end2
end1
```

In Sect. 2, the enhancement of tagging data source, overriding, grouping and sorting by key with the enhanced reducer-side join algorithm is discussed. In the method, the values in the foremost tag are followed by the values of the subsequent one. The tag is added to both a key and a value in reducer-side join. As the tag is added to a key, to split the nodes, the portions must be overridden by the key. In this case for only one of input sets, the buffering is required. As in case of reduce-side join, this algorithm also needs two phases- one for pre-processing the data and the other for direct joining of the data. In the Reducer-side join, additional information regarding the source is not required because the data comes to the reducer same time.

4 Results and Comparison Analysis

4.1 Operational Environment

Experiments were conducted on 3 different kinds of clusters. There were two types of machines used that were slightly different from another and details are given in Table 1.

File structure used in all the experiments was HDFS. For conducting all experiments cluster setup (Table 2) was designed with help of different data node, name node and job tracker.

Experiments were conducted to compare the existing Map/Reduce algorithms and proposed algorithms. The values of the different data sets are compared by using tuple values given in Table 3. The results of existing and proposed results are compared.

The values given in the comparison chart shows that the execution time has been reduced by using proposed algorithms Fig. 2.

Table 1 Different machines used in experiments

	Type1	Type2
Cores	4	4
Memory	4 GB	4 GB
Cache size	2 MB	1 MB
OS	Scientific linux 5_5	Scientific linux 5_5
Linux kernel	26-18-194	26-18-194

Table 2 Different types of clusters used for the experiments

	Data node (s)	Name nodes(s)	Job trackers (s)	Total nodes (s)
Type 1	1	1	1	1
Type 2	3	1	1	5
Type 3	6	1	1	8

Table 3 Distribution values over algorithms

Tuples (x100000)	Existing map reduce algorithm (s)	Proposed map reduce algorithm (s)
1	21	9
2.5	30	13
5	43	25
7.5	55	36
10	64	47
25	142	125
50	275	267

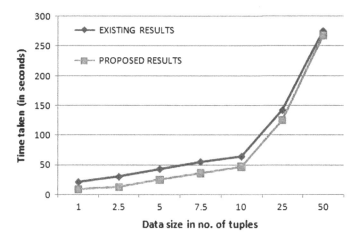

Fig. 2 Comparison analysis

5 Conclusion

In this paper, the basic join algorithms are discussed and analyzed. To overcome the drawbacks of the existing basic join algorithms, new algorithms are proposed which can improve the performance and optimization of the Joins in Map/Reduce. Hadoop environment makes the processing of large data sets in a massively parallel and distributed way on large clusters of nodes with higher scalability, reliability and fault-tolerance. Moreover, this paper compares and analyzes pair of elementary algorithms as presented above. The results are relied upon to give an insight of how Map/Reduce is a good fit for evaluating Joins.

References

1. J. Dean, S. Ghemawat, Mapreduce: simplified data processing on large clusters, in *Design and Implementation 6th Symposium on Operating Systems, ACM*, pp. 137–150 2004
2. Y. Mao, R. Morris, F. Kaashoek, *Optimizing MapReduce for Multicore Architectures* (Massachusetts Institute of Technology, Cambridge)
3. Thesis on Performance Analysis and Optimization of Left Outer Join on Map Side, Ming Hao, Stavanger, 15th June 2012
4. S. Blanas, J.M. Patel, V. Ercegovac, J. Rao,E.J. Shekita, Y. Tian, A comparison of joinalgorithms for log processing in MaPreduce, in *Proceedings of the 2010 International Conference on Management of Data* (2010) pp. 975–986
5. A. Abouzeid, K. Bajda-Pawlikowski, D. Abadi, A. Silberschatz, A. Rasin, Hadoopdb, An architectural hybrid of MapReduce and dbms technologies for analytical workloads, in *VLDB*, 2009

6. K.H. Lee, Y.J. Lee, H. Choi, Y.D. Chung, parallel Data Processing with MapReduce: a Survey, *Department of Computer Science, Department of Computer Science and Engineering* (Korea University in *KAIST*)
7. V. Jadhav1, J. Aghav, S. Dorwani2, Join algorithms using mapreduce a surveyn, in *International Conference on Electrical Engineering and Computer Science*, 21 Apr 2013
8. Binary Theta-Joins using MapReduce: Efficiency Analysis and Improvements, Ioannis K. Koumarelas, Athanasios Naskos, Anastasios Gounaris, Dept. of Informatics, Aristotle University
9. J. Dean, S. Ghemawat, Mapreduce: simplified data processing on large clusters. Commun. ACM **51**(1), 107–113 (2008)
10. Thesis in Implementation and Analysis of Join Algorithms to handle skew for the Hadoop MapReduce Framework, Fariha Atta, University of Eidenburgh 2010
11. Minimal MapReduce Algorithms, Yufei Tao, 1Chinese University of Hong Kong, Hong Kong, Wenqing Lin, Korea Advanced Institute of Science and Technology, Korea, Xiaokui Xiao, Nanyang Technological University, Singapore
12. K. Palla, A comparative analysis of join algorithms using the hadoop MapReduce framework. Master's thesis, MSc Informatics, School of Informatics, University of Edinburgh (2009)

Adaptive Spatio-Temporal Filtering with Motion Estimation for Mixed Noise Removal and Contrast Enhancement in Video Sequence

S. Madhura and K. Suresh

Abstract Naturally available noises in the videos are complex but fortunately they can be broadly classified as Gaussian and Impulse noises. Most of the available models for noise removal emphasize on any one kind of noise removal thus an optimum model of mixed noise removal is still a challenge. This paper describes about removal of video flickering and artifacts due to sensor motion, unprofessional recording behaviors, device defects, poor lighting conditions and high dynamic exposure. The adaptive spatio-temporal filter gives excellent result for mixed (Gaussian and Impulse) noise removal. Dense optical flow is introduced to reduce the motion blur and enhance the video. The analysis of PSNR and SSIM values were compared with existed method like Non-local Means and BM3D approach and results are tabulated. The Histogram graph gives the better intensity distribution in frames thus the proposed method even works good for low illumination or night vision surveillance videos.

Keywords Video enhancement · Noise reduction · Optical flow · Adaptive filters

1 Introduction

Video recording has become a trend in the existing rapidly developed digital advancement aided by availability of high focal length and high definition cameras. In this scenario videos are not only limited to professional application but have stepped into surveillance and customized video making since the capturing videos has become easy. Hence the artifacts such as blur, blocking, contrast distortion and

S. Madhura (✉)
Dayanandasar College of Engineering, Bangalore, India
e-mail: madhu4tulip@gmail.com

K. Suresh
SDMIT, Ujire, Karnataka, India
e-mail: ksece1@gmail.com

© Springer Nature Singapore Pte Ltd. 2017   501
S.C. Satapathy et al. (eds.), *Proceedings of the 5th International Conference on Frontiers in Intelligent Computing: Theory and Applications*, Advances in Intelligent Systems and Computing 516, DOI 10.1007/978-981-10-3156-4_52

ill-illumination would occur due to movement of the recording device, unprofessional recording behaviors, device defects, poor lighting conditions, high dynamic exposure, transmission loss and so forth. Even though most of the noises are random and complex they can be broadly classified as Impulse noise and Gaussian noise [1, 2]. During video acquisition additive Gaussian noise would get introduced and generally assumes zero mean distribution whereas impulse noise would assume uniform distribution and is produced due to transmission errors. It is still a challenge to develop a common model that deal equally well with different kinds of distortions. Most of the video filtering techniques defined in recent works mainly focuses on removal of impulse noise only [3].

For instance, order statistic filters work very well in removal of impulse noise which distinguishes fine features and image pixel but this method fails for shot noise or Gaussian noise [4–6]. Gaussian noise removal is better with bilateral filter [7], wavelet transforms [8] and anisotropic diffusion [9]. However some of the most effective Gaussian noise removal algorithm like Non-Local Means [10–12] and BM3D [13, 14] cannot remove impulse noise contamination. BM3D algorithm is DCT based and is generally used for Additive Gaussian White Noise (AGWN) removal. This algorithm is generally used for noise removal in still images but now it is extended to videos.

Along with the noisy frames if it is associated with blur due to camera movement or scene motion then there should be another algorithm which can normalize the frames and produce a desirable quality image. The motion blur of the image with the low temporal resolution can be directly measured by the images with high temporal resolution [8, 15]. Motivated by the above problems a novel algorithm has been developed which enhances the video which was subjected various kinds of noises and also which contain motion blur. For noise removal Adaptive Spatio-Temporal Filters would be used and for removal of blur motion estimation from optical flow would be considered. Moreover to make the algorithm faster and robust modified neighborhood relation would be introduced in the temporal domain.

2 Proposed Method

The proposed algorithm is schematically represented in Fig. 1. The entire process can be modeled into two parts: video denoising and deblurring. Firstly frames are extracted from the input video which is distorted by mixed [Gaussian and Impulse] noise and associated with motion blur. The spatially filtered frame would undergo forward and backward temporal filtering. Temporal filtering would be done more exclusively then spatial filtering [16].

Secondly for every three frames the first frame is tracked using the optical flow algorithm and hence their location in third frame would be known. With the assumption that the location of second frame is midway between first and third, the motion vectors of each pixel of first frame are divided by two which provides the

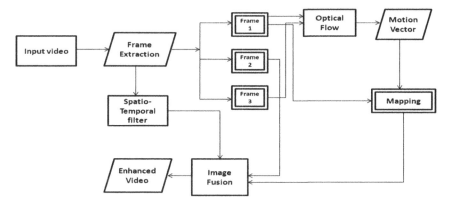

Fig. 1 Schematic of proposed algorithm

magnitude of the motion vector from first frame to second frame. However the motion vector directions remain the same as shown in Eq. (1).

$$I_1(X_1, Y_1) = I_2(X_2, Y_2) \tag{1}$$

where $I_1(X_1, Y_1)$ and $I_2(X_2, Y_2)$ are the intensities of first and second frame respectively. Finally image fusion would be simple logarithmic averaging operation between the frame reconstructed from optical flow and second original frame.

3 Adaptive Spatio-Temporal Filtering

The Adaptive Spatio-Temporal filter depends on the following points:

1. Number of pixels that it wants to combine and whether these pixels are in the area of the frame with motion.
2. The parameters for local illumination are chosen based on transition between spatial-only and temporal-only bilateral filtering.

To remove the spatial blur, blocking and temporal fluctuations within a frame sequence each frame is treated with spatial and temporal filter by proposed statistical spatio-temporal filter. Each frame estimation within the same frame is obtained by their neighboring pixels as shown in Eq. (2) which is a recursive updating function.

$$\theta_c^{(\omega+1)} = \theta_c^{(\omega)} - g\left(\theta_c^{(\omega)}\right) \tag{2}$$

where $\theta_c^{(\omega)}$ is the ω th estimation iterate of θ_c and $g(.)$ is a gradient function given by:

$$g\left(\theta_c^{(\omega)}\right) = \alpha^\omega \sum_{m \in M_c} \psi\left(\theta_c^{(\omega)} - \theta_m\right) \tag{3}$$

where m is the pixel index of current element, M_c is a set of neighboring pixels, $\psi(.)$ is the influence function representing the sensitivity to the difference $(\theta_c - \theta_m)$ and α is a step size parameter.

The blocking and ringing effect are reduced by processing the above iterative method. The temporal fluctuations are suppressed by making changes in Eq. (3) as follows:

$$g\left(\theta_c^{(\omega)}\right) = \alpha^\omega \left(\sum_{p \in P_c} \psi\left(\theta_c^\omega - \theta_p\right) + \sum_{p \in Q_c} \psi\left(\theta_c^{(\omega)} - \theta_q\right)\right) \tag{4}$$

where p and q are pixel indexes, P_c is the immediate previous frame and Q_c is the next frame sets of neighboring pixels $(\theta_c^{(\omega)} - \theta_p)$ are the forward temporal estimation and $(\theta_c^{(\omega)} - \theta_q)$ are the backward temporal estimation.

4 Motion Estimation by Optical Flow

Optical Flow has gained lot of momentum towards much computer vision study and significant amount of research is being carried out on the same. In proposed method Lucas-Kanade approach is used.

- The pixels within the small window possess same motion.
- The offset vector from first order approximation of Taylor expansion since the motion is very small.

Consider a window slice 8 × 8, each one with $p = 8^2$ pixels. An overconstrained system with p equations and two variables are formed using local constraint movement.

$$\begin{aligned}
I_{X1}u + I_{y1}v + I_{t1} &= 0 \\
I_{X2}u + I_{y2}v + I_{t2} &= 0 \\
&| \\
I_{Xp}u + I_{yp}v + I_{tp} &= 0
\end{aligned} \tag{5}$$

Equation (5) can be solved by using Least Mean Square (LMS) method for estimating optical flow vector. The estimated optical flow for each 8 × 8 window corresponds to the optical flow vector of all the pixels in the related window.

5   Results and Discussion

Various video sequences contaminated with mixed noise and blur were analyzed. Block-Matching and 3D filtering-BM3D [14] and two dimensional Non-local Means Denoising methods were taken as reference filters. Initially our proposed algorithm was tested on an image contaminated by additive Gaussian noise and impulse noise (Fig. 2).

Objective qualitative measure such as Peak Signal to Noise Ratio (PSNR), Structural Similarity (SSIM) and Improved Minimum Mean Squared Error (IMMSE) are shown in Table 1. PSNR is calculated for RGB component while SSIM and IMMSE would be calculated only for Luminance component.

Test video that was corrupted by various noise scenarios and motion blur was tested and compared with different filters with reference to PSNR and SSIM values. Figure 3 shows the results of Non-Local Mean filter, BM3D filter and our proposed filter results for nature video sequence. The objective qualitative measures are depicted in Table 2.

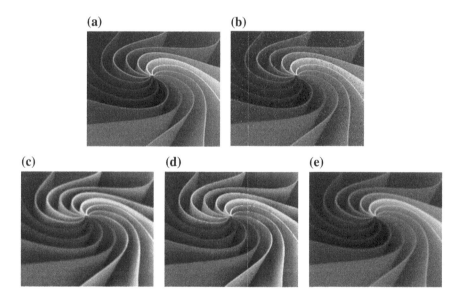

Fig. 2 Sample of **a** input image, **b** image corrupted by mixed noise, **c** non-local mean filter, **d** BM3D filter and **e** adaptive spatio-temporal filter

Table 1 Comparative qualitative measures of various methods for color image

Filters	PSNR (dB)	SSIM	IMMSE
Non-local mean	28.10	0.67	2.112
BM3D	26.07	0.71	2.330
Proposed adaptive spatio-temporal filter	33.48	0.88	2.664

Table 2 Comparative qualitative measures of various methods for nature video sequence

Filters	PSNR (dB)	SSIM	IMMSE
Non-local mean	23.50	0.23	2.912
BM3D	24.42	0.19	2.830
Adaptive spatio-temporal filter	29.48	0.816	2.564

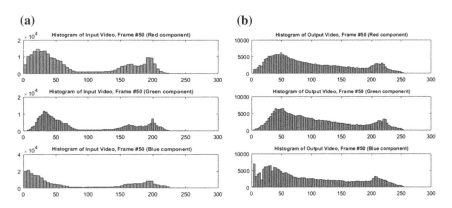

Fig. 3 Video sequence of Nature: **a** input image, **b** image corrupted by mixed noise, **c** non-local mean filter, **d** BM3D filter and **e** adaptive spatio-temporal filter

Fig. 4 Histogram graph of RGB component **a** before and **b** after execution of 50th frame of nature video sequence

The corresponding histogram of the RGB component before and after the execution is as shown in the Fig. 4. It is noted that the BM3D gives better results for Gaussian noise but the computational complexities of this filter gives limited application and also the filter requires the knowledge of noise model for execution.

Our proposed method even though it gives poor result for Gaussian noise it is simpler and gives good result for mixed noise. Even the Non-Local Mean filter does poor filtering of mixed noise.

It can be noted that the BM3D gives better results for Gaussian noise but the computational complexities of this filter gives limited application and also the filter requires the knowledge of noise model for execution. Our proposed method even though it gives poor result for Gaussian noise it is simpler and gives good result for mixed noise. Even the Non-Local Mean filter does poor filtering of mixed noise.

6 Conclusion

A universal noise removal system and video enhancement system has been presented which uses optical flow to reduce the motion blur. The adaptive Spatio-temporal filter switches from spatial to temporal based on the noise intensity which makes the model more robust. The approach in this paper removes the various scenarios of noises in video sequences and it even removes the motion artifacts caused due to undesirable camera movement. It is observed from the objective qualitative analysis of PSNR and SSIM that this proposed method gives desired result and is computationally simpler. However the execution time of this method can be reduced further if this algorithm is implemented on DSP processor with real time implementation.

References

1. P. Kisilev, S. Schein, Real-time video enhancement for high quality video conferencing. IEEE Trans. Commun. **31**(4), 532–540 (2010)
2. R. Garnett, T. Huegerich, C. Chui, W. He, A universal noise removal algorithm with an impulse detector. IEEE Trans. Image Process. **14**(11), 1747–1754 (2005)
3. M. Szczepanski, Fast spatio-temporal digital paths video filter. Real-Time Image Process. (2016). doi:10.1007/s11554-016-0561-7. Springer, Berlin, Heidelberg
4. J. Astola, P. Haavisto, Y. Neuovo, Vector median filters. IEEE Proc. **78**, 678–689 (1990)
5. B. Smolka, Peer group switching filter for impulse noise reduction incolor images. Pattern Recogn. Lett. **31**(6), 484–495 (2010). doi:10.1016/j.patrec.2009.09.012
6. B. Smolka, K. Plataniotis, A. Chydzinski, M. Szczepanski, Self-adaptive algorithm of impulsive noise reduction in color images. Patt. Recogn. **35**(8), 1771–1784 (2002)
7. C. Tomasi, R. Manduchi, Bilateral filtering for gray and color images, in *Proceedings of the 6th IEEE International Conference on Computer Vision (ICCV'98)*, Bombay, India, January 1998, pp. 839–846
8. M. Ben-Ezra, S. Nayar, Motion deblurring using hybrid imaging, in *Proceedings of CVPR 2003* (2003), pp. I-657–664
9. P. Perona, J. Malik, Scale-space and edge detection using anisotropic diffusion. IEEE Trans. Pattern Anal. Mach. Intell. **12**(7), 629–639 (1990)

10. K. Buades Dabov, A. Foi, V. Katkovnik, K. Egiazarian, Image denoising by sparse 3-d transform-domain collaborative filtering. IEEE Trans. Image Process. **16**(8), 2080–2095 (2007), doi:10.1109/TIP.2007.901238

11. K. Radlak, B. Smolka, B, Trimmed non-local means technique for mixed noise removal in color images, in *2013 IEEE International Symposium on Multimedia (ISM)* (2013), pp. 405–406

12. A. Buades, B. Coll, J.M. Morel, A non-local algorithm for image denoising, in *Proceedings of CVPR 2005* (2005), pp. II-60–65

13. K. Dabov, A. Foi, V. Katkovnik, K. Egiazarian, Image denoising by sparse 3-d transform-domain collaborative filtering. IEEE Trans. Image Process. **16**(8), 2080–2095 (2007), doi:10.1109/TIP.2007.901238

14. M. Maggioni, V. Katkovnik, K. Egiazarian, A. Foi, Nonlocal transform-domain filter for volumetric data denoising and reconstruction. IEEE Trans. Image Process. **22**(1), 119–133 (2013). doi:10.1109/TIP.2012.2210725

15. Y.W. Tai, H. Du, M. Brown, S. Lin, Image/video deblurring using a hybrid camera. Proc. CVPR **2008**, 1–8 (2008)

16. C. Wang, L.-F. Sun, B. Yang, Y.-M. Liu, S-Q Yang, Video enhancement using adaptive spatio-temporal connective filter and piecewise mapping. EURASIP J. Adv. Signal Process. 13 pages (2008). Article ID 165792, doi:10.1155/2008/165792

17. J. Portilla, V. Strela, M.J. Wainwright, E.P. Simoncelli, Image denoising using scale mixtures of Gaussians in the wavelet domain. IEEE Trans. Image Process. **12**(11), 1338–1351 (2003)

Social and Temporal-Aware Personalized Recommendation for Best Spreaders on Information Sharing

Ananthi Sheshasaayee and Hariharan Jayamangala

Abstract As the growth of online information sharing and online shopping is tremendous, the social networking sites and Online Shops (OSs) have become the potential information sources to the Recommender System (RS). The RS provides either services or products to the users based on their preferences. An Online Social Network (OSN) enables the users to share information with their social neighbors. An OS furnishes the products to the customers or users based on their requirements. The conventional context-aware recommender systems are inept at predicting the new user's preferences, and user's recent preferences. Usually, customers like to drift their preferences over time due to the evolution of the products in the OS. Hence, together considering of the key parameters of social popularity and temporal dynamics is crucial for modeling the RS. This paper presents Social and Temporal-Aware personalized Recommendation for the best Spreaders (STARS) approach which recommends the products based on the social influence and recent context information about the user. It employs the collaborative filtering and incorporates the three phases such as influence user identification using OSN, user's preference identification using OS, and recent preference based recommendation using temporal dynamics. Initially, the STARS identifies the best spreader in OSN by applying Eigen Vector Centrality (EVC) measurement in the k-shell structure. Secondly, it analyzes the customer's explicit as well as implicit feedback information using a user-item matrix factorization and Pearson correlation measurement. Finally, the STARS recommends the appropriate products to the users by predicting the user's recent preferences reading it from the context-aware explicit and implicit feedback information. The experimental results show that the STARS significantly outperform the conventional context-aware recommender systems.

A. Sheshasaayee (✉)
Department of Computer Science, Quaid-e-Millath Government College
for Women (Autonomous), Chennai 2, Tamilnadu, India
e-mail: ananthi.research@gmail.com

H. Jayamangala
Department of Computer Science,
Periyar University, Salem, Tamilnadu, India
e-mail: jayamangala.research@gmail.com

© Springer Nature Singapore Pte Ltd. 2017   509
S.C. Satapathy et al. (eds.), *Proceedings of the 5th International Conference on Frontiers in Intelligent Computing: Theory and Applications*, Advances in Intelligent Systems and Computing 516, DOI 10.1007/978-981-10-3156-4_53

Keywords Centrality · Social influence · User's preference · User-item matrix · Explicit · Implicit · Context-aware · Recommendation

1 Introduction

Nowadays, the emerging many Online Social Networks (OSNs), such as Facebook, Twitter, and LinkedIn, attracts a vast number of internet users. The OSNs allow the internet users to share their updates with family, friends, and acquaintances within a social network [1]. According to their identical interests, many users affiliate and establish the relationship in the social network. The Social influence of the user is exploited to leverage the viral marketing, which improves the product's popularity level within the allocated advertisement budget. Due to the dynamic growth of newly joining users in the social networks and their advent usage of OSNs, the social networks inundate abundant information creating the information overload. The information sharing mechanism of OSN accelerates the propagation of information to many users [2]. However, this information overload is very inconvenient for users who have several neighbor links while searching the useful social updates from the enormous information. Hence, to retain the customers of Online Shopping (OS), the RS is necessary for the social networks to recommend the product relevant information.

Recommender System (RS) [3] reduces the information overload in the social networks by providing personalized recommendations on different items such as music, movie, news, web pages, and books. Amazon and Netflix are the popular recommender systems which advertise the products to the customers to gain a place in the e-commerce field. The RS recommends either products or services to the users based on the user's preference by explicitly analyzing the user's feedback information or implicitly analyzing the user's behavior in OSN and OS [4]. The user's preference based recommendation facilitates the user to find the useful social updates and the required products from the abundant information. Usually, if a user is the best spreader in OSN, the RS commands popularity. The conventional recommendation systems employ the two types of algorithms, namely content-based filtering method [5] and collaborative filtering [6, 7] method. In general, a user's interest in the specific item changes over time, hence focusing on the temporal factor is crucial in RS. Temporal dynamics is time-interval based rating. Most of the existing recommender systems underpin the collaborative filtering algorithms to recommend the relevant items by using the corresponding user's feedback [8]. However, it lacks in accurately recommending the items, since the contextual factors indicate that a single user is providing the different ratings on an item [9]. Thus, both the social influence and accurate context of user's feedback, especially temporal information consideration are necessary for the RS.

2 Proposed Methodology

Most of the conventional Recommender Systems still possess two major short-comings, namely cold-start and data sparsity due to the lack of personalization of the users in a social network. The STARS approach focuses on maximizing the profitability of the advertiser within their budget by temporally identifying the context-aware user's preferences. The dynamic user's intentions are abruptly generated and rapidly propagated through social networks. Hence, the STARS employs

---

Input: OSN, and OSs website
Output: List of items to be recommended
//Identifying information hubs in online social network
for each user i **do**
    for each neighbor **do**
        Select the best spreader using Eigen score of user's in OSN
    endfor
endfor
//Analyzing user (re)views on a specific item in OS
while $U = \{U_1, U_2,...U_i\}$; $I = \{I_1, I_2,...I_j\}$; $R = \{R_1, R_2,...R_i\}$ **do**
    for all best spreaders or influential nodes **do**
        Construct the user-item matrix based on the user's ratings in OS
            if(U_i =new user) **then**
                Measure $Sim(I_1, I_2)$ using Pearson correlation co-efficient
                Predict r_{ij} using $F_{U,I}^{Exp}$
elseif
                Calculate $F_{U,I}^{Imp}$ using $f(I^P)$, S_t and $f(I_j^L)$
            endif
//Context-aware personalized recommendation
        while all social users with their preferences **do**
            for t=$\{t_1,t_2,...t_n\}$ **do**
                Filter the items based on the brand using $F_{U,I}^{Exp}$and $F_{U,I}^{Imp}$
                    if($N(U_{ij}^R)$>1) **then** (i.e.$N(U_{ij}^R)$-number of user's ratings)
                        Determine the recent preference of ith user on j^{th}
                item
                        Calculate $W(F_{U,I}^{Exp})_t$ and $W(F_{U,I}^{Imp})_t$
                        Identify R_{Pred} using $W(F_{U,I}^{Exp})_t$ and $W(F_{U,I}^{Imp})_t$
                  endif
                    if(R_{Pred}>η) **then**
                        Select item set to recommend i^{th} user
                    endif
            endfor
        endwhile
    endfor
endwhile

Fig. 1 Algorithm 1: the STARS algorithm

OSN and OS sources to recognize the social user's preferences. The STARS incorporates three phases such as identifying information hubs in the online social network which employs Eigen score measure to find influential node, analyzing user (re)views on a specific item in the online shop, and context-aware personalized recommendation which includes brand and time (Fig. 1).

3 Experimental Evaluation

This paper evaluates the STARS with the conventional recommendation method of Context-aware movie recommendation (CAMR) approach [10] to illustrate the prediction and recommendation accuracy.

The STARS approach runs the experiments on Linux Ubuntu 12.04 LTS 64-bit machine with a 2.9 GHz Intel CPU and 32 GB memory. To evaluate the performance of recommended items, the STARS employs the Rival Java evaluation toolkit which provides the results of the assessment by comparing across different recommendation frameworks.

3.1 Results and Analysis with Metrics

3.1.1 Precision

Figure 2 reveals the precision of both the STARS and CAMR approach while varying the percentage of new users and New Items (NI) in OS. Precision is the ratio between the number of relevant items that are recommended and the number

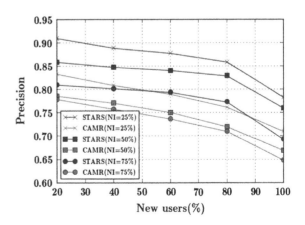

Fig. 2 Precision versus new users

of predicted items to be recommended. On the whole, the precision value decreases while increasing the percentage of new users in OS. At the point of 20 % of new users and NI = 25 %, the precision of the STARS has increased by 9.24 % more than the CAMR approach. While forecasting the missing rating of new users, the increasing percentage of new users degrade the performance of rating prediction and recommendation due to the lack of their historical rating information. Though the STARS approach maintains the precision value with a slight deviation of 1.9 % until reaching 80 % of the new users than the CAMR approach, it suddenly drops the precision value, because, the STARS approach also focuses the recent context-aware rating prediction while recommending the items to new users.

3.1.2 Recall

The Recall results of both the STARS and CAMR approach is illustrated in Fig. 3. Recall is the ratio between the number of relevant items that are recommended and the total number of relevant items that are to be recommended. At the point of new users is 20 % and NI = 25 %, the STARS approach attains 6.58 % improvement in the recommendation accuracy than that on the CAMR approach. Additionally, increasing from 20 to 80 % of the new users, the STARS approach keeps the recall value at 5 % depletion when NI = 25 %. However, the CAMR approach depletes the recall value by 8 %, since the model-based collaborative filtering method, i.e. neural network and machine learning requires the network model based training information to predict the missing rating. Moreover, the CAMR approach is a strenuous process in the sense of training all the rating information to the system, and maintaining the trade-off between the scalability and prediction performance.

Fig. 3 Recall versus new users

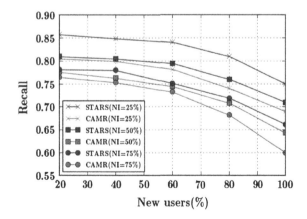

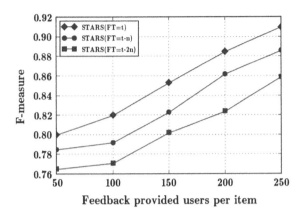

Fig. 4 F-measure versus feedback provided users per item

3.1.3 F-Measure

F-Measure or F-score is the discrepancy and balance between precision and recall ie. F-measure = 2 * ((P * R) / (P + R)). Figure 4 indicates the F-measure of the STARS approach while varying the number of feedback provided users on an item and Feedback Time (FT). FT is the time of providing either explicit or implicit feedback information in which 't' implies the recent time, 't-n' and 't-2n' denotes the preceding time. If many users rate a specific item, the F-measure value increases, since the number of feedback information provides the inherent importance of the specific item. The STARS approach recommends the items based on assigning the higher weight to the recent FT to accurately forecast the user's area-of-interests. Hence, the F-measure at time 't' is higher than the time 't-n' and 't-2n'. It reflects that the customer's ancient preferences are deviated from the recent preferences due to the dynamic item's popularity and user's taste. When increasing the number of customers who have provided the feedback information on an item from 50 to 250, the STARS approach achieves 13.74 % of increasing F-measure value when FT = t.

3.1.4 Profit

Figure 5 exhibits the profit of the recommender system for both the STARS and CAMR approach while increasing the number of ISH per item and NN. The ISH represents the Interesting Social influence Hubs which indicates the total number of best spreaders interested in an item. The NN denotes the Number of Neighbors of each ISH in OSN. The percentage of profit indicates the percentage of reaching the target budget while recommending the items to the social users. The STARS approach gains 96 % profit when NN = 30 and the number of ISH are 50, but when NN = 20, it earns 92 % profit due to the limited diffusion level. At the point of the number of ISH is 50, the STARS approach increases the profit value by 6 % than

Fig. 5 Profit versus number of ISH per item

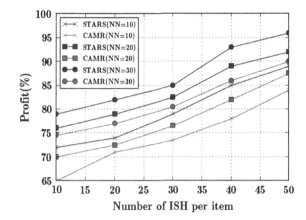

the CAMR approach when NN = 30, because, the STARS approach efficiently recognizes the best spreader using k-shell structure and recommends the relevant items to that user by combining both the temporal-aware explicit and implicit feedback information.

4 Conclusion

In this paper, the STARS approach has been proposed to recommend the relevant items to the social users. Intuitively, it targets to increase the advertiser's profit within the advertisement budget. The STARS approach incorporates the contextual information into the recommendation process and recommends the items based on the user's implicit and explicit feedback information. To estimate the missing or unknown rating, it handles the memory-based traditional collaborative filtering method along with temporal dynamics. The STARS approach incorporates three phases such as influence user identification using OSN, user's preference identification using OS, and recent preference based recommendation using temporal dynamics. Eventually, the experimental results empirically demonstrate that the STARS approach substantially outperforms the conventional recommendation method.

References

1. H. Quan, J. Wu, J. Shi, *Online Social Networks and Social Network Services: A Technical Survey. Handbook of Pervasive Communication* (CRC Press, Boca Raton, 2011)
2. E. Bakshy, I. Rosenn, C. Marlow, L. Adamic, The role of social networks in information diffusion, in *ACM Proceedings of the 21st International Conference on World Wide Web* (2012), pp. 519–528

3. J. Bobadilla, F. Ortega, A. Hernando, A. Gutiérrez, Recommender systems survey. Elsevier Trans. Knowledge-Based Syst. **46**, 109–132 (2013)
4. S.M. Kywe, E.P. Lim, F. Zhu, A survey of recommender systems in twitter. Springer Trans. Social Inf. **7710**, 420–433 (2012)
5. P. Lops, M. De Gemmis, G. Semeraro, Content-based recommender systems: state of the art and trends, in *Springer Transaction on Recommender Systems Handbook* (2011), pp. 73–105
6. X. Su, T.M. Khoshgoftaar, A survey of collaborative filtering techniques. ACM Proc. Adv. Artif. Intell. **4**, 1–19 (2009)
7. Y. Shi, M. Larson, A. Hanjalic, Collaborative filtering beyond the user-item matrix: a survey of the state of the art and future challenges. ACM Comput. Surv. (CSUR) **47**(1), 3 (2014)
8. X. Yang, Y. Guo, Y. Liu, H. Steck, A survey of collaborative filtering based social recommender systems. Elsevier Trans. Comput. Commun. **41**, 1–10 (2014)
9. G. Adomavicius, A. Tuzhilin, Toward the next generation of recommender systems: a survey of the state-of-the-art and possible extensions. IEEE Trans. Knowl. Data Eng. **17**(6), 734–749 (2005)
10. C. Biancalana, F. Gasparetti, A. Micarelli, A. Miola, G. Sansonetti, Context-aware movie recommendation based on signal processing and machine learning, in *ACM Proceedings of the 2nd Challenge on Context-Aware Movie Recommendation* (2011), pp. 5–10

A Combined System for Regionalization in Spatial Data Mining Based on Fuzzy C-Means Algorithm with Gravitational Search Algorithm

Ananthi Sheshasaayee and D. Sridevi

Abstract The proposed new hybrid approach for data clustering is achieved by initially exploiting spatial fuzzy c-means for clustering the vertex into homogeneous regions. Further to improve the fuzzy c-means with its achievement in segmentation, we make use of gravitational search algorithm which is inspired by Newton's rule of gravity. In this paper, a modified modularity measure to optimize the cluster is presented. The technique is evaluated under standard metrics of accuracy, sensitivity, specificity, Map, RMSE and MAD. From the results, we can infer that the proposed technique has obtained good results.

Keywords Spatial data mining · Clustering · Fuzzy c-means · Gravitational search algorithm · Regionalization · Metrics

1 Introduction

In spatial data mining, knowledge discovery denotes the unearthing of implied, previously unknown and interesting knowledge from spatial databases. The important task in spatial data mining is spatial clustering. It aims to group similar spatial objects into classes or clusters. The objects in a cluster have high similarity in estimation to one another and are distinct to objects in other clusters [1]. A prime application area for spatial clustering algorithms in geography is towards social and economical issue. In the scope a classical methodical problem of social geography is "regionalisation". In general it is known as a classification procedure applied to spatial objects with an area representation. It groups them into contiguous regions of homogeneous nature [2, 3]. The main objective of regionalization is to find

Ananthi Sheshasaayee (✉)
Quaid-E-Millath Goverment College for Women (Autonomous), Chennai, Tamil Nadu, India
e-mail: ananthi.research@gmail.com

D. Sridevi
Computer Science Department, Sri Chandrasekharendra Saraswathi Viswa Maha Vidyalaya, Kanchipuram, Tamil Nadu, India

© Springer Nature Singapore Pte Ltd. 2017
S.C. Satapathy et al. (eds.), *Proceedings of the 5th International Conference on Frontiers in Intelligent Computing: Theory and Applications*, Advances in Intelligent Systems and Computing 516, DOI 10.1007/978-981-10-3156-4_54

517

spatial regions of not a specific shape (arbitrary) with a homogeneous internal distribution of non-spatial variables with respect to compactness and density [4]. In this paper, we suggest a new hybrid approach for data clustering making use of FCM and gravitational search optimization with the results obtained for the different clusters.

2   Literature Survey

In the literature survey, several methods have been proposed for the Regionalization in spatial data mining. Among the most recently published works are those presented as follows:

Niesterowicz and Stepinski [5] has explained the regionalization of multi-categorical landscape. The land cover pattern is based on the principle of machine vision rather than clustering of landscape metrics. The NLCD 2006 shows spatially varying pattern of Land Use/Land Categories of maps. Using an LULC map as an input of their method locates and maps different types of landscapes as LTs, based on the characteristic of LULC pattern domination.

Additionally, Xie et al. [6] has explained the Regionalization of Chinese Medicinal Plants with climate factors based on Spatial Data Mining. They obtained data of nearly 100 varieties of herb plants and performed clustering which results as eco-environment has outcome on geo-authenticities.

In, Assuncao et al. [7] has explained the regionalization techniques using graph and neighbourhood by minimum spanning tree. The result was the division of the spatial objects into connected by regions that have more inner homogeneity.

3   Problem Definition

Clustering which is basically a process of partitioning data objects into subsets, is a critical task in data mining. Spatial clustering is one of the important components of spatial data mining. Most of the clustering technique has its own drawbacks. The regionalization can be solved with the employment of clustering techniques.

4   Proposed Methodology

In this paper, the proposed data clustering algorithm consist of various stage of processing. At each iteration, the proposed system initializes the cluster center and fuzzy membership matrix is calculated. Then, the fitness for each candidate solution is computed. This work uses a modularity based fitness measure to improve the accuracy. In addition to this, a measure μ is added with fitness measure in order to gain high clustering accuracy. After this, update the global best and worst fitness among them. Further calculate the mass and acceleration for each candidate solution. For each candidate solution update the velocity and position. Finally best candidate solution is obtained as output at $t + 1$ iteration.

4.1 Preliminaries

4.1.1 Fuzzy C-Means Clustering

Fuzzy C mean is the advanced version of k-means, where it needs to do a full inverse-distance weighting [8]. Fuzzy c-means is comparatively take less time and iterations than the popular k-means algorithm.

The step by step procedure of the fuzzy c-means clustering algorithm is

Step 1: Selection of cluster centre.
Step 2: Initialize U.
Step 3: Centre Vectors are calculated.
Step 4: Update the matrix with the calculated value.
Step 5: Check if the distance between the absolute value of the U with the existing, if it is less than ε of the matrix then stop. Otherwise go to step 2.

4.1.2 Gravitational Search Algorithm

One of the heuristic algorithms is gravitational search algorithm. It works by the law of gravity and also their mass interactions. Various application problems are solved using it like travelling salesman problem, image upgrading and filtering problem and designing DNA problem by sequencing. First the initial population is generated and is evaluated using the fitness function. Gravitational constant is later found out. Then, compute the gravitational masses, acceleration and the force. Update the particle velocity and position. Continue the process till the stopping criteria is met.

4.2 Proposed Modularity Based Hybrid Clustering Algorithm

The proposed modularity based hybrid clustering algorithm consists of various stages of processing. They are,

(i) Input data: The n inputs with features are clustered into k groups. The input centroid is optimized using algorithm.

(ii) Pre-processing: The second stage consist of the analysis of data to find the missing and unwanted data.

(iii) Modularity based Clustering: With the input data, membership matrix is calculated with the Euclidean distance. Further the fuzzy membership matrix and the centre is evaluated with our proposed fitness function.

The step by step procedure of our proposed modularity based hybrid clustering algorithm is given below,

Step 1: Initially fix the centroid C and initialize cluster centre.
Step 2: Next calculate the fuzzy membership matrix based on the equation.
Step 3: Intialize a group of solution as cluster center.
Step 4: Modularity measure for each agent is found and fitness is computed.
Step 5: Update the global best and worst solution based on fitness.
Step 6: Calculate the mass and acceleration for each solution.
Step 7: The velocity and position for each solution is updated.
Step 8: Continue steps 4–8 till stopping criteria are met.

5 Results

5.1 Experimental Setup with Dataset Description

We have implemented the proposed method using MATLAB as tool. The configuration was with 6 GB RAM and 2.6 GHz in Intel i-7 processor. To evaluate the performance of the proposed technique, Localization Data for Posture Reconstruction from the repository of UCI machine learning is considered [9]. Additionally, the data was tag readings of the people for the different moments. The total number of instance found in this dataset is 164860 and the total number of attributes found in this dataset is 8. The activities used in this dataset are walking, lying, sitting, standing and falling. The characteristics of the attribute are real, uni—variate, sequential and time-series. The paper is focussed with cluster size of 3, 4 and 5 which can be changed.

5.2 Evaluation Metric

The evaluation metrics used here for the purpose of performance evaluation prediction system are, (a) accuracy, (b) sensitivity and (c) specificity.

(a) **Accuracy**: It means probability that our proposed system can correctly predict both positive and negative examples.

$$Accuracy = (TP + TN)/(TP + TN + FP + FN)$$

(b) **Sensitivity**: It means probability that the algorithms can correctly predict only positive examples.

$$Sensitivity = TP/(TP + FN)$$

(c) **Specificity**: It means probability that the algorithms can correctly predict only negative examples

$$Specificity = TN/(TN + FP)$$

where, TP is true positive, TN is true negative, FP is false positive and FN is false negative.

(d) **Mean Absolute Percentage Error (MAPE)**: It is,

$$MAPE = (100/length(Obtain\ op) \times \sum \frac{(Obtain\ op - Orginal\ op)}{Orginal\ op}$$

(e) **Root Mean Square Error (RMSE)**: It is

$$RMSE = \sqrt{\frac{1}{length(Obtain\ op)}} \times \sum \frac{(Obtain\ op - Orginal\ op)}{Orginal\ op}$$

(f) **Mean Absolute Deviation (MAD)**: The MAD of an intrusion detection system is

$$MAD = \frac{1}{length(Obtain\ op)} \times \sum (Obtain\ op - Orginal\ op)$$

(g) **Mean Square Error (MSE)**: The MSE of an intrusion detection system is,

$$MSE = Mean((Original\ op - Obtain\ Op^2)$$

where Original op is Original output and Obtain op is obtained output.

5.3 Output

(a) **Accuracy**: From the above graph we found that the accuracy of our proposed method increases with cluster size (Fig. 1).

(b) **Sensitivity**: The sensitivity obtained for our proposed modularity based hybrid clustering algorithm is shown in the figure below (Fig. 2),

Fig. 1 Accuracy

Fig. 2 Sensitivity

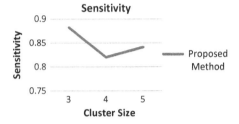

Fig. 3 Specificity

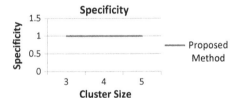

Fig. 4 MAPE

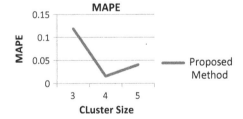

(c) **Specificity**: The specificity obtained for our proposed modularity based hybrid clustering algorithm is given below (Fig. 3),

(d) **Mean Absolute Percentage Error (MAPE)**: The obtained Mean Absolute Percentage Error for our proposed method is shown in the graph below (Fig. 4),

(e) **Root Mean Square Error (RMSE)**: The obtained Root Mean Square Error for our proposed modularity based hybrid clustering algorithm is shown below (Fig. 5),

Fig. 5 RMSE

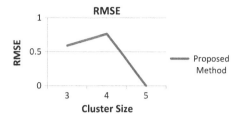

Fig. 6 MAD

Fig. 7 MSE

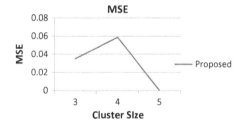

(f) **Mean Absolute Deviation (MAD)**: The obtained Mean Absolute Deviation of a proposed modularity based hybrid clustering algorithm is shown in the figure below (Fig. 6),

(g) **Mean Square Error (MSE)**: The obtained Mean Square Error for our proposed hybrid clustering algorithm is shown in the figure below (Fig. 7),

6 Conclusion

In this paper the designed hybrid clustering algorithm for optimal clustering is evaluated with the standard metrics. The fuzzy algorithms are mainly used as problem solver for resources consumption like oil, water, electricity power consumption, meteorological data's, climatic changes, image segmentation and in medical diagnosis for thyroid, tumour, cancer and heart diseases. Although it works well in real-world applications the extended versions are supportive to other areas like telecommunication, networking, Geographical Information System. Thus this combined system has a better performance when fuzzy c means is in conjunction with gravitational search algorithms.

References

1. P. Srinivas, S.K. Satpathy, L.K Sharma, A.K. Akasapu, regionalisation as spatial data mining problem: a comparative study. Int. J. Comput. Trends Technol. May to June Issue (2011)
2. C.D. Juan, R. Raul, S. Jordi, Supervised Regionalization Methods: a Survey, Res. Inst. Appl. Econ. (2006)
3. R.M. Assuncao, M.C. Neves, G. Câmara, C.C. Freitas, Efficient regionalization techniques for socio-economic geographical units using minimum spanning trees. Int. J. Geogr. Inf. Sci. **20** (7), 797–811 (2006)
4. J. Christina, Dr.K. Komathy, Analysis of hard clustering algorithms applicable to regionalization, in *Proceedings of 2013 IEEE Conference on Information and Communication Technologies* (ICT 2013)
5. Jacek Niesterowicz, Tomasz F. Stepinski, Regionalization of multi-categorical landscapes using machine vision methods. Appl. Geogr. **4**, 250–258 (2013)
6. C. Xie, S. Chen, F. Suo, D. yang, C. Sun, Regionalization of chinese medicinal plants based on spatial data mining, *2010 Seventh International Conference on Fuzzy Systems and Knowledge Discovery* (FSKD 2010)
7. R.M. Assuncao, M.C. Neves, G. Camara, C. Da Costa Freitas, Efficient regionalization techniques for socio-economic geographical units using minimum spanning trees. Int. J. Geogr. Inf. Sci. **20**(7), 797–811 (2006)
8. Y. Kumar, G. Sahoo, A review on gravitational search algorithm and its applications to data clustering and classification, I. J. Intell. Syst. Appl. **6**, 79–93 (2014)
9. http://archive.ics.uci.edu/ml/datasets/Localization+Data+for+Person+Activity

Design and Functional Verification of Axi2OCP Bridge for Highly Optimized Bus Utilization and Closure Using Functional Coverage

N. Shalini and K.P. Shashikala

Abstract Given the density of current SOC's, Bridge design is used to connect interconnects working on different frequencies, protocols and bus widths. AXI and OCP are very commonly used protocols in industry given the fact that they support wide range of features. AXI2OCP bridge is used to connect 2 interconnects, one working on AXI protocol another on OCP protocol. Expected functioning of the Bridge design can be obtained by the process of verification, without proper verification the system may show unexpected behavior. Performance of the bus can be increased with effective bus utilization. Building verification environment for AXI2OCP Bridge using System verilog, Generating and simulating the test cases for various features of AXI and OCP, Measuring the Bus utilization parameter for the AXI 3.0 protocol, implementing Functional coverage and assertions with the proposed integrated verification environment using Questa—sim tool is main idea of the paper.

Keywords AXI 3.0 protocol · System Verilog · Functional coverage · AXI2OCP bridge · Bus utilization

1 Background

In the past decade, advanced peripheral bus and the advanced high-performance bus of AMBA are the most popular communication architectures for SOC designs. Advanced extensible Interface (AXI) and open core protocol (OCP) are some advanced communication protocols that facilitate parallel communication are thus proposed. Verification of these complex protocols is bit challenging. It can be easily verified by building verification environment. The Advanced extensible interface 4

N. Shalini (✉) · K.P. Shashikala
Dayananda Sagar College of Engineering, Bangalore, India
e-mail: shalureddy911@gmail.com

K.P. Shashikala
e-mail: kp.shashikala@gmail.com

© Springer Nature Singapore Pte Ltd. 2017
S.C. Satapathy et al. (eds.), *Proceedings of the 5th International Conference on Frontiers in Intelligent Computing: Theory and Applications*, Advances in Intelligent Systems and Computing 516, DOI 10.1007/978-981-10-3156-4_55

(AXI4) update to AMBA AXI3 includes the following: updated write response requirements support for burst lengths up to 256 beats. AXI4 protocol supports 16 masters and 16 slaves interfacing. Verifying on-chip communication properties is the challenge of integration. In study of high performance protocols like AXI and OCP, in [1] an efficient read and writes channel for memory interface in SOC is described for AXI. Burst based transaction in AMBA AXI for SOC integration [2] is Designed and some AXI protocol rules such as read, write, and handshake mechanisms are simulated in modelsim. Design of AXI4 slave interface is verified and environment built on system verilog [3] different peripherals are connected to into AMBA based processors using slave interface without any bridge. Communication between multiple masters and multiple slaves is described in [4] the interconnect can works at 100 MHz frequency at Vertex E as target Device (Synthesized by Xilinx ISE 8.2i).

Verification Environment for AMBA axi protocol is build using system verilog, functional coverage and assertions are implemented [5] this proposed environment can improve the coverage and time spending in the verification can be reduced. Design and verification of OCP protocol using property specification language (PSL) assertions [6] and communication between AXI and OCP is been established. It includes developing FSM's for OCP and AXI using VHDL. AXI2OCP bridge supporting out of order transactions for achieving deadlock free communication [7] is Designed and verified and this address a deadlock problem in an on chip bus system which supports out of order transactions and also presents a graphical model which represent a status of bus system. In [8] using system verilog they focused on functional verification of AXI2OCP bridge for read, write, write_read test cases generation and bus utilization is calculated for a AMBA AXI 3.0. Practical waveforms are generated by the Mentor Graphics QUEST A-SIM EDA tool are explored.

2 Architecture of AXI and OCP

2.1 Advanced Extensible Interface Protocol

AXI is targeted for high frequency and high performance system designs which include wide range of features such as aligned, unaligned, burst, out of order, protected, overlapping kind of transfers.

Every transaction in AXI includes address and control information. Data is transferred to slave through write data channel and master through read data channel. Figure 1 shows AXI bus Architecture it has five independent channels used for write and read transfers consist of information signals and Two-way READY and VALID handshake mechanism to exchange information between sender and receiver. Write address channel and write data channel are used for write request. Read address channel is used for read request. For write response and read

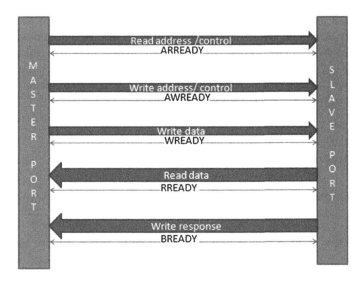

Fig. 1 AXI bus architecture

response, the write response channel and read data channel is used respectively. Acknowledgement for write request given by, AWREADY, WREADY signals. Read request given by ARREADY signal. Master acknowledges read response with RREADY and write response with BREADY.

2.2 Open Core Protocol

Basic block diagram of open core protocol is shown in the Fig. 2 consist of two cores one core (CORE 1) is having OCP master and another core (CORE 2) is having OCP slave.

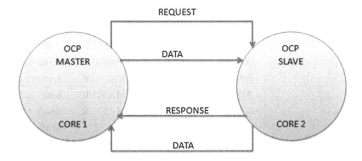

Fig. 2 OCP block diagram

By presenting request command on the request line to the slave the master starts either read or write request. Master also puts the request information on the data lines to the slave with the request command. Slave will acknowledge the master by registering the request information, slave waits for the response from the CORE2 after presenting request to CORE2. Later slave sends response information to master through response and data lines (Table 1).

3 Proposed Verification Method

The components that are included in the above verification environment shown in Fig. 3 are generator, bus functional model, AXI and OCP monitor, coverage module, reference module, Design under verification (DUV), OCP slave and checker. All these modules are designed according to their respective functionalities. For different user application generator will generates test cases using constrained randomization and through mailbox it will connect to the bus functional model (BFM), here mailbox acts as a synchronizer between two modules and BFM

Table 1 Signal descriptions of AXI and OCP

Signal naming	Width	Driver	Function
AWADDR	5	MASTER	Write address
AWID	3	MASTER	Write address ID
WDATA	5	MASTER	Write data
WID	3	MASTER	Write ID tag
ARADDR	5	MASTER	Read address
ARID	3	SLAVE	Read address ID
AWEADY	1	SLAVE	Write addressreadv
WREADY	1	SLAVE	Write ready
BRESP	2	SLAVE	Write response
BID	3	SLAVE	Response ID
AREADY	1	SLAVE	Read address readv
RRESP	2	SLAVE	Read response
EDATA	5	SLAVE	Read data
RID	3	SLAVE	Re ID tag
MAddr	5	MASTER	Transfer address
MData	5	MASTER	Write data
MCmd	3	MASTER	Transfer command
MBurstlength	3	MASTER	Burst length
MITagid	3	MASTER	Request tag ID
SData	5	SLAVE	Read data
SResplast	1	SLAVE	Last response in burst
STagid	3	SLAVE	Response tag ID

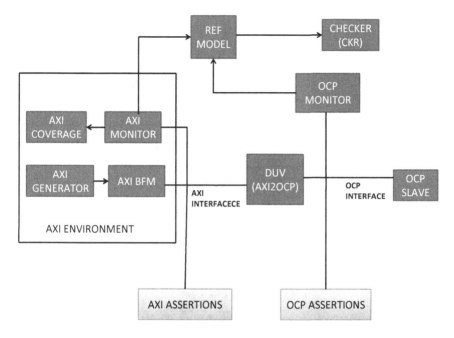

Fig. 3 Verification environment of AXI2OCP bridge

plays major role in driving transactions to the AXI interface and DUV converts all AXI format Signals to OCP format and communicate with OCP slave based on read or write request. Transactions at AXI and OCP side are monitored by AXI and OCP monitor respectively and both monitored transactions are given to the reference module which converts transactions to common format and gives to checker so checker will compare the AXI and OCP data and displays the results. Assertions are implemented to check the functionality of design. Every TB has a top most module and top most module has lot of initial used to generate clock, apply reset and we instantiate program block below top most module.

4 Results Analysis

4.1 Simulation Results for Write Operation

As shown in Fig. 4 data are writing in address location AWADDR [31:0] in AXI and same data appearing in OCP side through signal MData [63:0]. Test for transactions of different burst length AWLEN [3:0] is shown in Fig. 5, Test for unaligned transactions are shown in Fig. 6, Test for out of order transactions are shown in Fig. 7 Such that data will be written in the different order in which they are issued. Figure 8 shows test for overlapped transactions.

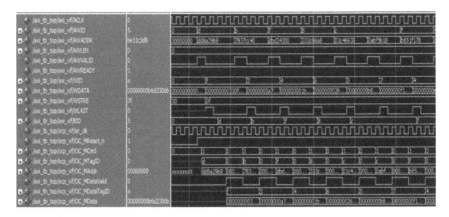

Fig. 4 Test for writing data's each carries different address

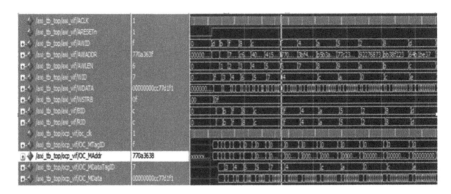

Fig. 5 Test for writing transactions with different burst lengths

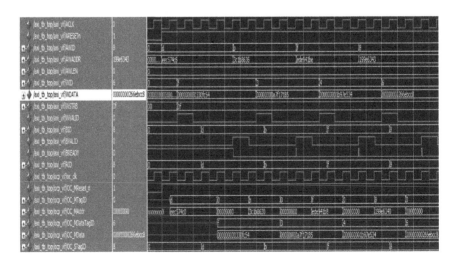

Fig. 6 Test for unaligned transactions

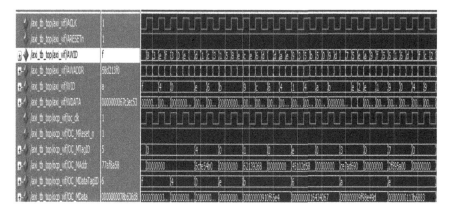

Fig. 7 Test for out of order transactions

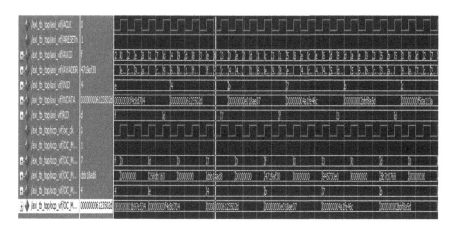

Fig. 8 Test for overlap transactions

4.2 Simulation Results for Read Operation

As shown in Fig. 9. Data are reading address location ARADDR [31:0] in AXI through signal and same data appearing in OCP side through signal SData [63:0]. Test for transactions of different burst length ARLEN [3:0] is shown in Fig. 10 and test for write_read operation is shown in Fig. 11.

Here in Fig. 12 read transaction took totally 18 cycles out of that only 16 cycles we are doing data transfer

Bus utilization = (16/18) % = close to 89 %

Here in Fig. 13 write transaction took totally 19 cycles out of that only 16 cycles we are doing data transfer

Bus utilization = (16/19) % = close to 84 % (Table 2)

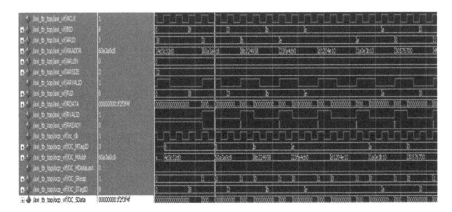

Fig. 9 Test for reading data's each carries different address

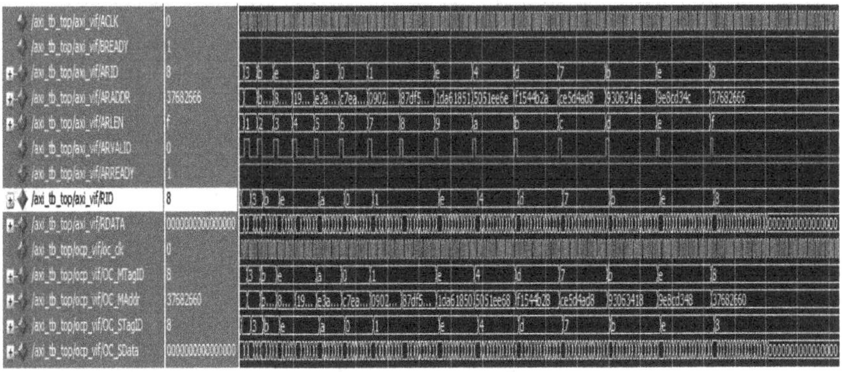

Fig. 10 Test for reading transactions with different burst lengths

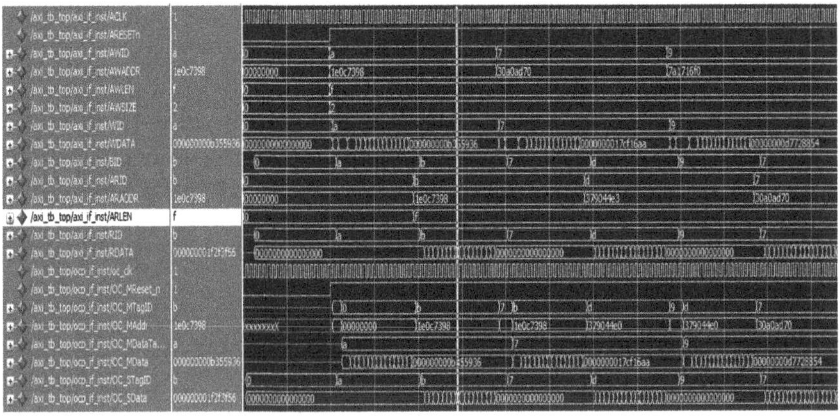

Fig. 11 Test for write_read operation

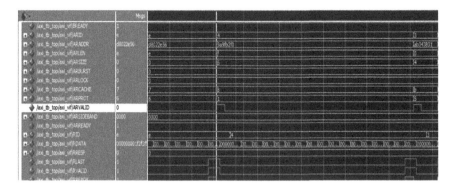

Fig. 12 Bus utilization for read phase

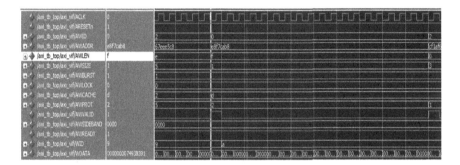

Fig. 13 Bus utilization for write phase

Table 2 Inference

Test case	Data cycles	Total cycles	Bus utilization (%)
Read phase	16	18	88.89
Write phase	16	19	84.21

4.3 Coverage Report

Coverage report shown in Fig. 14 depicts that 100 % Functional coverage obtained with 147 Covergroups and 3 Directive assertions hits for simple write and read transactions.

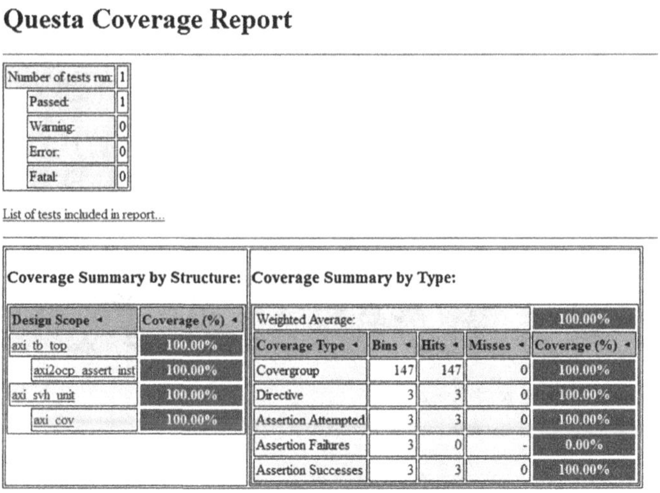

Questa Coverage Report

Number of tests run:	1
Passed:	1
Warning:	0
Error:	0
Fatal:	0

List of tests included in report...

Coverage Summary by Structure:		Coverage Summary by Type:				
Design Scope	Coverage (%)	Weighted Average:				100.00%
axi_tb_top	100.00%	Coverage Type	Bins	Hits	Misses	Coverage (%)
axi2ocp_assert_inst	100.00%	Covergroup	147	147	0	100.00%
axi_svh_unit	100.00%	Directive	3	3	0	100.00%
axi_cov	100.00%	Assertion Attempted	3	3	0	100.00%
		Assertion Failures	3	0	-	0.00%
		Assertion Successes	3	3	0	100.00%

Fig. 14 Functional coverage (assertions)

5 Conclusion and Future Work

Test cases are generated for some of the features of AXI2OCP Bridge like simple write and read transfers, burst transfers for read, write and write-read tests, aligned and unaligned transfers, out of order and overlapped transfers. Practical waveforms are obtained by Mentor Graphics QUEST A-SIM EDA tool without any change in the behavior of the design. The bus utilization parameter with proper inference is obtained and produced for AXI3.0 protocol and maximum coverage report is obtained for AXI2OCP Bridge.

Future work is to show some more unique features of AXI like protected and cache type of transactions. Code improvement can be done to set up regression to run all test cases as a group which can be carried out by PEARL script language and also to achieve 16 out of order transactions. We can include assertions at OCP interface to check with more functionality.

Acknowledgments Authors are thankful to department of electronics and communication, Dayananda Sagar College of Engineering for their encouragement and help for the execution of the project.

References

1. A. Tiwari, J. Nagar, An efficient axi read and write channel for memory interface in system-on-chip. Int. J. Digital Appl Contem. Res. **3**(3) (2014)
2. V.N.M.K. Brahmanandam, C. Monohar, Design of burst based transactions in AMBA-AXI protocol for soc integration.Int. J. Sci. Eng. Res. **3**(7), 2229–5518 (2012)

3. B. Krithi, S. Bhat, Y. Panchaksharaiah, Verification of AMBA based AXI4 slave interface. Int. J. Comput. Sci. Inf. Technol. **6**(3), 3135–3137 (2015)
4. M. Rai Nigam, S. Bande, AXI interconnect between four master and four slave interfaces. Int. J. Eng. Res. Gen. Sci. **2**(4), 2091–2730 (2014)
5. B.M. Mahendra, A.C. Ramachandra, Bus functional model verification ip development of AXI protocol. Int. J. Innovative Res. Sci. Eng. Technol. **3** (2014)
6. T. Naga Prasad Reddy, K. Avinash, Design and PSL verification of SoC interconnect using open core protocol (OCP). Int. J. Comput. Trends Technol.**4**, 2231–2803 (2013)
7. S.K. Roopa, Design and verification of AXI-OCP bridge supporting out-of-order transactions for achieving dead lock free communication. Int. J. Res. Eng. **01**, 2348–7852
8. Golla Mahesh., S.M. Sakthi Vel, Functional Verification of the Axi2Ocp Bridge using System Verilog and effective bus utilization calculation for AMBA AXI 3.0 Protocol in *IEEE Sponsored 2nd International Conference on Innovations in Information, Embedded and Communication systems* (2015)

Person Recognition Using Surf Features and Vola-Jones Algorithm

S. Shwetha, Sunanda Dixit and B.I. Khondanpur

Abstract Face recognition is one of the prominent biometric software applications, which can identify specific person in a digital image by analysing few parameters and comparing them. These type of recognitions are commonly used in security systems but are used increasingly in variety of other applications. Few non static conditions like facial hair can make recognition system a serious problem. The three stages of face recognition system are facing detection, feature extraction and classification. For enhancing the face recognition from video successions against dissimilar occlusion invariant and posture is proposed by using a novel approach. This face identification system made use of Viola and Jones algorithm for face detection and SURF (Speed Up Robust Feature) for feature extraction. Classifications of these face images are done using RBF (Radial Basis Function kernel) SVM (Support Vector Machine) classifier.

Keywords Face detection · Face recognition · SURF · SVM · Vola-jones algorithm

1 Introduction

State-of-Art in constrained environments the algorithms of face recognition demonstrate reliable plus satisfactory performance example, undeviating illumination and stationary frontal postures. They have mainly developed for still images.

S. Shwetha (✉) · S. Dixit · B.I. Khondanpur
Department of Information Science and Engineering,
Dayananda Sagar College of Engineering, Bangalore, India
e-mail: shwetha.s308@gmail.com

S. Dixit
e-mail: sunanda.bms@gmail.com

B.I. Khondanpur
e-mail: bi.khodanpur@gmail.com

© Springer Nature Singapore Pte Ltd. 2017
S.C. Satapathy et al. (eds.), *Proceedings of the 5th International Conference on Frontiers in Intelligent Computing: Theory and Applications*, Advances in Intelligent Systems and Computing 516, DOI 10.1007/978-981-10-3156-4_56

In video sequences the recognition of face is contribute in the fields of applications significantly, where the important attribute is considered as motion plus the performance of recognition growths as the data increases. In addition with video arrangement following these frameworks additionally require face acknowledgment that can incorporate data acquired for the complete video. Recognizing the face is one of the prominent biometric software applications. Which can identify specific person in a digital picture by examining few parameters and comparing them. These type of recognitions are commonly used in security systems but are used increasingly in variety of other applications.

Video observation frameworks show off a situation where the routine of the recorded people and the outcomes of environment fluctuate limitlessly. Abundantinvestigation endeavors have shown that recognizing of face algorithms that function great in oversaw situations are slanted to endure in observation connections. These issues have stimulated the improvement of face recognition algorithms that make utilization of the bounteous understanding offered by means of videos, to make amends for the terrible review conditions experienced in uncontrolled survey situations. In any case, video based recognition of face postures is a challenge task. Low quality facial pictures, brightening modifications, posturevariations, and occlusions are the factors which has been existing for the face detection and recognition algorithms.

Hence several researches on this is carried out. Natarajan and Selvaganesan [1] proposed an efficient video based face recognition system. With the help of SURF descriptor a wide-ranging POOF (Part-Based One-versus-One Features) feature set is created and to discover the similarity of face picture is done by using weighted holo entropy calculation. The active appearance model (AAM) is used to find the appearance based feature of the faces. For the choice of key frames POIF is used based on the supervised learning called as Fuzzy Clustering that uses Bat algorithm (FC-Bat). Using which the keyframes dictionary is created. Using the dictionary features face recognition from video are done. Dalal et al. [2] proposed an efficient face identification technique in a group photographs. To identify individuals in a group photographs SURF technique is used. Every time the extracted SURF features from the test image and train images are matched. If the feature matches then the classified result is obtained saying the person is identified.

Cui et al. [3] projected a new methodology for extracting region of face descriptor. They separated each frame from the video arrangement into a number of blocks and each block is symbolized by sum pooling the inadequate codes which are non-negative of position free fixes examined inside the block. To diminish the feature dimension Whitened Principal Component Analysis (WPCA) is utilized with the assistance of Spatial Face Region Descriptor (SFRD). Pairwise-constrained Multiple Metric Learning (PMML) is made used to integrate the face region descriptors resourcefully of all blocks. Researches in [4–7] also proposed a significant approaches on this context.

A novel methodology for improving the face recognition approach from video sequences against different stance and occlusion invariant is proposed. This face identification system made use of Viola and Jones algorithm for face detection and SURF for feature extraction. Classifications of these face images are done using RBF SVM classifier.

2 Methodology

The proposed procedure is as appeared in the Fig. 1. The framework comprises of two phases called testing stage and training stage. The video is takes as input for the testing phase and passed to frame generation block. Here the video is divided into frames for further operation. These generated frames are then passed to pre-processing block. In the pre-processing block different pre-processing steps like resizing and color conversions are done. The pre-processed image is then passed to face detection block.

Here only the face area in the entire frame is detected and cropped using Vola–Jones algorithm for additional processing. This step is followed by SURF feature extraction block for feature extraction. Extracted feature is then passed to RBF SVM classifier. In the training phase face features from all the frames are extracted using SURF and is stored in the knowledge base. The features that are obtained during testing stage are compared with the features that already stored in the knowledge base to give classified output.

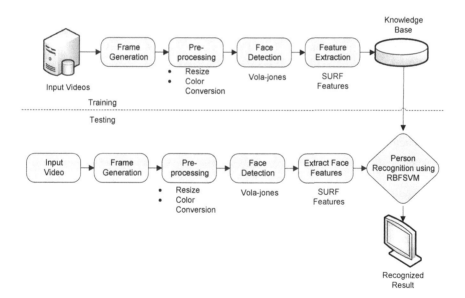

Fig. 1 Architecture of proposed system

(a) **Vola-Jones**

According to the research this algorithm made use of AdaBoost learning algorithm developed by Freund and Schapire [8]. This algorithm is made use of classifier for the selection of subset of visual features from larger set of potential features. In order to focus on the regions of face and eliminate the background portion of the image viola-jones combines classifiers which are known as cascade. The image representation called the integral image is furthermore used by the face detection algorithm. Face detection relies on the fact that objects of similar type will share some common visual features. Different properties like region of nose and cheeks are often brighter than the region of the eyes, bridge of the nose is brighter compared to eyes are considered to crop only face area [9].

(b) **Surf Features**

This feature is a enplane rotation and scale invariant feature. It contains descriptor and detector in the point of interest. The detector is used to detect the point of interest in the image and the descriptor is used to describe the features in the point of interest and develops the feature vectors of the interest focuses. The Hessian matrix $H(x, \sigma)$ in a image I at a point $X = (x, y)$ in x at σ scale is given by,

$$H(X, \sigma) = \begin{bmatrix} L_{xx}(x, \sigma) & L_{xy}(x, \sigma) \\ L_{xy}(x, \sigma) & L_{yy}(x, \sigma) \end{bmatrix} \tag{1}$$

where $L_{xx}(x, \sigma)$, $L_{xy}(x, \sigma)$, $L_{xy}(x, \sigma)$ and $L_{yy}(x, \sigma)$ are the convolution values of the Gaussian second order partial derivatives in the image I at point x. The scale and the area of the point of interest are chosen by depending on the quality given by the Hessian matrix. Point of interest restricted in a picture space and scale by applying non maximum suppression in a $3 \times 3 \times 3$ neighbourhood [2, 10].

(c) **RBFSVM**

We made use of RBFSVM as component classifier for feature matching obtained from SURF. RBF is one of the popular kernels used in the SVM based classification. This has a parameter known as Gaussian width with relatively large value. The Function RBF kernel is given by,

$$K(w_i, w) = e^{-\gamma |w_t - w|^2} \tag{2}$$

where w_i are the feature vectors for $i = 1 \ldots, M$ from the training set and w is nothing but the feature vector of the image to be classified. C and γ are the regularization parameter obtained from the training data. The brief flow for the SURF feature extraction is given in Fig. 2. The flow depicts calculation of calculating adoptive histogram equalization, location, orientation and descriptor calculation. Every time the distance between features in frames are calculated. The feature with more match are considered. Plot the locations where best match is found.

Fig. 2 Flow chart of
proposed system

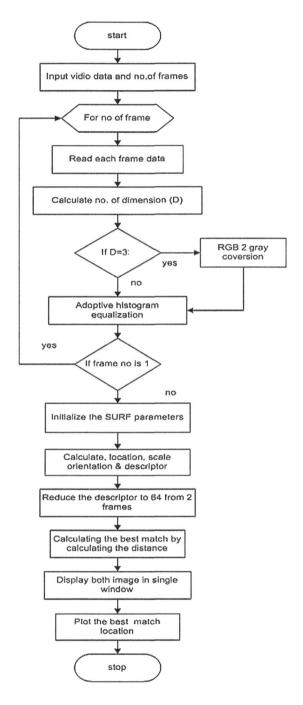

Input Frame	Detected Face area in Gray Image	Face Recognition

Fig. 3 Proposed system results

3 Experimental Results

Results obtained in each stage are explained briefly in this section. Figure 3a is the input frame of video 1, 2 and 3 with 2 different Frame different persons after certain pre-processing steps. (b) Depicts the Face detected image using Vola-Jones algorithm This image shows only detected face area of the person in the input frame. (c) Shows output image with the Recognized persons in the Frame of all the three videos. Each of these videos is having two different persons. This classification is done by RBFSVM using the extracted SURF features.

4 Conclusion

In the last few decades Recognition of face has concerned significant consideration because of its wide number of utilizations like observation and access control. Although considerable accomplishments have been achieved with 2D face recognition its exactness is still a challenge task. The proposed system discusses an efficient face detection and recognition system in a video. This methodology could give good accuracy by recognizing all the frames in the video efficiently. The method used in this paper detects the human face in all the generated frames. For facial detection Vola-jones algorithm is used. And used a SURF feature descriptor for extracting facial features which is later used by RBFSVM during classification.

References

1. K. Natarajan, J. Selvaganesan, Robust face recognition from video based on extensive feature set and fuzzy_bat algorithm. Indian J. Sci. Technol. **8**(35) (2015)
2. P. Singh, J. Dalal, M.S. Meena, Person identification in a group photograph using surf features. Int. J. Innovations Adv. Comput. Sci. (IJIACS) **4**(5) (2015)
3. W. Li, S. Shan, Z. Cui, X. Chen, in *Fusing Robust Face Region Descriptors via Multiple Metric Learning for Face Recognition in the Wild* (IEEE, 2013), pp. 3554–3561
4. S. Shankar Sastry, Z. Zhou, A.Y. Yang, L. Zhuang, in *Single-Sample Face Recognition with Image Corruption and Misalignment via Sparse Illumination Transfer* (IEEE, 2013), pp. 3546–3553
5. Y. Tian, Y. Shi, Z. Qi, Robust twin support vector machine for pattern classification. Elsevier **46**, 305–316 (2013)
6. Y. Guo, M. Hayat, Y. Lie, M. Bennamoun, An efficient 3D face recognition approach using local geometrical signatures. Elsevier (2013)
7. G.J. Bala, S.L. Femandes, A novel technique to detect and recognize faces in multi-view videos. Recent Adv. Comput. Sci. (2015)
8. J. VijayaBarathi, E. Gurumoorthi, N. Nirosha, P. Sasikala, Identification of gender and face recognition using adaboost and SVM classifier. Int. J. Eng. Comput. Sci. **3**(11) (2014)
9. P. Khatri, S. Agrawal, in *Facial Expression Detection Techniques: Based on Viola and Jones algorithm and Principal Component Analysis* (IEEE, 2015), pp. 108–112
10. Y. Liu, F. Guo, X. Luo, Research on feature extraction and match method based on the surf algorithm for mobile augmented reality system. Int. Ind. Inf. Comput. Eng. Conf. (IIICEC) (2015)

2D Shape Representation and Analysis Using Edge Histogram and Shape Feature

G.N. Manjula and Muzameel Ahmed

Abstract To identify the images, the images have so many components which will give the visual information of the image. Shape diagram are characterized that has to be described the shape features and the properties. The important properties to represent the image are shape property which represented in 2D or 3D in Euclidean plane. To represent the shape there are many methods and techniques are available like canny edge. The major aim of this paper is to find out the shape of the object by comparing with the mathematical formulas and properties of the 2D shapes with different orientation.

Keywords Image retrieval · Canny edge · Shape feature · Euclidean plane · SVM classifier · Bounding box

1 Introduction

Image recovery system is a computer system for finding, penetrating and recovering pictures from a enormous database of digital libraries [1]. The utilization of metadata such as headline, keywords or explanation. The images which are stored in the record along with the low level component removed from the image like shape, color, quality etc. it has been used till today for the purpose of image retrieval through the existing search engine. If a user describing some request

G.N. Manjula (✉)
Department of Information Science and Engineering,
Dayananda Sagar College of Engineering, Bangalore, India
e-mail: gnmanjula6@gmail.com

M. Ahmed
Jain University, Bangalore, India
e-mail: muzchk@yahoo.com

S.C. Satapathy et al. (eds.), *Proceedings of the 5th International Conference on Frontiers in Intelligent Computing: Theory and Applications*, Advances in Intelligent Systems and Computing 516, DOI 10.1007/978-981-10-3156-4_57

usually has in mind just one topic, while the Consequences produced to fulfill the request may belong to various topics. Hence only parts of the search results are applicable for a user.

Most original methods of retrieval image will make use of few methods by adding metadata likely captioning, keywords, or descriptions to the images so that retrieval can be performed over the explanation of words [2]. The disadvantages of text-based method are that it is subject to human awareness and problem of characteristics of images.

Basically, the geometrical functional are space, edge, radii of the marked and bounded circles, and the minimum and maximum diameters which allows thirty one shape diagrams to be built. Almost all of these shape diagrams can also been applied to more universal condensed sets than condensed curve sets. Starting from these six classical geometrical functional, a complete qualified study has been performed in order to examine the demonstration importance and judgment power of these thirty one shape diagrams.

2 Related Work

Serverine Rivollier et al. in his paper, the non empty compressed rounded sets in the Euclidean 2D plane are taken into account. Several arithmetical efficient like area, perimeters etc. are computed in order to differentiate the sets. They are connected by the so called geometric inequalities, which permit essential morphological functional and hereafter determining diagrams.

Densheng et al. Zhang reviewed for images using figured facial appearance has involved much awareness. There are several illustration and explanation techniques in the text. This part of the work categorizes and examines these significant techniques.

Yoshiki Kumagai et al. proposes [3] the clarification by sketch image retrieval using border relation histogram(ERH) using local and universal feature. ERH emphasizes on the relation among edge pixels, and ERH is shift, scale, rotation and balance invariant feature. This philosophy was tested 20,000 images in Corel photo gallery. New outcome show that the proposed technique is efficient in finding out the images. Rai et al. [4] projected a image retrieval using consistency, color and shape feature.

M. Suman et al. explains the procedure of recognizing and locating sharp discontinuities in an image. In his paper, edge recognition for image processing using several types of techniques is examined.

An edge can be explained as the border among an article and background in an image [5]. Curve recognition technique contains image smoothing and image discrimination plus a post processing for curve labeling.

3 Proposed System

The proposed system is divided into testing and training phase. In testing phase input image is applied with image pre-processing like noise removal, gray conversion and resizing. For the pre-processed image canny edge detection is applied. Shape features and edge histogram features are extracted and trained using SVM. Trained features are stored in a knowledge base.

In the testing phase same image processing techniques like pre-processing, edge detection, feature extraction is applied on query image. With help of result of training phase stored in knowledge base, classifier (SVM) recognizes the shape image (Fig. 1).

A. **Edge Detection**

It's a very frequent that all pictures taken from a camera will hold a few quantity of noise. To avoid that noise is incorrect for edges, should reduce noise. To reduce the noise in an image the Gaussian filter technique is used. Henceforth firstly image is smoothed by implementing a Gaussian filter. Gray scale image is taken as an input. Firstly by using Gaussian convolution image is smoothened.

A straightforward 2-D first derivative operator is applied to the smoothed picture to emphasize districts of the image with high first spatial derivatives. Gray scale image is a input. The result of the canny operative is firm by 3 parameters the width of the Gaussian kernel used in the smoothing the phase, By the tracker upper and lower thresholds is used. Gaussian kernel reduces the detectors sensitivity noise by increasing the width, at the cost of losing some of the finer detail in image. The center fault in detected edges also increases slightly as the Gaussian width is increased.

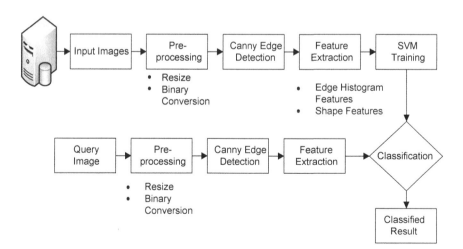

Fig. 1 Architecture of proposed system

B. Feature Extraction

In the proposed method shape feature and edge histogram features are extracted.

- **Edge histogram features**: Consider the edge detected image a L1 in which pixels are represented in binary. On the other hand, a gradient is applied to input gray image and the inverse tangent is applied (L2). Edge detected pixel in L1 image is replaced by '0' value in L2 and other values are remain the same. Histogram of resultant image is obtained.
- **Shape feature**: Shape is the core point of visual feature analysis and representation. The image content also can be determined on the basis of the characteristics of the image. Figure 2 shows the working flow of the shape feature. The shape character or features are center of gravity, Mass, ratio, angle, number of edges etc.

C. SVM Classifier

SVM is suitable to characterize the Color and shape includes precisely and effectively [6]. It is superintend learning process which break down and perceive examples, for example, surface, color, shape and it is utilized for classification and regression methods.

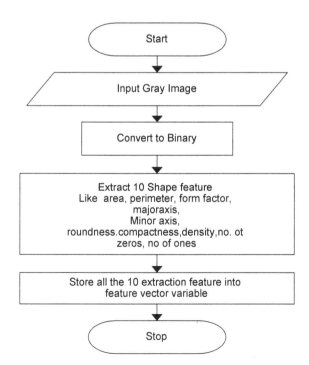

Fig. 2 Flowchart of shape feature

The grouping is associated by hyper plane which has outsized separation to nearest separation training data of any class (shading, surface, shape). In this manner the perception is bigger the margin, bring down the generalization error of classifier. The arrangement isolates information into preparing tests and testing tests. The goal of SVM is to foresee the objective estimations of test information by giving just the test information qualities.

SVM stands for "SUPPORT VECTOR MACHINE" which analyses the task of classification and regression of shapes of two dimensional objects. In classification SVM there are two types: Type 1 is known as "C-SVM classification". Type 2 is known as "nu-SVM classification". Regression SVM has two types: Type 1 is known as "epsilon-SVM regression". Type 2 is known as "nu-SVM regression".

4 Experimental Result

In the proposed algorithm some of the different shape images are a set of 25 images of 100 * 100 pixels and where 10 pixels are eliminated on four sides of an matrix and drawn two dimensional objects like circle of 5 different angles, triangle of 5 different angles, rectangle of different angles, square of 5 different angles and star shaped image of 5 different angles saved in a .jpg format which are used as input images. In the algorithm edge information is obtained using canny edge detection method the result of canny edge of query image and also with the help of shapes the image is retrieved accurately compared to traditional method.

The proposed system of an algorithm is compared with the two dimensional images of three primary colors as Red, Green and Blue (RGB), in which one images consists if number of objects like circle, triangle, square and rectangle [6, 7] (Table 1).

SVM classifier is a comparatively efficient classifier which is used in proposed system for the better retrieval of shape image.

Table 1 Results

Input image	Bounding box image	Canny edge image	Shape classification
			Belongs to Shape 2 OK
			Belongs to Shape 3 OK
			Belongs to Shape 4 OK

5 Conclusion

This paper deals with 2D shape diagram which is represented by set of points whose co-ordinates are the morphological functions. Past and proposed visual features image retrieval methods are reviewed. This proposed method is a Efficient image retrieval technique using which good accuracy (0.83) is achieved compared to other methods. For 2D shape image representation and analysis mathematical properties are very important which has to be well defined.

The limitation of the proposed system is that the two dimensional objects does not match with the three dimensional objects. Only the specific dataset can be used.

References

1. C.S. Gode, A.N. Ganar, Image retrieval by using colour, texture and shape features. Int. J. Adv. Res. Electr.
2. H.B. Kekre, P. Mukherjee, S. Wadhwa, Image retrieval with shape features extracted using gradient operators and slope magnitude technique with BTC. Int. J. Comput. Appl. (0975–8887) 6(8), September (2010)
3. Y. Kumagai, T. Arikawa, G. Ohashi, Query-by-sketch image retrieval using edge relation histogram MVA2011, in *IAPR Conference on Machine Vision Applications*, June 13–15, 2011, Nara, Japan
4. Y. Rai, T.S. Huang, S.F. Chang, Image retrieval: current technique, promising direction and open issues
5. M. Suman, M. Pawan, A survey on various methods of edge detection. Int. J. Adv. Res. Comput. Sci. Softw. Eng. 4(5) (2014)
6. S.V. Chhaya, S. Khera, S. Pradeep Kumar S, Basic geometric shape and primary colour detection using image processing on matlab. IJRET 04(05) (2015)
7. S. Rege, R. Memane, M. Phatak, P. Agarwal, 2d geometric shape and colour recognition using digital image processing. IJAREEIE 2(6) (2013)
8. S. Rivollier, J.-C. Pinoli, J. Debayle, Shape representation and analysis of 2D compact sets by shape diagrams, in HAL Id: hal-00509444. 3rd ed. by J. Clerk Maxwell, A Treatise on Electricity and Magnetism, vol. 2 (Oxford, Clarendon, 1892), pp. 68–73
9. D. Zhang, G. Lu, in *Review of Shape Representation and Escription Techniques*, ed. by K. Elissa, "Title of Paper if Known," unpublished. Accepted16 July 2003

Automatic Classification of Lung Nodules into Benign or Malignant Using SVM Classifier

B. Sasidhar, G. Geetha, B.I. Khodanpur and D.R. Ramesh Babu

Abstract Carcinoma of lungs is allied to the cancers that are causing the highest number of deaths all over the world. It is very important to improvise the detection methods so that the rate of survival can be increased. In this paper, new algorithm has been proposed to segment the lung regions using Active Contour method. Once the detection of nodules is through and Gray level Co-occurrence Matrix (GLCM) is used to calculate the texture features. HARALICK texture features are calculated and dominant features are extracted. Support Vector Machine (SVM) Classification of the nodules is done using SVM classifier. Satisfactory results have been obtained. Lung CT scan images are taken from LIDC-IDRI database.

Keywords Active contour · HARALICK texture features · GLCM

1 Introduction

Lung cancer is an solitary type of cancer with highest transience rate in the world. According to American cancer society around 1,58,040 people are dying every year, which means around 18 people die every hour. This mortality rate is alarming. If the mortality rate has to be reduced the prior exposure of lung cancer is vital. There have been various techniques researched and tried to detect cancer as early as possible. In India, Mizoram and Manipur have the highest incidence rate. Other

B. Sasidhar (✉) · G. Geetha · B.I. Khodanpur · D.R. Ramesh Babu
Dayananda Sagar College of Engineering, Bangalore, India
e-mail: bolasasi@gmail.com

G. Geetha
e-mail: geetha071992@gmail.com

B.I. Khodanpur
e-mail: bi.khodanpur@gmail.com

D.R. Ramesh Babu
e-mail: bobrammysore@gmail.com

© Springer Nature Singapore Pte Ltd. 2017
S.C. Satapathy et al. (eds.), *Proceedings of the 5th International Conference on Frontiers in Intelligent Computing: Theory and Applications*, Advances in Intelligent Systems and Computing 516, DOI 10.1007/978-981-10-3156-4_58

551

highest incident sites in India are Bangalore, Chennai, kollam, Kolkata, Tripura and Thiruvananthapuram.

There are many screening techniques available but Computed Tomography (CT) screening technique has stood the test of time. In CT scan, images are captured at various angles and they produce several slices of images thereby giving a detailed and accurate view of the internal parts of the body. In CT images artifacts are reduced to a great extent, giving a clear picture of the body part. CT images can be reconstructed to obtain a different volume of data too. In CT images overlapping anatomical structures are completely eliminated, these images will appear as though an actual cross section of the organ is presented for us to view and analyze. CT images gives information about the size, extent, texture, shape and the exact location. This information is very valuable for the analysis of the condition of the disease.

Computer Aided Detection (CAD) systems are computer programs written to detect and classify a disease. CAD systems help the radiologist in making a decision by giving a affirming opinion about the patient. CAD systems assist radiologists in case if any suspicion or confusion arises. CAD systems are developed for better diagnosis and classification of the disease. In simple terms, CAD systems are used to avoid oversights of radiologists and to interpret what is actually seen.

While building a CAD system, classifier plays an important role. Choosing a suitable and appropriate classifier for the CAD system is very crucial to determine the accuracy and efficiency of the system. Support Vector Machine (SVM) is one such binary classifier which is used to classify linearly separable data. Linearly separable means that the data can be separated into two partitions with a clear cut line of difference. Once the data is partitioned into two groups, the data to be tested is evaluated and is assigned to a particular group, thereby classifying it. SVM gives nearly accurate results even if the training samples are less. It classifies very well even when the features selected are more than the samples itself. SVM can handle the problems of over fitting. Training of SVM classifier is easier and faster. Error rate produced by SVM classifier is very less. Compared to other classifiers SVM is always proven to be a good one.

This paper is organized into the following sections. Section 2 addresses the related work. Section 3 depicts about the proposed methodology. Section 4 addresses the results and discussion. Section 5 depicts about conclusion.

2 Related Work

Elizabeth et al. [1] have suggested a new CAD system which is competent of selecting a trivial slice for the scrutiny of each nodule from a set of slices of a CT scan. CT image is pre-processed by segmenting the region of interest (ROI) using greedy snake algorithm, and then radial basis function neural network (RBFNN) for classification of the nodules. Thomas and Kumar [2] have made a assessment among classifiers like SVM, Minimum distance and k-nearest neighbour.

Morphological Operators is used for pre-processing of the images and gray level co occurrence matrix is used for the feature extraction. Narayanan and Jeeva [3] have proposed a CAD system which uses morphological operators and artificial neural networks for classifying the cropped lung regions as benign or malignant.

El-Bazl et al. [4] have proposed a CAD system which uses surface features and K-nearest classification of nodules. Kumar et al. [5] have proposed a system which uses autoencoder features to classify lung nodules. Zhang et al. [6] have proposed an automatic method which uses weighed Clique Percolation Method (CPMw) for classification. Punithavathy et al. [7] have used Contrast Limited Adaptive Histogram Equalization (CLAHE) and Fuzzy c means clustering for classification.

El-Baz et al. [8] have proposed a method which uses visual appearance features and rotation invariant second order Markov-Gibbs random field for classification than the conventional growth factor. Kaya and Can [9] have proposed a CAD system which uses radiographic descriptors for classification. Mukherjee et al. [10] have proposed a CAD system which uses bilateral filtering for noise removal, thresholding for segmentation and feature extraction for classifying the lung nodules into benign or malignant.

Kaur et al. [11] have proposed a CAD system which uses thresholding and textural and statistical features for classification using artificial neural networks.

Vivekanandan and Raj [12] have proposed a CAD system which uses Snake algorithm for preprocessing, Grey Level Co-occurrence Matrix (GLCM) for extraction of ROI and Nearest Neighbour (NN) for classification. Han et al. [13] have used 2D and 3DHaralick texture feature model for texture feature analysis and SVM for classification. Farag et al. [14] they have used active appearance models (AAM) to detect nodules and K-NN classifier for classification. There are various method used by researchers across the globe to identify a better technique for earlier lung carcinoma detection. However there are few challenges that are being faced by them always.

3 Methodology

The proposed methodology tells about the techniques used to segment and detect the lung nodules (Fig. 1):

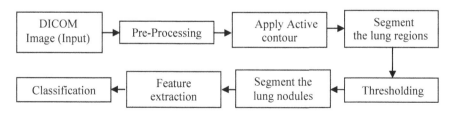

Fig. 1 Flow diagram of the proposed methodology

DICOM image is taken as input. This image is pre-processed to remove the noise and unwanted anatomical structures. To segment the lung nodules Active Contour algorithm is used. After pre-processing the image Active contour is applied to obtain an initial boundary of the lung regions. Then area is calculated to identify the largest connected components. Then the connected components with very large area only is retained and other connected components with lesser area is discarded. Once this is obtained morphological operators are used to fill the holes in the lung region. Now this acts like a mask. This mask is superimposed on the original DICOM image to segment out the lung regions. Now the original lung regions with the nodules are segmented accurately.

Nodules are extracted by calculating eccentricity. Blood vessels and other anatomical structures other than the nodules are discarded. Now Thresholding is done to discard the non nodules. Finally we obtain a mask for nodules, This mask for the nodules is superimposed on the segmented original lung regions. After this we obtain the original lung nodule.

For the nodule obtained, Haralick features are calculated and the texture features are extracted. Dominant features are suitably selected and the SVM classifier is feeded with the values of the dominant features. After the SVM classifier is trained the SVM is subjected to classify the nodules into benign or malignant

Algorithm 1-To segment lung region using Active contour and morphological operators:

Input: DICOM image of a lung.
Output: Segmented lung region.
Step 1: Input a DICOM format lung CT scan image.
Step 2: Perform edge detection using canny edge detector
Step 3: Use Active Contour to obtain a initial boundary of the lungs.
Step 4: Clear image borders and find the connected components.
Step 5: Identify the connected components with large area and fill the holes using morphological operators
Step 6: Superimpose the obtained mask over the original image to obtain the original lung regions with the nodules.

Algorithm 2-To segment the nodules from the lung regions:

Input: Segmented lung region
Output: Segmented lung nodules
Step 1: Calculate eccentricity for the components inside the lung region.
Step 2: Remove the components whose eccentricity is greater than and lesser than that of the nodules.
Step 3: Superimpose this nodule mask on the segmented lung region to get the original nodules.

Once the nodules are obtained using Gray level Co-occurrence Matrix (GLCM), calculate the HARALICK texture features. Dominant feature values are selected and then SVM classifier is trained to classify the nodules into benign or malignant.

4 Results and Discussion

The CT scan images are taken from LIDC-IDRI database. The obtained results are shown below:

No	Input image	Output image	Details
1	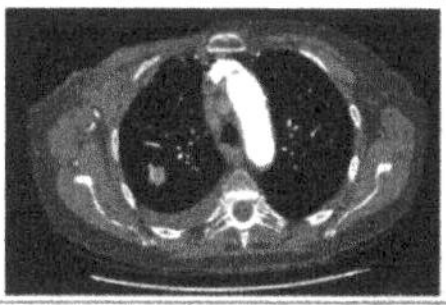	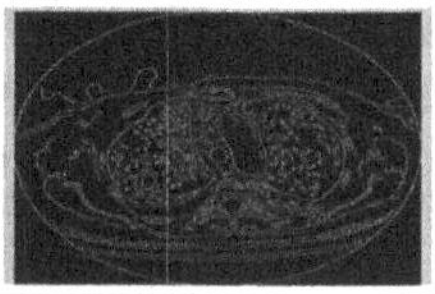	Output obtained is an edge detected image
2	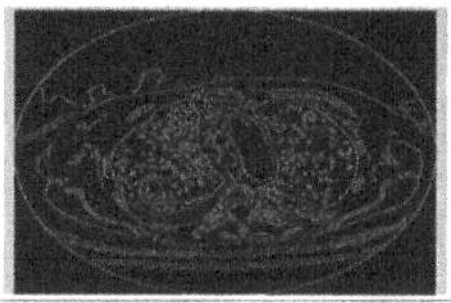	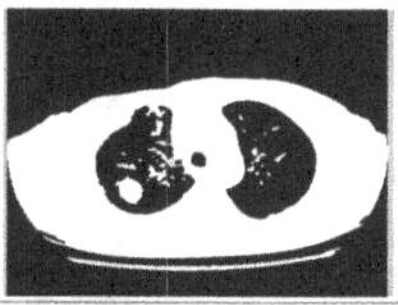	Output obtained is initial boundary of the lung regions
3	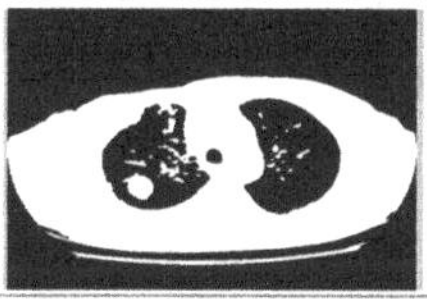	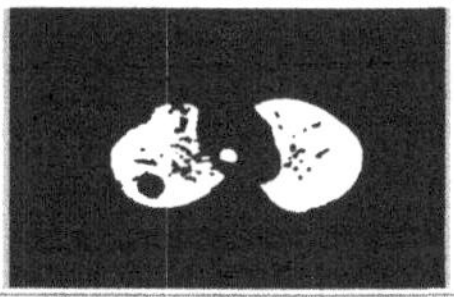	Output obtained is connected components in the lung region
4	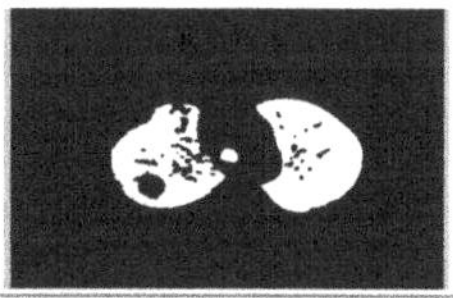	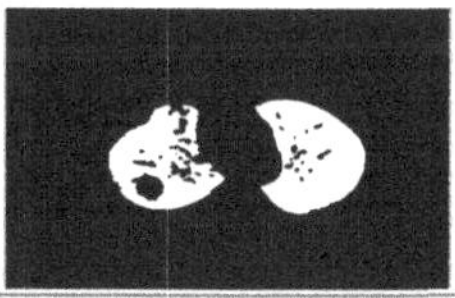	Output obtained is the components with large area
5	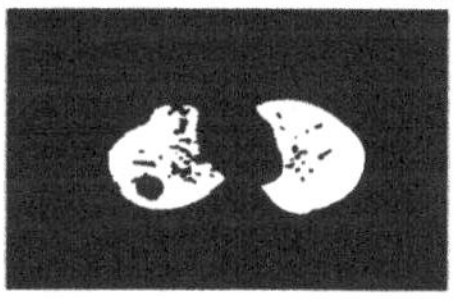	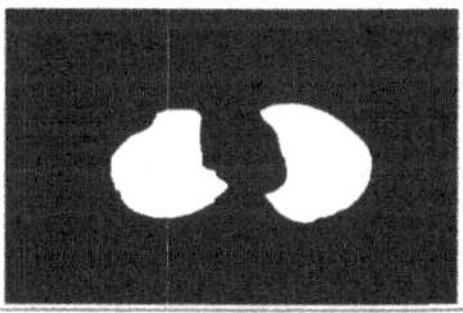	Output obtained is a holes filled image
6	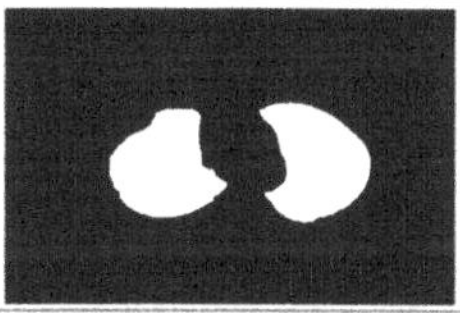	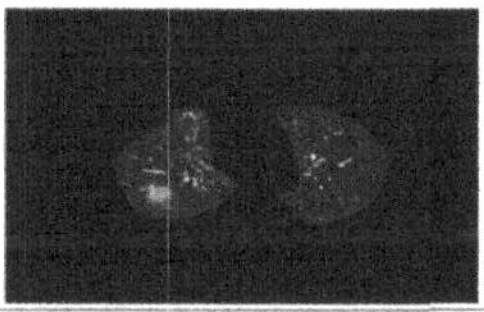	Output obtained is original lung regions
7	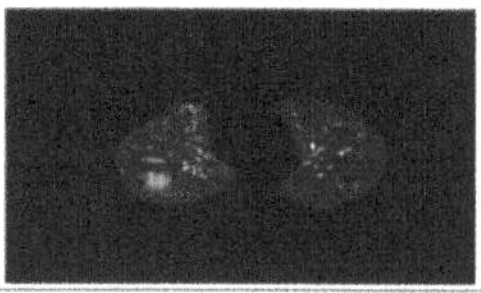		Output obtained is anatomical components inside lungs

(continued)

(continued)

No	Input image	Output image	Details
8			Output obtained is nodules and non nodules
9			Output obtained is just nodule's mask
10			Output obtained is original lung nodule

4.1 Feature Selection

HARALICK texture features gives us information about the texture of the cancerous lung regions. GLCM is used for the calculation of texture features. Features are calculated and dominant 5 features have been selected. The graphs for the selected features are as shown below (Graphs 1, 2, 3, 4 and 5):

The below formula is used to calculate the accuracy of the system:

$$accuracy\ (acc.) = \frac{TP + TN}{TP + TN + FP + FN} \qquad (1)$$

where

TP =   True positive
TN =   True negative

Graph 1 Homogeneity values

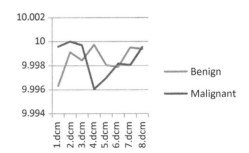

Graph 2 Correlation values

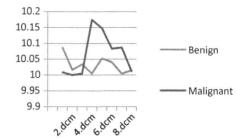

Graph 3 Maximum
probability values

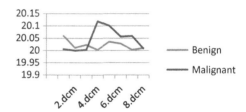

Graph 4 Sum average
values

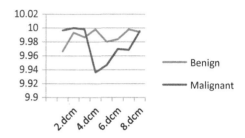

Graph 5 Variance values

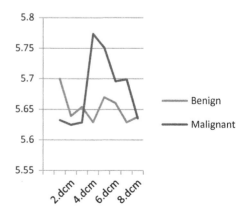

Fig. 2 ROC curve of
existing system

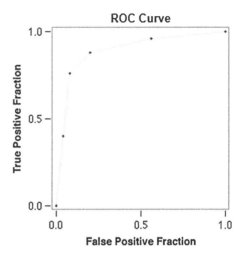

Fig. 3 ROC curve of the
proposed system

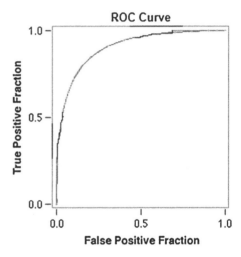

FP =   False positive
FN =   False negative

ROC curve for the proposed and existing system is as given below (Figs. 2 and 3):

The ROC curve of the existing system does not have a smooth curve; this is one of the indications that the system provides lesser accuracy. The ROC curve of the proposed system is more inclined towards the upper left corner; this means that the proposed system has better accuracy than the existing system.

5 Conclusion

This paper uses Active contour method to segment the lung regions and HAR-ALICK texture features. Support Vector Machine is used for classifying the nodules into benign or malignant. The proposed algorithm gives better classification comparatively. Segmentation produces results accurately and detection of nodules is very accurate. SVM classification gives nearly accurate results. The algorithm can be further improvised to get improved results. The results obtained are comparatively good and the accuracy obtained is around 92 %.

References

1. D.S. Elizabeth, H.K. Nehemiah, C.S. Retmin Raj, A. Kannan, Computer-aided diagnosis of lung cancer based on analysis of the significant slice of chest computed tomography image. IET Image Process. (2011)
2. R.A. Thomas, S.S. Kumar, *Automatic Detection of Lung Nodules Using Classifiers* (IEEE, 2014)
3. A.L. Narayanan, J.B. Jeeva, *A Computer Aided Diagnosis for Detection and Classification of Lung Nodules* (IEEE, 2015)
4. A. El-Bazl, M. Nitzkenl, E. Vanbogaertl, G. Gimel'jarb, R. Falfi, M. Abo El-Ghar, *A Novel Shape-Based Diagnostic Approach for Early Diagnosis of Lung Nodules* (IEEE, 2011)
5. D. Kumar, A. Wong, D.A. Clausi, *Lung Nodule Classification Using Deep Features in CT Images* (IEEE, 2015)
6. F. Zhang, W. Cai, Y. Song, M.-Z. Lee, S. Shan, D.D. Feng, *Overlapping Node Discovery for Improving Classification of Lung Nodules* (IEEE, 2013)
7. K. Punithavathy, M.M. Ramya, S. Poobal, Analysis of statistical texture features for automatic lung cancer detection in PET/CT images. Int. Conf. Robot. Autom. Control Embed Syst.—RACE (2015)
8. A. El-Baz, G. Gimel'farb, R. Falk, M. El-Ghar, *Appearance Analysis for Diagnosing Malignant Lung Nodules* (IEEE, 2010)
9. A. Kaya, A.B. Can, in *Characterization of Lung Nodules* (IEEE, 2013)
10. J. Mukherjee, A. Chakrabarti, S.H. Skaikh, M. Kar, *Automatic Detection and Classification of Solitary Pulmonary Nodules from Lung CT Images* (IEEE, 2014)
11. J. Kaur, N. Garg, D. Kaur, *An Automatic CAD System for Early Detection of Lung Tumor Using Back Propagation Network* (IEEE, 2014)
12. D. Vivekanandan, S.R. Raj, *A Feature Extraction Model for Assessing the Growth of Lung Cancer in Computer Aided Diagnosis* (IEEE, 2011)
13. F. Han, G. Zhang, H. Wang, B. Song, H. Lu, D. Zhao, H. Zhao, Z. Liang, *A Texture Feature Analysis for Diagnosis of Pulmonary Nodules Using LIDC-IDRI Database* (IEEE, 2013)
14. A. Farag, A. Ali, J. Graham, A. Farag, S. Elshazly, R. Falk, *Evaluation of Geometric Feature Descriptors for Detection and Classification of Lung Nodules in Low Dose CT Scans of the Chest* (IEEE, 2011)

Automatic Detection of Diabetic Retinopathy Using Two Phase Tophat Transformations—A Novel Approach

A.S. Akshaya, Sunanda Dixit and Ngangom Priyobata Singh

Abstract Diabetes is the most common disease which occurs when the pancreas fails to produce enough insulin. It gradually affects the retina of the human eye. As this disease aggravates, the vision of the patient starts deteriorating which ends up in Diabetic Retinopathy (DR). 80 % of all the patients who have had diabetes for 10 plus years are affected by this DR disease which can also lead to the vision loss. In this regard, the early detection of DR is hoped to help the patients from vision loss. In this paper, an attempt is made to propose a system for automatic classification of normal and abnormal retinal fundus images by detecting exudates and microaneurysms. Some other features like area of exudates, number of microaneurysms, entropy, homogeneity, contrast and energy are also calculated. The extracted features are fed to SVM classifier for automatic classification. The paper is based on secondary data gathered from different sources.

Keywords Diabetic retinopathy · DR · Tophat · Optic disk · Exudates · Microaneurysm

1 Introduction

Diabetes is a disease that occurs when the pancreas does not secret enough insulin. According to the World Health Organisation (WHO), 135 million people have diabetes worldwide and the number of people with diabetes increases to 300 million by the year 2025 [1].

A.S. Akshaya (✉) · S. Dixit · N.P. Singh
Department of Information Science and Engineering, Dayananda Sagar College of Engineering, Bangalore, Karnataka, India
e-mail: akshayamgd@gmail.com

S. Dixit
e-mail: sunanda.bms@gmail.com

N.P. Singh
e-mail: priyobatang@gmail.com

© Springer Nature Singapore Pte Ltd. 2017
S.C. Satapathy et al. (eds.), *Proceedings of the 5th International Conference on Frontiers in Intelligent Computing: Theory and Applications*, Advances in Intelligent Systems and Computing 516, DOI 10.1007/978-981-10-3156-4_59

Diabetes is the major cause of blindness. DR being one of the complications caused by diabetes, affects different parts of the retina such as blood vessels, optic disk, macula and fovea as shown in Fig. 1. DR is a medical condition where the retina is damaged because fluid leaks from blood vessels into the retina [2]. Due to this, new features such as micro aneurysms, exudates, and hemorrhages appear in back of the retina as shown in Fig. 2.

- **Micro aneurysms (MA)**—MA are the small enlargement of minute blood vessels of eye. They are the first clinical signs of DR.
- **Exudates** is a fluid rich in protein that leaks out of blood vessels into the back of the retina and is deposited in nearby tissues.

Until the DR has affected a large area on the retina, the patients do not notice any visual problems. So there is a need for mass-screening of diabetic patients' eyes to

Fig. 1 Normal retinal fundus image

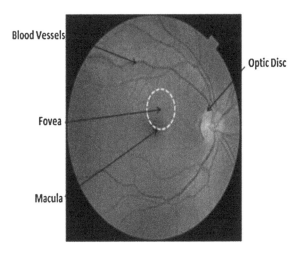

Fig. 2 Retinal image with DR lesions

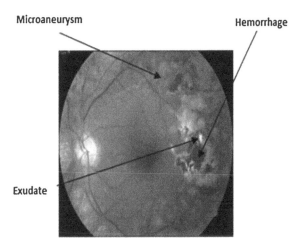

detect the DR as early as possible. Although the process of manual diagnosis is possible, it is very cumbersome and limits the mass-screening process [2]. Therefore the automatic method is very much needed.

2 Literature Survey

Agrawal et al. [1] have done a survey on automated MA detection in DR retinal images. Here the authors review and analyse techniques and methodologies used for the detection of MA from DR present retinal fundus images.

Mahendran and Dhanasekaran [3] have proposed a method on Morphological Process and SVM Classifier for diagnosis of DR. It focuses on automatic detection of DR through detecting exudates.

Mahendran et al. [4] have proposed Morphological Operations based Segmentation for the Detection of Exudates from the retinal. First it detects the DR through identifying exudates by Morphological process and then segregates the severity of the lesions by Cascade Neural Network (CNN) classifier.

Mahendran et al. [5] have proposed a method based on Morphological Process and PNN Classifier for the identification of Exudates for DR. It uses morphological process to detect exudates. Canny edge detector is used to remove the OD. Probabilistic Neural Network (PNN) classifier is used to classify the DR.

Sri et al. [6] have proposed a method where in mathematical and statistical approach is used to identify different stages of DR. After segmenting the blood vessels from the fundus images, it calculates the area and perimeter of the blood vessels which helps in identifying the severity of the DR.

3 Methodology

To understand the detection of DR process, some of the concepts are very much required. A brief discussion is carried out here.

(a) **Morphological Operations**

Image Morphology offers a unified and powerful method to many image processing problems. Usually it involves special mechanisms of merging two sets of pixels. Out of the two sets, one set consists of image to be processed and the other set consists of Structuring Element (SE) or also called as the kernel. In all the operations of morphology, SE is very important. It is matrix of 0 s and 1 s that have any arbitrary shape and size. This SE controls the operation on the image.

Morphological operations mainly involve Dilation, Erosion, Opening and Closing.

Morphological Dilation is the process of thickening the image which is controlled by SE. It is denoted by

$$A \oplus B = \left\{ z \in E \mid (B^s)_{\approx} \cap A \neq \varnothing \right\} \qquad (1)$$

Here B is the SE.

Morphological Erosion is the process of thinning the image which is controlled by SE. it is denoted by

$$A \ominus B = \left\{ z \in E \mid B_{\approx} \subseteq A \right\} \qquad (2)$$

Generally Morphological Opening smoothes the contour of an object, breaks narrow strips and eliminates thin projections. It is denoted by

$$A \circ B = (A \ominus B) \oplus B \qquad (3)$$

Morphological Closing tend to smooth the section of contour, but it generally fuses narrow breaks and long thin gulfs, eliminates small holes and fills gaps in the contour. It is denoted by

$$A \bullet B = (A \oplus B) \ominus B \qquad (4)$$

(b) TopHat Transformation

It extracts the small elements and details from the given image. It also eliminates the uneven illumination in an image. There are two kinds of TopHat Transformation. One is White TopHat transformation and the other is Black TopHat transformation. White transformation returns the image which contains those objects of input image that are smaller than the SE and brighter than their surroundings. It is denoted by

$$T_w(f) = f - f \circ b \qquad (5)$$

Here b is the SE.

Black TopHat transformation returns the image which contains those objects of input image that are smaller than the SE and darker than their surroundings.

It is denoted by

$$T_b(f) = f \bullet b - f \qquad (6)$$

4 Extraction of Retinal Features and the Proposed Method

(a) Extraction of Blood Vessels (BV)

The extraction of BV is very much needed in identification of both exudates and MA using image processing. The contrast between the BV and the background is very high in green component of the original fundus image. So this green component is preprocessed to get the high quality image. Now the Black TopHat transformation which follows Eq. (6) is applied on the preprocessed image. Then the image is adjusted to improve the contrast and to remove the uneven illumination in the image. The BV are extracted by converting image into binary image and inverting the intensity of the output image.

(b) **Detection of Exudates**

Exudates are the fluids rich in protein that are leaked from the BV. They are deposited in nearby tissues. They appear as a bright white-yellow spots. In order to detect the exudates, the retinal/fundus images are pre-processed. To do so, first, the images are converted into green component as the contrast between the objects and the background is very high in green component. Then to remove the speckle and other noise present in the image the median filter is applied. Finally to maintain a standard, the images are resized to 1500×1152.

Now to detect exudates, the TopHat transformation is used in two phases. In the first phase the White TopHat transformation concentrates on the actual size of exudates and the second phase of TopHat transformation concentrates on removing the Optic Disk (OD). For this, two SE (say SE1 and SE2) are used. SE1 is ball shaped with radius 5 and height 6. Using this SE1 the TopHat transformation is applied on the preprocessed image. Then contrast stretching is done. This results in good contrast between exudates and the background. As the intensity values of OD and exudates are almost similar, the part of OD is also present with exudates in the output image. However the TopHat transformation applied on preprocessed image adds some noise nearby the strong BV. Then the output image is converted into binary image. To get the actual size of exudates, morphological opening is done with disk as the SE with radius 3 and height 6.

In the second phase of TopHat transformation the ball shaped SE2 with radius 50 is used to remove the OD. Here the TopHat is applied on the preprocessed image with SE2. The output of this is contrast stretched and thresholding is done to convert it into binary image. The binary image contains the major portion of OD along with the exudates and other unwanted objects. To find the rough location of OD the morphological opening is done with disk of size 19 as SE is applied on this binary image. Then to get the coordinates of OD, the centroid is calculated. After that the OD mask of radius 85 (because the size of OD ranges from 80–83 pixels in the fundus image) is created. This mask is applied on the output of first phase TopHat transform to remove the OD. To remove the noise added during the first phase of TopHat transformation, the already extracted BV image is subtracted from the mask applied first phase TopHat transformation output. So the final image contains only the exudates with their actual size.

The process of detection of Exudates can be represented as shown in Fig. 3.

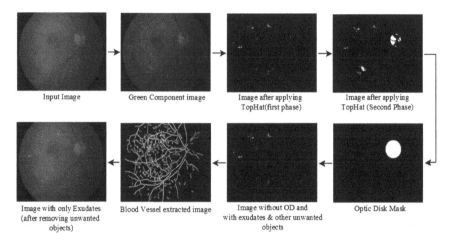

Fig. 3 Process of detection of exudates

(c) Detection of Microaneurysm (MA)

MA are the small enlargement of blood vessels of retina. They are the first clinical signs of DR. MA appear as a small dark spot. They are very difficult to detect because the size of MA ranges from 15 to 60 μ.

To detect MA, first the intensity of fundus image is adjusted and they are resized to 1500 × 1152 to pre-process them. The contrast of preprocessed image is stretched by applying Adaptive Histogram Equalization before applying canny edge detection method to detect the minute edges of the retina. The boundary is detected by filling the holes by using disk shaped SE of size 6 for morphological opening. To get the image without boundary, the edge detected image is subtracted from boundary detected image.

Then the holes are filled to get the MA and other unwanted objects. The extracted BV image is subtracted from this image to remove the unwanted objects. The process of extraction of MA is as shown in Fig. 4.

(d) Classification of DR

After detecting the exudates, the area of exudates is calculated. At the same time when the MA is detected, the number of MA present in the image is calculated. To classify the DR, SVM classifier is used. For the SVM to classify, some of the inputs are needed. Those inputs are the features extracted from the exudates detected image. These features are entropy, homogeneity, contrast and energy. They are extracted using Gray Level Co-occurrence Matrix (GLCM). It contains information about the position of the pixels having most similar gray levels. The GLCM which can represented as P[i, j] is used to calculate all pairs of pixels separated by distance vector having gray levels i and j.

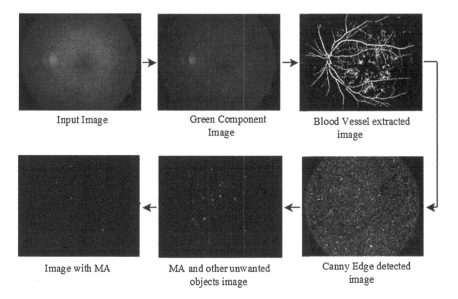

Fig. 4 Process of extraction of MA

The formulae for these features are:

$$Entropy = - \sum_{i,j} p(i,j) \log(p(i,j)) \tag{7}$$

$$Homogeneity = \sum_{i,j} \frac{1}{1 - (i-j)^2} p(i,j) \tag{8}$$

$$Contrast = \sum_{i,j} |i-j|^2 p(i,j) \tag{9}$$

$$Energy = \sum_{i,j} p(i,j)^2 \tag{10}$$

After extracting all these required features, they are given as input to SVM classifier to train and classify the DR as normal or abnormal (Fig. 5).

5 Experimental Results

After testing the proposed method on 136 retinal images, the following results are obtained.

True Positive (TP)—59 Images
True Negative (TN)—55 Images

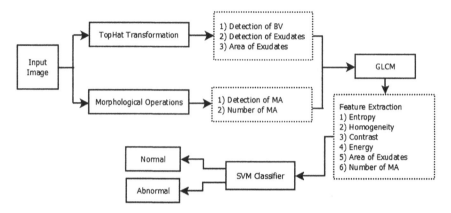

Fig. 5 Overall process of the proposed method

Table 1 Performance of the proposed method

Performance metric	Proposed method outcome (Values in %)
Sensitivity	72.84
Specificity	100
Accuracy	83.83

False Positive (FP)—00 Images
False Negative (FN)—22 Images

The Table 1 shows the performance of the proposed system as according to the formulae given below.

$$\text{Sensitivity} = \text{TP*100} /(\text{TP} + \text{FN}) \tag{11}$$

$$\text{Specificity} = \text{TN*100} /(\text{TN} + \text{FP}) \tag{12}$$

$$\text{Accuracy} = (\text{TP} + \text{TN})\text{*100} /(\text{TP} + \text{FN} + \text{TN} + \text{FP}) \tag{13}$$

6 Conclusion

Automatic Detection of DR presents many of the challenges. The size and the color of micro aneurysms are very similar to that of the blood vessels. Its size is variable and small. So it can be easily confused with the noise present in the image. In this paper a novel approach is presented to detect and classify the disease DR based on TopHat transformations. However the TopHat transformation adds some unwanted objects nearby strong blood vessels. These are removed in this proposed method.

Based on the study the researchers are able to get a precise perception of automatic detection of DR which can go a long way to enable them to develop more effective and better methods for diagnosis of DR.

References

1. A. Agrawal, C. Bhatnagar, A.S. Jalal, in *IEEE Conference on A Survey on Automated Micro Aneurysm Detection in Diabetic Retinopathy Retinal Images* (2013)
2. E.M. Shahin, T.E. Taha, W. Al-Nuaimy, S El Rabaie, F. Osama, in *IEEE Conference on Automated Detection of Diabetic Retinopathy in Blurred Digital Fundus Images* (2012)
3. G. Mahendran, R. Dhanasekaran, in *Diagnosis of Diabetic Retinopathy Using Morphological Process and SVM Classifier. IEEE* International Conference on Communication and Signal Processing (India, 2013)
4. G. Mahendran, R. Dhanasekaran, K.N. Narmadha Devi, in *Morphological Process based Segmentation for the Detection of Exudates from the Retinal images of Diabetic Patients.* IEEE International Conference on Advanced Communication Control and Computing Technologies (ICACCCT) (2014)
5. G. Mahendran, R. Dhanasekaran, K.N. Narmadha Devi, in *Identification of Exudates for Diabetic Retinopathy based on Morphological Process and PNNClassifier.* International Conference on Communication and Signal Processing (India, 2014)
6. R.M. Sri, M. Raghupathy Reddy, K.M.M. Rao, Image processing for identifying different stages of diabetic retinopathy. ACEEE Int. J. Recent Trends Eng. Technol. **11** (2014)

Evaluation of Heart Rate Using Reflectance of an Image

Neha Deshmukh, Sunanda Dixit and B.I. Khondanpur

Abstract Observing heart rate by electrocardiogram and oximetry sensors may cause skin irritation to some patients. In order to avoid this heart rate evaluation from face reflectance analysis is carried out. Initially procedure adapted is by taking face reflectance and details analysis is carried out. This results in changes in hemoglobin of blood vessels. The green channel is selected because hemoglobin observes only green colour. Hilbert-Huang transform is utilized to procure heart rate by reducing the light variations. Next to reduce noise EMD is applied. The proposed methodology is to get most accurate heart rate with some variations. With this approach, it is now possible to get heart rate accurate up to 70 %.

Keywords Heart rate · Hilbert huang transform · Reflectance · Empirical mode decomposition

1 Introduction

The demise rates are expanding step by step because of cardiovascular assault, there are such variety of strategies executed in our medicinal science. Checking heart rates utilizing traditional electrocardiogram hardware and industrially accessible heartbeat oximetry sensors may bring about skin bothering and distress for patients.

For, this issue an assessment of heart rate utilizing reflectance of a picture is taken into consideration for assessment of heart rate variety. Face reflectance ini-

N. Deshmukh (✉) · S. Dixit · B.I. Khondanpur
Department of Information Science and Engineering, Dayananda Sagar College
of Engineering, Bangalore, India
e-mail: deshmukhneha1008@gmail.com

S. Dixit
e-mail: sunanda.bms@gmail.com

B.I. Khondanpur
e-mail: bi.khodanpur@gmail.com

© Springer Nature Singapore Pte Ltd. 2017   571
S.C. Satapathy et al. (eds.), *Proceedings of the 5th International Conference on Frontiers
in Intelligent Computing: Theory and Applications*, Advances in Intelligent Systems
and Computing 516, DOI 10.1007/978-981-10-3156-4_60

tially disintegrated against solitary picture afterward heart signal assessment is performed as indicated by the intermittent variety of reflectance quality coming about because of changes in absorptivity of hemoglobin as heartbeats aim alteration in blood volume on the face.

Hilbert–Huang change is used to acquire the key heart rate although decreasing impact of changes in light. Firstly, video as an information test is considered, produce and pre-handled casings for that video test, among RGB design just green channel picture for further process, from considered green channel picture to get heart rate.

Face reflectance ought is initially deteriorated by applying Empirical Mode Decomposition calculation, from this, any confused information can be disintegrated into a limited, frequently little count of part. From the after effect of EMD the crest purpose of the sign is considered and investigate the heart rate by considering the reflectance picture.

2 Related Work

Smith and Waterman [1] has proposed A Robust Local Image Descriptor components. They exhibited test results on human face location which likewise demonstrates a promising execution practically identical to the best published results.

May et al. [2] has proposed a strategy for gender acknowledgment from face pictures with neighborhood wld descriptor, in this they reported the best blend of these parameters and that our proposed spatial WLD descriptor with the least difficult classifier gives better precision i.e. 99.08 % with lesser algorithmic multifaceted nature than that of best in class gender acknowledgment approaches.

Foster and Kesselman [3] has proposed Fingerprint Liveness Detection taking into account Weber Local Image Descriptor, in this paper they have demonstrated diverse sensors show WLD to perform positively contrasted with the cutting edge strategies in unique mark liveness identification. Also, by consolidating Weber Local Descriptor with Local Phase Quantization which comes about further enhance essentially.

Czajkowski et al. [4] has proposed Design and Development of a Heart Rate Measuring Device utilizing Fingertip, in this paper they have contrasted the execution of HRM gadget and Electrocardiogram reports and manual heartbeat estimation of pulse of 90 human subjects of various ages. In this manner, results demonstrated that the blunder rate of the gadget is irrelevant.

Foster et al. [5] has proposed Remote Heart Rate Measurement From Face Videos Under suitable conditions, in this paper they have exhibit that our technique considerably beats every past strategy in remote heart rate estimation from face recordings.

3 Architecture

Figure 1 demonstrates the square outline of proposed framework, which take into account the assessment of heart rate variety with no physical contact with patient. For this implementation one video is taken as input. Face reflectance disintegrated against a solitary picture first afterward heart rate assessment is performed as per the intermittent variety of reflectance quality coming about because which causes changes in hemoglobin absorptivity over the noticeable light range as heartbeats aim changes to blood volume on the face by utilizing Weber Local Descriptor. To accomplish a vigorous assessment, in this paper we are utilizing outfit Empirical Mode Decomposition (EMD) of the Hilbert–Huang change in order to accomplish the essential heart rate signal although lessening impact due to surrounding light. Firstly, we consider video as an info test, create and pre-handled casings for that video test, among RGB design we are considering just green channel picture for further process, from considered green channel picture to give powerful heart rate checking, the face reflectance ought to last initially deteriorated by applying Empirical Mode Decomposition calculation, from this, any confounded information set is decayed into limited and frequently little number of segments, whichever is gathering of IMF. From the consequence of EMD we can assess the crest point signal.

(a) **Green Chanel Intensity Image**

After eras of casings from the information video, for reflectance deterioration a picture covering the subject's temples territory is initially caught and afterward the green channel is chosen on the grounds that oxygenated hemoglobin assimilates green light. Effective reflectance quality of green channel is demonstrated by watching the variety of oxygen levels in blood. the reflectance quality changes with each pulse. This time sign is excessively boisterous, making it impossible to recognize every pulse period. Sometimes can't recognize every heart beat period as there is a lot of clamor in the created time signal, with a specific end goal to

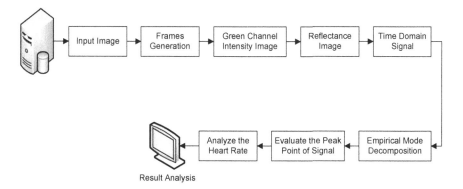

Fig. 1 Architecture of proposed system

evacuate ecological commotion, EEMD is utilized to part the first run through sign against the face reflectance to more than one natural technique capacity.

(b) **Reflectance Image**

RGB shading sensors get signals comprise enlightenment against natural light sources and reflectance caught from targets. Face reflectance ought to be initially disintegrated with a specific end goal to screen the hearty heart rate. The differential excitation segment is component of proportion between two conditions, one is relative power contrasts between current pixel against its neighbors, another is force of present pixel. The introduction segment is the slope introduction of present pixel. For a given picture, we utilize two segments just before developing WLD histogram.

Weber's Law is as follows:

$$\frac{\Delta I}{I} = K \tag{1}$$

Weber Local Descriptor comprises of two parts, they are differential excitation and introduction. In the begining compute the contrasts between the inside point and its neighbor utilizing the channel f_{00}

$$V_S^{00} = \sum_{I=0}^{P-1} (\Delta x_i) = \sum_{i=0}^{P-1} (x_i - x_c) \tag{2}$$

where x_i ($i = 0, 1,...$ p−1) signifies the i-th neighbors of x_c and p is the quantity of neighbors. Taking after clues in Weber's Law, we then process the proportion of the distinctions to the power of the present point by consolidating the yields of the two channels.

f_{00} and f_{01} (whose output v_s^{01} is the original image in fact):

$$G_{ratio}(x_c) = \frac{v_s^{00}}{v_s^{01}} \tag{3}$$

We then utilize the arctangent function on $G_{ratio}()$

$$G_{arctan}[G_{ratio}(x_c)] = \arctan[G_{ratio}(x_c)] \tag{4}$$

Combining (2), (3), (4), we have

$$G_{arctan}[G_{ratio}(x_c)] = \gamma_s^0 = \arctan\left[\frac{v_s^{00}}{v_s^{01}}\right] = \arctan\left[\sum_{i=0}^{p-1}\left(\frac{x_i - x_c}{x_c}\right)\right] \tag{5}$$

So, the differential excitation is a follows as:

$$\varepsilon(x_c) = \arctan\left[\frac{v_s^{00}}{v_s^{01}}\right] = \arctan\left[\sum_{i=0}^{p-1}\left(\frac{x_i - x_c}{x_c}\right)\right] \tag{6}$$

(c) **Empirical Mode Decomposition**

For disintegrating a sign into natural mode capacities (IMF) and to acquire quick recurrence information will utilize the Hilbert–Huang change (HHT) calculation. Any confounded information set be able to be decayed within a limited and little count of segments, whichever is accumulation of intrinsic mode function by utilizing the exact mode disintegration (EMD) strategy. EMD has no predetermined premise. Its premise is adaptively created relying upon the sign itself, which results in high decay productivity as well as sharp recurrence and time restriction.

Two diverse stoppage criteria are utilized. Firstly, the primary foundation depends on the S-number, which is characterized as the quantity of sequential shiftings when the quantity of zero-intersections and extrema are equivalent or at most vary by one. In particular, a S-number is pre-chosen. The moving procedure will discontinue just if quantities of zero-intersections and extrema are equivalent or at most vary by one for S continuous cycles and second is the filtering process stops if the standard deviation SD is littler than pre-characterized esteem.

SD is characterized by

$$SD = \sum_{t=0}^{T} \left[\frac{|h_{k-1}(t) - h_k(t)|^2}{h_{k-1}^2(t)} \right] \tag{7}$$

Where h_{k-1} and h_k are the $k-1$th component and kth component respectively.

$$r_1 = x(t) - c_1. \tag{8}$$

The r_n is defined by

$$r_n = r_n - 1 - c_n. \tag{9}$$

From the above comparisons, we get x(t) by

$$x(t) = \sum_{k=1}^{n} c_k(t) + r_n(t) \tag{10}$$

In this manner, a disintegration of the information into n-observational modes is accomplished.

4 Experimental Results

Beneath figures demonstrate the exploratory after effects of our proposed work. we consider video as an information test appeared in Fig. 2a, produce and pre-handled casings for that video test, among RGB design we are considering just green channel picture for further process appeared in Fig. 2b, c speak to green channel force picture, from considered green channel power picture to give hearty heart rate observing, the face reflectance will create squares which is appeared in Fig. 2d and proselyte those qualities into time domain signals which is appeared in Fig. 2e, that reflectance picture ought to be initially deteriorated by applying Empirical Mode Decomposition calculation, from this, any confounded information set can be disintegrated within a limited and frequently little count of segments, whichever make sense of IMF (characteristic mode capacities). From the after effect of EMD qualities appeared in Fig. 2f, we can assess the purpose of the sign and investigate the heart rate by considering the reflectance picture in Fig. 2g.

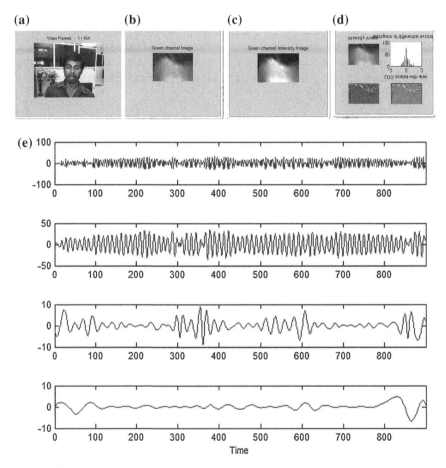

Fig. 2 **a** Video frames; **b** Green channel image; **c** Green channel intensity image; **d** Block generations; **e** Time domain signal; **f** EMD values; **g** ensemble empirical mode decomposition peak signal results

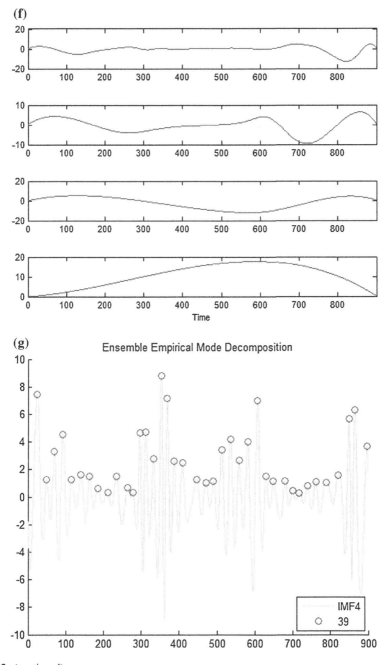

Fig. 2 (continued)

5 Conclusion

In this paper, the proposed assessment of the heart rate variety with no physical contact. Heart rate assessment is first directed when the face reflectance disintegration from a solitary picture, from successive edges as indicated by the occasional variety of reflectance quality coming about because of varieties in hemoglobin absorptivity over noticeable light range with every pulse expands/diminishes the blood volume in the veins of the face. Group Empirical Mode Decomposition (EEMD) of the Hilbert-Huang Transform (HHT) has been utilized to accomplish a strong assessment and to get the essential heart rate signal although lessening the impact of modification to surrounding light, with reference to a littler level of fluctuation it will give exact estimation, therefore exhibiting its relevance in true situations. Test results exhibit the productivity of the proposed idea when contrast with other customary strategies.

Limitations The limitations are as follows:

* Sometimes the heart rate may differ due to real world environment such as light changes.
* Not much helpful in emergency conditions.

References

1. T.F. Smith, M.S. Waterman, Identification of common molecular subsequences. J. Mol. Biol. **147**, 195–197 (1981)
2. P. May, H.C. Ehrlich, T. Steinke, ZIB Structure Prediction Pipeline: Composing a Complex Biological Workflow through Web Services, in *Euro-Par 2006*, vol. 4128, LNCS, ed. by W.E. Nagel, W.V. Walter, W. Lehner (Springer, Heidelberg, 2006), pp. 1148–1158
3. I. Foster, C. Kesselman, *The Grid: Blueprint for a New Computing Infrastructure* (Morgan Kaufmann, San Francisco, 1999)
4. K. Czajkowski, S. Fitzgerald, I. Foster, C. Kesselman, in *Grid Information Services for Distributed Resource Sharing*. 10th IEEE International Symposium on High Performance Distributed Computing (IEEE Press, New York, 2001), pp. 181–184
5. I. Foster, C. Kesselman, J. Nick, S. Tuecke, *The Physiology of the Grid: an Open Grid Services Architecture for Distributed Systems Integration* (Technical report, Global Grid Forum, 2002)

Analysis of Logs by Using Logstash

Sushma Sanjappa and Muzameel Ahmed

Abstract The key functionality of this proposed system is its ability to handle, collect and analysis huge volume of different kinds of log data. When deployed in a network would facilitate collection of logs from different nodes across the network. This paper explains the proposed system which collects the logs using Logstash which is having a capability of handling the many types of Logs data which helps to identify the malicious activity in the network.

Keywords Logstash · Nx-log · Raw logs · Security information event management · Syslog

1 Introduction

In the field of computer security, **Security Information and Event Management** produces and facilities combine security information management (SIM) and security event management (SEM). A SEM system merges a storage and explanation of logs and in the real-time analysis [1]. The system collects the information into a vital source for analysis and provides manual writing for submission and centralized reporting. The combining both SIEM systems provide documentation, analysis and retrieval of security events. A SIEM system collects the logs and security-related documentation for analysis.

Security information and event management (SIEM) systems are to collect security log events for huge amount of data. By passing the logs data, SIEM products allow central analysis and reporting for an organization's security events

S. Sanjappa (✉)
Department of Information Science and Engineering, Dayananda Sagar
College of Engineering, Bengaluru, India
e-mail: sahanasush027@gmail.com

M. Ahmed
Jain University, Bengaluru, India
e-mail: Muzchk@yahoo.com

© Springer Nature Singapore Pte Ltd. 2017 579
S.C. Satapathy et al. (eds.), *Proceedings of the 5th International Conference on Frontiers in Intelligent Computing: Theory and Applications*, Advances in Intelligent Systems and Computing 516, DOI 10.1007/978-981-10-3156-4_61

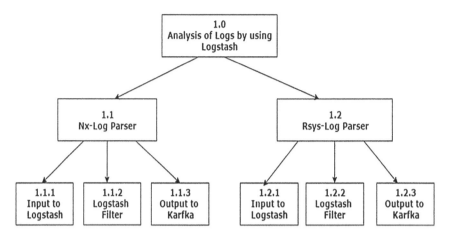

Fig. 1 Flow diagram of analysis of logs

[2]. The analysis may result in the detection of attacks and some SIEM products have the capabilities to attempt to stop attacks in the network.

Logstash is a tool for managing events and logs in a larger system of log collection, processing, storage and searching activities. Logstash alone does not have exact meaning and centralized management of log files.

Collection is accomplished via configuration are Input plugin. Once input plugin has collected data it can be processed by number of bits which modify and annotate the event data [3]. Logstash it has three stages: Input, Filter, Output.

Inputs logs can be collected from windows, Linux, TCP, UDP, Filter is used to modify the inputs data transformable can be done easily. Any modification can be done by using the configuration, not required the code [4]. Outputs can be written in file, Database, Karfka. It is also having an Extra features that logs data can also write in own custom modification in filter stage.

The entire analysis of logs can be shown below figure (Fig. 1).

The logs are collected from windows by using Nxlog and in Linux machine Rsyslog. Each block plays a very vital role in collecting the logs namely input, filter and output to obtain the desired log parser.

2   Related Work

Ariel Rabkin et al. [1] proposed the system in which it collect, analysis the data and process the data with MapReduce. The system analysis would need to build a infrastructure to connect data sources with processing tools. MapReduce is designed is used to store the data in a distributed filesystem like Hadoop's HDFS.

Robert Bonham et al. [3] proposed that data analyzed from the systems through the parsing and analysis of various log files such as server logs, syslogs, and cisco firewall events logs generated by both SAS and third-party resources in a SAS deployment.

Brad Hale et al. [5] proposed that log management is having the ability to approximation the group of Logs. This is required for generally two primary reasons to estimate the amount of Log data required storage and to estimate the cost of model.

Anton Chuvakin et al. [2] proposed analyze the relationship between these two technologies SIEM and log management meeting not only on the technical differences and different uses for these technologies, but also on architecting their joint deployments.

Mahmoud Awad et al. [4] proposed to explore and implement the various components of the framework, mainly focus on automated workload characterization using system log analysis, which is the primary function of the Workload Model Analyzer component.

3 Proposed System

The proposed system is divided into testing and training phase. In testing phase the input is taken from both windows and Linux logs. The collection of logs from different system like windows, syslog, event log, cisco firewall logs, DHCP logs.

Window log mainly holds records of login/logout activity and other security-related events measured by means of the system's audit policy [6, 7].

The Security Log is one of the prime tools cast off by Administrators to detect and investigate attempted and successful illegal activity and to troubleshoot difficulties. Event logs record events on system when user logs encounters an error. DHCP logs provide dynamic host configuration protocol service to a network.

The logs are collected from Logstash configuration from both Linux and windows machine. The main function is to ensure that the log collector publishes event logs. This event are not guarantee to be deliver the log data [8–10].

3.1 Log Collector

The logs are collected and then it is parsed to then log events for both windows and Linux machine. Log that holds records of login/logout activity and security-related events. In the Security Log it helps to configure the windows for operating system activities. After collecting logs from multiple nodes normalizes the logs and publishes them and the log collector must have the ability to handle high volume, high rate and a variety of logs.

3.2 Log Stash

A remote shipper component function is to send their logs to central Logstash server. Web interface can search and the log details at specified time range Indexer in storage. It will receive the log data from remote agents and will file it, so that it is necessary for storage them late. Broker it will collect the log event data from different agents, from servers. Any changes can be done in filter phase according to the logs.

3.3 Karfka

The desired logs are can be stored in database like Cassandra, NOSQL, and File. But in this paper logs data are stored in the database for processing. It helps to allow a single group to serve as the central for a large organization. Data streams are split and send over a cluster of machines to allow data streams.

After collecting the logs from various systems by using Logstash, the logs may be success by parsing or it may be failure. Success login is one which the logs parsed without any malicious data Failure is one in which it checks all the last log data then it parses the data.

```
define ROOT C:\Program Files (x86)\nxlog

Moduledir %ROOT%\modules
CacheDir %ROOT%\data
Pidfile %ROOT%\data\nxlog.pid
SpoolDir %ROOT%\data
LogFile %ROOT%\data\nxlog.log

<Extension json>
    Module        xm_json
</Extension>

<Input internal>
    Module        im_internal
</Input>

<Input file>
    Module  im_file
    File    "C:\\MyApp\\Logs\\mylog.json"
</Input>

<Input eventlog>
    Module        im_msvistalog
# Uncomment if you want only specific logs
#    Query         <QueryList>\
#                      <Query Id="0">\
#                          <Select Path="Application">*</Select>\
#                          <Select Path="System">*</Select>\
#                          <Select Path="Security">*</Select>\
```

The proposed system is used to collect the logs from windows and Linux machine algorithm as follows:

For syslog configuration is given by

```
<86>May 17 20:53:32 server sshd[14330]: Accepted password for darshan from 192.168.0.103 port 63742 ssh2
<30>May 17 20:53:32 server systemd: Started Session 28 of user darshan.
<38>May 17 20:53:32 server systemd-logind: New session 28 of user darshan.
<30>May 17 20:53:32 server systemd: Starting Session 28 of user darshan.
<86>May 17 20:53:32 server sshd[14330]: pam_unix(sshd:session): session opened for user darshan by (uid=0)
<30>May 17 20:53:35 server dhclient[10497]: DHCPREQUEST on eth0 to 192.168.0.1 port 67 (xid=0x4830dcac)
<85>May 17 20:55:36 server unix_chkpwd[14579]: password check failed for user (darshan)
<85>May 17 20:55:36 server sshd[14573]: pam_unix(sshd:auth): authentication failure; logname= uid=0 euid=0 tty=ssh ruser= rhost=192.168.0.103  user=darshan
<86>May 17 20:55:37 server sshd[14573]: Failed password for darshan from 192.168.0.103 port 63756 ssh2
<86>May 17 20:55:47 server sshd[14573]: Accepted password for darshan from 192.168.0.103 port 63756 ssh2
<86>May 17 20:55:51 server sshd[1457β]: pam_unix(sshd:session): session closed for user darshan
```

4 Experimental Results

In the proposed method raw logs are collected and configured the Logstash on both windows and Linux machines. The desired data are taken from the Logstash Data (Figs. 2, 3 and 4).

Finally the out is written in file, console. The Logstash send the log events to Logstash, gathers the message converts them into Json format and stored in karfka. The parsed logs are able to check whether it is a failed login or success login.

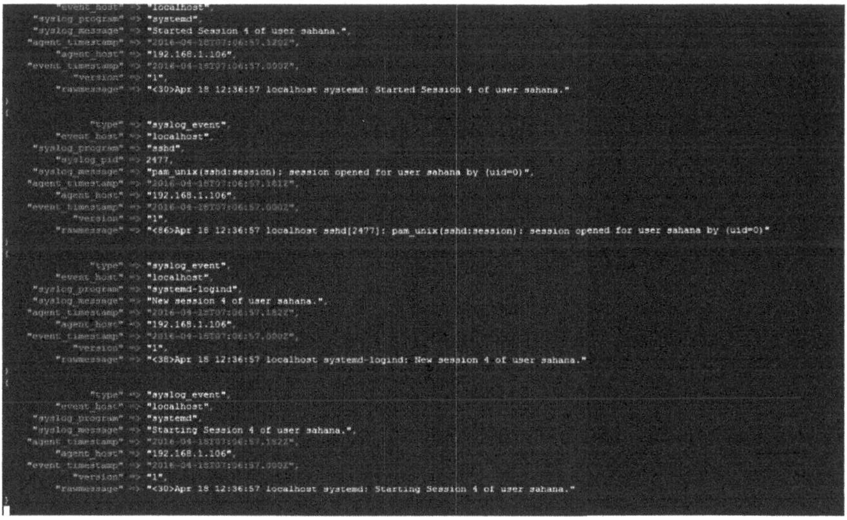

Fig. 2 Windows output for raw logs

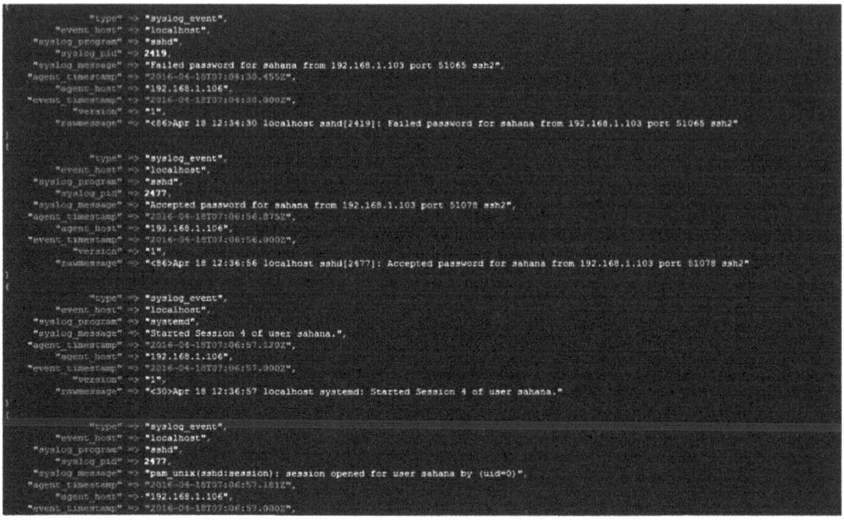

Fig. 3 Linux raw logs

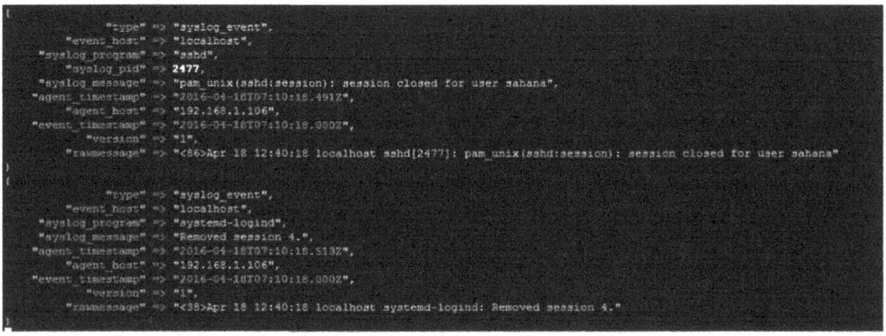

Fig. 4 Logstash output

5 Conclusion

One of the vital conclusions from this work is to remember that everybody has logs and that means that everybody ultimately needs log management. It contains all the details of logs and the proposed system has a capability to handle large volumes of Log messages. Different systems of logs are collected and it forward to Logstash. It is a log forwarder and has multiple features and it handles many raw log input, filters and outputs.

To run the windows raw Logs virtual box is used and Syslog in putty software and the language used is Java, Testing was completed on numerous datasets and results were as expected.

References

1. The Foundation for Federal Security and Compliance "Log Management". http://www.mcafee.com
2. Dr. Anton Chauvin, The wide guide to log and event management. Int. J. Future Comput. Commun. (2013)
3. V. Vianello, V. Gulisano, Ricardo, A scalable SIEM systems and applications, in *International Conference on Availability, Reliability and Security* (2013)
4. Mahmoud Awad, D.A. Menasce, Logging system for automatic process, in *Computer Measurement Group Conference*, Sanantonio, TX, Nov 2015
5. Bernard Jasen, Log analysis by semantic search. Elsevier Inf. Res. (2015)
6. Sudeheendra, Suhas, Mitesh Patel, Pratik Kumar, On the predictive properties of performance models derived through input-output relationships, in *Proceeding of 32nd International Computer Measurement Group Conference* (2006)
7. R. Vaarandi, M. Pihelgas, Using security logs for collecting and reporting technical matrix. IEEE Trans. (2014)
8. T.M. Vaarandi, Imagining event log clusters, in *2015 11th International Conference Network and Service management (CNSM)*
9. David Swift, *Security Log Management Policies for Appraisal and Compliance.* SAAN Institute Infosec Reading Room, Nov 2010
10. Afsaneh Madanis, Classification of security operation centers. IEEE Trans. (2013)

Estimation of Degree of Connectivity to Predict Quality of Software Design

S.D. Thilothame, M. Mashetty Chitralekha, U.S. Poornima
and V. Suma

Abstract The main goal of Object Oriented Methodology is to deliver software which is maintainable. *Post_Development_Quality_Requirements* such as software maintainability is depending on design quality. Coupling and Cohesion (C&C) are two design quality factors which are measurable. C&C are influenced by structure of a class which is a basic unit of Object Oriented design. Defining the class structure and their relationships measures the design quality which in turn is indicator for quality requirements such as maintainability, reusability and scalability. This paper explores how different dependency types between classes adds on to design complexity and hence software quality by proposing a model which calculates Degree of Connectivity (DC) between the classes and Coupling Index (CI) of overall software. Thus, it is now possible to infer that design quality not only depends on class structure, but also upon the level of relationships such as inheritance, aggregation, composition and association present in software.

Keywords Design quality · Class relationships · C&C · Degree of Connectivity (DC) · Coupling Index (CI) · Post_Development_Quality_Requirements

1 Introduction

Software Engineering domain provides a platform for programmers and researchers to develop quality software, tools and process methodologies. Object Oriented Programming is a popular methodology which facilitates to develop complex software. Complexity increases due to bad design. Researchers are working on

S.D. Thilothame · M.M. Chitralekha · U.S. Poornima (✉) · V. Suma
Dayananda Sagar College of Engineering, Bangalore, India
e-mail: uspaims@gmail.com

V. Suma
e-mail: sumadsce@gmail.com

U.S. Poornima
Raja Reddy Institute of Technology, Bangalore, India

© Springer Nature Singapore Pte Ltd. 2017
S.C. Satapathy et al. (eds.), *Proceedings of the 5th International Conference on Frontiers in Intelligent Computing: Theory and Applications*, Advances in Intelligent Systems and Computing 516, DOI 10.1007/978-981-10-3156-4_62

assessing the design quality either at code level (Post Assessment) or during High Level Design (Pre Assessment) through UML diagrams. Research has been going on to measure internal binding between attributes, methods and attributes_methods [1, 2]. Cohesion reflects such binding at code level, namely Low Level Design (LLD). Author of [3] formed metric to measure the cohesion at High Level Design (HLD) using UML diagrams. Many such metrics are proposed and mathematically validated for both LLD and HLDs.

It is difficult to anticipate the complexity only on strength of internal dependency in a class. Design complexity also depends on external binding of classes. Class relationships such as inheritance, aggregation, composition and association have their own impact on design quality. Each class relationship coins at data and method level. This work considers (1) *Data Binding, when two classes binds each other through object data*, (2) *Method Binding, when two classes binds each other through methods*, (3) *Data_Method Binding, when two classes binds each other through both data and method. Further, the work considers that aggregation and composition relationships proposes Data Binding, Dependency relationship proposes Method Binding and Inheritance proposes Data_Method Binding.*

This paper proposes a model to measure the Degree of Connectivity (DC) of different relationships and their individual contribution to design complexity. The model then calculates Coupling Index (CI) of overall project which is an indicator of design complexity during maintenance. This part of the research has used a Java project with 6 classes and different coupling types to calculate DC and CI.

This paper is organised as below, Sect. 2 is on related work, Sect. 3 proposes a model, Sect. 4 is about the measuring degree of connectivity using matrix representation, Sect. 5 is on calculating Coupling Index and Sect. 6 concludes the paper.

2 Related Work

This section summarises the survey of related work on coupling and cohesion.

Author [4] discusses software coupling approach on Object Oriented dependencies between classes mainly concentrated on types of coupling that are unavailable until after the completion of implementation. The coupling measure from source code has the advantage of having quantitative and more specific measure, but information is not available before implementation.

Author [5] made survey on design pattern. The pattern contains the dynamic and static behaviour of different types of entities which can be traced as functional diagram with dependency between them. The coupling factor represents the degree of relationship which the industries perceive and measures for quality of software design. The main principle of design of pattern is to reduce the coupling index for minimizing the complexity of design. The entire quality of software design is thus based on complexity of relationship between modules.

Author [6] discusses the significance of cohesion and coupling on design quality. Objects and classes are the entities in solution space and software quality directly depends on design quality of such logical entities. C&C are the two prime factors in object oriented design, measuring them can become an indicator to reduce the complexity. In complicated software, architecture of design required to be flexible and maintainable.

Author of [7] discussed that there are many number of techniques and tools are available to perform metric analysis on such code or software. The entire software modularization is partitioned into three main components (1) Use of API (2) Use of non API (3) Use of shared variable. This study provides a conceptual and practical framework for measurement of various factors like polymorphism, inheritance, coupling and cohesion and depth of inheritance. They used "step-in out" technique to get their functioning therefore increasing the entire quality of software and productivity.

Author of [8] discusses that earlier coupling measures consider only the static coupling but they do not consider dynamic coupling because of polymorphism and may usually deprecate the software complexity and miscalculate the need for coding inspection, testing and debugging. The proposed method consists of three steps such as introspection procedure, post processing and coupling measure. Finally metrics of coupling are evaluated for dynamic coupling. The development result represent that propose system will accurately evaluate the metrics of coupling dynamically. Finally author recommended dynamic coupling evaluation techniques which contain introspection procedure, including trace events into functions of all classes and anticipating dynamic behaviour at the time of execution of source code.

Author of [9] suggested the basic metric of coupling for object-oriented systems. In that metric, they stated CBO (Coupling between Objects) metric as number of non-inheritance dependent couples with other remaining classes.

Author of [10] states that measure the dynamic coupling at phase of analysis only. They define that dynamic coupling depends on the frequency with which classes communicate at runtime. They suggested Dynamic or run time Clustering Mechanism (DCM) that performs by capturing the circumstances for dynamic coupling at analysis phase.

Author of [11] recommended a dynamic method to calculate coupling index of software systems. It gives final result that generally used analysis of the and gives the partial dynamic behaviour of the system.

3 Algorithmic Representation of Proposed Model

The proposed model works as below.

Step 1: Accepts a Java Project
Step 2: Calculates the DC which is overall connection between classes
Step 3: Calculates types of dependency of individual class
Step 4: Calculates Coupling Index of each relationship
Step 5: Reports severity of coupling using severity index table.

4 Measuring the Degree of Connectivity Using Matrix Representation of Coupling

To assess the complexity, it is better to have quantitative information on different types of connectivity between the classes. Hence, this part of the work dealt with designing a model which takes a moderate size Java project with 6 classes, namely, A, B, C, D, E and F. The intension is to get a quantitative measure on different types on connectivity between the classes and to calculate Coupling Index which is an indicator for project maintainability. The connectivity is represented in matrix form to identify the degree of connectivity as in Table 1.

In the above Table 1, the value 0 denotes connectivity present between two classes and 1 denotes no connectivity. The degree of connectivity is formulated as below.

$$Degree\ of\ Connectivity\ (DC) = \frac{Number\ of\ connectivity}{Total\ Number\ of\ Connectivity} \times 100 \qquad (1)$$

Result: DC for sample project = (13/36) × 100 = 36 %.

The sample project has 36 % of overall connectivity between the classes which is a first level indicator of representing design complexity for maintenance group.

After finding the degree of connectivity, the model determines which type of dependency exits between two classes. The dependency between classes can be

Table 1 Matrix representation of coupling

Name of classes	A	B	C	D	E	F
A	0	1	0	1	1	0
B	1	0	1	1	1	1
C	0	0	0	1	1	0
D	1	1	0	0	0	0
E	0	0	1	0	0	0
F	1	0	0	0	0	0

Table 2 Matrix representation of class relationships of a sample project

Name of classes	A	B	C	D	E	F
A	No dependency	Inheritance	No dependency	Aggregation	Association	No dependency
B	Composition	No dependency	Inheritance	Composition	Aggregation	Association
C	No dependency	No dependency	No dependency	Inheritance	Inheritance	No dependency
D	Inheritance	Aggregation	No dependency	No dependency	No dependency	No dependency
E	No dependency	No dependency	Aggregation	No dependency	No dependency	No dependency
F	Inheritance	No dependency	No dependency	No dependency	No dependency	No dependency

Table 3 Types of dependency for class A

Name of class	Inheritance	Aggregation	Association	Composition
A	1	1	1	0

Table 4 Types of dependency for class B

Name of class	Inheritance	Aggregation	Association	Composition
B	1	1	1	1

Data Binding (Aggregation, Composition), Method Binding (Dependency) or Data _Method Binding (Inheritance) as shown in Table 2.

Table 2 Presents relationships between various classes present in project which is an input to identify individual class relationships to find CI as in Tables 3 and 4.

Similarly, dependency types are calculated for classes C, D, E and F.

5 Calculating the Coupling Index

After finding types of dependency for each class relationship with each class is tabularised as in Table 5. This is used to calculate CI of a project.

Hence the CI represents strength of each relationship in a project which can be calculated using the formula as shown below.

$$Coupling\ Index\ (CI) = \sum \frac{number\ of\ relation\ present}{total\ number\ relations} \times 100 \qquad (2)$$

Result: The following Table 6 lists CI of each relationship in a sample project.

Using CI, severity index of each class coupling relationship is calculated as shown in Table 6. The range of 1–4 is taken to fix the severity of each coupling type (Table 7).

Using the severity index table, the sensitivity of coupling for the sample is calculated which shows that Inheritance coupling in a project is extremely coupled,

Table 5 Class with total dependency

Name of classes	Inheritance	Aggregation	Association	Composition
A	1	1	1	0
B	1	1	1	2
C	2	0	0	0
D	1	1	0	0
E	0	1	0	0
F	1	0	0	0

Table 6 CI of each relationship

Name of relationship	Formula	CI in percentage (%)
Inheritance	6/14	42.85
Aggregation	4/14	28.57
Association	2/14	14.28
Composition	2/14	14.28

Table 7 Severity index of each coupling using CI

Coupling range	Percentage (%)	Severity of coupling type
More than 4	40	Extremely coupled
2–3	30	Tightly coupled
1–2	20	Moderately coupled
0–1	10	Loosely coupled

aggregation is tightly coupled and both association and composition are loosely coupled.

Thus, CI decides the severity of each relationship in software. Depending on CI value, the software maintainability can be identified as high risk, medium risk and low risk which mirrors the design complexity.

6   Conclusion

Software Engineering domain invites researchers and programmers to improve development process, Techniques and Tools to provide quality software. Improving design quality facilitates *Post_Development_Quality_Requirements* such as software maintainability, reusability at ease. Quality design is achieved through well defined class in Object Oriented Programming. Much research has been going on Coupling and Cohesion which are two design quality decisive factors represents dependency between classes and attributes within a class respectively. This paper proposes a model to calculate overall Degree of Connectivity (DC) and Coupling Index (CI) of a project to find the level of complexity which becomes an indicator for maintainability in future.

Data Binding (Aggregation, Composition), Method Binding (Dependency) or Data _Method Binding (Inheritance) are major coupling types exists between the classes. Since Coupling and Cohesion are closed knitted, the other features of Object Oriented Programming such as dynamic binding, polymorphism in coupling types would influences the cohesion and vice versa.

Acknowledgments The authors would like to sincerely acknowledge all the industry personnel for their valuable suggestions, help and guidance in carrying out this part of research. The complete work is undertaken under the framework of Non Disclosure Agreement.

References

1. Al Dallal, Jehad, L.C. Briand, A precise method-method interaction-based cohesion metric for object-oriented classes. ACM Trans. Softw. Eng. Methodol. (TOSEM) **21**(2) (2012)
2. Al Dallal, Jehad, Incorporating transitive relations in low-level design-based class cohesion measurement. Softw.: Pract. Exp. **43**(6), 685–704 (2013)
3. Al Dallal, Jehad, L.C. Briand, An object-oriented high-level design-based class cohesion metric. Inf. Softw. Technol. **52**(12), 1346–1361 (2010)
4. J. Offutt, A. Abdurazik, S.R. Schach, Quantitatively measuring object-oriented couplings. Softw. Qual. J. **16**(4), 489–512 (2008)
5. P. Wolfgang, *Design patterns for object-oriented software development* (Addison-Wesley, Reading, Mass, 1994)
6. U.S. Poornima, Factors modulating software design quality (2014). arXiv:1402.2374
7. Deepti Gupta, Coupling based structural metrics—an quality assessment of software modularization. Int. J. Adv. Res. Comput. Sci. Softw. Eng. **3**(6) (2013). ISSN: 2277 128X
8. S. Babu, Dr. R.M.S. Parvathi, Development of dynamic coupling measurement of distributed object oriented software based on trace events. Int. J. Softw. Eng. Appl. (IJSEA) **3**(1) (2012)
9. S.R. Chidamber, C.F. Kemerer, Towards a metrics suite for object-oriented design, in *Proceedings of the Conference on Object-Oriented Programming: Systems, Languages and Applications, (OOPSLA' 91)*, SIGPLAN Notices, vol. 26, no. 11 (1991), pp. 197–211
10. H. Paques, L. Delcambre, A mechanism for assessing class interactions using dynamic coupling during the analysis phase, in *Proceedings of XVIII Brazilian Symposium on Software Engineering—SBES'99*, Florianopolis, Santa Catarina, Brasil (1999)
11. E. Schikuta, *Dynamic Coupling Metrics* (1993)

Multipath Load Balancing and Secure Adaptive Routing Protocol for Service Oriented WSNs

C.P. Anandkumar, A.M. Prasad and V. Suma

Abstract Existing multipath routing methods in a Wireless Sensor Networks (WSNs) have presented the effective distribution of traffic over multipath to accomplish required quality of service (QOS). But, failure of the any link in WSN affects the data transmission performance, security of data, scalability and reliability of WSN. Hence, by considering the reliability, scalability, security and congestion for the multipath in a wireless sensor network, it is necessary to design and develop a service focused multi path routing scheme which should provide high failure tolerance and effective routing scheme. This paper put forth a Secure Multipath AODV (SMAODV) protocol in which RSA algorithm is used for secure data transmission and path vacant ratio is calculated to discover the link disjoint path to destination sensor node from source sensor node from all presented paths in a network. Load balancing metrics and technique like detection of congestion, congestion notification and congestion control is used to fine tune and balance the load over multi paths. Split the data packets into multiple fragments and sent it to the destination node through multipath based on the path vacant ratio.

Keywords Congestion detection and control · RSA algorithm · Multipath · AODV · Wireless sensor networks

C.P. Anandkumar (✉) · A.M. Prasad
Department of CSE, DSCE, Bangalore, India
e-mail: anandcp05@gmail.com

A.M. Prasad
e-mail: prasaddsce@gmail.com

V. Suma
Department of ISE, RIIC, DSCE, Bangalore, India
e-mail: sumavdsce@gmail.com

© Springer Nature Singapore Pte Ltd. 2017
S.C. Satapathy et al. (eds.), *Proceedings of the 5th International Conference on Frontiers in Intelligent Computing: Theory and Applications*, Advances in Intelligent Systems and Computing 516, DOI 10.1007/978-981-10-3156-4_63

1   Introduction

Wireless sensor networks (WSNs) consist of randomly deployed autonomous sensor nodes. Each node is having the ability of interacting or communicating with the physical word and gathers the information about the physical word. Most of the existing service oriented wireless sensor network is designed for a specific purpose. But these are lack in standard representation and standard process for the sensor data [1–3]. Interoperability between the different applications has been provided by developing the service oriented architecture for the WSNs in which functionality delivered wireless sensor network are taken as the services such as data processing, data localization and data aggregation service [4, 5]. Self-determining service provisioning and scalable wireless sensing technology are combined in Service oriented wireless sensor network (WSN) in which applications are taken as the services. This technique helps to support through resource management and protocol design [6]. Service oriented architectures advantageous for application oriented wireless sensor network and also generic WSNs. Services provided by the service oriented WSNs are having number of performance metrics. These performance metrics are delay, bandwidth, load balancing, and reliability that are contained within the service system. Quality of service (QOS) parameter associated with the services, for instance, load balancing, are provided by the sensor nodes [7–9].

Applications can be developed in Service oriented wireless sensor network (WSNs) based on the requirement of the services which is needed. The main challenge in the service oriented WSNs are to improve the throughput. Because large amount of data or packet traffic is swapped over the WSNs [3]. Security is also another challenge in service oriented wireless sensor network. To provide the quality of service requirement, it is necessary to design and develop the multipath and secure routing system. Design of the multipath from source to destination reduces the traffic over the single path and also provides the reliable paths. If the services want to run over the secure and reliable wireless sensor architecture, it is necessary to provide the assured bandwidth for each path in multipath routing scheme. In-service oriented WSNs, each node should have the capability to estimate the performance of its neighbors sensor node based on the path reliability [9–13].

In this paper has presented a multiple paths and secure routing method with the features (a) Independence of application (b) Adaptive congestion control (c) Secure data transmission.

2   Proposed Scheme

By considering the above mentioned challenges, this work has proposed a multipath load balancing and secure routing protocol based on the Ad Hoc On-Demand Distance Vector (AODV) which has the below features.

(a) RSA algorithm is used to secure the data in the network. Use of Multipath to send the packets, reduces the delay. This is first stage of SMAODV protocol. This improves the privacy and confidentiality of data in the service oriented WSNs.

(b) Path vacant ratio is calculated for the each path in the Wireless sensor network. Based on the path vacant ratio, load on the each path is evaluated. More data packets are send through the path which are having high path vacant ratio and less data packets. Path vacant ratio is calculated by considering the importance degree of the nodes in a network.

(c) The congestion detection and control method is adapted in SMAODV to detect and notify the congestion occurrence on each path. Each sensor node in the network, monitors its ongoing traffic. If the node notices the congestion existence, then it informs to its parent node by HELLO message. Parent node decreases the packet rate based on the level of congestion.

Figure 1 displays the architectural diagram of the proposed SMAODV protocol. The protocol uses disjoint multipath to form multipath from a given source node to destination node. Path vacant ratio is used as evaluation metric for load balancing and to form multipath. Path vacant ratio Vi for ith path is used for load balancing which is given by,

$$vi = \frac{\sum_{i=1}^{n} PI(\text{path}i) - PI(\text{path}i)}{\sum_{i=1}^{n} PI(\text{path}i)} \qquad (1)$$

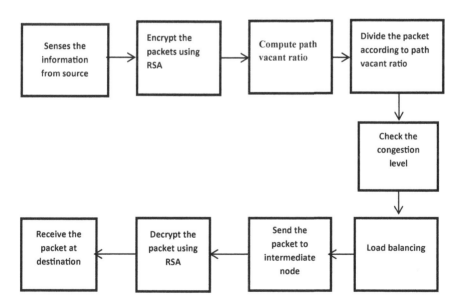

Fig. 1 Proposed SM AODV protocol architecture

where PI specifies the important path of the sensor node in a network and PI is calculated by taking the each node connectivity degree in a sensor network.

Each node in wireless sensor network continuously monitors its traffic load. If any of the nodes sense the incidence of congestion. After that node adds this congestion statistics in to the HELLO message. After adding the information regarding congestion into the Hello message, child node sends that updated Hello Message to its parent sensor node. The parent node verifies the congestion information and evaluates the congestion level. If the level of congestion is too high, then parent node regulates the delivery rate of data packet on the congested path. This increases the throughput.

The algorithm for the proposed SMAODV protocol multipath congestion control is as below.

Step 1: Detection of congestion by sensor node.
Step 2: Congestion notification by HELLO message.
Step 3: Congestion evaluation and adjust the load on congested path.

When a node detects the congestion, it sends that congestion information to the parent node by adding congestion information into the HELLO Message. Parent node evaluates the congestion level based on the congestion information. If the CONGESTION_VALUE is equal to one that means load on the path is normal so parent node no need to do any changes in packet delivery rate. If the CONGESTION_VALUE is equal to more than one then congestion is high so the parent node has to decrease the packet delivery rate to by its next lower level. Before detecting the congestion, the data packet is encrypted using RSA algorithm and encrypted data is spitted in to number of segments. Based on the path vacant ratio, the data packets are sent over multipath. Detection of congestion level and load balancing is done by considering the value of packet service ratio $r(i)$. These packets are then decrypted using RSA and received at the destination.

3 Simulation and Performance Evaluation

3.1 Simulation Setups

NS-2.3 is used for simulating the proposed work. Sixty sample sensor nodes had taken for the simulation and all these nodes are deployed randomly in 500 m × 500 m square area. Basic parameter for the simulated network is shown in Table 1. Data rate is set to 1 Mbps. For each sensor node in network, carrier sensing range is 550 m and radio transmission range is 250 m. Average delay and packet delivery percentage are the two performance metrics. For each setting 30 trails are simulated. Assuming that at multipath having at least 3 hops between sources to destination and topology of network is known.

Table 1 Parameters for simulation

Traffic flow	UDP
Simulation time	160 s
Transmission range	120 m
Simulation area	500 × 500 m
Packet size	512 bytes
Number of nodes	60
Buffer size	40 packets

3.2 Simulation Analysis

The packet delivery rate and the average delay are the two performance metrics which are used for performance evaluation of the proposed secure multipath AODV (SMAOVD) protocol by comparing with existing protocols.

(1) *Packet delivery ratio (Pr)*. Packet delivery rate is calculated based on the number of packets successfully received or delivered to the destination sensor node and the number of packets sent to the destination by source node.

$$Pr = \frac{P\text{succ}}{P\text{send}} \tag{2}$$

in which Pr *is the packet deliver rate,* Psucc is the packets received at destination and Psend is the packets sent from the source node.

(2) *Average end to end delay*. Average delay is calculated by considering time taken for path discovery phase, at MAC layer packet retransmission time, at interfaces packet queuing time and packet transmission time from source sensor node to destination node.

$$T\text{delay} = T\text{disv} + T\text{queu} + T\text{tran} + T\text{prop} \tag{3}$$

where Tdelay is the average end to end delay and Tprop is packet propagation time.

3.3 Simulation Results

The performance of the SMAODV is simulated with different node speed like 0, 5, and 10, 30 m/s. Figure 2 Represents the packet delivery ratio (PDR) of the secure multipath AODV protocol along with AOMDV and CRAODV protocol.

As in above figure, the secure multipath AODV (SMAODV) is having the higher packet delivery ratio and less packet loss than the other protocol. This is because of more stable routes and less overhead in route discovery.

Figure 3 shows the comparison of average packet delay of the proposed SMAODV, AODV and AOMDV protocols. As in the figure, SMAODV is having

Fig. 2 Packet delivery ratio

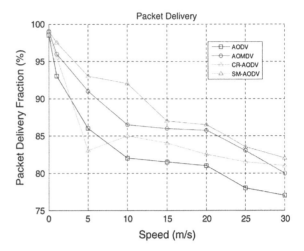

Fig. 3 Average end to end delay

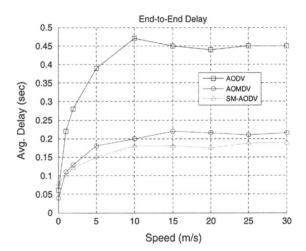

lower average end to end delay up to 6 % than existing protocol like AODV and AOMDV. Because, the proposed SMAODV uses the path vacant rate to schedule the load on the multi path which reduces the congestion during data transfer.

4   Conclusion

The proposed SMAODV protocol for wireless sensor networks (WSNs) is an enhanced AODV protocol for efficient data transfer between source and destination nodes. Since fault tolerance is one of the challenging issues in all networks, it needs to be addressed in WSN as it has direct affect on sensor node battery life time.

SMAODV proposes a technique for load balancing to secure and reliable transfer of packets from source to destination on multi path with the uses of RSA and congestion control algorithm. This improves sensor nodes' capability in applications and also provides optimization choices for secured transferring the data over network which are not present in the exiting protocol.

References

1. S. Li, X. Wang, S. Zhao, Multipath routing for video streaming in wireless mesh networks. Ad Hoc Sensor Wirel. Netw. (½), 73–92 (2011)
2. T. Zhao, K. Yang, H.H. Chen, Topology control for service oriented wireless mesh networks. IEEE Wirel. Commun. **16**(4), 64–71 (2009)
3. K. Lin, J.J.P.C. Rodrigues, H. Ge, N. Xiong, X. Liang, Energy efficiency QoS assurance routing in wireless multimedia sensor networks. IEEE Syst. J. **5**(4), 495–506 (2011)
4. K. Wang, X. Bai, J. Li, C. Ding, A service-based framework for pharmacogenomics data integration. Enterp. Inf. Syst. **4**(3), 225–245 (2010)
5. D. Chiang, C. Lin, M. Chen, The adaptive approach for storage assignment by mining data of warehouse management system for distribution centers. Enterp. Inf. Syst. **5**(2), 219–234 (2011)
6. A. Rezgui, M Eltoweissy, μRACER: a reliable adaptive service driven efficient routing protocol suite for sensor-actuator networks. IEEE Trans. Parallel Distrib. Syst. **20**(5), 607–622 (2009)
7. L. Li, Introduction: advances in E-business engineering. Inf. Technol. Manage. **12**(2), 49–50 (2011)
8. S. Xu, L.D. Xu, Management: a scientific discipline for humanity. Inf. Technol. Manage. **12** (2) (2011)
9. L.D. Xu, C. Wang, Z. Bi, J. Yu, AutoAssem: an automated assembly planning system for complex products. IEEE Trans. Ind. Inform. **8**(3), 669–678
10. X. Zhang, H. Su, Network-coding-based scheduling and routing schemes for service-oriented wireless mesh networks. IEEE Wirel. Commun. **16**(4), 40–46 (2009)
11. L.D. Xu, Enterprise systems: state-of-the-art and future trends. IEEE Trans. Ind. Inform. **7**(4), 630–640 (2011)
12. L. Duan, W. Street, E. Xu, Healthcare information systems: data mining methods in the creation of a clinical recommender system. Enterp. Inf. Syst. **5**(2), 169–181 (2011)
13. E. Xu, M. Wermus, D. Bauman, Development of an integrated medical supply chain information system. Enterp. Inf. Syst. **5**(3), 385–399 (2011)

Visual Based Information Retrieval Using Voronoi Tree

Megha Biradar and Muzameel Ahmed

Abstract Content retrieval from large databases needs an efficient approach due to the increasing growth in the digital images. Especially content based image retrieval is an extensive research area. This mainly includes retrieving similar images from the large dataset based on the extracted features. The extracted feature content can be texture, colour, shape etc. Efficient method for image recuperation is proposed in this paper based on shape feature. Shape features like computing Boundary, mode using morphological operations and Harris corner detector and Voronoi diagram are proposed. These matching decisions can be made by different classification models. SVM classifier is used in this research work to get the best matched images during image retrieval. The proposed algorithm is evaluated on JPEG images to get accuracy of about 90 %.

Keywords Voronoi tree · Support vector machine (SVM) classifier · Boundary and node detection

1 Introduction

With the speedy development of the web and the falling cost of storage instruments, it becomes more and more general to store texts, graphics, images, video, and audio in digital format. This raises the challenge of designing methods that support strong search and navigation through the contents of huge digital archives. As a part of this common predicament, image recuperation and indexing had been active research areas for more than a decade. This aims at recuperation images efficiently from the

M. Biradar (✉) · M. Ahmed
Department of Information Science and Engineering, Dayananda Sagar College
of Engineering, Bangalore, India
e-mail: Meghasbiradar93@gmail.com

M. Ahmed
e-mail: Muzchk@yahoo.com

M. Biradar · M. Ahmed
Jain University, Bangalore, India

© Springer Nature Singapore Pte Ltd. 2017   603
S.C. Satapathy et al. (eds.), *Proceedings of the 5th International Conference on Frontiers
in Intelligent Computing: Theory and Applications*, Advances in Intelligent Systems
and Computing 516, DOI 10.1007/978-981-10-3156-4_64

huge database using the extracted features. The extracted features can be obtained from shape, colour and texture properties of the image in the stored dataset [1].

Houssem Chatbri et al. [2] proposed an efficient method content based image retrieval based on skeletons and contours shape representation. This paper gives an empirical study of these two shape representations by comparing their performance to various categories of binary image. Different categories include nearly thin, thick and elongated images. Image variations include addition of noise, reduction of size and image blurring. Ritika Hirwane et al. [3] proposed an efficient method for solving semantic gap problem in CBIR approach. This paper mainly aimed at introducing the challenges and problems concerned with the creation of CBIR systems.

Vikrant Gunjal et al. [4] proposed an efficient method for Content Based Visual Information Retrieval (CBVIR). Which defines effective set of irrelated and related images regard to a query image by evaluating three criteria's like density, diversity and unreliability. This aims at selecting the most informative images in the image collection. Apeksha Chaudhari et al. [5] proposed an efficient method for content based image categorisation using SVM classifier. Using this in this work, an attempt to construct an efficient mapping between the semantic and low level is made in order to make a decision on which category an image belongs to. Researchers in [1, 6] all have proposed an efficient methods for shape based retrievals.

An efficient method for CBIR based on shape feature is proposed in this work. Different shape parameters like Boundary, Node and Voronoi tree [7] which are then used as the feature extraction during classification. SVM classifier is thus used in this work for Image classification to get the proper matched Image Retrieval. The detail description of the proposed system is shown in the paper below.

2 Methodology

The proposed system is as shown in the Fig. 1. The proposed system consists of two phased called testing phase and testing Phase. In the testing phase the input image with only boundary of particular image is passed, followed by pre-processing block. In this step different pre-processing steps like image resizing and binary conversion of boundary image is done if the boundary image is not in binary format. The pre-processed image is then passed to Node detection block. Here the nodes in the Boundary image are calculated using Harris Corner Detector. The obtained nodes are then passed to feature extraction block. Here the Voronoi Tree algorithm makes use of these nodes for Voronoi tree creation, which is taken as the unique feature for this particular test image. Features diameter, area and thinness are obtained. Like The extracted feature is then passed to SVM classifier [8] for further classification.

In the training phase the Different images with single objects are passed. Pre-processing steps like image resizing, binarizing as in testing phase is carried

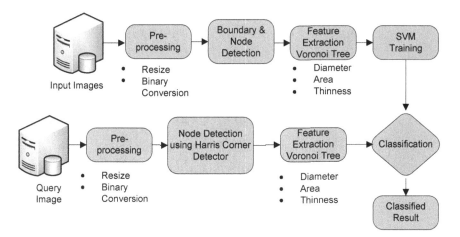

Fig. 1 Architecture of proposed system

out. Once the binary image is obtained, the boundary and nodes of this image is calculated before feature extraction. The obtained nodes of the generated boundary image is then passed to feature extraction block, where the Voronoi Tree for these nodes are calculated and stored in the knowledge base as a feature for all the trained images. The features obtained in the testing phase are then compared with the features stored in the knowledge base using SVM classifier to get the proper matched image.

2.1 Harris Corner Detector

As this approach is one of the popular algorithms for corner detector, this approach is used in this work for consequence point detection because to its powerful performance even in rotation, enlightenment and scale variation. This method is exits on the auto relation property (function). This function calculate the local changes of the signal with the shifted patches by a small amount in various path (directions). The approach is explained in detail with following steps,

Considering a shift $(\Delta x, \Delta y)$ and a point(x, y) the defined function for auto correlation is given by,

$$C(x, y) = \sum_{w} [I(x_i, y_i) - I(x_i + \Delta x, y_{i+} \Delta y)]^2 \tag{1}$$

where the image function is assigned by $I(.)$ and the points in the Window W is denoted by (x, y). To the first order terms Taylor expansion is done to approximate the shifted image.

$$I(x_i + \Delta x, y + \Delta y) = I(x_{i,} y_i) + \left[I_x(x_i y_i) I_y(x_i, y_i) \right] \begin{bmatrix} \Delta x \\ \Delta y \end{bmatrix} \quad (2)$$

where I_x and I_y represents the partial derivative in both x and y, Eq. (3) is obtained by substituting Eq. (2) in Eq. (1).

$$C(x, y) = \sum_w [I(x_{i,} y_i) - I(x_i + \Delta x, y_{i+} \Delta y)]^2 \quad (3)$$

$$= \sum_w (I(x_i, y_i) - I(x_i, y_i) - [I_x(x_i, y_i) I_y(x_i, y_i)] \begin{bmatrix} \Delta x^2 \\ \Delta y \end{bmatrix} \quad (4)$$

$$= \sum_w (-I_x(x_i, y_i) I_y(x_i y_i) \begin{bmatrix} \Delta x \\ \Delta y \end{bmatrix}^2 \quad (5)$$

$$= \sum_w ([I_x(x_i, y_i) I_y(x_i, y_i)] \begin{bmatrix} \Delta x \\ \Delta y \end{bmatrix}^2 \quad (6)$$

Intensity structure of the all the local neighbourhood is captured by the matrix C (x, y). γ_1 and γ_2 denotes the Eigen values of the matrix C(x, y). If these two values are very small then the local auto correlation function becomes flat. If these values are high then the auto correlation function becomes ridge shaped. If both the Eigen values are high then the auto correlation function becomes shapely peaked and this indicated a corner [01].

2.2 Voronoi Tree

The Voronoi tree is one of the efficient algorithms for computing the shape features of the obtained curvature extrema along the border. Voronoi diagram is calculated using the corner points obtained. Set of limited no of points are considered and is called as the generator points in the Euclidian plane in this work. Association of all these points in the plane to their nearest generator is done. Generator is represented by S. The location set is then assign to each generator forming a area called as Voronoi cell or Voronoi polygon [5]. Association of all the generators with set of Voronai polynomial forms the Voronoi diagram. The Voronoi polygon and Voronoi diagram is given by the sum of the generators set given by, $P = \{p_1, \ldots, P_n\} \in R^2$, where $2 < n < \infty$ and it is given that for $i \neq j$, $i, j \in I_n = \{1, \ldots, n\}$. The Voronoi Polynomial is given by the equation given below,

Fig. 2 The Voronoi diagram with its polygons and generators

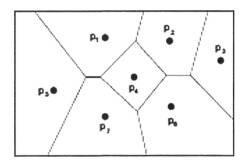

$$Voronoi\,Polynomial\,(p_i) = \{p|d(p,p_j)\}\,for\,j \neq i, j \in I_n \tag{7}$$

Minimum distance between p and p_i i.e. the length of the line connecting p and p_i in the Euclidean distance is called the Voronoi Polynomial (VP) and the associated Voronoi Diagram is given by Eq. (8). Figure 2 depicts the Voronoi Diagram with its polygons and generators [9].

$$Voronoi\,Diagram = \{VP(p_1), \ldots, VP(P_n)\} \tag{8}$$

2.3 SVM Classifier

Collecting similar type of images in large dataset is a difficult task when it comes to manual browsing. As there is a large amount of database searching identical images would be a complicated task. Hence to improve this kind of classification using the extracted feature using efficient classification algorithm is very necessary. In this proposed work, SVM classifier is used to get the best match to obtain accurate image retrieval system. Nonlinear support vector machine approach is further used in this research. This approach maps all the data into other space with the help of nonlinear mapping function. This function is called as a kernel function and the space used here is nothing but a Euclidian distance. By training the several binary classifiers it is possible to obtain the multi class SVM classifier [5].

3 Experimental Result

This section depicts the result obtained at each stage of the proposed system. Initially when the Query boundary image is given it undergoes certain pre-processing and the obtained image is as in the Fig. 3a. The pre-processed image is then subjected to Harries algorithm to get the corner points as in (b). This step is

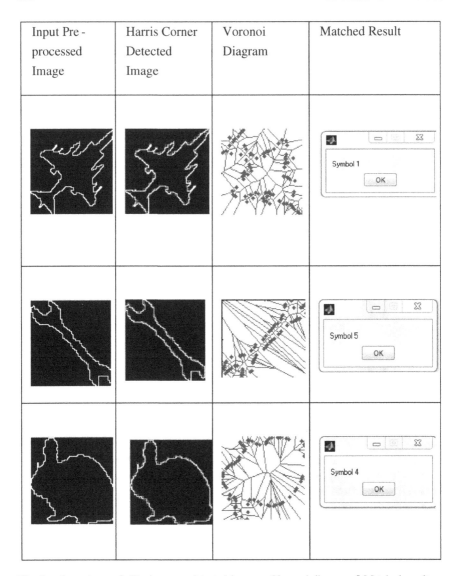

Input Pre - processed Image	Harris Corner Detected Image	Voronoi Diagram	Matched Result

Fig. 3 **a** Input image. **b** Harris corner detected image. **c** Voronoi diagram. **d** Matched result

followed by Voronoi tree based feature extraction to get the Voronai Diagram as in (c). These extracted features are then passed to SVM classifier to get the matched result. When the match is found the dialogue box saying the folder name in displayed regardless of change in orientation in the image. This folder consists of the trained images which are matched with the query image as in (d). Total of 6 training folders are taken with 6 different images. Each folder consists minimum of 4 images of same type with different orientation.

4 Conclusion

The method which is used in this research is most efficient to match the images. The Harris corner detection will give the corner detected point of the image. The special practice here is use of the Voronoi diagram from that it is possible to draw the Voronoi tree to get the feature of the image. The feature which gets from this tree is trained with the system that is stored in the storage. The SVM will give the result of match or not match. The method which is used in this work is the most accurate and useful when compared to other method. The accuracy of the result will be 90 % or more than that. Here the one limitation is if more Voronoi point will create ambiguity while drawing the tree.

References

1. D. Li, S. Simske, *Shape Retrieval Based on Distance Ratio Distribution* (2002)
2. H. Chatbria, K. Kameyamab, P. Kwanc, *A Comparative Study Using Contours and Skeletons as Shape Representations for Binary Image Matching* (Elsevier, 2015), pp. 1–8
3. R. Hirwane, Content based image retrieval: feature extraction, current techniques, promising directions. Int. J. Adv. Res. Comput. Sci. Softw. Eng. **5**(5) (2015)
4. V. Gunjal, V. Baranwal, Relevance feedback for content-based visual information retrieval using active learning method. Int. J. Adv. Res. Comput. Eng. Technol. **4**(5) (2015)
5. A. Chaudhari, P.K.S. Bhagat, An overview of content based image categorization using support vector machine. Int. J. Innov. Sci. Eng. Technol. **1**(10) (2014)
6. M. Alkhawlani, M. Elmogy, H. El Bakry, Text-based, content-based and semantic-based image retrievals: a survey. Int. J. Comput. Inf. Technol. **04**(01) (2015)
7. M.A. Migut, M. Worring, C.J. Veenman, *Visualizing Multi-Dimensional Decision Boundaries in 2D* (Springer, 2015)
8. S. Bansal, Er. R. Kaur, A review on content based image retrieval using SVM. Int. J. Adv. Res. Comput. Sci. Softw. Eng. **4**(7) (2014)
9. M. Kolahdouzan, C. Shahabi, Voronoi-based K nearest neighbor search for spatial network databases, in *VLDB Conference* (2004)
10. S.Y. Pattar, Study of corner detection algorithms and evaluation methods. Int. J. Innov. Res. Sci. **4**(5), (2015)
11. X. Wang, X. Ying, Y.-J. Liu, S.-Q. Xin, W. Wang, X. Gue, W. Mueller-Wittig, Y. Hea, *Intrinsic Computation of Centroidal Voronoi Tessellation (CVT) on Meshes*, vol. 58 (Elsevier, 2015), pp 51–61
12. Y. Song, *Visual Information Retrieval from 2D Shapes by Bipolar Matching* (IEEE, 2010)

Gradient Magnitude Based Watershed Segmentation for Brain Tumor Segmentation and Classification

Ngangom Priyobata Singh, Sunanda Dixit, A.S. Akshaya
and B.I. Khodanpur

Abstract MRI is one of the tool for detecting the tumor in any part of the body. But precise tumor segmentation from such Magnetic resonance imaging (MRI) is difficult and also time consuming technique. To overcome such difficulty, this work proposes a very simple, efficient and automatic segmentation and classification of brain tumor. The proposed system is composed of four stages to segment, detect and classified tumor as benign and malignant. Pre-processing is carried out in the first stage after which watershed segmentation technique is applied for segmenting the image which is the second stage. The segmented image undergo for post processing to remove the unwanted segmented image so as to obtain only the tumor image. In the last stage, gray-level co-occurrence matrix (GLCM) is used to extract the feature. This feature is given as input to Support Vector Machine (SVM) to classify the brain tumor. Results and experiment shows that the proposed method accurately segments and classified the brain tumor in MR images.

Keywords Gradient image · GLCM · MRI · SVM · Watershed segmentation

1 Introduction

Biomedical image processing may be defined as a practice that undertakes medicine engineering and technology jointly. Computer process techniques become a lot of powerful through many benefits in non-invasive medicine imaging and image

N.P. Singh (✉) · S. Dixit · A.S. Akshaya · B.I. Khodanpur
Dayananda Sagar College of Engineering, Bangalore, India
e-mail: priyobatang@gmail.com

S. Dixit
e-mail: sunanda.bms@gmail.com

A.S. Akshaya
e-mail: akshayamgd@gmail.com

B.I. Khodanpur
e-mail: bi.khodanpur@gmail.com

© Springer Nature Singapore Pte Ltd. 2017   611
S.C. Satapathy et al. (eds.), *Proceedings of the 5th International Conference on Frontiers in Intelligent Computing: Theory and Applications*, Advances in Intelligent Systems and Computing 516, DOI 10.1007/978-981-10-3156-4_65

processing [1]. Image processing becomes one of the important role in the field of Medical Imaging. MRI becomes an important tool for most of the researcher as it produces high quantized image giving minute details regarding delegate structure within human body. The results obtained from analysis are used to guide for the treatment.

The cell is considered to be one of the primary structural unit in living organisms. There are millions of cells in human body. Each of the cell has its own unique functions. For the proper functioning of the body, these cells have to be divided to create new cells in a very governable manner. But typically, they split and grow unconditional to form new cells. This ends up in a mass of unwanted tissue that is outlined as a tumor [1]. Studies have found that the rate of brain tumor cases is increased drastically [2]. Scientist made a conclusion that the cause is due to exposure to ionizing radiation such as radiation therapy where the machine aims for the head region and even caused due to family history. So it becomes an important to detect tumor in early stage so as to give early treatment.

There are different imaging techniques existing such as MRI, CT etc. for brain tumor examination. MRI is considered to be one of the recommended machine as it does not use radiation which is harmful to human body [3]. Also it gives a clear distinction between soft and hard tissue.

Getting segmented brain tumor from MRIs become very important for neurosurgeon, oncologist and radiotherapy which is accountable to measure tumor responses for treatment. Segmentation can be done manually but it takes considerable time and is prone to error. So automatic detection and segmentation [4] is highly desirable. Automatic segmentation and detection is highly challenging task. For example, tumor can vary in size, shape and location [5]. So this research proposes a fully automatic brain tumor segmentation which combine two efficient technique i.e. gradient magnitude and watershed segmentation.

2 Methodology

The proposed method consist of two phase: training and testing phase. The planned technique of tumor segmentation encompasses four steps. The first step is pre-processing followed by gradient magnitude based watershed segmentation method which is the second step. Third step is post processing which is applied to obtain the region of interest (ROI). Lastly, GLCM is used to extract the feature for further classification using SVM. Figure 1 shows the overview of the above said plan technique.

2.1 Data Acquisition

Implementation of the proposed system starts with collection of the MRI brain image from the Brain Web Database at the McConnell Brain Imaging center of the

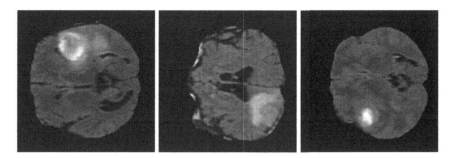

Fig. 1 Brain MRI images with tumor

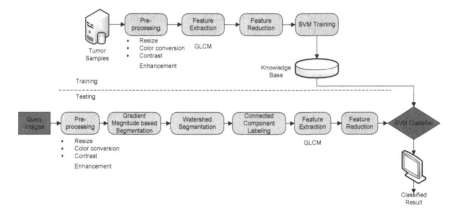

Fig. 2 Overview of proposed system

Montreal Neurological Institute (MNI), McGill University [6]. Brain image can get from link (http://www.bic.mni.mcgill.ca/brainweb) [7]. The following figure shows the sample brain MRI (Fig. 2).

2.2 Pre-processing

The MRI that is obtained for segmenting the tumor is in .mha format. The images can not feed directly for the process. These image contain noise. Therefore the image has to undergo for pre-process technique. The pre-processing stage comprise of different steps as follow.

(1) RGB to Gray-level conversion: The first step in pre-processing is to convert the obtain image into gray-level image.
(2) Image resize: The gray-level image is resized to 260 × 260 for providing uniformity.

(3) Median filter: It is employed to remove the noise so as to boost the results of later process. It uses the concept of sliding window which is also known as kernel. The main idea behind the median filter is slide the kernel over the entire image and replace the middle value in the kernel with the median of all the values in the kernel.

(4) Binary image conversion: The enhanced image is then converted from gray-level to binary image for further processing.

2.3 Gradient Based Watershed Segmentation

Image segmentation is the process of dividing or partitioning an image into identical explicable regions. The division of an image into meaty structures is usually a vital step in image processing, object illustration, visualization, and lots of different image process tasks. The definitive point in an exceedingly sizeable measure is to concentrate important alternative from the image learning, from that a framework, translation, or comprehension of the setting will be given by the machine. It is vital for identifying tumors, edema and necrotic tissues. In short, segmentation induce the Region of Interest (ROIs).

Watershed segmentation is an established calculation utilized for division, that is, for isolating diverse objects in an image. Watershed implies that space of area wherever all the water channels off it and goes into indistinguishable spot. In earth science, watershed line is plot as the line isolating two development basins. The rains that fall on either facet of the watershed line can flow into a similar lake. This idea is utilized in image processing as a method for taking care of numerous segmentation issues. Figure 3 shows the two dimension of the above said segmentation technique. To comprehend the watershed segmentation, a grayscale image is seen as a topological surface, where the estimations of f(x, y) compare to statures [8].

Fig. 3 Watershed segmentation simplified to 2D

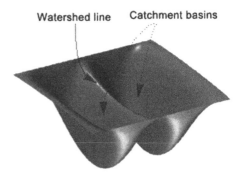

The algorithm of watershed segmentation is worked as follows:

1. Assume a hole is made at each of the catchment basin and the whole geography is overflowed from the beneath by giving the water a chance to ascend through the opening at a uniform rate.
2. Pixels underneath the water level at a given time are labeled as flooded.
3. When the water level rises at uniform rate, the overflowed regions will develop in size. In the end, the water will raise uniformly to a certain point where two overwhelmed areas from isolated catchment basins will combine.
4. When this happens, the algorithm builds a one-pixel dam that isolates the two areas.
5. The overflowed proceeds until the whole image is portioned into independent catchment bowls isolated be watershed edge line.

Normal watershed segmentation has drawback of over segmentation. To defeat this issue, this study introduced gradient magnitude based watershed segmentation approach. Before applying watershed segmentation technique on the image, the image into gradient image is first converted. In such image, pixels associated with the object edges are of high values and that of other pixels are of low values. Let the image be denoted by f(x, y). Then the gradient vector magnitude and the point at which most extreme rate of progress of intensity level happens for the image can be measured using two mathematical statement which is given below [3]:

$$g(x, y) = \sqrt{\{g_1^2(x, y) + g_2^2(x, y)\}} \tag{1}$$

$$\alpha(x, y) = \tan^{-1}\left(\frac{g_2(x, y)}{g_1(x, y)}\right) \tag{2}$$

where

$g_1^2(x, y)$ gradient in x direction
$g_2^2(x, y)$ gradient in y direction.

The degree of the gradient can be accomplished by utilizing sobel mask. If the gradient image is given as input to the watershed function, over segmentation will reduce.

2.4 Post-processing

The image obtain after segmentation contains unwanted pixel information which is out of tumor regions. To remove such unwanted pixel information, this study aims to apply morphological operation. Different types of morphological operation such as dilation, erosion and analysis of connected pixels were performed. Morphological technical uses the concept of structuring element. Structure element is taken which is of appropriate shape and size that will remove the unwanted pixel information.

2.5 Feature Extraction

It is a system to locate the important features from the images, which are utilized to comprehend the images easier. In this proposed method, GLCM is used to extract the important features for classification. The ROI is tumor which is given as input to the GLCM method to extract the feature. Some of the relevant features are Auto-correlation, Correlation, Entropy, Energy, Contrast, and Homogeneity [9].

Contrast: It gives back a measure of the intensity difference between a pixel and its neighbor.

Homogeneity: It gives back a quality that measures the closeness of the conveyance of component in the GLCM to the GLCM diagonal.

Energy: Gives back the aggregate squared components in the GLCM.

Entropy: It is a measure of randomness.

2.6 Classification

SVM is one of the supervised learning strategy. It is a descent tool for data examination and classification. It is used for classification for two class problem. It depends on the idea of decision planes. A decision plane may be define as a line (2D) or a plane (3D) that isolates between a groups of items having distinctive class membership [9]. In this proposed method, classification is done using SVM technique. SVM with 6 images is trained: 3 benign and 3 malignant image. In SVM, the classes are accept to be recognize as $+1$ or -1. The decision boundary is estimated as $y = 0$ using the formula

$$y = \sum_{i=1}^{n} w_i x_i + b \qquad (3)$$

where

w_i weight
x_i input pattern.

The two equation of the line partitioning the classes is given below

$$x_i w_i + b \geq 1 \quad \text{when } y = +1 \qquad (4)$$

$$x_i w_i + b \leq 1 \quad \text{when } y = -1 \qquad (5)$$

Fig. 4 Separating margin
between two classes

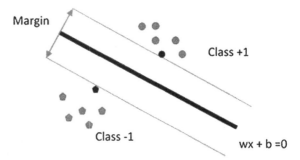

The distance to the origin from the hyperplane ($x_i w_i + b = 0$) is $\frac{-b}{w}$, where w is the norm of w. Again the distance to the origin from the hyperplane is $M = \frac{2}{w}$, where M is the margin. So the largest margin is obtain by maximizing w so as to increase accuracy of classification (Fig. 4).

3   Results and Analysis

The proposed method is reenacted in MATLAB 2010a. A sample MRI brain is the input image which contain tumor. The simulated output of pre-processing stage, segmentation, detection and finally classification of tumor is given in Fig. 5.

3.1   Performance

(A) Performance measure:

Performance of the proposed system is tested and analysed using the parameters such as "sensitivity", "specificity" and "accuracy". These parameters are computed using the given equations:

$$\text{Sensitivity} = TP/(TP + FN) * 100\,\% \tag{6}$$

$$\text{Specificity} = TN/(TN + FP) * 100\,\% \tag{7}$$

$$\text{Accuracy} = (TP + TN)/(TP + TN + FP + FN) * 100\,\% \tag{8}$$

where

TP   Abnormal brain precisely identify as abnormal.
TN   Normal brain precisely identify as normal.
FP   Abnormal brain wrongly identify as normal (Table 1).

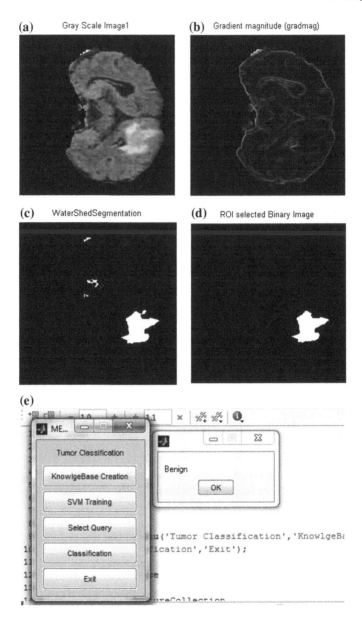

Fig. 5 a Grey level image. **b** Gradient image. **c** Watershed segmentation. **d** ROI (tumor). **e** Classification of MRI brain image

Table 1 Performance analysis

Parameter	Values in %
Accuracy	90
Specificity	100
Sensitivity	87.75

4 Conclusion

In this paper, the brain tumor segmentation and classification is implemented effectively. The performance of this system is analysed based on sensitivity, specificity and accuracy. The method achieves the accuracy rate of 90 %. It has been established in the environment of MATLAB 2013a platform. Optimization can be performed and kept for future work.

References

1. Joseph, R. Paul, M. Manikandan, C. Senthil Singh, Brain tumor MRI image segmentation and detection in image processing. Int. J. Res. Eng. Technol. **03**(01 | NC-WiCOMET-2014) (2014)
2. Cancer Facts & Figure 2015, American Cancer Society
3. K. Ain, R. Rulaningtyas, Edge detection for brain tumor pattern recognition. Physics Department, University of Airlangga Surabaya
4. W. Yang, M. Huang, W. Chen, Y. Wu, J. Jiang, Brain tumor segmentation based on local independent projection-based classification. IEEE Trans. Biomed. Eng. **61**(10) (2014)
5. H. Tang, W. Liu, H. Lu, X. Tao, *Tumor Segmentation from Single Contrast MRI Images of Human Brain* (Healthcare Philips Research China, IEEE, 2015)
6. R. Helen, N. Kamaraj, CAD Scheme to detect brain tumour in MR images using active contour models and tree classifiers. J. Electr. Eng. Technol. **10**(2), 742 (2014)
7. Amritpal Singh, Parveen, Detection of brain tumor in MRI images, using combination of fuzzy C-Means and SVM, in *2015 2nd International Conference on Signal Processing and Integrated Networks (SPIN)*
8. Dr. J. Jayakumari, K.S. Angel Viji, Automatic detection of brain tumor based on magnetic resonance image using cad system with watershed segmentation, in *Proceedings of 2011 International Conference on Signal Processing, Communication, Computing and Networking Technologies (ICSCCN 2011)*
9. K. Amshakala, K.B. Vaishnavee, An automated MRI brain image segmentation and tumor detection using SOM clustering and proximal support vector machine classifier, in *2015 IEEE International Conference on Engineering and Technology (ICETECH)*, 20 Mar 2015, Coimbatore, TN, India

Information Retrieval Through the Web and Semantic Knowledge-Driven Automatic Question Answering System

Ananthi Sheshasaayee and S. Jayalakshmi

Abstract The rising popularity of the Information Retrieval (IR) field has created a high demand for the services which facilitates the web users to rapidly and reliably retrieve the most pertinent information. Question Answering (QA) system is one of the services which provide the adequate sentences as answers to the specific natural language questions. Despite its importance, it lacks in providing the accurate answer along with the adequate, significant information while increasing the degree of ambiguity in the candidate answers. It encompasses three phases to enhance the performance of QA system using the web as well as the semantic knowledge. The WAD approach defines the context-aware candidate sentences by using the query expansion technique and entity linking method, second, Ranks the sentences by exploiting the conditional probability between the query and candidate sentences and the automated system, third, identifies the precise answer including the reasonable, adequate information by optimal answer type identification and validation using conditional probability and ontology structure. The WAD methodology provides an answer to a posted query with maximum accuracy than baseline method.

Keywords Question answering · Ontology · Query expansion · Semantic · Information retrieval

A. Sheshasaayee (✉)
PG & Research Department of Computer Science, Quaid-e-Millath Government College for Women (Autonomous), Chennai, Tamilnadu, India
e-mail: ananthi.research@gmail.com

S. Jayalakshmi
Periyar University, Salem, India
e-mail: jayalakshmi.research@gmail.com

© Springer Nature Singapore Pte Ltd. 2017
S.C. Satapathy et al. (eds.), *Proceedings of the 5th International Conference on Frontiers in Intelligent Computing: Theory and Applications*, Advances in Intelligent Systems and Computing 516, DOI 10.1007/978-981-10-3156-4_66

1 Introduction

The usage of WWW increases day by day and people submit numerous questions on the web search engine which retrieves the query-relevant web documents. As the growth of the web searching information is rapid, the users demand an efficient system that provides the relevant information with ease, is also high [1]. Natural Language processing plays an important role in automatic evaluations and extracting a relevant document for the posted query. In Information Retrieval (IR) field, Question Answering (QA) system satisfies the desires of the users, provides the precise textual answers with adequate sentences to natural language questions posed instead of a set of relevant web documents [2]. QA system incorporates the three main modules such as question analysis module, relevant documents retrieval module, and candidate answer extraction module. The Sematic and Syntactic features are used to define the degree of relationship between the words and define the grammar, rearrange the sentences in an appropriate manner.

The QA system employs both the structured and unstructured collection of data to determine the correct answers. Based on these information sources, QA system is categorized into structured data of knowledge-based and unstructured data of the corpus-based. The existing QA systems exploit the set of retrieved web documents to answer the similar questions of the others [3]. The large-scale information sources such as DBpedia [4], Freebase [5], YAGO [6], Microsoft's Satori, and Google's knowledge graph, store the globally valuable relationship of information in the form of triplets. QA systems utilize the semantic web sources such as ontology that includes concepts and their relationships. Ontology is one of the main blocks of the semantic web, plays a significant role in providing the background knowledge of the QA system in IR process. Most of the conventional QA systems generate the candidate answers by focusing on the web-based information sources. However, it lacks in providing the sufficient knowledge of the candidate answer, since these QA systems validate only the candidate answer with the response type. Thus, providing the candidate answer including the concise, relevant information is imperative for QA system. The proposed QA system concentrates on the web-based QA system with ontology relationship structure while generating the candidate answers to achieve this goal.

The main contributions of Web And semantic knowledge-Driven automatic question answering system (WAD) includes, to provides the appropriate answer including the adequate, and significant information for the posted query, expands the queries using both the web search engine and ontology and retains the snippets of the documents, Exploiting conditional probability based semantic relevance score ranks the context-aware candidate answer sentences, determines the answer type of each query based on the probability of the query and candidate answer sentences.

2   Related Work

The existing QA systems are primarily based on the following two information.

2.1   Web-Based Question-Answering Systems

The existing QA systems provide the answers to the users using the rich web corpus. Ask the QA system [7] to provide answer to the users' natural language questions. It contains only the dynamic information source to provide the accurate result of the questions in a timely fashion. While missing entity relationship in the large-scale knowledge bases, the web-based searching is used to generate the appropriate candidate answers [8]. Google search engine [9] defines the complex mappings between the surface form users questions and entities from the Google API. The work in [10] presents the framework for a question-answering system that contains question analysis, search, hypothesis generation, and hypothesis scoring. An automatic QA system in [11] exploits the search engine results and n-gram co-occurrence statistics to answer the complex and non-factoid questions by implementing the dynamic programming algorithm.

To predict the appropriate answer type of each question, the existing QA systems contemplate on the question analysis regarding question classification, and context based answer type validation [12, 13]. The precise context information on the users' questions in [14] facilitates the QA system to determine the corresponding answer type which is retrieved by employing the target types in DBpedia ontology [15]. An approach [16] exploits the correlation based typing scores and Wikipedia resource based type information. To assign the rank to the candidate answers, the QA system employs the probabilistic graphical model. Its prediction model is based on the joint probability which estimates the correctness of individual answers and its correlations [17].

2.2   Ontology-Based Question-Answering Systems

The information retrieval system utilizes the ontology services in ONKI selector widget with functions to expand the input queries before integrating the questions with a search interface [18]. Conceptual semantic space [19] is used to expand the users' questions by exploiting the WordNetontology-based semantic relations of the keywords appearing in the questions. Template based QA system generates an SPARQL template to identify the internal structure of the question by parsing a question. The system uses integer linear programming over knowledge bases on the web of linked data to translate the natural language questions into structured SPARQL queries [20]. The semantic relation of natural language question that are submitted to the search engine is matched with the subgraph in the knowledge base to answer the natural language questions [21]. A Freebase based QA system [22] extracts the information in the sense of question patterns with answer patterns using Freebase in which information is retrieved from the web-scale corpus.

Relational Pattern based Semantic Question Answering system over Linked Data (RP-SQALD) approach retrieves the direct answer from the knowledge bases by mapping the users' questions to ontology. It maps the natural language questions in terms of Resource Description Framework (RDF) triplet relational patterns. The research work addresses the problem of entity linking with multiple knowledge bases by utilizing textual and knowledge base features, ontology modularization, and collective inference.

3 An Overview of the WAD Approach

The WAD approach aims to identify the appropriate, concise candidate answers automatically for the posted.

3.1 Generating Context-Aware Candidate Sentences

The WAD approach applies preprocessing procedure on the natural language factoid questions submitted by the users. Then, it expands the input questions using the snippets retrieved from web source and the semantic relation of the query terms retrieved from WordNet ontology.

3.1.1 Query Expansion

In QA system, the numerous web users submit variety of requests as questions on the web search engine. The WAD approach implements the preprocessing method of the natural language questions to evacuate the most unimportant terms regarding the context of the question which restricts the processing time. To accurately retrieve the relevant documents from the search engine, the WAD approach handles the two levels of query expansion method based on the web source. The first level query expansion method involves snippets selection, and relevant terms included in the original query terms. In the second level query expansion method, the expanded query terms processed from the first level expansion are penetrated to the WordNet ontology to obtain the expanded query terms with the meta-features which are the semantic query terms.

3.1.2 Determining the Relevant Snippets Through Entity Linking

The WAD approach exploits entity linking method between the web corpus and the ontology to disambiguate the corresponding entity of each term appearing in the snippet. The input of the entity linking system is the set of retrieved documents' snippets, and the output of the system is the list of matched entities including confidence value in the ontology graph structure. Then, the list of entities is processed together with the words or phrases in the snippets to identify the question's intention based specific entity.

$$\text{Entity\_Link}\left(W_i^{NP}, S_i\right) = \begin{cases} \text{Score\_Link}\left(E_i, W_i^{NP}\right), & \text{if } 1 \leq i \leq n \\ 0, & \text{Otherwise} \end{cases} \tag{1}$$

From Eq. (1), $i = \{1, \ldots, n\}$ represents the entities appearing in ontology. If an entity is not included in the ontology, the distance between the new term and query term in a snippet is used as the score of the snippet.

$$\text{Link}\left(E_i, W_i^{NP}\right) = \arg\min_{E_i \in E_n} \left(\alpha\left(\text{score}\left(E_i | W_i^{NP}, S_i\right)\right) + \beta\left(\text{dist\_}\left(E_i, W_i^{NP}\right)\right)\right) \tag{2}$$

$$p\left(E_i | W_i^{NP}, E_n\right) = \frac{p\left(E_i | W_i^{NP}\right)}{\sum_{E_i \in E_n} p\left(E_i | W_i^{NP}\right)} \tag{3}$$

Equation (2) represents the measurement of recognized entity score based on the context of the snippet, the Eq. (3) to measure the value of the probability in which the conditional probability ($p(E_i | W_i^{NP})$) is measured using the summation of distance between the entity, and the NP and other words of a snippet in the ontology.

3.2 Ranking Candidate Answer Sentences

To identify the relevance of candidate answer sentence to a corresponding question, the non-trivial semantic relevance measurement is applied. Semantic relevance reveals the level of intimacy between the user's query and the candidate answer sentences.

$$\text{Rel}(Q, A) = \begin{cases} P_{ij}, & \text{if } Q_i = \text{new\_}Q_i \\ (P_{ij}) * \text{freq\_}(Q_i, A_j), & \text{if } Q_i \neq \text{new\_}Q_i \end{cases} \tag{4}$$

$$P_{ij} = \left[\sum_{K=1}^{N} \text{Sem\_sim}\left(Q_i, A_j\right)\right] * \left(1 / \text{dist\_}(Q_k, A_l)\right) \tag{5}$$

The semantic relevance measurement is obtained using Eqs. (4) and (5) (Fig. 1).

Algorithm 1: The WAD algorithm

WAD Algorithm

Input: User's question (Q_i),
Output: Corresponding answer for a given question (A_j)

While $Q_i \in Q_n$ **do**
//**Generating context-aware candidate sentences**
for each Q_i **do**
Q_i --> Preprocessing
Preprocessed Q_i --> {set of keywords} i.e. Q_i^s
Q_i^s --> Web search engine
$S_i(Q_i^s)$ <-- Web search engine
if($S_i(Q_i^s)_{novel\ terms} \leq \alpha$) **then**
Select S_i including novel terms and add that novel terms into Q_i^s
endif
for all expanded(Q_i^s) **do**
Expanded(Q_i^s) = $E(Q_i^s)$ = $S_i(Q_i^s)_{novel\ terms} + Q_i^s$
$E(Q_i^s)$ -->WordNet ontology -->Semantic_$E(Q_i^s)$
endfor
Semantic_$E(Q_i^s)$ --> Web search engine
Web search engine --> D_s including S_i
if($S_i(D_s)_{novel\ terms} \leq \beta$) **then**
Select set of S_i and S_i -->Yago ontology
endif
Yago ontology --> recognized entities of S_i
foreach NP **do**
Identify entity link
Endfor
//**Ranking candidate answer sentences**
foreach S_i based sentence **do**
for $Q_k \in Q_i$ and $A_l \in A_j$ **do**
if(Q_i=new_Q_i) **then**
Measure Rel(Q,A) using P_{ij}
else
Measure Rel(Q,A) using P_{ij} and freq_(Q_i,A_j
endif
endfor
//**Predicting precise answer based on the answer type**
foreach scored candidate sentence **do**
if(Q_i=WH-operator) **then**
Identify $A_t(Q_i)$ using the WH-word and ontology
else
Identify $A_t(Q_i)$ to predicate in ontology
endif
foreach $A_t(Q_i)$ **do**
Validate $A_t(Qi)$ with $A_j^1(Q_i$
Provide appropriate A_j to a given Q_i

endfor
endfor
endfor

endfor
endwhile

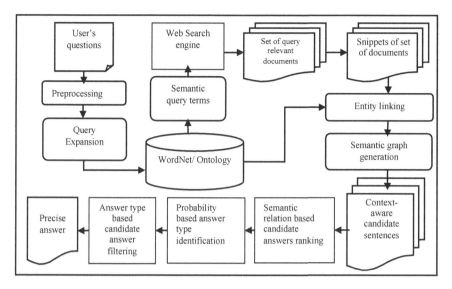

Fig. 1 The WAD methodology

3.3 Forecasting Accurate Answer Based on the Answer Type

The WAD approach generates the precise answer by identifying and examining the answer type in accordance with the user's question. Initially, it determines the answer type for each question by estimating the probability of the set of query terms with all the suitable answer types. Then, validates this answer type with the candidate answers to pinpoint the answers accurately.

3.3.1 Answer Type Identification and Validation

To predict the appropriate candidate answer, the proposed QA system recognizes the original intention of the questions through determining the answer type. The WAD approach contemplates the queries and candidate answer sentences as the vectors of the set of word stems. i.e. $Q = \{q_1^w, q_2^w, ... q_m^w\}$ and $A = \{a_1^w, a_2^w, ... a_n^w\}$.

Usually, in the case of informal questions and complex questions, the headword is insufficient to understand the exact answer type. Hence, the WAD approach exploits the set of query words along with the WH-operator appearing in a question.

4 Research Evaluations

The WAD and RP-SQALD systems are evaluated on the Question-Answer dataset [23]. This dataset contains the question-answer pairs from the year of 2008, 2009, and 2010.

4.1 Evaluation Metrics

Precision: The ratio between the number of correct answers and the number of questions to answer returned.

Recall: The ratio between the number of correct answers and the total number of questions.

F-measure: F-measure is defined as the combination of precision and recall value

$$F-\text{measure} = 2 * \frac{\text{Precision} * \text{Recall}}{\text{Precision} + \text{Recall}}$$

4.2 Results and Analysis

Describes the implementation results of both the WAD and RP-SQALD approach when evaluating the system on the QA dataset. It shows the performance in terms of precision, recall, F-measure, and response time.

4.2.1 Precision Versus Number of Questions

Figure 2 reveals the precision of both the WAD and RP-SQALD approach on the QA dataset while varying the number of input questions and Answer Ambiguity (AA). The AA is the ratio between the number of candidate answers that has a predicted answer type and the total number of candidate answers, represented in percentage. On the whole, the precision value decreases while increasing the number of questions in QA system. When increasing the number of questions from 200 to 1000 and AA is at 10 %, the RP-SQALD approach rapidly decreases its precision value by 2.66 %, but the WAD approach marginally decreases by 1.86 %. It is because, by exploiting entity linking method, which enables the QA system to disambiguate the candidate sentences. The performance of the WAD approach (when AA = 30 %) is high due to the consideration of snippets as well as the questions.

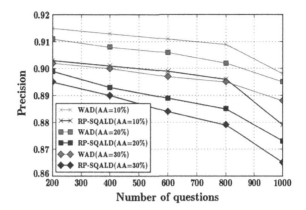

Fig. 2 Precision versus number of questions

4.2.2 Recall Versus Number of Questions

Figure 3 illustrates the comparative result of the WAD and RP-SQALD approaches. While varying the number of questions from 200 to 1000 and AA from 10 to 30 %. From the figure, the recall value of both the WAD and RP-SQALD approach decreases while increasing the number of questions.

4.2.3 F-measure Versus Number of Snippets

Figure 4 indicates F-measure of the WAD approach while varying the number of retrieved snippets from 100 to 500 and RS(S, Q). The RS(S, Q) represents the Relevance Score between the snippets and the corresponding query. The results presented in

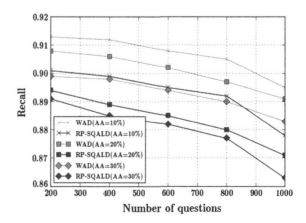

Fig. 3 Recall versus number of questions

Fig. 4 F-measure versus
number of snippets

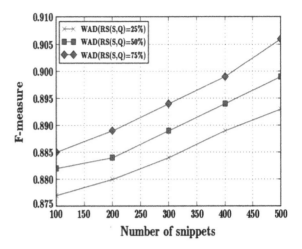

4.2.4 Response Time Versus Question Length

Figure 5 indicates the comparative results of the WAD and RP-SQALD approach. It reveals the response time while varying the length of the input questions in which length represents the number of words in a question. If a load or complexity of a question increases, the response time of the QA system also increases due to the gaining loads of the input.

The WAD approach balances the response time until reaching the certain limit on the question length, and then it slightly delays the response time.

Fig. 5 Response time versus
question length

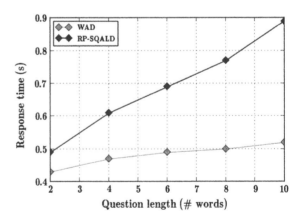

5 Conclusion

The WAD approach is used to accomplish both the web and semantic knowledge. It encompasses three important phases are query expansion and entity linking method which incorporate the semantically related terms into query terms and disambiguate the candidate answer. The candidate answers are ranked using conditional probability based semantically relevance between the query and the answer type. This approach has further improved the QA system, including relevant answer type and ontology structure. It accurately validates the standard answer with corresponding list of candidate answers and defines the exact answer with relevant information. The WAD approach has achieved a high performance while comparing with the baseline QA system.

References

1. O. Etzioni, Search needs a shake-up. Nature **476**(7358), 25–26 (2011)
2. O. Kolomiyets, M.-F. Moens, A survey on question answering technology from an information retrieval perspective. Elsevier Trans. Inf. Sci. **181**(24), 5412–5434 (2011)
3. V. Singh, S.K. Dwivedi, Question answering: a survey of research, techniques and issues. Elsevier Trans. Procedia Technol. **10**, 417–424 (2013)
4. C. Unger, L. Buhmann, J. Lehmann, A.-C. Ngonga Ngomo, D. Gerber, P. Cimiano, Template-based question answering over RDF data. ACM Proc. WWW, 639–648 (2012)
5. K. Bollacker, C. Evans, P. Paritosh, T. Sturge, J. Taylor, Freebase: a collaboratively created graph database for structuring human knowledge. ACM Proc. SIGMOD, 1247–1250 (2008)
6. F.M. Suchanek, G. Kasneci, G. Weikum, Yago: a core of semantic knowledge. ACM Proc. WWW, 697–706 (2007)
7. E. Brill, S. Dumais, M. Banko, An analysis of the AskMSR question-answering system, in *EMNLP* (2002), pp. 257–264
8. R. West, E. Gabrilovich, K. Murphy, S. Sun, R. Gupta, D. Lin, Knowledge base completion via search-based question answering, in *ACM Proceedings of the 23rd International Conference on World Wide Web* (2014), pp. 515–526
9. S. Monahan, J. Lehmann, T. Nyberg, J. Plymale, A. Jung, Cross-lingual cross-document coreference with entity linking, in *TAC 2011 Workshop* (2011)
10. D. Ferrucci, E. Brown, J. Chu-Carroll, J. Fan, D. Gondek, A.A. Kalyanpur, A. Lally, J.W. Murdock, E. Nyberg, J. Prager, Building watson: an overview of the deepqa project. AI Mag. **31**(3), 59–79 (2010)
11. R. Soricut, E. Brill, Automatic question answering using the web: Beyond the factoid. J. Inf. Retrieval-Spec. Issue Web Inf. Retrieval **9**(2), 191–206 (2006)
12. A. Lally, J.M. Prager, M.C. McCord, B. Boguraev, S. Patwardhan, J. Fan, P. Fodor, J. Chu-Carroll, Question analysis: how watson reads a clue. IBM J. Res. Dev. **56**(3.4), 2–1 (2012)
13. J.W. Murdock, A. Kalyanpur, C. Welty, J. Fan, D.A. Ferrucci, D. Gondek, L. Zhang, H. Kanayama, Typing candidate answers using type coercion. IBM J. Res. Dev. **56**(3.4), 7–1 (2012)
14. X. Luo, H. Raghavan, V. Castelli, S. Maskey, R. Florian, Finding what matters in questions, in *HLT-NAACL* (2013), pp. 878–887
15. K. Balog, R. Neumayer, Hierarchical target type identification for entity-oriented queries. ACM Proc. CIKM, 2391–2394 (2012)

16. A. Grappy, B. Grau, Answer type validation in question answering systems, in *Proceedings RIAOVAdaptivity, Personalization and Fusion of Heterogeneous Information* (2010), pp. 9–15
17. J. Ko, E. Nyberg, L. Si, A probabilistic graphical model for joint answer ranking in question answering, in *ACM Proceedings of the 30th Annual International SIGIR Conference on Research and Development in Information Retrieval* (2007), pp. 343–350
18. J. Tuominen, T. Kauppinen, K. Viljanen, E. Hyvönen, Ontology-based query expansion widget for information retrieval, in *Proceedings of the 5th Workshop on Scripting and Development for the Semantic Web, 6th European Semantic Web Conference*, vol. 449 (2009)
19. J. Zhang, B, Deng, X. Li, Concept based query expansion using WordNet, in *Proceedings of International e-Conference on Advanced Science and Technology* (2009), pp. 52–55
20. M. Yahya, K. Berberich, S. Elbassuoni, M. Ramanath, V. Tresp, G. Weikum, Natural language questions for the web of data, in *EMNLP-CoNLL* (2012), pp. 379–390
21. L. Zou, R. Huang, H. Wang, J.X. Yu, W. He, D. Zhao, Natural language question answering over RDF: a graph data driven approach. ACM Proc. SIGMOD, 313–324 (2014)
22. X. Yao, B. Van Durme, Information extraction over structured data: question answering with freebase, in *ACL* (2014)
23. http://searchdocs.net/

Level Set Based Liver Segmentation and Classification by SVM

Mallikarjun Kesaratti, Sunanda Dixit and B.I. Khodanpur

Abstract Liver segmentation from CT image is the key exploration works in representing a liver, which has incredible effect on the examination of liver issue. Hence, numerous computer-aided segmentation approaches have been proposed to partition liver locale from medical image automatically in the past numerous years. A method for liver segmentation system is proposed by consolidating level set based method with Pseudo Zenerike moment and GLDM Features. The objective of proposed algorithm is to solve the segmentation issue which is created by indistinguishable intensities between liver region and its adjacent tissues. Radial Basis Function SVM is used in this work to classify the type of the tumor.

Keywords Level set segmentation · Zenerike and GLDM features · Principal component analysis · RBF-SVM classifier

1 Introduction

Liver which is one of the organ of human being requires minimally invasive surgeries. According to the 'cancer facts and society', it is reported that the death rate due to the liver cancer increases rapidly. It also becomes one of the risk factor of death in the world. So it becomes very important to detect liver cancer in early stage to safe the precious human life.

Finding the exact location of liver tumor is the first task before going for surgery. Getting such location requires segmentation of the liver tumor. An accurate analysis of such segmented tumor gives important information such as tumor staging or whether surgery or therapy has to provide or not. It also gives the best approach for

M. Kesaratti (✉) · S. Dixit · B.I. Khodanpur
Department of ISE, Dayananda Sagar College of Engineering, Bengaluru, India
e-mail: mallikarjun.sk@outlook.com

S. Dixit
e-mail: sunanda.bms@gmail.com

B.I. Khodanpur
e-mail: bi.khodanpur@gmail.com

© Springer Nature Singapore Pte Ltd. 2017   633
S.C. Satapathy et al. (eds.), *Proceedings of the 5th International Conference on Frontiers in Intelligent Computing: Theory and Applications*, Advances in Intelligent Systems and Computing 516, DOI 10.1007/978-981-10-3156-4_67

treatment. Tumor segmentation also plays an important role in the development of 3D image that helps and guide the surgeon to remove the tumor easily.

Since CT is one of the tool for diagnosis of liver tumor, segmentation becomes essential. Manual segmentation can be done. But they more prone to error and also it is very time consuming process. So it becomes an important for the scientist to developed and automatic system which will segment and detect the liver tumor. Error given by system is less as compared to that of error given by human. So we proposed a method which has two parts, one Training SVM and other Testing. For training part tumor samples are taken as the input, then pre-processing is performed, the pre-processed image is taken as a input for feature extraction and feature reduction for SVM training and knowledge is stored.

The second part is testing, input liver image is taken and pre-processed step is performed, then for segmentation the level set is used, from the segmented image features are extracted using Pseudo Zernike moment and GLDM, feature reduction step is performed by PCA and at last SVM classifier will recognise and produce the result based on training data and testing data.

Segmentation of an image is a major step in image analysis process. Classical approaches of Image segmentation such as thresholding, splitting and merging, and region growing. Many others were largely studied by way of researchers and a lot of versions had been proposed. Usually many segmentation ways were developed. the region other than the ROI are marked in the same way, with a difficulty of purpose to solve this issue. Here first we are segmenting the region of image using level set methods and edges are well outlined and classification of tumor is done by SVM.

2   Related Work

Xinjian Chen et al. [1] proposed a method which is based on the active appearance which are strategically combined model (AAM), live wire, and graph cuts for Segmentation. Which consists of three main parts they are Model building, Object recognition and delineation. In Model building, they have constructed an AAM and then train the LW function, and GC parameters. In the second part they proposed an algorithm for improving the conventional AAM, AAm and LW are combined effectively, which results in oriented AAM. For object initialization a multi object strategy is adapted. Pseudo 3D initialization and segmentation of organs slices with the help of OAAM. For delineation part, 3D shape constrained GC is proposed.

BalaKumar and Mohamed Syed Ali [2] proposed a segmentation method of liver CT scan which is fully automatic which used a fuzzy c-means clustering and level set. Boundaries are clearer by improving the difference of unique image. Then anatomical previous information and the spatial fuzzy c-means clustering are used to extract liver automatically.

Fernández-de-Manuel [3] proposed a system to segment liver volumes from abdominal by using active surface method. It finds the surface that minimizes energy function by combining the inside and outside intensities of the surface.

Sajith and Hariharan [4] proposed a system, which is simple and clinically useful for segmenting liver tumor form CT scan. In image processing the level set method is widely for segmentation. Firstly fuzzy c-means algorithm is used and then for fine delineation a level set method is used. Segmentation performed by this method the liver region and tumor is very well defined.

3 Proposed System

In our approach, we divided the work in 2 phases, i.e. testing and training phase. In the first phase, that is training phase we take segmented tumor part of images and we pre-process those images which includes re-sizing, color conversion followed by contrast enhancement. Then features such as Pseudo Zenerike moment and GLDM of the image will be extracted followed by feature reduction using principal component analysis. Save those trained images in knowledge base. In the second phase, i.e. testing phase, this research takes a CT image and pre-process the image, in order to detect the tumour region we apply segmentation using level set algorithm and classify the tumor type using RBF-SVM classifier (Fig. 1).

A. **Pre-processing**

Pre-processing is a stage where the input image is converted to gray scale and resized to 256 × 256, increase the contrast enhancement.

B. **Segmentation**: **Level Set Algorithm**

Level set is a nonstop deformable model procedure with certain representation. It's fundamental thought is to insert the de-formable model in a d + a dimensional space, to fragment iteratively an item in a d dimensional space, making utilization

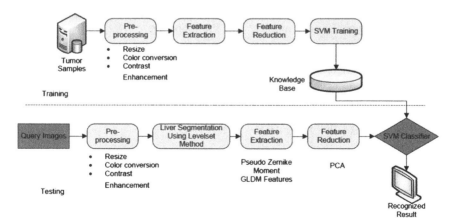

Fig. 1 Proposed architecture

of halfway differential equation. The fundamental point of preference of level sets is that it permits changes of surface topology verifiably.

The function for standard level set is characterized as Eq. (1)

$$\begin{cases} \frac{\partial \varnothing}{\partial t} + F|\nabla_\phi| = 0 \\ \phi(0, x, y) = \phi_0(x, y) \end{cases} \tag{1}$$

where $|\nabla_\phi|$: normal direction

$\phi_0(x, y)$: initial contour

F: comprehensive-forces. $\phi_0(x, y)$ is usually defined as in the Eq. (2)

$$\phi_0(x, y) = \begin{cases} -C_0 & if\ (x, y)\ is\ inside\ \phi_0 \\ C_0 & otherwise \end{cases} \tag{2}$$

Here $C_0 > 0$ is constant.

Here the level set needs re-initialization because the level set function (LSF) normally creates abnormalities in its evolution which causes numerical error and lastly break the stability of the level set evolution. Despite the fact that re-initialization as a numerical solution is able to preserve the regularity of the LSF, it is going to move incorrectly the zero level set away from the normal liver segmentation method.

C. Feature Extraction

Pseudo-Zernike moments have been shown to be superior to Zernike and Hu moments as feature descriptors in terms of their exhibited error rate [4]. However due to their costly computation and high dynamic range, pseudo-Zernike moments haven't seen wide use in image recognition yet. Pseudo-Zernike moments of order $p + q$ contain twice as much information as Zernike moments, because the restriction that $p - |q|$ is even, does not exist.

Pseudo-Zernike moments can be thought of as an augmented version of Zernike moments. They inherit the same orthogonality property, but instead of the radial polynomials pseudo-Zernike moments use the following,

$$S_{nm}(r) = \sum_{s=0}^{n-|m|} \frac{r^{n-s}(-1)^s(2n+1-s)!}{s!(n-|m|-s)!(n+|m|+1-s)!} \tag{3}$$

We can write them in the same form as Zernike moments using the pseudo-Zernike polynomial $V_{pq}(r, \varphi)$

$$Z_{pq} = \frac{(p+1)}{n} \int_0^{2\pi} \int_0^1 \tilde{V}_{pq} f(r, \varphi) r dr d\varphi, \quad r \geq 1 \tag{4}$$

where

$$\tilde{V}_{pq} = s_{nm}(r)e^{iq\varphi}$$

GLDM: One of the least complex techniques for representing texture is to make use of the statistical moments of the intensity histogram from an image or region. Use of only histogram in calculation will result in texture measurement that convey the most effective information about distribution of intensities, however not concerning the relative position of pixels as for each other in that texture. By making use of statistical technique like difference matrix will give helpful information concerning the relative position of the neighboring pixels in an image.

With a given image I of size N × N, the co-occurrence matrix P is calculated by using the following Eq. (5)

$$p(i,j) = \sum_{x=1}^{N} \sum_{y=1}^{N} \begin{cases} 1, & if\ I(x,y) = i\ and\ I(x+\Delta_x, y+\Delta_y) = 1 \\ 0, & otherwise \end{cases} \tag{5}$$

D. **Feature Reduction**: **PCA**

Principal component analysis method is used as a prerequisite for feature reduction in classification for CT image. PCA is additionally utilized as a part of conjunction with DWT in object classification issues like face recognition. Here, PCA reduces the feature vector measurement acquired from GLDM of CT pictures.

E. **Classification**: **SVM**

Support vector is the self-learning feature and it is considered as one of the most accurate and efficient classifier, it is a linear classifier, the training time for SVM is high, and is independent of dimensionality and feature space. Free forward network along with the nonlinear units and classification are highly accurate. The working principal of the SVM is by minimizing the bound on the mistakes which is done by the machine learning on the test dataset which are not used for the purpose of the training. Objective function of training is not minimized by the SVM. Henceforth, the SVM can perform flawlessly over the images that are not used while training. Concentrating and learning the datasets while training process. Such a learning categorized datasets in training are known as Support vectors.

4 Results and Discussion

Result of our proposed work is shown in the Fig. 2, where (A) Input image and (B) Curve evolution image. Level set technique is used to detect the segmented part of an image which is shown in the figure (C). Noise removed segmented image is shown in the figure (D) and classified result is shown in the figure (E).

Fig. 2 Results of proposed work

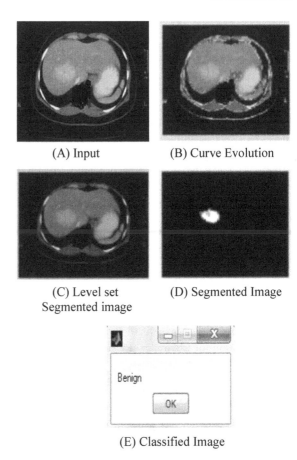

(A) Input   (B) Curve Evolution

(C) Level set   (D) Segmented Image
Segmented image

(E) Classified Image

Performance Measure of the proposed system is as shown below

Total no of images = 15   TP = 7 FP = 2   FN = 0 TN = 6
Accuracy = 86.7 %   Specificity = 75 %   Recall = 100 %

5   Conclusion

Proposed a novel methodology to identify and classify the tumor in liver from CT images. In order to achieve better classification accuracy level set approach for segmentation is used. Pseudo Zenerike moments Features extracted from GLDM and SVM classifier is used for the classification of tumor. The result of proposed method shows that it detects and classifies tumor as benign or malignant accurately. The accuracy rate of the proposed system is 86.7 %.

References

1. K. Jayaram, X. Chen, U. Bagci, Y. Zhuge, J. Yao, Medical image segmentation by combining graph cuts and oriented active appearance models. IEEE Trans. Image Process. **21**(4) (2012)
2. A. Mohamed Syed Ali, B. BalaKumar, Integrating spatial fuzzy clustering with level set methods for liver segmentation from computed tomography scans. Comput. Appl. Int. J. (CAIJ) **1**(1) (2014)
3. L. Fernández-de-Manuel, J.L. Rubio, J. Pascau, M.J. Ledesma-Carbayo, E. Ramón, M. Desco, J.M. Tellado, 3D liver segmentation in preoperative CT images using a level-sets active surface method, in *Annual International Conference of the IEEE EMBS* (2009)
4. S. Hariharan, A.G. Sajith, Medical image segmentation using CT scans-a level set approach. Int. J. Innovative Technol. Exploring Eng. **2**(6) (2013)
5. R. Poongodi, N. UmaDevi, Integration of spatial fuzzy clustering with level set for efficient image segmentation. Int. J. Comput. Sci. Commun. Netw. **3**(4), 296–301
6. X. Li, S. Luo, J. Li, Review on the methods of automatic liver segmentation from abdominal images. J. Comput. Commun. (2014)
7. S. Luo, J. Li, X. Li, Liver segmentation from CT image using fuzzy clustering and level set. J. Signal Inf. Process. (2013)
8. P. Kasza, Pseudo-Zernike Moments for Feature Extraction and Chinese Character Recognition
9. R. Rajagopal, P. Subbaiah, A survey on liver tumor detection and segmentation methods. ARPN J. Eng. Appl. Sci. **10**(6) (2015)
10. G. Chen, L. Gu, L. Qian, J. Xu, An improved level set for liver segmentation and perfusion analysis in MRIs. IEEE Trans. Inf. Technol. Biomed. **13**(1) (2008)
11. Z. Chen, S. Nie, L. Qian, Z. Chen, J. Xu, Automatic liver segmentation method based on a Gaussian blurring technique for CT images, in *The 2nd International Conference on IEEE Bioinformatics and Biomedical Engineering, 2008. ICBBE 2008*, 18 May 2008
12. S. Luo, J.S. Jin, S.K. Chalup, G. Qian, A liver segmentation algorithm based on wavelets and machine learning, in *IEEE 2009 International Conference on Computational Intelligence and Natural Computing*, 7 June 2009
13. S. Sangewa, A.A. Peshattiwar, V. Alagdeve, R. Balpande, Liver segmentation of CT scan images using k-means algorithms, in *2013 International Conference on Advanced Electronic System (ICAES)*
14. O. FekryAbd-Elaziz, M. SharafSayed, M. Ibrahim Abdullah, Liver tumors segmentation from abdominal CT images using region growing and morphological processing, in *2014 International Conference on IEEE Engineering and Technology (ICET)*, 19–20 Apr 2014

An Exploratory Study of RDF: A Data Model for Cloud Computing

A. Clara Kanmani, T. Chockalingam and N. Guruprasad

Abstract Semantic web is an extension of the web which focuses on the meaning of data content rather than the structure of the data content. It promotes many standards by the world wide web consortium (W3C). RDF is a data interchange standard widely used by semantic web community. Cloud computing is a computing paradigm which involves outsourcing of computing resources with the capabilities of resource scalability, on demand provisioning with little or no up-front IT infrastructure investment costs. Resource Description framework is a semantic data model for cloud computing. This paper analyze RDF in terms of its present status, comparison between RDF and traditional data models, its usage in semantic web data management, overview of semantic web rule languages and finally its limitations in representing concepts are examined.

Keywords Semantic web · Cloud computing · RDF · Data model

1 Introduction

Semantic web hush-up the vision of web enriched by annotations which allows both software and humans to use the content by making the web to some degree machine comprehensible. One of the key technologies used is resource description framework. It is a basis for dealing out XML based metadata which provides interop-

A.C. Kanmani (✉) · T. Chockalingam
Department of Computer Science and Engineering, Visvesvaraya
Technological University, Research Resource Center, Belagavi 590018, India
e-mail: claracalton.joseph@gmail.com

T. Chockalingam
e-mail: tchocks@gmail.com

N. Guruprasad
Department of Computer Science and Engineering, New Horizon College
of Engineering, Bengaluru 560103, India
e-mail: nguruprasad18@gmail.com

© Springer Nature Singapore Pte Ltd. 2017
S.C. Satapathy et al. (eds.), *Proceedings of the 5th International Conference on Frontiers in Intelligent Computing: Theory and Applications*, Advances in Intelligent Systems and Computing 516, DOI 10.1007/978-981-10-3156-4_68

641

erability between applications that exchange machine understandable information on the web.

Semantic Web technologies has wide range of applications in data integration, resource discovery and classification, which provides a chance to integrate the data in different locations and to perform an efficient search. It further can be used in cataloging, which describes the content and content relationships, available at a particular web site and knowledge sharing and exchange.

1.1 Architecture of the Semantic Web Stack

The semantic web stack [1], encompasses three layers, as depicted in Fig. 1.

Lower layers includes IRI, Internationalized Resource Identifier (IRI), which, identifies semantic web resources and Unicode serves to represent and manipulate text in many languages. XML is a markup language that makes possible creating documents consisting of structured data.

Middle layers includes RDF, a graph data model which represents web resources as subject, object and predicate, together known as triples. RDF vocabulary is well described in RDFS. Advanced constructs to express RDF statements are available in web ontology language (OWL). Users who want to query any kind of RDF data can utilize the RDF query language SPARQL. There is standardization required in the upper layers.

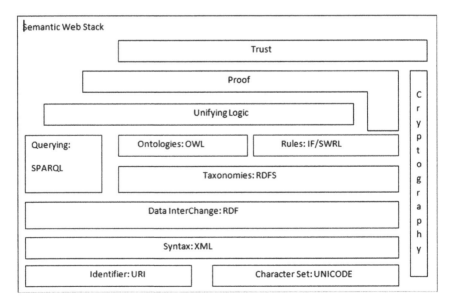

Fig. 1 Semantic web stack

1.2 Semantic Web and Cloud Computing

Cloud computing is a budding paradigm in data processing communities. Businesses store and access data at remote locations in the cloud. Data is the most vital asset the application generates over time and is essential for continued functioning. As the popularity of cloud computing grows, there are many challenges to be addressed by service provider. They have to maintain huge quantities of heterogeneous data while providing efficient information retrieval.

The effect of an inefficient data model may reduce an organization business performance on cloud environment. The change might even required to carry across different data models. At the same time semantic web is also an emerging area to augment human reasoning. A competent data model for cloud computing can be constructed using RDF.

This paper is organized as follows. Section 2 gives an introduction to RDF, design goals and evolution, Sect. 3 makes an attempt to perform an analysis taking into account of some of the vital criteria pertaining to it and finally Sect. 4 ends up with analysis and discussion followed by conclusion.

2 Resource Description Framework

Resource Description framework is an infrastructure that enables the encoding, exchange and reuse of structured metadata. This session gives a brief of evolution of RDF, introduction to RDF and design goals of RDF.

2.1 Evolution of RDF

Content rating has information about the content of web pages. It also gives an idea about whether this information is given by a qualified researcher. In this context, PICS (Platform for Internet Content Selection) is a metadata framework which provides content rating. RDF as a general metadata framework and knowledge representation mechanism is the inspiration obtained from PICS.

2.2 Introduction to RDF

The Resource Description Framework (RDF) is a framework for expressing information about resources. Resources can be anything, including documents, people, physical objects, and abstract concepts, as shown in Fig. 2.

Fig. 2 RDF graph

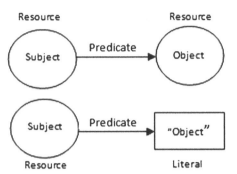

It is a simple language description model that makes statements about a resource by either defining its relationships with other resources or by defining its attribute. The resource is the subject, the relationship is the predicate, and the attribute value is the object.

2.3 Design Goals of RDF

The design of RDF is proposed to meet the following goals, to have a simple model, formal semantics and provable reference. It focuses on using an extensible URI-based vocabulary and XML-based syntax. It supports the use of XML schema data types and allows anyone to make statements about any resource.

The aim of RDF is to have a mechanism for expressing resources where no assumptions can be made about a particular application domain, nor the semantics of any domain is defined.

3 RDF: An Analysis

An analysis about RDF is carried in different ways. To begin with Sect. 3.1 makes a comparison between RDF and traditional data model to verify whether RDF becomes a unique ontological framework for conceptual representation compared to an ER model or an Object oriented data model. Section 3.2 look at the present status of RDF as accepted by semantic web community. Section 3.3 deals with RDF usage in semantic web data management. Finally, Sect. 3.4 examines the limitations of current RDF in representing concepts followed by analysis and discussion and finally ends up with conclusion.

3.1 Comparison Between RDF and Traditional Data Models

RDF is a model of entities and relationships [2]. In ER model there are entity types, and there is a set of relationships defined for it. The RDF model is the same as ER model but the relationships are of prime importance, they are identified by a URI. ER model is a static data model representing raw data, entities and relationships. Raw data is inferred. ER model is suited for structured data modeling. It becomes a messy when dealing with semi-structured content.

The RDF model is a dynamic data model. It captures the perceived semantics and dynamically relations can be added in RDF. Any web resource is expressed in the form subject, predicate and object called triples The RDF model does not have methods and all parts of the RDF graph are public.

There is a direct connection between a model of relational databases with semantic data model [2]. Any relational database has tables, which possess rows, or records. Each record has a set of fields. Any record is a RDF node, column name denotes a predicate or property and record field is object which can be literal.

RDF data is managed by Relational database systems, in a different way. Though the relational data model is the most prevalent one, it has become unsuitable for data federation purpose.

At whatever point any sort of new information with various attributes arrive, it can undoubtedly be attached to the current model without changing the current schema. This flexibility nature of RDF could be seen as an information driven configuration. This very flexibility nature of RDF makes one to view it as information driven outline. This information driven configuration makes RDF an exceptional ontological system contrasted with other conventional information models. Figures 3 and 4 shows the dynamic nature of RDF about how relations can be added in RDF.

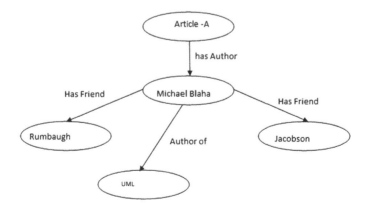

Fig. 3 RDF schema

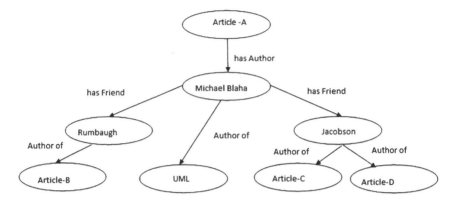

Fig. 4 Modified RDF schema

3.2   RDF Present Status

There is an RDF interest group [3] who provides a forum for discussion of a wide range of RDF related activities and issues. Another adaptation of RDF may incorporate changes regarding components, semantics, and serialization punctuations, in reverse similarity is of a central significance. In fact, RDF has been conveyed by apparatuses and applications, and the most recent couple of years have seen a noteworthy uptake of Semantic Web innovations and production of billions of triples coming from open databases (e.g. the Linked Open Data people group).

It would be, accordingly, disadvantageous to this advancement if RDF was seen as shaky and if the legitimacy of current application would be made due by a future development. As a result, with any progressions of RDF [4], in reverse similarity prerequisites ought to be formalized along the accompanying bearings, any substantial RDF graphs as far as the RDF 2004 adaptation ought to stay legitimate regarding another rendition of RDF, change or changes on the RDF (S) semantics.

There is a prerequisite for clarifications, blunders, updates on the RDF (S) specifications. There can be an upgrade in the zone of serialization formats, center RDF (S) highlights, vocabulary terms those applications require. Different ranges of advancement incorporates following changes in RDF graphs, how provenance fits in RDF and tending to best practices (use of rdf: value, reification).

3.3   Semantic Web Rule Languages

Haase et al. [5] makes a comparison of semantic web query languages. Harold Boley et al. gives an overview [6] of RuleML, a rule markup language for the semantic web. Modular syntax and semantics of RuleML and current RuleML 0.8 DTDs are presented. Negation handling, priorities and implementations of RuleML via XSLT is been analysed. Boley et al. [6] also provides the requirements for further RuleML versions.

Horrocks et al. (2004) puts forth a proposal [7] for SWRL: A Semantic Web Rule Language which combines OWL and ruleML. SWRL incorporates an abnormal state dynamic sentence structure for Horn-like guidelines in both the OWL DL and OWL Lite sublanguages of OWL. A XML language structure taking into account RuleML and the OWL XML Presentation Syntax and also a RDF solid linguistic structure in light of the OWL RDF/XML trade grammar are likewise determined, with a few illustrations.

3.4 RDF Usage in Web Data Management

The volume of RDF information keeps on becoming over the previous decade and numerous known RDF datasets have billions of triples. Expansive measure of semantic information are accessible in the RDF group in numerous fields of science, designing, and business, including bioinformatics, life sciences, business knowledge and interpersonal organizations. An incredible test of dealing with this tremendous RDF information is the means by which to get to this huge RDF information productively.

In this [8] a well known methodology is been proposed for tending to the issue which constructs a full set of permutations of (S, P, O) records. Despite the fact that this methodology has appeared to quicken joins by requests of size, the expansive space overhead constrains the adaptability of this methodology and makes it heavyweight. TripleBit [8] a quick and reduced framework for putting away and getting to RDF information is been planned.

A structure list for RDF is anticipated [9] which can be utilized for questioning RDF data for which the schema is fragmented or not accessible. In light of data caught by the structure file, comparably organized information components are physically assembled and put away adjacently on disk.

At questioning time, the list is utilized for "structure-level" handling to distinguish the gatherings of information that match the inquiry structure. Structure-level preparing is then consolidated with standard "data level" operations that include recovery and join systems executed against the information.

Proposed methodology mentioned in this [9] indicates adaptability results, i.e., execution change builds more than straightly with the span of the information. This is on the grounds that its execution does not entirely correspond with information measure but rather relies on upon the heterogeneity of structure examples displayed by the information.

3.5 Limitations of RDF

New serializations groups (e.g. Turtle) have picked up a huge backing by the group, while the entanglements in RDF/XML punctuation have made a few challenges in practice [10] and also in the acknowledgment of RDF by a bigger Web people group. At last, at present there is no standard programming API to oversee RDF information, the necessity may emerge to characterize such a standard either in a general, programming dialect autonomous path or for a portion of the essential dialects (Javascript/ECMAscript, Java, Python, ...).

There are a few confinements of current RDF diagram as for expressiveness of specific properties. These properties are depicted in detail [11] underneath.

- Local extent of properties

rdfs: range characterizes the scope of a property (e.g. eats) for all classes. In RDF Schema we can't announce range limitations that apply to a few classes only. E.g. we can't say that cows eat just plants, while different creatures may eat meat, as well.

- Disjointness of classes

Now and then we wish to say that classes are disjoint (e.g. tyke and adult).This property is hard to express in RDF.

- Boolean blends of classes

Now and then we wish to assemble new classes by consolidating different classes utilizing union, crossing point, and supplement. E.g. human is the disjoint of the classes youngster and grown-up.

4   Analysis and Discussion

RDF that begins as a simple graph data model further gets expanded into complex structures. It has more advantages compared to the conventional data models. Using RDF one can extend and adapt data and their organizational schema flexibly as per user requirements. There is a provision for exploration and analysis with this model which is not easily possible with other existing models. This model can be used without disturbing the pre-existing schema. Relational database has proved itself as a good database with structured data. It stores data in tables which saves space, ideal for the enterprises etc.

5   Conclusion

RDF is new and still emerging. There are lot of issues which are not addressed in RDF. Once those issues are identified, it could be represented with ease which is very much intricate in the in present scenario. Looking at the present status of RDF by semantic web community, lots of development is still to be done in this area and the main goal is to reuse, as much as possible, existing data in its existing form minimizing the RDF data which has to be created manually.

References

1. Semantic Web Stack, https://en.wikipedia.org/wiki/Semantic_Web_Stack
2. W3C Semantic Web Activity, http://www.w3.org/2001/sw/
3. RDF Interest Group, https://www.w3.org/RDF/Interest/
4. W3C Workshop, https://www.w3.org/2009/12/rdf-ws/cfp
5. P. Haase, J. Broekstra, A. Eberhart, R. Volz, A comparison of RDF query languages, in *ISWC 2004 LNCS* 3298 (2004), pp. 502–517
6. H. Boley, S. Tabet, G. Wagner, Design rationale of RuleML, in *A Markup Language For Semantic Web Rules* (2001)
7. I. Horrocks, P.F. Patel-Schneider, F. Van Harmelen, A semantic web rule language combining OWL and RuleML, in *W3C submission*
8. TripleBit: A Fast and Compact System for Large Scale RDF Data, in *Proceedings of VLDB*, vol. 6(7) 2013
9. T. Tran, G. Ladwig, S. Rudolph, Managing structured and semi structured RDF data using structure Indexes
10. http://answers.semanticweb.com/questions/11557/what-are-the-most-severe-limitations-of-rdfxml
11. http://www2.cs.man.ac.uk/~raym8/comp38212/main/node185.html

Domain Independent Approach for Aspect Oriented Sentiment Analysis for Product Reviews

Nilesh Shelke, Shriniwas Deshpande and Vilas Thakare

Abstract The Sentiment analysis from text documents is emerging field for the research in Natural Language Processing (NLP) and text mining. Feature specific opinion matters more than the overall opinion. Given a collection of review texts, the goal is to detect the individual product aspects comments by reviewers and to decide whether the comments are rather positive or negative. In this research paper unsupervised approach for domain independent feature specific sentiment analysis has been proposed. SentiWordNet lexical resource has been used to determine the polarity of identified features. Research work has shown the promising results over the previously used approaches using SentiWordNet. Newly introduced Senti-WordNet 3.0 has been proved to be an important lexical resource.

Keywords Sentiment analysis · Noun phrase · Verb phrase · Synsets

1 Introduction

Sentiment analysis from text documents is emerging field for the research in Natural Language Processing (NLP) and text mining. Individuals as well as institutions are paying increasing attention to sentiment analysis. Companies are interested in what bloggers are saying about their products. Politicians wish to know how different media are portraying them. Even as a part of business intelligence and decision

N. Shelke (✉)
S.G.B.A.U, Amravati, MS, India
e-mail: nileshshelke08@gmail.com

S. Deshpande
P. G Department of Computer Science and Technology, DCPE HVPM,
Amravati, India
e-mail: shrinivasdeshpande68@gmail.com

V. Thakare
P. G Department of CSE, SGBAU, Amravati, India
e-mail: vilthakare@yahoo.co.in

© Springer Nature Singapore Pte Ltd. 2017
S.C. Satapathy et al. (eds.), *Proceedings of the 5th International Conference on Frontiers
in Intelligent Computing: Theory and Applications*, Advances in Intelligent Systems
and Computing 516, DOI 10.1007/978-981-10-3156-4_69

support system, sentiment analysis is carried out on the enormous volume of text data acquired from various sources.

Much of the existing work done in this direction is focused on knowing overall emotive content of a document. However, a review is generally expressed regarding all the features of products. For instance, a laptop review will probably discuss display, memory, size, price, camera, voice, processor etc.

Very first step involved in this process is to extract the features from the reviews. Much research has been done on extracting the features from domain dependent reviews. This research work addresses domain independent approach for feature extraction from the given reviews. Major steps for doing Aspect (Feature) Oriented Sentiment Analysis are the following:

I.   Identify product features.
II. Identify opinions regarding product features.
III. Determine the polarity of opinions.
IV. Rank opinions based on their strength.

This paper is organized as follows: Sect. 2 includes the literature survey and methodologies researchers have used. Section 3 tackles proposed system including architecture, Algorithm etc. Section 4 discusses Experimental setup and Senti-wordnet lexical resource. In Sect. 5, Result analysis has been discussed. Section 6 concludes proposed work.

2 Literature Survey

Mubarak Himmat et al. provide basic knowledge about scholarly papers on sentiment analysis. This survey paper has answered many queries that have been posed by many researchers on sentiment analysis [1].

A generative model for sentiment analysis that helps to shift from simple "negative or positive" classification toward a deeper comprehension of the sentiments in blogs has been proposed in [2]. Using SPLSA as a means of "summarizing" sentiment information from reviews, they have developed ARSA, a model for predicting sales performance based on the sentiment information and the product's past sales performance.

Researchers have used various methods for feature selection schemes like Gini Index, Information Gain etc. [3]. Many researchers exploited LSI to determine features as it is very helpful in lowering the dimensionality [4]. Among other feature selection methods is of Song et al. [5] who used Latent Dirichlet Allocation (LDA) in addition to Hidden Markov Model (HMM) for feature selection.

Researchers have used rule-based classifiers in which data space is considered for set of rules. With support and confidence parameters, rules are handcrafted during training phase [6].

Chaumartin et al. [7] has pioneered research work for rule based approach. Masaki Aono et al. [8] handcrafted hard-coded rules for all the possible structures.

But these strategies could not gain popularity because of several drawbacks like expensive and tedious work of rule creation and it being domain dependent, inconsistent and have weak reliability of coverage.

Common practice for finding the polarity of sentiments is to use dictionary approach. Most commonly used seed words with known emotion are collected. Then famous corpora such as WordNet, SentiWordNet are travelled for looking synonyms and antonyms for these seed words. Words thus found get appended to the seed word list and they become the new seed words and again search is started. After satisfactory rounds, manual intervention can be done to deleted unwanted/irrelevant words [9].

Existing research work have dominantly used machine classifiers like Naïve Bayes along with SNB, MNB, CNB as well as SVM, Maximum Entropy, Term count frequency etc They are limited with the accuracy and precision of below 75 %. Therefore there is a need to apply different approach for improving results.

3 Proposed System

Sentiment Analysis as a sub-field of text mining, analyzes opinions expressed regarding an object, a topic, or an issue. An opinion is expressed by a person using some opinion terms or phrases regarding issues or features. Proposed research work is domain independent. For illustration, camera reviews are taken. In Table 1, the bold words (atmosphere, location, drinks, esp lychee martini, food, fajita, mole sauce, ca non powershot g3, purchase, camera, price, olympus, user-interface, canon, battery life) are the product features, or issues about which the opinions are expressed. Underlined words (uncomfortable, substandard, love, very good, taste-less and burned, too sweet, extremely satisfied, fantastic, clumsy-looking, not as friendly) are semantic features that express opinions.

4 Dataset

Experiments on customer reviews on various domains like digital cameras, vehicles, restaurant etc are conducted. Reviews from Amazon.com are collected. Products on this site have a large number of reviews.

Table 1 Samples from data-sets

#	Review
1.	The **atmosphere** is attractive, but a little _uncomfortable_.
2.	This particular **location** certainly uses _substandard_ **meats.**
3.	I _love_ the **drinks, esp lychee martini**, and the **food** is also _VERY good._
4.	The **fajita** we tried was _tasteless_ and _burned_ and the **mole sauce** was way _too sweet._
5.	I recently purchased the **canon powershot g3** and am _extremely satisfied_ with the **purchase**.
6.	It is a _fantastic_ **camera** and well worth the **price**.

Table 2 The 10 top-ranked positive synsets and the 10 top-ranked negative synsets in SENTIWORDNET 3.0

Synsets	Positive	Negative	Objective
Good	0.75	0	0.25
Better_off	0.875	0	0.125
Divine	0.625	0	0.375
Elysian	0	0	1
Inspired	0.875	0	0.125
Unfortunate#1	0	0.125	0.875
Bad#1	0	0.625	0.325
Pitiful#2	0	1	0
Unfit	0	0.375	0.625
Unsound	0	0.875	0.125

SentiWordNet has been employed for the task of assigning polarity to the identified features. SentiWordNet is distributed as a single file and publicly available freely. SentiWordNet 3.0 is an advanced version of SentiWordNet 1.0. Lexicon contains Positive, negative and objective score and their total is always zero. Table 2 shows the sample score from SentiWordNet.

$$Pos.Score(term) + Neg.Score(term) + Objective\ Score(term) = 1 \qquad (1)$$

5 Experimental Setup

5.1 Architecture of Proposed System

Figure 1 gives an architectural overview for feature specific sentiment analysis system. Tokenization is the process of dividing up a text into separate grammatical segments, e.g. into individual tokens, or sentential elements. This segmentation into tokens is a pre-processing step for other annotations such as part-of-speech tagging. Lingo correction module have been introduced for spelling corrections and for replacement of abbreviated terms like gr8, gm, etc. After cleaning the review data, it is fed up to Stanford Part-Of-Speech Tagger. A Part-Of-Speech Tagger (POS Tagger) is a snippet of software that reads text in some language and assigns parts of speech to each word (and other token), such as noun, verb, adjective, etc. Output of the POS tagger is Noun Phrase and Verb Phrase. Out of these Noun Phrase and Verb Phrase, features are extracted as illustrated in the algorithm section and opinionated words are looked into Sentiwordnet for their sentiment scores. Sentiment scores are calculated according to the formula given in the algorithm. Many of the times it may happen that particular opinionated words many not occur in the sentiwordnet. Then its synonyms are searched sentiwordnet. Failure to get this synonyms in sentiwordnet, its synonym is again identified in WSD (Word Sense Disambiguation) and its score is checked in the looked in WSD. With the help of this synonym sentiment score are calculated.

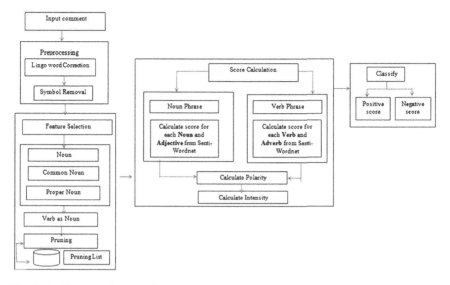

Fig. 1 Architecture of proposed system

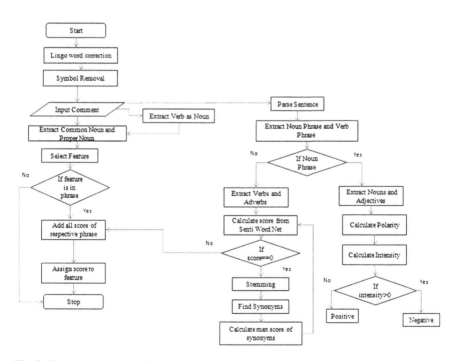

Fig. 2 Flowchart of proposed system

5.2 Flowchart of the Proposed System

See (Fig. 2).

5.3 Algorithms Developed

Algorithm for determining Polarity
Input : Single Comment.
Output: Polarity, Intensity.
Step 1: Select Comment.
Step 2: Apply POS tagging using Stanford POS tagger.
Step 3: Get proper Noun and common Noun.
Step 4: Take Noun phrase and verb Phrase using Stanford parser.
Step 5:

　　　If it is Noun phrase then for every Noun Phrase Noun and Adjective is checked. If it is a Noun then calculate the polarity with the formula

$$\text{PolarityN} = \sum\nolimits_{k=0}^{i}\ [(1-Ni)*Ni]$$

and if it is adjective then calculate it with the fomula

$$\text{PolarityA} = \sum\nolimits_{k=0}^{i}\ [(1-Ai)*Ai]$$

Setp 6:

　　　If it is a Verb phrase, then for every Verb phrase, check for verb and adverb. If it is a verb calculate the polarity as

$$\text{PolarityA} = \sum\nolimits_{k=0}^{i} [(1-Vi)*Vi]$$

and if it is adverb calculate it as

$$\text{PolarityA} = \sum\nolimits_{k=0}^{i} [(1-Advi)*Advi]$$

Step 7: Intensity= polarityN + polarityA;
Step 8: classification
　　　If intensity>0
　　　　　Positive;
　　　Else
　　　　　Negative;

Algorithm for feature calculation and score calculation
How N,A,V, Adv calculate
Step 1: Get N,A,V,Adv as word
Step 2: find word in to sentiwordnet
　　　If(found)
　　　　　　　　Score=(1/2)(1st score from
　　　　　　　　SWN)*(1/3)(2nd score from SWN)……….
　　　　　　　　Return score;
Step 3: Else
　　　　　　　　Apply stemming and get set of synonyms
　　　　　　　　from WSD
　　　　　　　　For each synonyms
　　　　　　　　　　Calculate score for each word.
　　　　　　　　End for.
　　　End else a
Step 4: Return max(score);
Step 5: end

6 Result Analysis

Although work is domain independent, for the illustration, only camera reviews are considered. For illustration first 100 reviews are considered only (Figs. 3, 4, 5 and 6; Tables 3 and 4).

Fig. 3 Commentwise evaluation parameters

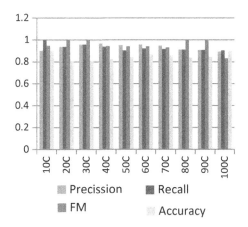

Fig. 4 F-Measure

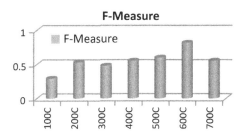

Fig. 5 Accuracy

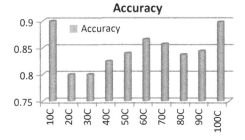

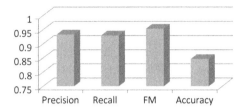

Fig. 6 Average precision, recall, FM and accuracy

Table 3 Evaluation parameters for camera reviews

Comments	TP	FP	TN	FN	Precision	Recall	FM	Accuracy
10	9	1	0	0	0.9	1	0.9473	0.9
20	15	1	3	3	0.9375	0.9375	1	0.8
30	23	1	1	5	0.958	0.958	1	0.8
40	31	1	2	6	0.96875	0.9393	0.9447	0.825
50	38	2	4	6	0.95	0.9047	0.9424	0.84
60	48	2	4	6	0.96	0.923	0.9411	0.866
70	55	3	5	7	0.9482	0.9166	0.932	0.8571
80	61	6	6	7	0.9104	0.9104	1	0.8375
90	69	7	7	7	0.9078	0.9078	1	0.844
100	75	9	8	8	0.8928	0.9036	0.83	0.8981
					0.933345	0.93009	0.95375	0.84677

Table 4 Average precision, recall, FM and accuracy

Precision	Recall	FM	Accuracy
0.9333	0.93	0.9537	0.8467

7 Conclusion

Researchers could not arrive with satisfactory solutions to many NLP problems and inherent NLP issues. Same are the issues with sentiment analysis. But sentiment analysis is in high demand in industry. Proposed system used SentiWordNet 3.0 for aspect level analysis. User interface has been developed using NetBeans IDE 8.0.2. Proposed system is domain independent. Experimentation work is shown with camera reviews. However, proposed work being domain independent it works same for the other domains like restaurants, vehicles, books, computer, printer etc.

Noun Phrase and Verb Phrase have been extracted from the output of the POS tagger. Product features can be extracted from noun phrases and some verb phrases. These are candidate product features. Unwanted product features are pruned using pre-determined list. SentiWordNet 3.0 has been exploited for computation of semantic score. Semantic score of the review is calculated by summing the semantic score of all the terms in the sentence. For deciding the final polarity formula is designed as shown in algorithm.

Proposed system gives an average accuracy of 0.8467 at the sentence level which is promising one compared with earlier research work.

References

1. M. Himmat, N. Salim, Survey on product review sentiment classification and analysis challenges, in *Proceedings of First International Conference on Advanced Data and Information Engineering (DaEng-2013)*. Lecture Notes in Electrical Engineering (Springer Science+Business Media publication, Singapore, 2014), pp 101–105
2. X. Yu, J.X. Huang, Mining online reviews for predicting sales performance: a case study in the movie domain. IEEE Trans. Knowl. Data Eng. **24**(4), 96–103 (2012)
3. J. Chenlo, A. Hogenboom, D. Losada, Sentiment-based ranking of blog posts using rhetorical structure theory, in *Presented at the 18th International Conference on Applications of Natural Language to Information Systems (NLDB'13)* (2013)
4. S. Deerwester, S. Dumais, T. Landauer, G. Furnas, R. Harshman, Indexing by latent semantic analysis. JASIS **41**, 391–407 (1990)
5. D. Adnan, S. Fei, Feature selection for sentiment analysis based on content and syntax models. Decis. Support Syst. **53**, 704–711 (2012)
6. L. Bing, H. Wynne, M. Yiming, Integrating classification and association rule mining. Presented at the ACM KDD conference (1998)
7. F.-R. Chaumartin, UPAR7: A knowledge-based system for headline sentiment tagging, in *Proceedings of 4th International Workshop on Semantic Evaluations, SemEval'07* (2007), pp 422–425
8. A. Neviarouskaya, M. Aono, Extracting causes of emotions from text, in *International Joint Conference on Natural Language Processing* (2013), pp. 932–936
9. S. Mohammad, C. Dunne, B. Dorr, Generating high-coverage semantic orientation lexicons from overly marked words and a thesaurus, in *Proceedings of the Conference on Empirical Methods in Natural Language Processing (EMNLP'09)* (2009)

Approach for Emotion Extraction from Text

Nilesh Shelke, Shriniwas Deshpande and Vilas Thakare

Abstract Emotion extraction from text is the categorization of given pieces of text (reviews/comments) into diffident emotions with NLP techniques. Now a days, internet is flooded with individual's social interaction. Also there are emotionally rich environments on the internet where close friends can share their emotions, feelings and thoughts. It has lots of applications in the next generation of human-computer interfaces. Experimentation aims at evaluating efficiency performance of proposed KEA algorithm for emotion extraction from text for ISEAR dataset as well as for any user defined comments. Fuzzy rules also have been incorporated in the algorithm.

Keywords Fuzzy rules · Key phrase · Pos tagger · Text mining · Emotion mining

1 Introduction

Human beings are sensitive to emotions and emotion mining is heavily used for decision making. Currently, World Wide Web is flooded with large datasets. These datasets may consist of numbers or texts. Recognizing emotion from text is a not an easy job. Existing methods attempted to address this issue and tried to get necessary information from these datasets. Research practitioners have made difference between emotion extraction and sentiment analysis. Although both are exploited on

N. Shelke (✉)
S.G.B.A.U, Amravati, MS, India
e-mail: nileshshelke08@gmail.com

S. Deshpande
Department of CST, DCPE, Amravati, India
e-mail: shrinivasdeshpande68@gmail.com

V. Thakare
Department of CSE, SGBAU, Amravati, India
e-mail: vilthakare@yahoo.co.in

© Springer Nature Singapore Pte Ltd. 2017   661
S.C. Satapathy et al. (eds.), *Proceedings of the 5th International Conference on Frontiers in Intelligent Computing: Theory and Applications*, Advances in Intelligent Systems and Computing 516, DOI 10.1007/978-981-10-3156-4_70

text data, scope of Sentiment analysis is restricted in knowing whether the given text is positive, neutral, or negative. However, emotion extraction from text deals with multi-emotion classification problem.

Emotion Extraction or Sentiment analysis aims to use automated tools to detect subjective information. It is the extraction of people's opinions, appraisals and emotions toward entities, events and their attributes [1].

Millions of comments are posted on YouTube, Twitter, Facebook and several social sites. Many research issues can be addressed by applying prototypes on these comments [2].

The best media for marketers wishing to imbibe an image or identity in the minds of their customers for their product, brand, or organization is the web. As a matter of fact the automatic emotion analysis of such online opinions requires a deep understanding of natural language text by machines, from which existing work is still very far [3].

Emotion detection and analysis has been investigated widely in neuroscience, psychology and behavior science, as they are important elements of human nature. In computer science, this task has also attracted the attention of many researchers, especially in the field of human computer interactions [4].

Simple topic classification and Emotional classification are two very different tasks. Topic classification can be performed by choosing seed words in a given review compare it against a lexicon of 'keywords'. Sentiment classification is to determine the semantic orientations of words, sentences or documents [5].

2   Literature Survey

Bincy Thomas et al. [6] attempted multi-classification of emotions from text. Multinomial Naïve Bayes (MNB) has been exploited for this task. Unigrams are used recognize emotions with the help of bag of words. More details emotive contents are trapped by the bigrams and trigrams. Prototype developed has been experimented on ISEAR dataset. They got approximately 77 % accuracy on unigram set.

Qiuhui Zheng et al. [7] found Latent Semantic Analysis Algorithm more promising than traditional Vector Space Model as LSA decreases search space by exploiting Singular Value Decomposition.

Much of the previous work like [8] attempted to recognize emotions with affective keywords. These existing works also handcrafted rules for recognizing the emotions. Problem with these rules are that they are dataset dependent.

Other prevalent existing work is development of text classifiers using various flavours of Naïve Bayes methods like CNB, SNB, MNB with a blend of Fuzzy approach. One of the approaches in this direction is by Surya Sumpeno et al. [9] for emotion classification from text. However Mean Precision and Mean Recall lies between 40 and 60 % for ISEAR Dataset.

Ameeta Agrawal et al. [10] experimented on Alm dataset. Semantic similarity is compared between strong emotive content words and relative emotion concepts. Syntactic dependencies with structure of comments are also verified for pruning.

Unsupervised methods are seldom used comparative to supervised methods. Frank et al. [11] used clustering technique to know structure pattern in unlabelled reviews. Once structure pattern is known emotions can be grouped into clusters.

Most of the existing research work in this domain has used machine classifiers like Naïve Bayes along with SNB, MNB, CNB as well as SVM, Maximum Entropy, Maximum Log-likelihood scores. They are limited with the accuracy and precision of below 75 % [6, 10]. Therefore there has been a need to apply different approach for improving results.

3 Dataset

ISEAR dataset is used for evaluation of proposed system. Many researchers have used this standard dataset as it is a strong emotive content corpus. Original ISEAR dataset have 7,666 text files. It has seven emotion classes and contains 4,40,060 running words [12]. Name of different emotions and their individual word contribution after processing on the dataset is shown in the Table 1.

4 Proposed System

Figure 1 shows architecture of proposed System. Initially pre-processing is done on ISEAR dataset and input comment. It includes POS tagging by Stanford POS Tagger. Lingo word corrections have been done. Because of removal of stop words and other unnecessary symbols, total words have been reduced from 449,060 to 66,836 (see Fig. 2). After POS Tagging sentences are segregated as adjective, adjective, noun, verb etc. Only these terms are considered for further analysis. Following example illustrate how POS tagger outputs the parse tree for the given comment:

"Would not return for the amount we paid."

Table 1 Keyword Contribution for different emotions in post-processed ISEAR dataset

Category	Sadness	Shame	Fear	Anger	Guilt	Joy	Disgust	Total
Total words in each category	8563	9272	10292	10567	9756	8708	9678	66836

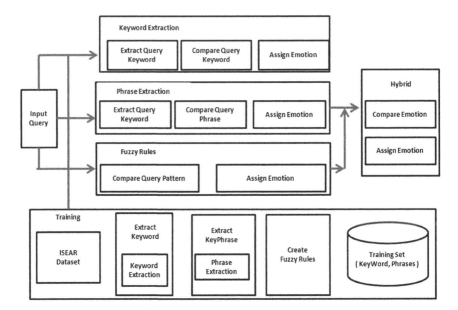

Fig. 1 Architecture of proposed system

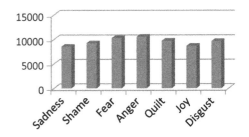

Fig. 2 Keyword contribution for different emotions in post-processed ISEAR dataset

```
(ROOT
 (SINV
  (VP (MD would) (RB not)
   (VP (VB return)
    (PP (IN for)
     (NP (DT the) (NN amount)))))
  (NP
   (NP (PRP we))
   (VP (VBN paid)))))
parsed==>(ROOT (SINV (VP (MD would) (RB not) (VP (VB return) (PP (IN for)
(NP (DT the) (NN amount))))) (NP (NP (PRP we)) (VP (VBN paid)))).
```

Keywords from the inputted query have been extracted. Then Key Phrases from the inputted query have been extracted. Figure 3 shows the details of this step.

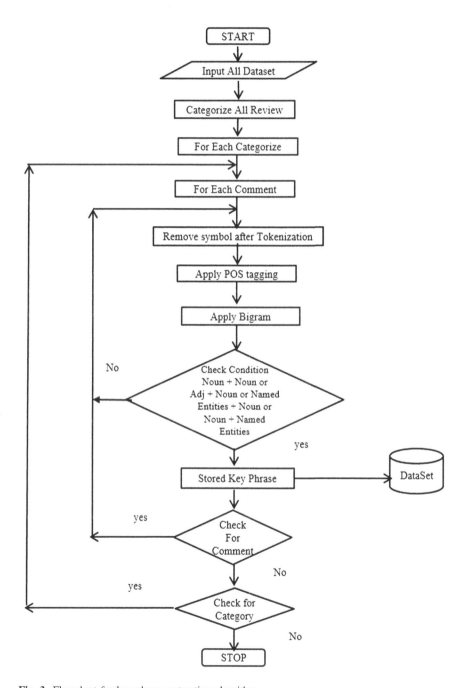

Fig. 3 Flowchart for key phrase extraction algorithm

Rest of the work is done in following three phrases.

(1) Extract the keywords for each emotion

All the sentences in the dataset belong to one of the seven emotions. First all the sentences are separated according to each emotion given in the dataset and grouped keywords according to its emotions. Table 1 shows the keywords contribution after post-processing.

(2) Extracting Key Phrases

Here bigram phrases have been extracted using KEA (Key Phrase Extraction Algorithm) from each emotion category. Then the bigram key phrase set has been created for each emotion. Table 2 shows bigram key phrases contribution after post-processing.

(3) Creating Rules

As mentioned in Sect. 3, only eight columns of ISEAR data have been considered for processing. Here fuzzy rules are developed. These rules will be based on the questionnaires asked to the users while giving the situation description. Each column in the dataset has its unique meaning which is described in the questionnaire document. Using the information from each column through the questionnaire document fuzzy rules has been created.

Prominent Implementation Steps

(1) Keywords from the inputted query have been extracted.
(2) Then Key Phrase from the inputted query has been extracted. Figure 3 shows the details of this step.
(3) Inputted query is compared with the keyword set and key phrase set.
(4) If no comparison is found then the inputted query is compared with the fuzzy rules of each emotion.
(5) After applying the above process the query is assigned to the respective emotion.

Table 2 Key phrases which are qualified as Noun + Noun or Adjective + Noun or Named Etities + Noun or noun + Named entities

Category	Sadness	Shame	Fear	Anger	Guilt	Joy	Disgust	Total
Total words in each category	3852	4195	4877	5000	4410	4122	4651	31107

5 Experimental Set up

ISEAR dataset is available in .mdb format. It has been converted into MySQL database as proposed work is experimented with Linux operating system. NetBeans IDE 8.0.2 has been used for user interfaces. There were two parts of the proposed work: one is for testing proposed algorithm with ISEAR dataset and another is for testing the same algorithm for online/user comments. Questionnaire has been created depending upon the eight columns of ISEAR dataset. Fuzzy rules are developed out of data of these eight columns and both ISEAR dataset and user comments are tested with fuzzy rules.

Questionnaire:

1. Did you hide your feelings?
2. What do you think was responsible for the event in the place?
3. Were you laughing/smiling?
4. Were you sobbing/crying?
5. Did you show aggregation?
6. Did it help hinder or help to follow your plans to achieve your aim?
7. Did you find the event itself pleasant or unpleasant?
8. How did this event affect your feelings about yourself, such as your self-esteem or your self-confidence?

Typical fuzzy rule is like:

1. If feeling is <u>not at all</u> hidden and cause is <u>not applicable</u> the person is <u>not smiling, crying</u> and is not showing <u>aggression</u> this event has <u>it didn't matter</u> the person and the event is <u>neutral</u> and it affected the person <u>not applicable.</u> Hence the emotion detected is <u>FEAR</u>.
2. If feeling is <u>not at all</u> hidden and cause is <u>not applicable</u> the person is <u>not smiling, not crying</u> and is showing <u>aggression</u> this event has <u>not applicable</u> the person and the event is <u>unpleasant</u> and it affected the person <u>not at all.</u> Hence the emotion detected is <u>DISGUST</u>.

6 Result Analysis and Discussion

ISEAR dataset consists of labeled emotion. Evaluation is done by comparing these labeled tags with the emotion tags computed by proposed system. KEA algorithm provided encouraging results for ISEAR dataset as depicted in Fig. 4. However, there is poor performance of fuzzy rules as shown in Fig. 5. It is because ISEAR dataset contains very surprising labels for the emotions. For example, some respondents have assigned happy emotions for someone's death also. Fuzzy rules have shown very promising performance for user defined emotions. However, user has to answer some questions for retrieving the right emotion.

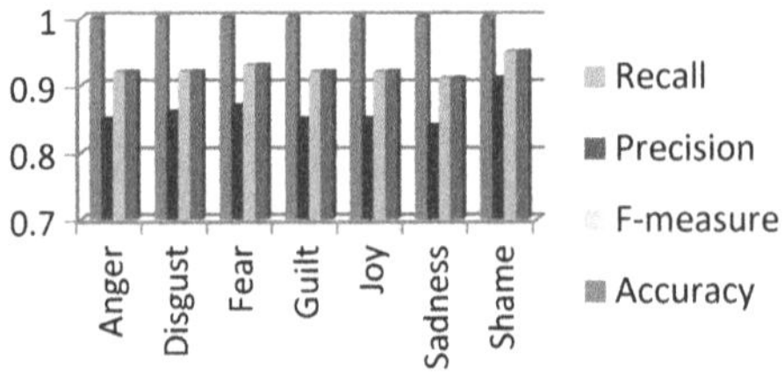

Fig. 4 Evaluation parameter
for emotion extraction by
KEA algorithm

Fig. 5 Evaluation parameter
for emotion extraction using
fuzzy rules

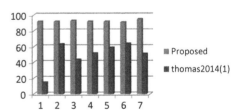

Comparison with other existing work is very tricky as existing work has not shown performance on all the parameters in one research paper. Accuracy of the proposed work has been compared with [6] in Fig. 6 and for precision with [12] in Fig. 7. Bincy Thomas et al. found good accuracy with Mutual Information (MI) as compared with WLLS and Normalized Google Distance (NGD). Alexandra Balahur et al. [12] developed core of EmotiNet, a new knowledge base (KB) for representing and storing affective reaction to real-life contexts.

Fig. 6 Accuracy comparison
with [6]

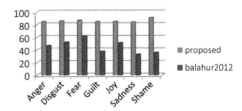

Fig. 7 Precision comparison
with [12]

7 Conclusions and Discussions

Different approach for emotion recognition system from natural text has been proposed. Contribution presents the simple but effective classifier using KEA algorithm and fuzzy rules. The results are very encouraging with KEA algorithm. Once system is trained, it responds to the user's online comments also. Fuzzy rules also have been exploited for this. Such extension in the earlier work has not been observed.

The work presented in this paper can be pursued further in several directions. Applying anaphora resolution, particularly important in cases when the stimulus is an agent; returning a pronoun as an emotion stimulus without anaphora resolution will not be as informative. Also, cases needs to be addressed when where there are more than one emotion expressed.

References

1. H. Chen, D. Zimbra, Artificial intelligence and opinion mining. IEEE Intell. Syst. **25**(3), 74–80 (2010). IEEE Computer Society
2. S. Bao, S. Xu, L. Zhang, R. Yan, Z. Su, D. Han, Y. Yu, Mining social emotions from affective text. IEEE Trans. Knowl. Data Eng. **24**(9), 1658–1670 (2012)
3. B. Liu, H. Wang, C. Havas, Knowledge-based approaches to concept level sentiment analysis. IEEE Intell. Syst. **28**(2), 12–14 (2013). IEEE Computer Society
4. E. Mower, Matarić, S. Narayanan, A framework for automatic human emotion classification using emotion profiles. IEEE Intell. Syst. **25**(3), 1057–1070 (2011). IEEE Computer Society
5. K. Nirmala Devi, V. Murali Bhaskar, Text sentiments for forums hotspot detection. Int. J. Inf. Sci. Tech. (IJIST) **2**(3), 53–61 (2012). Gandhigram Rural Institute, India
6. B. Thomas, K.A. Dhanya, P. Vinod, Synthesized feature space for multiclass emotion classification, in *Proceedings of First International Conference on Networks & Soft Computing (ICNSC)*, (IEEE Explore), pp 182–190, Aug 2014
7. Q. Zheng, X. Wang, Text emotion classification research based on improved latent semantic analysis algorithm, in *Proceedings of the 2nd International Conference on Computer Science and Electronics Engineering* (Atlantis Press, Paris, France, 2013)
8. A. Neviarouskaya, M. Aono, Extracting causes of emotions from text, in *International Joint Conference on Natural Language Processing* (2013), pp. 932–936
9. S. Sumpeno, M. Hariadi, M.H. Purnomo, Facial Emotional expressions of life-like character based on text classifier and fuzzy logic. IAENG Int. J. Comput. Sci. **38**(2), 25–30 (2011)
10. A. Agrawal, A. An, Unsupervised emotion detection from text using semantic and syntactic relations, in *Proceedings of the 2012 IEEE/WIC/ACM International Joint Conferences on Web Intelligence and Intelligent Agent Technology*, Vol. 01 (IEEE Computer Society Washington, DC, USA) ©2012 (2012), pp. 346–353
11. I.H. Witten, E. Frank, *Data Mining: Practical Machine Learning Tools and Techniques*, 2nd edn. Morgan Kaufmann Series in Data Management Systems. (Morgan Kaufmann Publishers Inc., San Francisco, CA, USA, 2005)
12. A. Balahur, J.M. Hermida, A. Montoyo, Building and exploiting EmotiNet, a knowledge base for emotion detection based on the appraisal theory model. IEEE Trans. Affect. Comput. **3**(1), 123–125 (2012). IEEE publications

Performance of Multiple String Matching Algorithms in Text Mining

Ananthi Sheshasaayee and G. Thailambal

Abstract Ever since the evolution of Internet Information retrieval is being made by surfers in large amount. The data gets increased everyday as the thirst of acquiring knowledge by the users gets increased day-by-day. The data which is raw needs to be processed for usage which increases the potential value in all major areas like Education, Business etc. Therefore Text Mining is an emerging area where unstructured information were made as relevant information. Text mining process can be divided into Information Extraction, Topic Tracking, Summarization, Categorization, Clustering, concept Linkage and Information visualization. Even though all other things can be applied to text only properly it is extracted from the web. Using Pattern matching or String matching algorithms to retrieve proper results from the Sea of information. In this paper we discuss the three types of algorithms Aho Corasick, Wu Manber and Commentz Walter. The performance of the algorithms are identified by implementing it in Python language. Finally the suitable algorithm for extracting information is found.

Keywords Text mining · Aho Corasick · Wu Manber · Commentz Walter · Boyer Moore · Python

1 Introduction

Web Mining is a term used in finding information from large repository of web. Web Mining is divided into three categories such as Web Content Mining, Web structure Mining and Web Usage Mining. In Web content mining the content can be either

A. Sheshasaayee
Quaide-Milleth College for Women, Chennai, India
e-mail: ananthi.research@gmail.com

G. Thailambal (✉)
Shri Chandrasekharendra Saraswathi Vishwa Maha Vidyalaya University,
Kancheepuram, India
e-mail: thaila.research@gmail.com

© Springer Nature Singapore Pte Ltd. 2017   671
S.C. Satapathy et al. (eds.), *Proceedings of the 5th International Conference on Frontiers in Intelligent Computing: Theory and Applications*, Advances in Intelligent Systems and Computing 516, DOI 10.1007/978-981-10-3156-4_71

Text, Image, Audio or Video. In Text Mining only the characters taken as input and in Image mining pictures were taken as input and in Audio Mining Audio files were taken as input and in Video Mining mpeg files were taken as input and processed for retrieving structured information. But the most popular transaction in web is through text and no programs can be used for information retrieval. Text Mining also known as Text Data Mining and Knowledge Discovery in Textual Databases [1].

String matching helps to find out pattern from given text where Symbols or characters are like {a, b} then abab is a string. The Pattern is denoted by P [1…m] and the text is denoted by T [1…n]. Two shifts Valid if P occurs with shift in T otherwise invalid [2]. String matching is further divided into Exact and Approximate in which Complete pattern is compared with the selected text window whereas only small portion of pattern is matched in approximate [3].

Multiple String pattern matching algorithm helps to identify more number of patterns from the text. It is used to filtering text from large number of text documents. This consists of two types one is Single String matching and Multiple String matching. In Single string matching only single occurrence of string is searched and in multiple string matching multiple occurrences of a string is searched. The commonly used algorithms of Single string matching is KMP, Boyer Moore, Naïve string matching [4]. The commonly used multiple string matching algorithms are Aho Corsaick, Wu Manber and Commentz Walter. This area is expected to be growing since longer strings create noise and random variations [5]. Two important categories available such as Left-to-Right, Right-To-Left. The Matching of string patterns with the text can be done by either Left-to-Right or Right-to-Left direction. Some of the algorithms in L-to-R is Brute Force and Knuth Morris Pratt algorithm. The algorithms in which processing is done by Right-to-Left is Boyer Moore and Turbo Boyer Moore [3]. Single Pattern matching helps to find a single occurrence of a Pattern in a text and Multiple Patterns matching gives more than one pattern matched with the text simultaneously. Multiple pattern matching have high performance and good Practicability [6].

1.1 Applications of String Matching Algorithms

Various applications where in text mining the String matching algorithms used are Spell Checkers, Spam Detection Systems, Intrusion Detection, Search Engines, Plagiarism Detections, and DNA Sequencing [7]. Spell Checkers helps in finding spellings using a trie of patterns are developed and in reaching its final states after matching is found the occurrence will be given [8]. Spam Detection Systems helps to filter spam emails. Spams are unwanted bulk e-mails increasing our memory size and bandwidth. They are checked with the suspected Signature Patterns by applying these algorithms in group of e-mails received [9] Intrusion Detection System helps to find the intruder of the information. All the malicious code will be stored in database and checked with the algorithms for finding the match. If the match is found that packet will be discarded [10] Search Engine is the place where our requested pages are searched and displayed. Web contains more data and the data can be accessed by the user with the help of these algorithms based on Categorization using Keywords [11] Plagiarism Detection is finding the place and the content which is duplicated. Copying the work of others and claiming as their own work which can be detected with the similarity of content using these algorithms [12]. String matching algorithms helps to find Genetic sequences and DNA patterns in bioinformatics which is a combination of Biology with application of computer science [13]. In this paper we present the three algorithms Aho Corasick, Wu Manber and Commentz Walter with pseudo-code and example. We apply for certain patterns all the three algorithms and the performance of each algorithm is found. Some of the related works of all the three algorithms are also discussed along with their pseudo code and example.

2 Multiple Pattern Matching Algorithms

Multiple pattern matching algorithms are divided into three types namely Prefix algorithms, Suffix Algorithms and Factor algorithms. Prefix algorithms the pattern searching is stored in a trie and a data structure where root node represents empty string and every node represents the prefix of one of the patterns. Best example is Aho Corasick algorithm. In Suffix algorithms the patterns are stored in backward of suffix automaton, a rooted directory tree with suffixes of all patterns. Best example is Commentz Walter algorithm. In Factor searching algorithms a trie with additional transition that can recognize any substring of the patterns. Best example is MultiDBM and Dawg-Match algorithms [14].

An algorithm IPMafC which is Index based Multiple Pattern Matching algorithms using Frequent Character Count and it will compare DNA and Protein Sequence which reduces character comparison effort at each attempt [15]. Regular expressions can be used in String searching methods combined with grep statements to search strings. A regular expressions matching method based on Network-on-Chip architecture to get a high matching rate is done by partitioning the regex into several parts to make the finite state machine simpler [16]. Usage of regular expressions in Natural Language Processing using validating data fields, Filtering texts, Identifying Particular strings in a text, Converting the output for one processing component into second [17]. In this paper discuss only the Prefix and Suffix type algorithms which includes Aho Corasick, Wu Manber and Commentz Walter only.

2.1 Aho Corsaick Algorithm

The first algorithm which uses automata approach to solve multiple string matching problems. Proposed by Alfred Aho and Corasick in 1975. Basically it consists of two parts in which first parts contains the keywords and in second part text string is given as input. A pattern matching machine for a set of strings is takes x as input string and produces the output of locations where the strings appear as substrings. This machine consists of set of states [18] Each state of the machine is given by an identification number. Three functions are used such as goto g, failure f and output. The output state specifies the keyword match is found and goto state either specifies a state or a failure state. Failure state specifies that the match is not found. Each step in pattern matching happens in constant time of O (n) running time. The space or memory requirements is needed quite large to accommodate the pattern and text input [19] Initially it combines all the patterns into non-deterministic Finite automaton (NFA) and finally into a Deterministic Finite automaton (DFA) [6] Many new algorithms have been proposed on the basis of Aho Corasick and implemented them with minor changes in the actual algorithm. Banded row format automaton and sparse row format automaton to construct two dimensional table dynamically which is used as Shift table. With this the author has proved that the space efficiency is improved without decreasing the time efficiency of Pattern matching [20]. This algorithm is combined with Parallel Failure less Aho Corasick algorithm is applied to protein sequence data sets which is 15 times faster to original algorithm [21]. Different algorithms can be used for string searching other than Aho Corasick, Wu Manber, Commentz Walter algorithm is Rabin-Karp, Needleman Wunsch, Knuth Morris Pratt, Smith Waterman, Brute Force, and Boyer Moore algorithm.

The various applications in which this algorithm used is Detecting Intrusions, Detecting Plagiarism, Bio-Informatics, Digital Forensics and Text Mining. This is an attractive algorithm for large number of keywords as it can matched in single pass [22] (Fig. 1) [18].

The following figure shows how the function of pattern matching machine occurs for the set of keywords {he, she, his, hers} (Fig. 2) [18].

2.2   Commentz Walter Algorithm

Commentz walter algorithm combines the functionality of Boyer Moore and Aho Corsaick with the only difference of reversed pattern. Two phases one is pre-computing and another one is matching Phase. Pre-computing phase the reversed tree pattern is created and the matching phase is divided into finite automaton technique and Boyer-Moore shifting techniques. The process is done by scanning backwards through the input string and when mismatch occurs the history of the matched characters are used as index which helps to shift before next match is attempted [19] (Fig. 3). The Fig. 3 gives python code implementation of Commentz Walter algorithm [23].

A Comparative analysis of five algorithms Aho Corasick, Rabin-Karp, Bit-Parallel, Commentz Walter and Wu Manber has been made and found that Commentz Walter is better comparatively to all other [23]. The following figure shows the reverse finite automaton structure for the set of keywords {he, she, his, hers} (Fig. 4) [24].

```
class ahoCorasick:
            for key in keywords.split(' '):
            j = 0
            state = 0
            current = self.root
            key = key.upper()
            while j < len(key):
                ch = key[j]
                j = j+ 1
                child = current.getTransition(ch)
                if child != None:
                    current = child
                else:
                    self.newstate = self.newstate +1
                    nd = State(self.newstate, ch)
                    current.tranList.append(nd)
                    current = nd
                    while j < len(key):
                        self.newstate = self.newstate +1
                        nd2 = State(self.newstate, key[j])
                        current.tranList.append(nd2)
                        current = nd2
                        j = j+1
                    break
            current.outputSet.add(key)
```

Fig. 1 Part of Python code of Aho Corasick algorithm [18]

Fig. 2 Automaton for Aho Corasick algorithm [18]

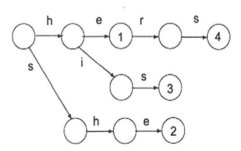

2.3 Wu Manber Algorithm

This algorithm is variant of Boyer-Moore and uses bad-Character shift rule of BM algorithm. This algorithm also combines the concept of UNIX grep tool in order to improve it working performance. The idea is looking through the text and patterns as blocks instead of single character matching. They have pre-processing as their first step in which three tables are constructed such as SHIFT, PREFIX and HASH

1: **procedure** CW($y, n, m, p, root$)
    ▷ Input:
    ▷ $y \leftarrow$ array of n bytes representing the text input
    ▷ $n \leftarrow$ integer representing the text length
    ▷ $m \leftarrow$ array of keyword lengths
    ▷ $p \leftarrow$ number of keywords
    ▷ $root \leftarrow$ root node of the trie

2:       $v \leftarrow root$ ▷ The current node
3:       $i \leftarrow \min\{m[0], m[1], ..., m[p-1]\}$ ▷ i points to the current position in y
4:       $j \leftarrow 0$ ▷ j indicates depth of the current node v

5:       **while** $i \leq n$ **do**   ▷ Matching

6:           **while** v has child v' labeled $y[i-j]$ **do**
7:               $v \leftarrow v'$
8:               $j \leftarrow j+1$
9:               **if** $out(v) \neq \emptyset$ **then**
10:                  **output** $i-j$ ▷ Path from v to root matches $y[i-j]$ to $y[i]$
11:              **end if**
12:          **end while**
                                                                     ▷ Shifting
13:          $i \leftarrow i+ \min \{ shift2(v), \max \{ shift1(v), char(y[i-j]) - j - 1 \} \}$
14:          $j \leftarrow 0$

15:      **end while**
16: **end procedure**

Fig. 3 Pseudo code for Commentz Walter algorithm [23]

Fig. 4 Example for
Commentz Walter algorithm
[24]

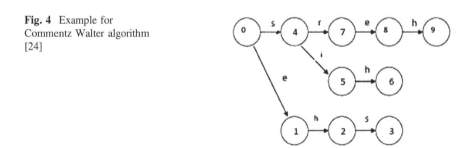

tables. During the second part of the phase the occurrence is checked with the previous three tables. The hash value is created and checked with the shift value and if the shift value is greater than zero then the shift is made according to the shift value. If it is less than zero then a match will be found and the tables HASH and PREFIX will be considered to check the matching [19]. By dividing the patterns into different pattern sets according to their length and processing debases the effect on Performance caused short patterns. If Independent data structures are used the high concurrence can be obtained [25]. This algorithm requires all patterns of same length as the distance of SHIFT table is limited with length of patterns.

If the shortest string is too small then the distance will be quite small influencing matching efficiency [26]. Some of the grep statements tool for searching patterns are GNUgrep, Nr-Grep, and Agrep. Agrep is a Fast approximate Pattern Matching tool Search for approximate patterns, Search for records rather than lines, Searching for multiple patterns with and, or logic queries [27]. Nrgrep is a pattern matching tool built over the BNDM algorithm (hence the name "nondeterministic reverse grep", since BNDM (Backward Non-deterministic DWAG matching) scans windows of the text in reverse direction) [28]. LZGREP is A Boyer Moore String matching tool to search text in decompressed form instead of decompressing and Searching [29]. After Wu Manber algorithm is developed many algorithms were developed with different variations to original algorithm. Quick Wu Manber is developed which uses Quick Search Algorithms. It introduces a head table with first two characters of the patterns. Maximum shift distance is reached by using mismatched details during searching Phase [30]. Improved Wu Manber is the algorithm proposed which uses two shift tables in the algorithm. As the number of patterns increases the values in shift table decreases [31]. Quick Multiple Matching is an aggressive algorithm that reduces the need of shift and Hash tables in the original algorithm which checks the character next to scan window for matching [32] (Fig. 5). Figure 5 gives python code implementation of Wu-Manber algorithm [25].

```
class WuManber:
  def search_text(self,nocase=True,verbose=False):
    s = time.time()
    if nocase:
      self.nocase = c_int(1)
    else:
      self.nocase = c_int(0)
    self.__search_init__()
    null_ptr1,null_ptr2=POINTER(c_int)()POINTER(c_int)()
    if not verbose:
      cb = null_ptr1
    else:
      cb = WM_CALLBACK(self.__callback__)
    wm_ret =
self.so.wm_search_text(self.wm,self.ctext,self.len_ctext,
      cb,null_ptr2)
    sys.stderr.write("search_text took %.2f
seconds\n"%((time.time()-s)))
    return wm_ret
```

Fig. 5 Part of Python code of Wu Manber algorithm [25]

Patterns	Length
STOP	4
STORM	5
MAD	3

Shift Key	Shift Value
ST	1
TO	0
MA	1
AD	0

Hash Key		
TO	STOP	STORM
AD	MADE	

Fig. 6 Patterns and their length, shift table and shift values, hash table

The Following figure shows the different SHIFT, PREFIX, HASH value tables for the set {STOP, STORM, MAD} (Fig. 6).

3 Python Language

C and C++ languages will have some advantages like abstraction, object Polymorphism, Automatic variable destruction, Templates. Java is also a Powerful language to implement String search algorithms. Many built-in functions are available to ease the usage. Python is an Open source high level language with decreasing lines of programming code comparatively to Java and C++. Python developed by Guido Van Rossum in 1996 which is a Structured, Object-Oriented and Structured programming language.

Pyahocorasick 1.0.0 is a module which uses two kinds of data structures trie and Aho Corasick automaton. Wu Manber is implemented in Python language. There is no python version for Commentz Walter and it is implemented in C language. Python has many packages related to string algorithms available which helps to use it efficiently and faster than other languages. Indentation is available in Python which is not available in other languages. Python has smaller runtime comparatively to other conventional languages like C, C++ and Java.

4 Performance Evaluation of Three Algorithms in Python and C Language

From this figure when pattern length increases Aho Corasick Performs well comparatively to Wu Manber and Commentz Walter in Python Language. Wu Manber has only slight variation since it is the reverse automaton of Aho Corasick algorithm. The time taken in Commentz Walter has only little variation with Aho Corasick and Wu Manber. Even though the three algorithms has only slight variation the Aho Corasick is best as it is the original algorithm developed before

Fig. 7 Time taken when pattern length increases

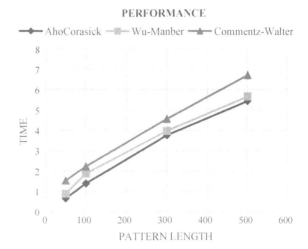

Commentz Walter and Wu Manber. Without combining other algorithms like Boyer Moore in other two algorithms Aho Corasick can perform well. The Performance of this algorithm can be improved to get efficient results (Fig. 7).

5 Conclusion

The Three algorithms can be used in string searching by many researchers and the pros and cons of them discussed in many papers. In future Commentz Walter can be applied in Python for getting better results. According to the implementation in python it proved that Aho Corasick is better performing than other two algorithms and the performance efficiency can be improved for further promising results. The algorithms are applied only to 500 patterns. In future this can be applied to web documents consisting of more than 1000 lines and it can be improvised according to its performance for web documents.

References

1. N. Zhong, Y. Li, S.-T. Wu, Effective pattern discovery for text mining. IEEE Trans. Knowl. Data Eng. **24**(1) (2012). 1041- 4347/12/$31.00 2012 IEEE
2. A. Rasool, A. Tiwari, G. Singla, N. Khare, String matching methodologies: a comparative analysis. (IJCSIT) Int. J. Comput. Sci. Inf. Technol. **3**(2), 3394–3397 (2012). ISSN: 0975-9646
3. I. Hussain, S. Kausar, L. Hussain, M. Asif Khan, Improved approach for exact pattern matching (bidirectional exact pattern matching). IJCSI Int. J. Comput. Sci. Issues **10**(3), No 1 (2013). ISSN (Print): 1694-0814 | ISSN (Online): 1694-0784

4. A. Sheshasaayee, G. Thailambal, A Comparative analysis of single pattern matching algorithms in text mining, in *IEEExplore Digital Library*, pp. 720–725. http://dx.doi.org/10.1109/ICGCIoT.2015.7380557
5. N. Singla, D. Garg, String matching algorithms and their applicability in various applications. Int. J. Soft Comput. Eng. (IJSCE) **1**(6) (2012). ISSN: 2231-2307
6. N.L. Dang, D.-N. Le, V.T. Le, A new multiple pattern matching algorithm for the network intrusion detection system. IACSIT Int. J. Eng. Technol. **8**(2) (2016)
7. K.K. Soni, R. Vyas, A. Sinhal, Importance of string matching in real world problems. Int. J. Eng. Comput. Sci. **3**(6), 6371–6375 (2014). ISSN: 2319-7242
8. A. Apostolico, ZviGalil, *Pattern Matching Algorithms*, 1st edn. (Oxford University Press, USA), 29 May 1997
9. C.-T. Wu, K.-T. Cheng, Q. Zhu, Y.-L. Wu, Using visual feature for anti-spam filtering, in *The Proceedings of IEEE International Conference on Image Processing (ICIP2005)* (2005), pp. 509–512
10. H. Kim, H.S. Kim, S. Kang, A memory efficient bit-split parallel string matching using pattern dividing for intrusion detection systems. IEEE Trans. Parallel Distrib. Syst. **22**(11), 1904–1911 (2011)
11. D. Sanchez, M.J. Martin-Bautista, I. Blanco, C. Torre, Text knowledge mining: an alternative to text data mining, in *The Proceedings of IEEE International Conference on Data Mining Workshops, ICDMW '08*, pp. 664–672, 15–19 Dec 2008
12. R.S. Aygün, Structural-to-syntactic matching similar documents. J. Knowl. Inf. Syst. ACM Digital Libr. **16**(3), 303–329 (2008)
13. L.-L. Cheng, D.W. Cheung, S.-M. Yiu, Approximate string matching in DNA sequences, in *Proceedings of the Eighth International Conference on Database Systems for Advanced Applications (DASFAA'03)*, pp. 303–310, 26–28 Mar 2003
14. C.S. Kouzinopoulos, P.D. Michailidis, K.G. Margaritis, Parallel Processing of Multiple Pattern Matching Algorithms for Biological Sequences: Methods and Performance Results, Systems and Computational Biology—Bioinformatics and Computational Modelling, ed. by Prof. N.-S. Yang (2011). ISBN: 978-953-307-875-5
15. S. Nirmala Devi, S.P. Rajagopalan, V. Anuradha, Index based multiple pattern matching algorithm using frequent character count in patterns. Int. J. Adv. Res. Comput. Sci. Softw. Eng. **3**(5), (2013). ISSN: 2277 128X
16. C. Linhai, An innovative approach for regular expression matching based on NoC architecture. Int. J. Smart Home **8**(1), 45–52 (2014). http://dx.doi.org/10.14257/ijsh.2014.8.1.06
17. G. Kaur, Usage of regular expressions in NLP. IJRET: Int. J. Res. Eng. Technol. eISSN: 2319-1163 | pISSN: 2321-7308
18. A.V. Aho, M.V. Coraisick, Efficient string matching: an aid to bibliographic search. Commun. ACM **18**(6), 333–340 (1975)
19. A.I. Jony, Analysis of multiple string pattern matching algorithms. Int. J. Adv. Comput. Sci. Inf. Technol. (IJACSIT) **3**(4), 344–353 (2014). ISSN: 2296-1739
20. S. Liangxu, L. Linlin, Improve Aho-Corasick algorithm for multiple patterns matching memory efficiency optimization. J. Convergence Inf. Technol. (JCIT) **7**(19) (2012). doi:10.4156/jcit.vol7.issue19.19
21. S. Soroushnia, High performance pattern matching on heterogeneous platform. J. Integr. Bioinform. **11**(3), 253 (2014). doi:10.2390/biecoll-jib-2014-253
22. S. Hasib, M. Motwani, A. Saxena, Importance of Aho Corasick string matching algorithm in real world applications. (IJCSIT) Int. J. Comput. Sci. Inf. Technol. **4**(3), 467–469 (2013). ISSN: 0975-9646
23. Z.A. Khan, R.K. Pateriya, Multiple pattern string matching methodologies: a comparative analysis. Int. J. Sci. Res. Publ. (IJSRP) **2**(7), 498–504 (2012)
24. S.M. Vidanagamachchi, S.D. Dewasurendra, R.G. Ragel, M. Niranjan, Commentz-Walter: any better than Aho-Corasick for peptide identification? Int. J. Res. Comput. Sci. **2**(6), 33–37 (2012). doi:10.7815/ijorcs.26.2012.053

25. B. Zhang, X. Chen, X. Pan, Z. Wu, High concurrence Wu Manber multiple patterns matching algorithm, in *Proceedings of the 2009 International Symposium on Information Processing (ISIP'09)*, Huangshan, P. R. China, pp. 404-409, 21–23 Aug 2009. ISBN: 978-952-5726-02-2 (Print), 978-952-5726-03-9 (CD-ROM)
26. Z. Zhang, X. Jin, Z. Hao, X. Guan, An improved Wu Manber multiple patterns matching algorithm for mobile internet content security audit in a Chinese environment. Int. J. Secur. Appl. **8**(3), 339–354 (2014). http://dx.doi.org/10.14257/ijsia.2014.8.3.34, ISSN: 1738-9976 IJSIA
27. S. Wu, U. Manber, Agrep- A Fast approximate Pattern Matching tool, Usenix Winter 92, pp. 153–162
28. G. Navarro, NR-grep: a fast and flexible pattern matching tool. Softw. Pract. Exp. **31**, 1265–1312 (2001)
29. G. Navarro, J. Tarhio, LZgrep: a Boyer-Moore string matching tool for Ziv-Lempel compressed text. Softw. Pract. Exp. **35**(12), 1107–1130 (2005)
30. Yang, S. Ding, An improved pattern matching algorithm based on BMHS, in *The Proceedings of 11th International Symposium on Distributed Computing and Applications to Business, Engineering & Science* (2012)
31. D.M. Sunday, A very fast substring search algorithm. Commun. ACM **33**(8), 132–142 (1990)
32. C. Zhen, W. Di, Improving Wu-Manber: a multi-pattern matching algorithm, in *The Proceedings of 2008 IEEE International Conference on Networking, Sensing and control (ICNSC)*, pp. 812–817, 6–8 Apr 2008

Author Index

Printed in the United States
By Bookmasters